Hierarchical Micro/ Nanostructured Materials

Fabrication, Properties, and Applications

Advances in Materials Science and Engineering

Series Editor

Sam Zhang

Aerospace Materials Handbook, *edited by Sam Zhang and Dongliang Zhao*

Biological and Biomedical Coatings Handbook: Applications, *edited by Sam Zhang*

Biological and Biomedical Coatings Handbook: Processing and Characterization, *edited by Sam Zhang*

Hierarchical Micro/Nanostructured Materials: Fabrication, Properties, and Applications, *Weiping Cai, Guotao Duan, Yue Li*

Hydroxyapatite Coatings for Biomedical Applications, *edited by Sam Zhang*

Nanobiomaterials: Development and Applications, *edited by Dong Kee Yi and Georgia C. Papaefthymiou*

Nanostructured and Advanced Materials for Fuel Cells, *edited by San Ping Jiang and Pei Kang Shen*

Micro- and Macromechanical Properties of Materials, *Yichun Zhou, Li Yang, and Yongli Huang*

Hierarchical Micro/ Nanostructured Materials

Fabrication, Properties, and Applications

Weiping Cai • Guotao Duan • Yue Li

CRC Press
Taylor & Francis Group
Boca Raton London New York

CRC Press is an imprint of the
Taylor & Francis Group, an **informa** business

CRC Press
Taylor & Francis Group
6000 Broken Sound Parkway NW, Suite 300
Boca Raton, FL 33487-2742

Printed on acid-free paper
Version Date: 20140203

International Standard Book Number-13: 978-1-4398-7682-4 (Hardback)

Library of Congress Cataloging-in-Publication Data

Cai, Weiping.
Hierarchical micro/nanostructured materials : fabrication, properties, and applications / Weiping Cai, Guotao Duan, Yue Li.
pages cm -- (Advances in materials science and engineering)
Summary: "Nanomaterials and nanotechnology have attracted much attention and been extensively studied for three decades due to their promising applications and scientific significance. Great progresses have been achieved in this field. The researchers in this area have the experiences from the creation of new structure in the early stage to the controlled grwoth, property and performance's study, and device fabrication and applications in the past decade"-- Provided by publisher.
Includes bibliographical references and index.
ISBN 978-1-4398-7682-4 (hardback)
1. Nanostructured materials. I. Duan, Guotao. II. Li, Yue, 1973- III. Title.

TA418.9.N35C3335 2014
620.1'15--dc23 2014000462

Visit the Taylor & Francis Web site at
http://www.taylorandfrancis.com

and the CRC Press Web site at
http://www.crcpress.com

Contents

Series Statement

Materials form the foundation of technologies that govern our everyday life, from housing and household appliances to handheld phones, drug delivery systems, airplanes, and satellites. Development of new and increasingly tailored materials is the key to further advancing important applications with the potential to dramatically enhance and enrich our experiences.

The Advances in Materials Science and Engineering series by CRC Press/Taylor & Francis is designed to help meet new and exciting challenges in materials science and engineering disciplines. The books and monographs in the series are based on cutting-edge research and development, and thus are up to date with new discoveries, new understanding, and new insights in all aspects of materials development, including processing, characterization, and applications in metallurgy; bulk or surface engineering; interfaces; thin films; coatings; and composites, just to name a few.

The series aims at delivering an authoritative information source to readers in academia, research institutes, and industry. The publisher and its series editor are fully aware of the importance of materials science and engineering as the foundation for many other disciplines of knowledge. As such, the team is committed to making this series the most comprehensive and accurate literary source to serve the whole materials world and the associated fields.

As series editor, I would like to thank all authors and editors of the books in this series for their noble contributions to the advancement of materials science and engineering and to the advancement of humankind.

Sam Zhang

Preface

Nanomaterials and nanotechnology have attracted much attention and been extensively studied for three decades because of their promising applications and scientific significance. Great progress has been achieved in this field. The researchers in this area have experienced the creation of new structures in early stages to controlled growth, property and performance studies, and device fabrication and applications in the past decade. Nanomaterial synthesis experienced changes from simple and unitary nanostructures to the complicated nanostructures and superstructures, from nanopowders to assemblies, from nanoparticles to hollow structures, from inorganic to organic materials, from zero to multidimension, from random to regular growth, from disordered arrangement to the periodic arrays, and so forth.

Hierarchical micro/nanostructured materials, which are composed of microsized objects with nanostructures, have been considered in the recent decade. Such structured materials show the surface activity and specific surface area of nanomaterials, and the structural stability and robustness of the bulk materials. They combine the advantages of both nanostructured and bulk materials. Hierarchical micro/nanostructured materials can exist in the forms of powders and regularly ranged arrays.

Micro/nanostructured object powders have large surface-to-volume ratios, high stability against aggregation, and are very easily separated from solution. These materials exhibit strong structurally enhanced properties, such as enhanced adsorption and catalysis performances compared with nanopowders or bulk ones, and hence could be good candidates for new environmental materials for high-efficient removal of contaminants in the environment. Further, if micro/nanostructured objects are arranged into a pattern on a substrate in some way, hierarchical micro/nanostructured arrays will be formed. Such arrays could be the important bases of the next generation of devices. There exist great potential applications in many fields, such as catalysis, integrated nanophotonics, optical devices, super-high-density storage media, sensors, nanobiotechnology, surface-enhanced Raman scattering (SERS) substrates, and so forth.

Although morphology- and structure-controlled growth and synthesis of these materials remain a promising challenge, a great deal of work in establishing parallel micro/nanofabrication techniques, performance exploration, and related applications has been done. Our group has also been in this field for nearly 10 years and focused on the development of new fabrication methods, exploration of structurally enhanced performance, surface properties, device applications, and so forth.

In this book, we will mainly present recent research progress of our group in the hierarchical micro/nanostructured materials, including two sections: Hierarchical Micro/Nanostructured Powders and Hierarchical Micro/Nanostructured Arrays in fabrication, properties, and applications, and hope to reflect and show the perspectives of the hierarchical micro/nanostructured materials in fundamental research and applications.

This book consists of 12 chapters. In addition to the general introduction of hierarchical micro/nanostructured materials in Chapter 1, we introduce, in detail, the mass production methods for hierarchical micro/nanostructured powders in Chapters 2 through 4, including solvothermal routes, template-etching strategies, and electrospinning technology, followed by structurally enhanced photocatalytic and adsorption performance in Chapters 5 and 6, respectively. Further, we introduce the modified colloidal lithography-based solution, electrodeposition strategies, and so forth, for fabrication of hierarchical micro/nanostructured object arrays and their devices in Chapters 7 and 8. In Chapters 9 through 12, we introduce and discuss the structure-dependent properties and performance of the micro/nanostructured arrays, including surface wettability, optical properties, gas-sensing performance, SERS performance, and detection applications. This book also introduces applications of hierarchical micro/nanostructured materials in environmental remediation and detection devices, and reviews the future trend of these materials in research and applications.

Acknowledgments

This book is financially supported by the National Key Basic Research Program of China (Grant No. 2013CB934303), the National Basic Research Program of China (973 Program, Grant No. 2011CB302103), Recruitment Program of Global Experts (C), the National Natural Science Foundation of China (Grant Nos. 11374303, 11174286, 51371165), and Anhui Provincial Natural Science Foundation for Distinguished Young Scholar (Grant Nos. 1108085J20, 1408085J10).

Authors

Weiping Cai earned a BS and MS in materials sciences from Northeast University in 1982 and 1984, respectively. In 1997, he earned a PhD from Huazhong University of Science and Technology in materials sciences. Since 1997, he has been a professor at the Institute of Solid State Physics, Chinese Academy of Sciences (CAS). His research interests include micro/nanostructured patterns and detection devices, and micro/nanomaterials for environmental applications.

Guotao Duan earned his PhD in 2007 from the Institute of Solid State Physics, CAS, and continued to work there. From April 2009 to March 2010, he did postdoctoral work in the National Institute for Materials Science in Japan. His current research interests focus on micro/nanofabrication, ordered micro/nanostructured arrays, and micro/nanodevices.

Yue Li earned his PhD in condensed matter physics at the Institute of Solid State Physics, CAS, in 2005. Later he worked as a postdoctoral fellow or visiting scientist in the Korea Advanced Institute of Science and Technology, National Institute of Advanced Industrial Science and Technology, and Max Planck Institute of Colloids and Interfaces. Since 2011, he has been working as a professor in the Institute of Solid State Physics, CAS. His research interests mainly focus on the fabrications, applications, and devices of micro/nanostructured arrays based on the colloidal monolayer template techniques.

1 General Introduction

Hierarchical micro/nanostructured materials are composed of microsized objects with nanostructures. Such structured materials possess not only the high surface activity and specific surface area of nanomaterials but also the structural stability and robustness of the bulk. Micro/nanostructured powders are of large surface to volume ratios, have high stability against aggregation, are very easily separated from solution, and also exhibit strong structurally enhanced adsorption and catalysis performances and, hence, can be used for the highly efficient removal of contaminants in the environment. Further, if the micro/nano-sized objects are arranged into a pattern on a substrate, hierarchical micro/nanostructured arrays are formed. These arrays have great potential applications in many fields, such as catalysis, integrated nanophotonics, optical devices, super-high-density storage media, sensors, nanobiotechnology, and surface-enhanced Raman scattering (SERS) substrates.

1.1 HIERARCHICAL MICRO/NANOSTRUCTURED POWDERS

The hazardous materials, such as organic pollutants and heavy metals from industrial production, released by human activities result in decreasing environmental quality or in environmental pollution [1–4]. Such environmental pollution has attracted more and more attention in recent years, especially in developing countries. The harmful materials, including heavy metal ions, organic pollutants have been released to air, water and soil which directly threaten the human health. Till now, many useful methods and techniques have been developed for pollution remediation [5–9]. Among them, removal of pollutants by the method of adsorption and/or catalysis using adsorptive and/or catalytic materials is important and effective [10]. In this case, the key issue is the efficiency of absorptive and/or catalytic materials. Materials with high surface activity, high specific surface area, good selective adsorption, strong structural stability, and excellent reusable performance are expected. Although nanosized particle powders can be candidates for such materials due to their high surface area and surface activity [5,6,11–16], they have no structural stability and are thus very easily aggregated, leading to unwanted reductions in active surface area and surface activity and, hence, bad reusable property. In addition, the used nanosized adsorbents or photocatalysts are suspended in solution and difficult to separate from the bulk solution. On the other hand, the microsized particle powders are of good structural stability and have antiaggregation property, in addition to being easily separated, but they are lacking in terms of surface area and activity.

If the microscaled building blocks and the nanostructure are combined together, hierarchical micro/nanostructured materials are formed [10,17]. Figure 1.1 shows the typical morphology of TiO_2 micro/nanostructured powders by the solvothermal

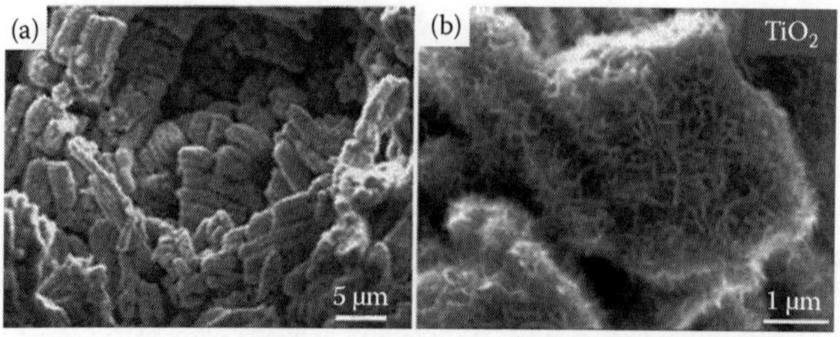

FIGURE 1.1 A typical morphology of TiO_2 microscaled particles with nanostructure by solvothermal route. (a) and (b) Field-emission scanning electron microscopic images with different magnifications.

route. The microsized objects resist aggregation, and the nanostructure supplies high surface area and surface activity during usage. Therefore, such materials with micro/nanoarchitectures not only possess large surface to volume ratios, have high structural stability against aggregation, and are very easily separated from solution during application in pollution remediation but also exhibit strong structurally enhanced adsorption and catalysis performances. They could overcome the aforementioned shortcomings of the nanoparticles and microscaled objects, and be good candidates of the environmental materials for high efficient removal of contaminants.

In recent years, many methods have been developed for controlled and mass production of such micro/nanostructured materials. Also, many progresses have been made in removal of pollutants by using these materials as adsorbents or catalysts.

1.2 HIERARCHICAL MICRO/NANOSTRUCTURED ARRAYS

If the randomly distributed nano-objects [such as zero-dimensional nanoparticles or nanodots, one-dimensional (1D) nano-objects, and two-dimensional (2D) nanoplates] or the microscaled objects with nanostructures are regularly arranged, in a certain way, on a substrate, a micro/nanostructured array is formed. Figure 1.2 shows some typical micro/nanostructured arrays, including the nanodot array, 1D nano-object array, and nano–hollow sphere array [18–20]. Such arrays are the important basis of next generation of devices and have important potential applications in areas such as catalysis, sensors, cells, SERS substrates, data storage, superhydrophobic or superhydrophilic films, photonic crystals, optoelectronics, microelectronics, and optical devices [21].

The properties of micro/nanostructured arrays are strongly correlated with their structural parameters, such as the size, shape, and interspacing of the building blocks in the array, and the supporting substrate, in addition to the intrinsic characters. Such arrays possess not only the properties of the individual building block but also some new performances due to the coupling effects between the building blocks. Usually, the supporting substrate is selected according to device realization and compatibility with desired materials. Building blocks should be controlled in both morphology and size for a favorable functionality. Arrangement or packing of the building blocks is

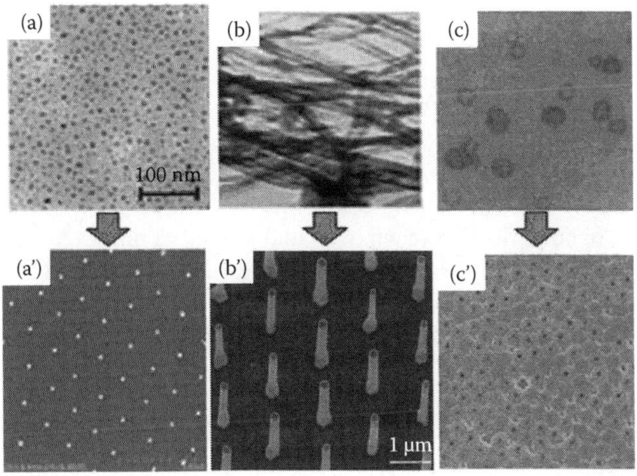

FIGURE 1.2 Morphologies of some typical nanoparticles and micro/nanostructured arrays: (a) nanoparticles, (b) one-dimensional (1D) nano-objects, (c) nano–hollow spheres, (a') nanodot array, (b') 1D nano-object array, and (c') nano–hollow sphere array. (With kind permission from Springer Science+Business Media: *Appl. Phys. B*, Laser morphological manipulation of gold nanoparticles periodically arranged on solid supports, 81, 2005, 765–8, Sun et al.; Reprinted with permission from Liu et al., 2006, 2375–8. Copyright 2006 American Chemical Society.)

designed according to practical applications. In many cases, an ordered arrangement of the building blocks is of high importance and allows a homogeneous surface characteristic in a large dimension on the array film, which is obviously advantageous to the designed devices and, thus, the subsequent stability of output functions.

It is well known that nanoparticles have many unique properties, which depend on the size, shape, and structure of the particles. Development of new synthesis methods for nanoparticles with controllable sizes, shapes, and structures and revealing the new functional performances of nanoparticles have been the hot spot of research activity in this field. For micro/nanostructured arrays, key issues include construction with low cost and according to need, reproducible fabrication and structural consistency, and finding new performances and their structural dependence as well as tunability. Aiming at the aforementioned key problems, extensive studies have been performed and big advances been made.

In general, micro/nanostructured arrays can be fabricated by photolithography [22–24], electron-beam lithography [25–28], microcontact printing [29,30], self-assembly techniques [31,32], and so on. In the past decade, using 2D colloidal crystals (i.e., the colloidal crystal with only few layers) as a template has shown great promise for the fabrication of micro/nanostructured arrays. It has been found that the monodispersed organic colloidal spheres can self-assemble into an ordered, hexagonally close-packed arrangement on a cleaned substrate driven by surface tension. Based on such ordered, arranged colloidal spheres and other assistant techniques, such as physical vapor deposition [33], sol gel [34], solution dipping [35], and electrodeposition [36,37], one can obtain various micro/nanostructured arrays after removal of the colloidal spheres. This strategy, what we call colloidal lithography, is of great advantages due to the material and substrate general, inexpensive, flexible in controlling surface morphology and size.

The aforementioned hierarchical micro/nanostructured materials have attracted much attention in recent years. Their properties depend on morphology and structure. Although the morphology and structure-controlled growth and synthesis of these materials remain a promising challenge, a great deal of work in establishing parallel micro/nanofabrication techniques and related applications has been done. In this book, we present recent advances in the field of hierarchical micro/nanostructured materials, in two parts, "Hierarchical Micro/Nanostructured Powders" and "Hierarchical Micro/Nanostructured Arrays," in terms of fabrication, properties, and applications.

This book introduces the new routes and technologies in the micro/nanostructured material field in detail, including structure-directed solvothermal routes, template-etching strategies, electrospinning and in situ conversion, and modified colloidal lithography–based solution and electrodeposition strategies. We introduce and discuss the correlative novel performances and property control arising from the micro/nanostructures. This book also introduces the applications of hierarchical micro/nanostructured materials in environmental remediation and detection devices and reviews the future trend of these materials in research and applications.

REFERENCES

1. Zhang, W. X. 2003. Environmental technologies at the nanoscale. *Environ. Sci. Technol.* 37: 102A–8A.
2. Ru, J.; Liu, H. J.; Qu, J. H.; Wang, A. M.; and Dai, R. H. 2007. Removal of dieldrin from aqueous solution by a novel triolein-embedded composite adsorbent. *J. Hazard. Mater.* 141: 61–9.
3. Schwarzenbach, R. P.; Escher, B. I.; Fenner, K.; Hofstetter,T. B.; Johnson, C. A.; Gunten, U.; and Wehrli, B. 2006. The challenge of micropollutants in aquatic systems. *Science* 313: 1072–7.
4. Shannon, M. A.; Bohn, P. W.; Elimelech, M.; Georgiadis, J. G.; Marinas, B. J.; and Mayes, A. M. 2008. Science and technology for water purification in the coming decades. *Nature* 452: 301–10.
5. Theron, J.; Walker, J. A.; and Cloete, T. E. 2008. Nanotechnology and water treatment: Applications and emerging opportunities. *Crit. Rev. Microbiol.* 34: 43–9.
6. Varanasi, P.; Fullana, A.; and Sidhu, S. 2007. Remediation of PCB contaminated soils using iron nano-particles. *Chemosphere* 66: 1031–8.
7. Peng, X. J.; Li, Y. H.; Luan, Z. K.; Di, Z. C.; Wang, H. Y.; Tian, B. H.; and Jia, Z. P. 2003. Adsorption of 1,2-dichlorobenzene from water to carbon nanotubes. *Chem. Phys. Lett.* 376: 154–8.
8. Kostal, J.; Mulchandani, A.; Gropp, K.; and Chen, W. A. 2003. A temperature responsive biopolymer for mercury remediation. *Environ. Sci. Technol.* 37: 4457–62.
9. Hu, J. S.; Zhong, L. S.; Song, W. G.; and Wan, L. J. 2008. Synthesis of hierarchically structured metal oxides and their application in heavy metal ion removal. *Adv. Mater.* 20: 2977–82.
10. Zhong, L. S.; Hu, J. S.; Liang, H. P.; Cao, A. M.; Song, W. G.; and Wan, L. J. 2006. Self-assembled 3D flowerlike iron oxide nanostructures and their application in water treatment. *Adv. Mater.* 18: 2426.

11. Ponder, S. M.; Darab, J. G.; and Mallouk, T. E. 2000. Remediation of Cr(VI) and Pb(II) aqueous solutions using supported, nanoscale zero-valent iron. *Environ. Sci. Technol.* 34: 2564–9.
12. He, P. and Zhao, D. Y. 2005. Preparation and characterization of a new class of starch-stabilized bimetallic nanoparticles for degradation of chlorinated hydrocarbons in water. *Environ. Sci. Technol.* 39: 3314–20.
13. Wang, C. B. and Zhang, W. X. 1997. Synthesizing nanoscale iron particles for rapid and complete dechlorination of TCE and PCBs. *Environ. Sci. Technol.* 31: 2154–6.
14. Keum, Y. S. and Li, Q. X. 2005. Reductive debromination of polybrominated diphenyl ethers by zerovalent iron. *Environ. Sci. Technol.* 39: 2280–6.
15. Shinde, V. R.; Gujar, T. P.; Noda, T.; Fujita, D.; Vinu, A.; Grandcolas, M.; and Ye, J. 2010. Growth of shape-and size-selective zinc oxide nanorods by a microwave-assisted chemical bath deposition method: Effect on photocatalysis properties. *Chem. Eur. J.* 16: 10569–75.
16. Yantasee, W.; Lin, Y. H.; Fryxell, G. E.; Alford, K. L.; Busche, B. J.; and Johnson, C. D. 2004. Selective removal of copper(II) from aqueous solutions using fine-grained activated carbon functionalized with a mine. *Ind. Eng. Chem. Res.* 43: 2759–64.
17. Wang, X. B.; Cai, W. P.; Liu, S. W.; Wang, G. Z.; Wu, Z. K.; and Zhao, H. J. 2013. ZnO hollow microspheres with exposed porous nanosheets surface: Structurally enhanced adsorption towards heavy metal ions. *Colloids Surf. A* 422: 199–205.
18. Sun, F.; Cai, W.; Li, Y.; Duan, G.; Nichols, W. T.; Liang, C.; Koshizaki, N.; Fang, Q.; and Boyd. I. W. 2005. Laser morphological manipulation of gold nanoparticles periodically arranged on solid supports. *Appl. Phys. B* 81: 765–8.
19. Liu, D. F.; Xiang, Y. J.; Wu, X. C.; Zhang, Z. X.; Liu, L. F.; Song, L.; Zhao, X. W.; Luo, S. D.; Ma, W. J.; and She, J. 2006. Periodic ZnO nanorod arrays defined by polystyrene microsphere self-assembled monolayers. *Nano Lett.* 6: 2375–8.
20. Li, Y.; Cai, W. P.; Duan, G. T.; Cao, B. Q.; and Sun, F. Q. 2005. Two-dimensional ordered polymer hollow sphere and convex structure arrays based on monolayer pore films. *J. Mater. Res.* 20: 338–43.
21. Cai, W. P.; Li, Y.; and Duan, G. T. 2007. *Advanced Materials Research Trends*. L. V. Basbanes (Ed.). Nova Science Publishers, Inc., New York, Chapter 2.
22. Howard, R. E.; Liao, P. F.; Skocpol, W. J.; Jackel, L. D.; and Craighead, H. G. 1983. Microfabrication as a scientific tool. *Science* 221: 117–21.
23. Ito, T. and Okazaki, S. 2000. Pushing the limits of lithography. *Nature* 406: 1027–31.
24. Pease, R. F. 1992. *Nanostructures and Mesoscopic Systems.* W. P. Kirk and M. A. Reed (Eds.). Academic Press, Boston, MA, pp. 37–50.
25. Pease, R. F. W. 1992. Nanolithography and its prospects as a manufacturing technology. *J. Vac. Sci. Technol. B* 10: 278–85.
26. Nakayama, Y. 1990. Electron-beam cell projection lithography: A new high-throughput electro-beam direct-writing technology using a specially tailored Si aperture. *J. Vac. Sci. Technol. B* 8: 1836–40.
27. Berger, S. D. 1991. Projection electron-beam lithography: A new approach. *J. Vac. Sci. Technol. B* 9: 2996–9.
28. Pfeiffer, H. C. and Stickel, W. 1995. PREVAIL—An e-beam stepper with variable axis immersion lenses. *Microelectron. Eng.* 27: 143–6.
29. Xia, Y.; Rogers, J.; Paul, K. E.; and Whitesides, G. M. 1999. Unconventional methods for fabricating and patterning nanostructures. *Chem. Rev.* 99: 1823–48.
30. Jeon, N. L.; Choi, I. S.; and Whitesides, G. M. 1999. Patterned polymer growth on silicon surfaces using microcontact printing and surface-initiated polymerization. *Appl. Phys. Lett.* 75: 4201.
31. Nagayama, K.; Takeda, S.; Endo, S.; and Yoshimura, H. 1995. Fabrication and control of 2-dimensional crystalline arrays of protein molecules. *Jpn. J. Appl. Phys.* 34: 3947–54.

32. Matsushita, S.; Miwa, T.; and Fujishima, A. 1997. Preparation of a new nanostructured TiO_2 surface using a two-dimensional array-based template. *Chem. Lett.* 309: 925–6.
33. Matsushita, S. I.; Miwa, T.; Tryk, D. A.; and Fujishima, A. 1998. New mesostructured porous TiO_2 surface prepared using a two-dimensional array-based template of silica particles. *Langmuir* 14: 6441–7.
34. Tessier, P. M.; Velev, O. D.; and Kalambur, A. T. 2000. Assembly of gold nanostructured films templated by colloidal crystals and use in surface-enhanced Raman spectroscopy. *J. Am. Chem. Soc.* 122: 9554–5.
35. Sun, F.; Cai, W.; Li, Y.; Cao, B.; Lei, Y.; and Zhang, L. 2004. Morphology-controlled growth of large-area two-dimensional ordered pore arrays. *Adv. Funct. Mater.* 14: 283–8.
36. Bartlett, P. N.; Baumberg, J. J.; Coyle, S.; and Abdelsalam, M. E. 2004. Optical properties of nanostructured metal films. *Faraday Discuss.* 125: 117–32.
37. Sun, F.; Cai, W.; Li, Y.; Cao, B.; Lu, F.; Duan, G.; and Zhang, L. 2004. Morphology control and transferability of ordered through-pore arrays based on electrodeposition and colloidal monolayer. *Adv. Mater.* 16: 1116–21.

Section I

Hierarchical Micro/ Nanostructured Powders

2 Solvothermal Routes

2.1 INTRODUCTION

The solvothermal/hydrothermal method is known for its low cost, ease of operation, and ease of control for the fabrication of micro/nanostructured materials. The general strategy is illustrated in Figure 2.1. Briefly, structural directors (surfactants or organic solvents) are added to precursor solutions and then maintained at 150°C–250°C for reaction. Novel micro/nanostructure materials are obtained after reaction for a certain time (about 10 hours) and/or subsequent heat treatment. The morphology of the products depends on the structural directors used. We can produce various micro/nanostructured materials, such as metal oxides or sulfides and even composites, by choosing different precursors and structural directors.

2.2 NOVEL MICRO/NANOSTRUCTURED ZnO

It is well known that ZnO is a promising candidate for photocatalysts [1,2], gas sensors [3,4], solar cells [5,6], and so on. There have been extensive reports in these fields. Also, ZnO is an environment-friendly material [7] and its surface can attach to many functional groups, such as hydroxyl groups, that can be active sites for adsorption and is thus a good candidate as an adsorbent for wastewater treatment [8,9]. The many hierarchical micro/nanostructured ZnO materials, with controlled sizes and morphologies, were prepared [1,10,11]. For instance, platelike micro/nanostructured ZnO, which has a high specific surface area, was fabricated by various methods, such as hydrothermal [12–17], solvothermal [18], chemical vapor deposition [19,20], electrochemical deposition [21–23], and microwave methods [3]. Comparatively, the hydrothermal or solvothermal route is quite a simple technique for the fabrication of ZnO micro/nanostructures and possesses several advantages, such as low cost, mass production capability, and a relatively low reaction temperature. Herein, we introduce the mass fabrication of micro/nanostructured porous ZnO nanoplates, nanoplate-built hollow spheres, and nanoplate-built core/shell-structured objects, which have a higher specific surface area, based on solvothermal methods, using different morphology directors, and a subsequent annealing process.

2.2.1 Micro/Nanostructured Porous ZnO Plates

According to the strategy shown in Figure 2.1, if ethylene glycol (EG) is used as the morphological director and a mixture consisting of $Zn(CH_3COO)_2{\cdot}2H_2O$, urea $[(NH_2)_2CO]$, and water with a volume ratio of $V_{EG}/V_{DIW} = 1{:}1$ (DIW refers to deionized water) is taken as the precursor solution, we can obtain platelike micro/nanostructured products after reaction at 160°C for 12 hours before the reactants are cooled to room temperature naturally [24], as shown in Figure 2.2a. The product

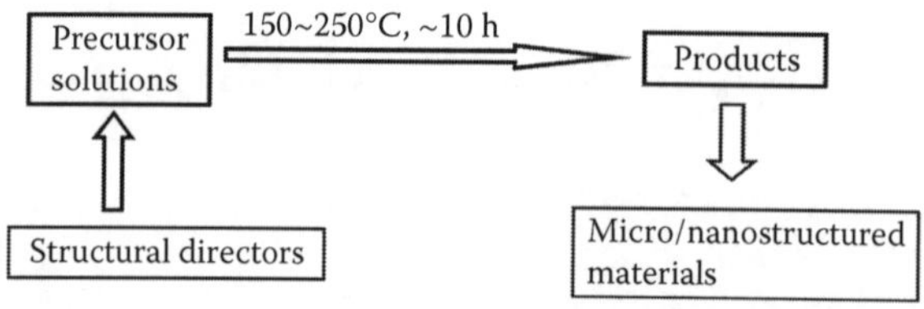

FIGURE 2.1 General synthetic route for solvothermal/hydrothermal preparation of micro/nanostructured materials.

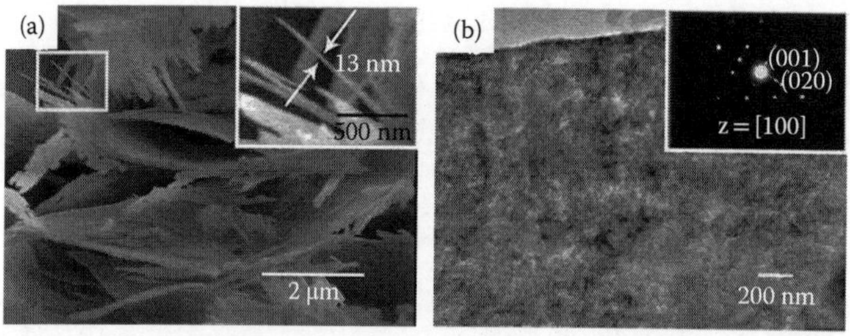

FIGURE 2.2 (a) Field emission scanning electron microscopy and (b) transmission electron microscopy images of the products ZnHC. Inset in (a) shows the local magnification corresponding to the frame area marked in (a), and inset in (b) shows the corresponding selected area electron diffraction pattern. (Wang et al., *J. Mater. Chem.*, 20, 8582–90, 2010. Reproduced by permission of The Royal Society of Chemistry.)

ZnHC consists of nanoplates with a smooth planar surface and irregular profiles. The plates are microsized in planar dimension and about 10–15 nm in thickness. X-ray diffraction (XRD) has revealed that such products are of monoclinic hydrozincite $Zn_5(OH)_6(CO_3)_2$ (or ZnHC for short) phase. Transmission electron microscopy (TEM) examination confirmed that the plates are of single crystalline structure with the (100) plane for the planar surface, as shown in Figure 2.2b; in the inset, the selected area electron diffraction (SAED) pattern is shown.

2.2.1.1 Morphology and Structure

The as-prepared products were subsequently annealed at 400°C, and the ZnHC was transformed to ZnO, which has been confirmed by XRD measurement [24]. Correspondingly, field emission scanning electron microscopy (FESEM) and TEM observations show that the annealed products consist of porous nanoplates with similar thickness to those shown in Figure 2.2a, as illustrated in Figure 2.3a and b. The pores fall in the range of 5–20 nm in size and disperse randomly in the plates. Figure 2.3c presents the high-resolution transmission electron microscopy (HRTEM) image and SAED pattern of an individual porous ZnO nanoplate. A plane spacing of 0.26 nm corresponds to the (0002) plane of the hexagonal ZnO phase. The ZnO plates are of single crystalline structure with the planar surface of the $(10\bar{1}0)$ or (100) plane. A crystal orientation relation exists between the ZnHC nanoplate and

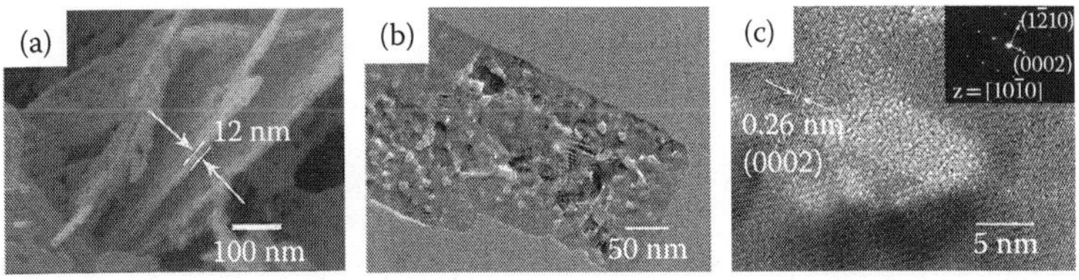

FIGURE 2.3 The morphology of the annealed nanoplates: (a) field emission scanning electron microscopy image; (b) transmission electron microscopy image of a single ZnO nanoplate; and (c) high-resolution transmission electron microscopy image of the local area in a single ZnO nanoplate, and the inset shows the corresponding selected area electron diffraction pattern. (Wang et al., *J. Mater. Chem.*, 20, 8582–90, 2010. Reproduced by permission of The Royal Society of Chemistry.)

the ZnO nanoplate, or ZnHC (100)//ZnO (100) or (10$\bar{1}$0). It means that the planar surfaces of such ZnO plates are nonpolar planes. In addition, the yield of ZnO porous nanoplates can be estimated to be more than 94% based on the Zn molar numbers in the precursor solution and the final product ZnO (by weighing). Obviously, the synthesis method in this study can realize mass production of porous ZnO nanoplates with high yield and low cost.

Further, the measurement of N_2 sorption for porous ZnO nanoplates has shown a type II isotherm [25], which means monolayer adsorption at low pressure and multilayer adsorption at high pressure, as shown in Figure 2.4. Based on the Brunauer–Emmett–Teller equation [26], the specific surface area was evaluated to be about 147 $m^2 \cdot g^{-1}$. In addition, it was estimated, from the density and thickness of the ZnO nanoplates, that the surface area of the exposed (10$\bar{1}$0) plane is about 30 $m^2 \cdot g^{-1}$. Therefore, 80% of the measured surface area is contributed by the pore walls in the nanoplates. The hysteresis loop in the isotherm clearly reveals the presence of mesopores in the plates. Most pores fall into the 5–20 nm range, as illustrated in the pore size distribution in the inset of Figure 2.4. This is consistent with FESEM observations (see Figure 2.3). The pores with larger size (>20 nm) should be mainly contributed from the interstitials induced by the random pileup of the nanoparticles during N_2 sorption measurement.

2.2.1.2 Influence of V_{EG}/V_{DIW} on Products

Further experiments have revealed that the EG in the precursor solution plays an important role in the formation of the platelike morphology of the products. With increasing EG content in the solution, the products' morphology evolves from nanosheets to plates, particles, and microspheres [24]. Without EG, the product consists of the scale-like ZnHC nanosheets. Too high V_{EG}/V_{DIW} (say, 7:3) produces nearly uniaxial particles 1 to 2 μm in size. Extremely, if EG is used for solvent (or without water in the solution), spherical particles can be formed with 2–4 μm in diameter. Only a proper ratio of V_{EG}/V_{DIW} (say, 1:1) leads to micro/nanostructured ZnHC nanoplates (Figure 2.2).

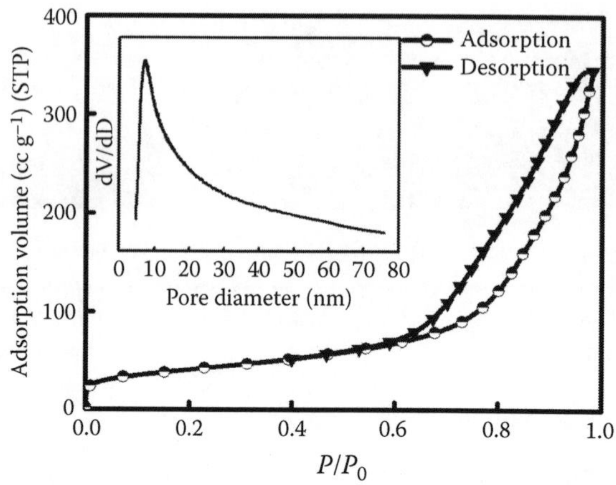

FIGURE 2.4 Nitrogen sorption isotherm of the porous ZnO nanoplates: inset shows pore size distribution. (Wang et al., *J. Mater. Chem.*, 20, 8582–90, 2010. Reproduced by permission of The Royal Society of Chemistry.)

2.2.1.3 Formation of Nanoplates

The formation of ZnHC can be described by the following reactions:

$$CO(NH_2)_2 + 3H_2O \rightarrow 2NH_3 \cdot H_2O + CO_2 \tag{2.1}$$

$$NH_3 \cdot H_2O \rightarrow NH_4^+ + OH^- \tag{2.2}$$

$$CO_2 + 2OH^- \rightarrow CO_3^{2-} + H_2O \tag{2.3}$$

$$5Zn^{2+} + 2CO_3^{2-} + 6OH^- \rightarrow Zn_5(OH)_6(CO_3)_2 \tag{2.4}$$

When the precursor solution is heated at 160°C, urea will release OH^- as it is a weak alkali (Reactions 2.1 and 2.2). Further, the CO_2 in the alkali solution produces CO_3^{2-} ions according to Reaction 2.3. The OH^- and CO_3^{2-} ions will further react with the Zn^{2+} ions released by $Zn(Ac)_2$ in the solution and form ZnHC (Reaction 2.4). As the reaction proceeds, the concentration of ZnHC in the solution will become saturated and, hence, ZnHC crystal nuclei will be formed in the solution. The formed crystal nuclei grow subsequently. Generally, EG prefers to adsorb on a certain crystal plane, leading to preferential growth along the plane and a platelike morphology due to the anisotropic growth [27–29]. In our case, EG could adsorb on the (100) plane of the hydrozincite during nucleation. This adsorption leads to preferential growth along [001] and [020] within the (100) plane, and the formation of nanoplates, as shown in Figure 2.2.

For porous ZnO nanoplates, due to the layered structure, ZnHC nanoplates with the (100) planar surface are assembled by zinc octahedrons $Zn_3(OH)_6O_2$ and tetrahedrons $Zn(OH)_3O$, which are held together by CO_3 groups, as illustrated in

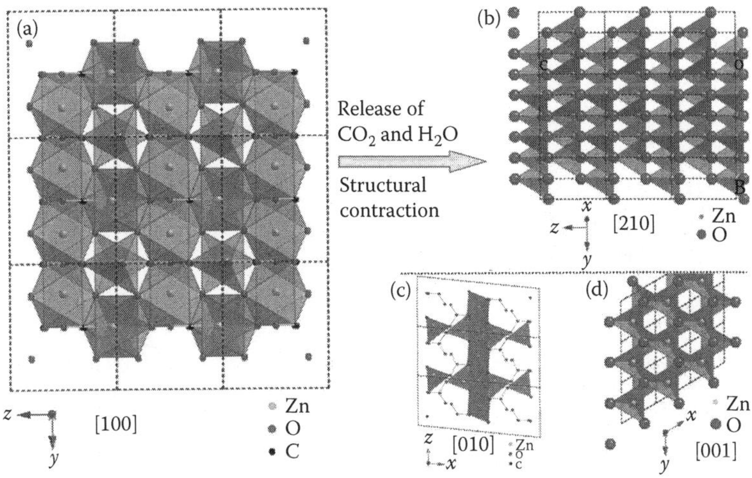

FIGURE 2.5 Structural illustration of transformation from hydrozincite to ZnO: (a) and (c) view along [100] and [010], respectively, for hydrozincite; and (b) and (d) view along [210] and [001], respectively, for ZnO. (Wang et al., *J. Mater. Chem.*, 20, 8582–90, 2010. Reproduced by permission of The Royal Society of Chemistry.)

Figure 2.5a and c. During subsequent annealing at 400°C, such ZnHC sheets will decompose and release CO_2 and H_2O or the following reaction will occur:

$$Zn_5(CO_3)_2(OH)_6 \rightarrow 5ZnO + 3H_2O + 2CO_2 \quad (2.5)$$

The octahedrons and tetrahedrons in the ZnHC sheets will thus collapse and contract to ZnO tetrahedrons, forming ZnO nanoplates with (100) or $(10\bar{1}0)$ planar surface, as shown in Figure 2.4b and d. It means that the crystal orientation relation of ZnHC (100)//ZnO (100) comes from the structural contraction induced by the release of CO_2 and H_2O. Obviously, the formation of pores or voids in the ZnO nanoplates should be attributed to topotactic transformation or a contraction of the original structure, that is, the Zn number density in wurtzite is higher compared to the precursor ZnHC, from which CO_2 and H_2O are released. Such a contraction of the original structure must lead to voids.

Further, differential thermal analysis and thermal gravimetric (TG) measurements were conducted for the ZnHC products. The corresponding results are given in Figure 2.6. The weight loss before 100°C (exothermic peak at 31°C) ascribes to the removal of a small amount of adsorbed water. The weight loss from 100°C to 310°C (endothermic peak at 290°C) corresponds to the decomposition of ZnHC nanoplates, and the release of carbon dioxide and water, leaving the porous ZnO (Reaction 2.5). The last stage of weight loss occurs from 310°C to 500°C (endothermic peak at 360°C) owing to the complete pyrolysis of hydrozincite. Further, the weight loss can be estimated, from Reaction 2.5, to be 25.87%, which is in good agreement with our TG experiment (26.1%) shown

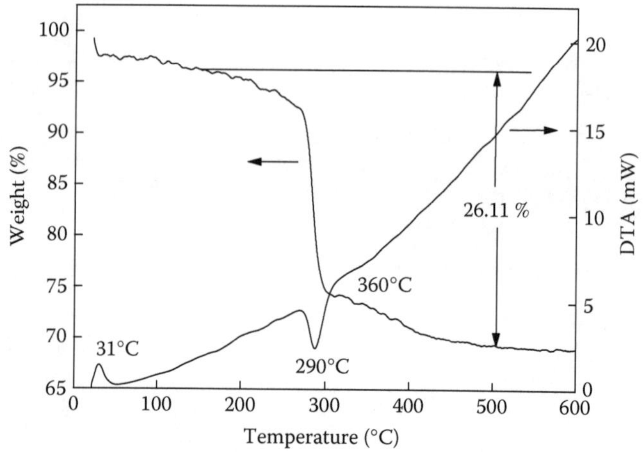

FIGURE 2.6 Thermal gravimetric and differential thermal analysis (DTA) curves of the hydrozincite nanoplates. (Wang et al., *J. Mater. Chem.*, 20, 8582–90, 2010. Reproduced by permission of The Royal Society of Chemistry.)

in Figure 2.6. The porosity of the porous nanoplates can thus be estimated to be about 40% from the TG curve and the densities of ZnHC (4.42 g·cm^{-3}) and ZnO (5.61 g·cm^{-3}).

2.2.2 Standing Porous Nanoplate-Built Hollow Spheres

As mentioned in the first paragraph of Section 2.2, ZnO with a micro/nanostructure could be a good adsorbent for wastewater treatment. The porous micro/nanostructured ZnO nanoplates with high specific surface areas have shown excellent adsorption performance to Cu(II) ions in aqueous solution [24]. However, these porous nanoplates were easy to stack together or overlap, due to their planar geometry, which decreased the surface area exposed to the solution. The structurally enhanced adsorption performance of the micro/nanostructured material was only partially exhibited. Obviously, if we assemble these porous nanoplates into a micro/nanostructure and all the porous nanoplates are vertically standing and cross-linked, stacking or pileup of the nanoplates will not take place, or the surface area within the porous nanoplates will be sufficiently exposed to the solution. In this case, the structurally enhanced adsorption performance would be sufficiently exhibited.

Based on the modified hydrothermal route shown in Figure 2.1, ZnO hollow microspheres with exposed porous nanosheets' surface could be produced using citrate as the structural director, and subsequent annealing treatment [30]. We used the same precursor as that for the aforementioned porous nanoplates, or an aqueous solution with zinc acetate dehydrate [$Zn(CH_3COO)_2 \cdot 2H_2O$] and urea [$(NH_2)_2CO$], but adopted trisodium citrate dihydrate ($C_6H_5Na_3O_7 \cdot 2H_2O$) (or TC for short) as the structural director (the final TC concentration was 1.2 g·L^{-1}).

2.2.2.1 Morphology and Structure

After reaction of the precursor solution at 160°C for 12 hours, white products were obtained. The corresponding XRD pattern revealed the formation of monoclinic hydrozincite ZnHC phase [30]. FESEM observation has shown that the ZnHC products consist of microsized spherical particles with rough surfaces, and these spheres are 5–20 μm in size centered around 12 μm, as illustrated in Figure 2.7a and b. Further, high-magnification examination has revealed that the spherical particles are honeycomb like in morphology (Figure 2.7c) or built of cross-linked nanoplates. The nanoplates are nearly vertically standing and cross-linked around the spherical part. Local magnification shows that the nanoplate is approximately 25 nm in thickness, as illustrated in Figure 2.7d. By crushing the spheres with agate mortar, we found that they were hollow in structure with approximately 5 μm hollow size (the inset of Figure 2.7c). The SAED pattern indicates that the standing nanoplate is of single crystalline structure with an exposed plane of (100) [30].

After subsequent annealing at 400°C for 2 hours, the phase structure of the products was changed from ZnHC to ZnO, which is in good agreement with other reports [3,24,31]. In addition, it should be mentioned that the yield or productivity of our product was quite high and estimated to be more than 70% based on the molar

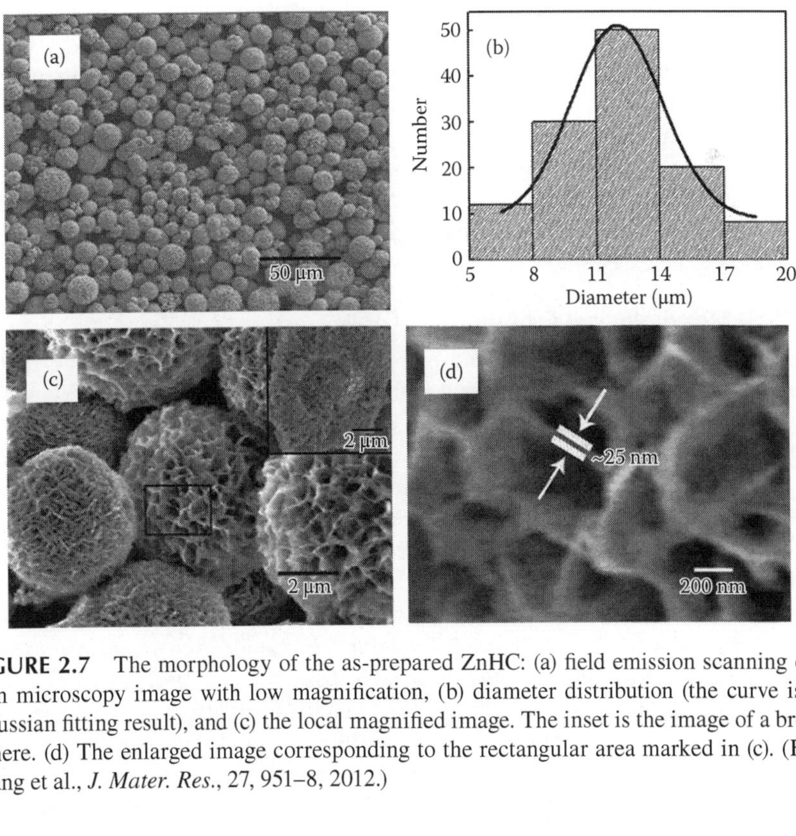

FIGURE 2.7 The morphology of the as-prepared ZnHC: (a) field emission scanning electron microscopy image with low magnification, (b) diameter distribution (the curve is the Gaussian fitting result), and (c) the local magnified image. The inset is the image of a broken sphere. (d) The enlarged image corresponding to the rectangular area marked in (c). (From Wang et al., *J. Mater. Res.*, 27, 951–8, 2012.)

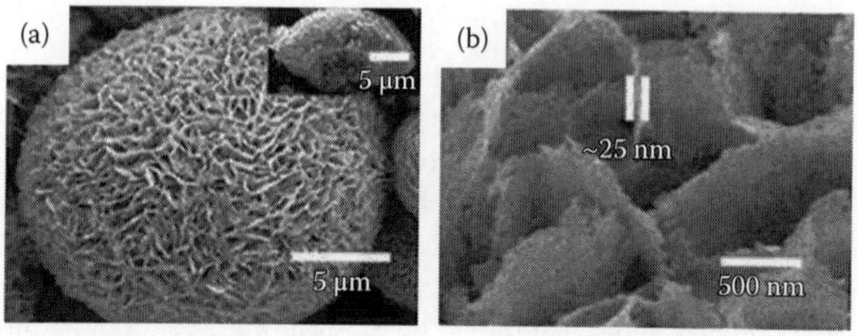

FIGURE 2.8 Morphology of the product after annealing at 400°C: (a) field emission scanning electron microscopy image of a single sphere. The inset shows a broken sphere. (b) The local enlarged image of (a). (From Wang et al., *J. Mater. Res.*, 27, 951–8, 2012.)

numbers of zinc acetate dihydrate in the reaction solution and ZnO in the final products. It is easy to realize mass production by using a big enough autoclave.

The morphology of the annealed product is similar to that of the honeycomb-like ZnHC hollow spheres. It also shows a hollow structure, as shown in the inset of Figure 2.8a, corresponding to a crushed sphere. However, a close observation reveals that there exist a large number of pores distributed uniformly in the nanoplates within the honeycomb-like microspheres (see Figure 2.8b). Further, HRTEM examination shows that the nanoplate is of single crystalline structure, similar to that shown in Figure 2.3c. It is indicated that the ZnO nanoplates grow, along with [0001] and [$1\bar{2}10$] axes, within the plane of ($10\bar{1}0$) [or (100)]. There also exists a crystal orientation relation between the ZnHC nanoplate and the ZnO nanoplate in the honeycomb-like structure, or ZnHC (100)//ZnO (100) or ($10\bar{1}0$) [30]. It means that such a ZnO plate is of nonpolar planar surface. In addition, the pores within the nanoplates are clearly seen in Figure 2.8b. The pores are around 10–15 nm in size. All these are similar to the pure nanoplates mentioned in Section 2.2.2. Nitrogen sorption isothermal measurement has indicated a specific surface area of about 46 $m^2 \cdot g^{-1}$ for such hollow spheres [30].

2.2.2.2 Morphological Evolution

To reveal the formation of honeycomb-like micro/nanostructured porous ZnO hollow spheres, their morphological evolution with reaction time was examined. Figure 2.9 gives the corresponding results.

In the initial reaction state (say, in 5 minutes), a few spheres with smooth surfaces and nearly 3–5 μm in diameter are formed very quickly (Figure 2.9a). TEM and XRD have confirmed that such smooth microsized spheres are amorphous in structure (see the inset of Figure 2.9b). Further reaction (say, 10 minutes or longer), on the one hand, leads to the appearance of more smooth (or amorphous) spheres and, on the other hand, induces the outward growth of *floc* on the surface of the preformed smooth spheres, as shown in Figure 2.9c and d. We can also see some spheres with a broken surface layer, which exhibit partially dissolved core parts or a core/shell-like structure. The broken surface layer should be induced during product collection and FESEM sample preparation. Correspondingly, XRD shows that a diffraction peak at $2\theta = 34°$ emerges, indicating that the spherical surface layer is crystalline. This thin layer is crystallized and thus fragile. It means that the crystallization of

FIGURE 2.9 Morphologies of the products (ZnHC) after reaction at 160°C for different times: (a) 5 minutes; (b) transmission electron microscopy image of a single sphere shown in (a), and the inset shows the corresponding selected area electron diffraction; (c) and (d) 10 minutes (in different fields), and the inset in (d) shows the field emission scanning electron microscopy image of a sphere with a partially dissolved core part; (e) 1 hour; and (f) 2 hour, and the inset shows a broken sphere. (From Wang et al., *J. Mater. Res.*, 27, 951–8, 2012.)

preformed smooth spheres takes place in the surface layer after reaction for 10 minutes. Subsequent reaction induces continuous outward growth of the floc, leading to vertically standing nanoplates. Meanwhile, the amorphous core parts dissolve and finally form a hollow structure within 2 hours, as shown in Figure 2.9e and f. The hollow size is comparable to that of the smooth amorphous spheres (3–5 μm in diameter). Obviously, this process (dissolution and crystal growth) is relatively slow compared with the formation of the amorphous smooth spheres. After reaction for 1 or 2 hours, almost all diffraction peaks of the phase ZnHC can be detected in the XRD measurement. When the reaction time is longer than 2 hours, the vertically standing nanoplates remain ever growing and keep the hollow cores unchanged in size. When the reaction time reaches 12 hours, much larger micro/nanostructured honeycomb-like hollow spheres are obtained, as seen in Figure 2.7.

In summary, the aforementioned results indicate that when the reaction time is 5–10 minutes amorphous ZnHC spheres with smooth surfaces and around 3–5 μm in diameter are formed. After that, crystallization of the preformed smooth spheres starts in the surface layer. Then crystalline nanoplates grow vertically outward on the crystallized surface layer of the spheres; meanwhile, the amorphous core part underneath the crystallized surface layer dissolves continuously and forms a hollow structure within 2 hours.

2.2.2.3 Influence of Trisodium Citrate Dihydrate on Morphology

Further experiments have revealed that the structural director TC in the precursor solution plays a dominant role in the formation of micro/nanostructured honeycomb-like porous ZnO hollow spheres. Figure 2.10 shows the morphological dependence on the TC content in the precursor solution. Without TC, only crystalline nanoplates were obtained and no amorphous phase was found during initial reaction, as shown in Figure 2.10a. The nanoplates are much thicker than those shown in Figure 2.7d. At a low TC concentration (say, 0.6 $g{\cdot}L^{-1}$), the nearly spherical aggregation formed is accompanied by some small nanoplates (Figure 2.10b). Only when the TC concentration is about 1.2 $g{\cdot}L^{-1}$ are the regular honeycomb-like micro/nanostructured hollow spheres formed (Figure 2.7). When more TC is added, the morphology of the microspheres remains similar; but the thickness of the vertically standing nanoplates becomes thinner and the number density higher, as shown in Figure 2.10c and d, corresponding to the TC contents of 2 and 4 $g{\cdot}L^{-1}$, respectively. Further, if the TC content increases to a very high level (say, >24 $g{\cdot}L^{-1}$), no product is obtained.

2.2.2.4 Formation of Porous Nanoplate-Built Honeycomb-Like ZnO Hollow Spheres

During preparation of the precursor solution, the zinc acetate dihydrate and TC would dissolve in water and the Zn^{2+} cations and $C_6H_5O_7^{3-}$ anions were thus be formed in the solution. When the precursor solution is heated at 160°C, the urea molecules in the solution would hydrolyze [32] and release $NH_3{\cdot}H_2O$ and CO_2 (see Reaction 2.1).

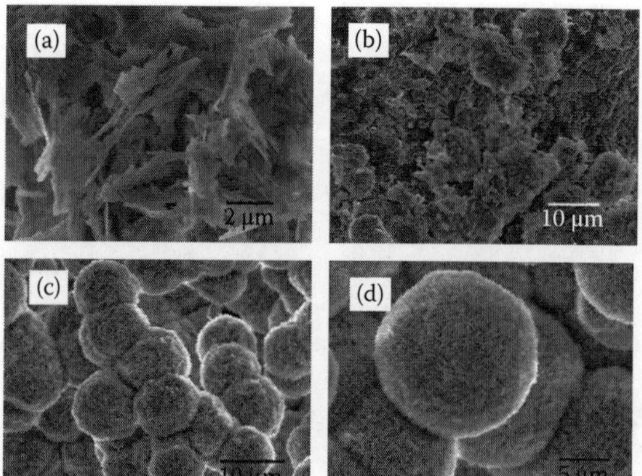

FIGURE 2.10 Field emission scanning electron microscopy images of the products (ZnHC) obtained from the precursor solution with different trisodium citrate dihydrate concentrations: (a) 0, (b) 0.6, (c) 2, and (d) 4 $g{\cdot}L^{-1}$. (Reaction temperature: 160°C; reaction time: 12 hours.) (From Wang et al., *J. Mater. Res.*, 27, 951–8, 2012.)

The $NH_3{\cdot}H_2O$ can release NH_4^+ and OH^- ions in solution (Reaction 2.2) and then react with CO_2 to produce CO_3^{2-} anions, or

$$2NH_3{\cdot}H_2O + CO_2 \rightarrow NH_4^+ + CO_3^{2-} \quad (2.6)$$

Finally, the formed OH^- and CO_3^{2-} ions would further react with Zn^{2+} to produce ZnHC molecules in the solution (Reaction 2.4).

2.2.2.4.1 Citrate Adsorption–Induced Formation of Amorphous ZnHC Spheres

The concentration of ZnHC molecules in the solution increases when the reaction proceeds, and when the concentration reaches supersaturation solid-phase ZnHC colloids will precipitate and grow. Correspondingly, the total free energy, G, for a spherical ZnHC solid phase should be the summation of its bulk and surface free energies [33] and can be written as a function of its size, or

$$G = f(R,\sigma) = \frac{4}{3}\pi R^3 G_v + 4\pi R^2 \sigma \quad (2.7)$$

where R is the radius of the spherical phase, and G_v and σ are the free energy in unit volume and the specific surface energy for the sphere, respectively. Usually, G_v and σ for amorphous ZnHC are bigger than those of the crystal one, since the crystal ZnHC is more stable than the amorphous one in the normal case, that is, the total free energy, G_a, of an amorphous sphere is bigger than that of a crystalline phase, G_c, with the same volume, or $G_a > G_c$ for all volumes, as schematically illustrated in Figure 2.11 (curve 1 is higher than curve 2). It means that precipitated ZnHC colloids should be crystalline in the usual case. However, the opposite is true in our case when the colloidal size is below approximately 3–5 μm (Figure 2.9a). This could be attributed to the existence of $C_6H_5O_7^{3-}$ in the solution.

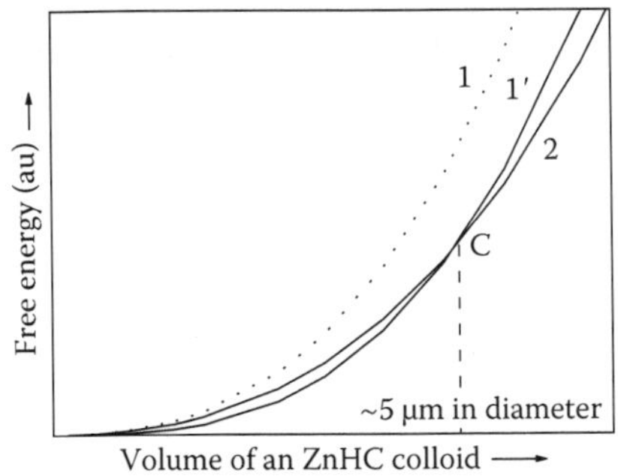

FIGURE 2.11 Schematic illustration for the free energy of a solid ZnHC sphere as a function of its volume: curves 1 and 1′ correspond to an amorphous ZnHC sphere without and with $C_6H_5O_7^{3-}$ adsorption, respectively. Curve 2 is for a crystal ZnHC sphere. Point C is the intersection point of curves 1′ and 2. (From Wang et al., *J. Mater. Res.*, 27, 951–8, 2012.)

The $C_6H_5O_7^{3-}$ in the solution will adsorb on the ZnHC colloids, due to the affinity between them, which was previously reported in the literature [34,35]. This adsorption will decrease the colloidal specific surface energyσ. When R is very small, the second term in the right of Equation 2.7 is dominant. If there is a significant decrease in σ value, curve 1 could decrease and be lower than curve 2 when the size is small enough, as schematically shown in curve 1′ of Figure 2.11. In this case, there should exist a critical size, C_s. When the colloidal size is smaller than C_s, the amorphous phase is more stable than the crystallized phase. Otherwise, when the size is larger than C_s, the opposite is true. On this basis, in our case also we believe that there exists a C_s, corresponding to point C in Figure 2.11 falling in the range of 3–5 μm, due to the adsorption of $C_6H_5O_7^{3-}$. When the colloidal size is smaller than 3–5 μm, the amorphous ZnHC phase is stable. Only when the size is bigger (>3–5 μm) will the crystallized ZnHC phase be more stable or the crystallization take place and start in the surface layer of the amorphous spheres (Figure 2.9c and d). Meanwhile, crystal growth will also occur outward on the crystallized layer due to attachment of the ZnHC molecules in the solution to form the rough surface morphology (Figure 2.9c and d).

2.2.2.4.2 *Kinetically Controlled Amorphous Spheres' Dissolution and Nanoplates' Growth*

Once the surface-crystallized layer of the spheres is formed, however, further inward crystallization will induce a significant increase in elastic or distortion energy due to crystallization-induced structural contraction [36]. It means that crystallization of the amorphous part underneath the surface-crystallized layer should be difficult and stop. In this case, dynamically, the dissolution of the amorphous part should be easier, since the chemical potential of ZnHC molecules is higher in the core part than in the crystallized surface layer. So, after the surface layer of amorphous spheres was crystallized, the ZnHC molecules in the inner parts would gradually dissolve into the solution, and the hollow structure was thus formed, which was finished within 1–2 hours during reaction (see Figure 2.9c through f). Meanwhile, the ZnHC molecules in the solution, due to reaction and dissolution, will diffuse and attach to the outward-growing crystal phase. Since crystal ZnHC is of a layered structure along the plane (100) [3], it tends to follow platelike growth with exposed plane of (100), leading to vertically standing nanoplates and formation of final honeycomb-like hollow spheres (Figure 2.7). Figure 2.12 schematically shows the formation process of the honeycomb-like ZnHC hollow spheres (stages I–VI).

2.2.2.4.3 *Effects of Trisodium Citrate Dihydrate*

Based on the discussion in Section 2.2.2.4.1, citrate anions $\left(C_6H_5O_7^{3-}\right)$ released from TC in the precursor solution is crucial to the formation of honeycomb-like micro/nanostructured ZnHC hollow spheres. It means that the amorphous spheres in the initial stage will not be formed during reaction without the addition of TC in the precursor solution, as there is no $C_6H_5O_7^{3-}$ adsorption. This has been confirmed, as shown in Figure 2.10a. Only crystal nanoplates were found and no amorphous phase was observed during initial reaction. Furthermore, after the addition of TC, since

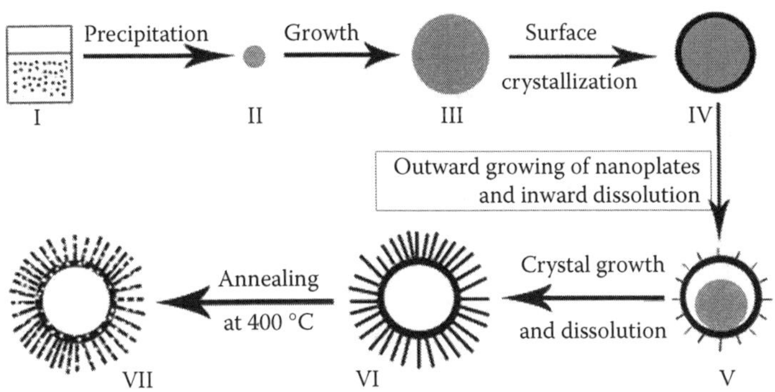

FIGURE 2.12 The schematic illustration of formation of a standing porous nanoplate-built ZnO hollow microsphere. Stage I: supersaturated solution. Stage II: a precipitated amorphous ZnHC colloid with small size. Stage III: the amorphous ZnHC sphere with a big enough size (say, ~5 μm) due to growth. Stage IV: the amorphous sphere with a crystallized surface layer (shell). Stage V: the sphere with outward-growing nanoplates on the crystallized surface layer and partially dissolved core part. Stage VI: the standing nanoplate-built ZnHC hollow sphere. Stage VII: the standing porous nanoplate-built ZnO hollow sphere. Gray color corresponds to the amorphous phase and black color to the crystallized phase. (From Wang et al., *J. Mater. Res.*, 27, 951–8, 2012.)

$C_6H_5O_7^{3-}$ adsorption will also occur on the nanoplates, an increase in TC content leads to the formation of thinner and denser nanoplates on the hollow spheres, as seen in Figure 2.10c and d. Furthermore, citrate is a chelating ligand, which can react with metal ions, form a complex, and control the release of metal ions [37–39]. So, if we add too much TC (say, 24 g·L^{-1} or higher) to the precursor solution a large amount of zinc cations would combine with citrate anions, or the reaction

$$Zn^{2+} + C_6H_5O_7^{3-} \rightarrow Zn(C_6H_5O_7)^-_{aq} \quad (2.8)$$

takes place and, hence, no product would be formed due to there being no adequate free Zn^{2+} ions in the solution.

Finally, the formation of porous honeycomb-like ZnO hollow spheres is similar to that of the pure ZnO nanoplates mentioned in Section 2.1 due to the decomposition of ZnHC during annealing at 400°C. Obviously, the formation of pores or voids in the ZnO nanoplates should be attributed to the topotactic transformation or a contraction of the original structure, that is, the Zn number density in wurtzite is higher compared to the precursor ZnHC, from which CO_2 and H_2O are released. This contraction of the original structure must lead to voids (see stage VII in Figure 2.12).

2.2.3 Nanoplate-Built Core/Shell-Structured ZnO Objects

Similarly, using ethylenediamine (EDA) ($NH_2CH_2CH_2NH_2$) as the structural director, and adopting Zn foil as the Zn source, we can obtain another newly structured ZnO hierarchical micro/nanoarchitecture via a facile solvothermal approach [1].

Typically, an aqueous solution of EDA was prepared with DIW and EDA according to the volume ratio 1:7. A piece of zinc foil was put into the solution together with a small amount of NaOH before reaction in an autoclave at 160°C for 12 hours. The products were then collected from the Zn foil.

2.2.3.1 Morphology and Structure

XRD pattern has indicated that the as-synthesized products scraped off from the Zn foil are wurtzite ZnO [1].

Figure 2.13 demonstrates the corresponding morphology of the as-synthesized ZnO. The products consist of a large number of conicle-like microsized particles (see Figure 2.13a). High-magnification FESEM image shows that the conicles are of netlike or gridlike morphology (see Figure 2.13b). It seems to be built from numerous nanosheets that stand almost vertically on the conicles' surface and alternately connect with each other to form networks or grids. The ZnO nanosheets are about 10 nm in thickness, and hundreds of nanometers in planar size, as shown in Figure 2.13c. A close examination of these nanosheet-built networks reveals that there is nearly a 60° angle between adjacent connected nanosheets for many nanosheets and the networks are about 100 nm in grid size. The slight deviation from 60° should mainly be attributed to the tilted observation

FIGURE 2.13 Field emission scanning electron microscopy images of the products prepared by solvothermal reaction at 160°C for 12 hours, V_{DIW}/V_{EDA} = 1:7. (a) A general-view image of products; (b) the local magnification image of (a); (c) a further enlarged top-view image of (b); and (d) an enlarged local image at the edge of the networks, incomplete nanosheet-built networks. The arrows in (c) indicate the distortion resulting from the jostling action between the densely arranged nanosheets. (Lu et al., *Adv. Funct. Mater.* 2008. 18. 1047–56. Copyright Wiley-VCH Verlag GmbH & Co. KGaA. Reproduced with permission.)

and the distortion due to jostle among the densely arranged nanosheets (see the arrow marks in Figure 2.13c). At some regions of incomplete nanosheet-built networks, where jostle exists relatively weakly, more adjacent nanosheets are easily observed jointing each other with the angle of about 60° (as shown in Figure 2.13d).

Figure 2.14a presents a TEM image for a typical isolated ZnO nanosheet obtained by ultrasonic dispersion of the as-prepared sample in ethanol. The corresponding SAED pattern (the inset in Figure 2.14a) indicates the single crystalline nature of the isolated nanosheet. The HRTEM image exhibits well-resolved two-dimensional lattice fringes with spacing of 2.6 and 2.8 Å, which are in good agreement with the interplanar spacing of {0002} and $\{01\bar{1}0\}$ planes, respectively, as shown in Figure 2.14b. It indicates that the nanosheet is of single crystalline structure with sheet-planar surfaces $\{2\bar{1}\bar{1}0\}$. One end of the nanosheet, marked by an arrow in Figure 2.14a, seems to be an interface disconnected from another sheet.

Further, the specific surface area was evaluated to be 185 $m^2 \cdot g^{-1}$ for this structured ZnO by measurement of the nitrogen sorption isotherm [1]. Obviously, this hierarchically structured ZnO with a high specific surface area will assume stability against aggregation, which could exhibit the potential application in catalysis and adsorption (see Chapters 5 and 6).

2.2.3.2 Morphology Evolution with Reaction Time

To understand how the ZnO hierarchical structures are formed, their time-dependent morphological evolution process was examined by FESEM. Figure 2.15 shows the morphologies of the sample after different reaction times. After reaction for 1 hour, no ZnO structures with discernible morphology are formed on the surface of the zinc foil, as shown in Figure 2.15a. After 2 hours, a few hexagonal prism–like microcrystals with a pyramidal top are observed, which aggregate together on the substrate

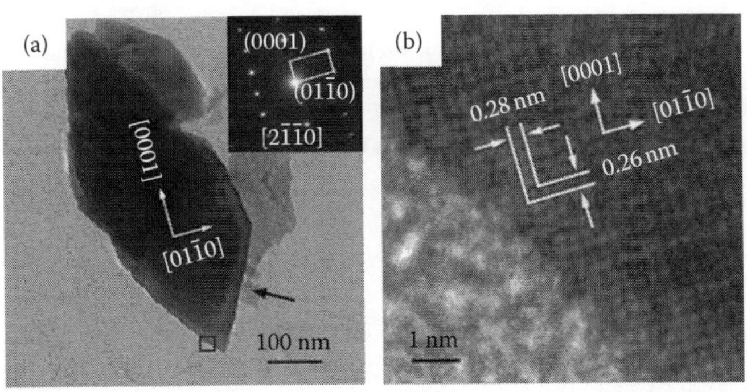

FIGURE 2.14 (a) Transmission electron microscopy (TEM) image of a typical nanosheet from the as-prepared sample (inset: the selected area electron diffraction pattern corresponding to the small frame area marked in the nanosheet, recorded along the $[2\bar{1}\bar{1}0]$ zone), and (b) the corresponding high-resolution TEM image. The arrow marked at one end of the nanosheet indicates an interface disconnected from another sheet. (Lu et al., *Adv. Funct. Mater.* 2008. 18. 1047–56. Copyright Wiley-VCH Verlag GmbH & Co. KGaA. Reproduced with permission.)

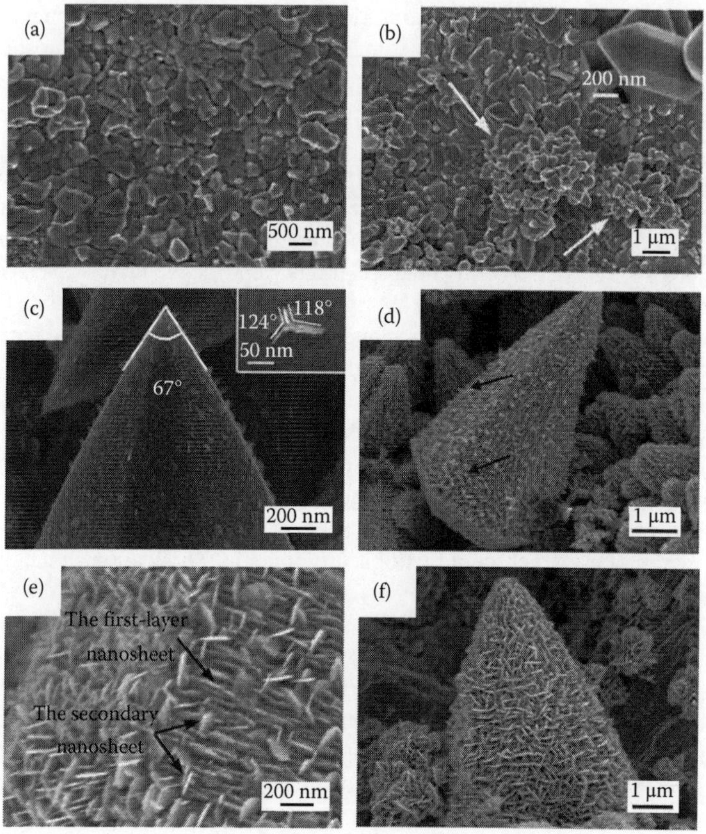

FIGURE 2.15 Field emission scanning electron microscopy images of the samples after different reaction times at 160°C for the solution with V_{DIW}/V_{EDA} = 1:7 and Zn foil: (a) 1 hour, (b) 2 hours, (c) 4 hours, (d) and (e) 8 hours, and (f) 16 hours. The inset in (c) shows a local top view exhibiting that partial nanosheets are connected in every three sheets, with the angle between two adjacent sheets being about 120°. The arrows in (d) indicate the outmost bladelike nanosheets standing astride on the bottom rows. (Lu et al., *Adv. Funct. Mater.* 2008. 18. 1047–56. Copyright Wiley-VCH Verlag GmbH & Co. KGaA. Reproduced with permission.)

(see the arrow marks), as shown in Figure 2.15b. The surfaces of these crystals are clean, and there is no covering on the surface (see the inset in Figure 2.15b). When the reaction time reaches up to 4 hours, some small sheets are found to stand nearly vertically on the surface of the microsized hexagonal pyramids, as shown in Figure 2.15c. The angle between two opposite edges at the tip of the micropyramid measures about 67°, which is close to the angle between two opposite edges at the tip of the hexagonal pyramid composed of the $\{01\bar{1}1\}$ planes as the side surfaces (63.94°). The deviation would result from a slightly tilted view of the shooting angle. The local high-magnification FESEM image (the inset in Figure 2.15c) shows that partial sheets have been connected, in every three sheets, with the angle between two adjacent sheets being nearly 120°. When the reaction is prolonged further, a significant change occurs in the morphology of the ZnO microcrystals. They would evolve from micropyramids to a more complex micro/nanoarchitecture with a delicate surface structure. After 8 hours, all the surfaces of the micropyramids become rougher, as displayed in Figure 2.15d. Two layers of vertically arranged nanosheets seem to be formed on the surfaces of the micropyramids. The top layer is incomplete (see the arrow-marked areas in Figure 2.15d). These two layers of vertically arranged nanosheets can be seen more clearly in Figure 2.15e, which corresponds to the local area of Figure 2.15d. Most of the nanosheets on the surface of the pyramid-like microcrystal (or the first layer) are arranged into dense parallel rows (nonnetwork pattern), whereas the nanosheets on the top layer vertically stand and show a netlike morphology from the top view (see Figure 2.15e), which is similar to that shown in Figure 2.13. As the reaction time reaches up to 12 hours, the complete top layer is formed with the nanosheets standing nearly vertically and interlacing into a netlike structure, as shown in Figure 2.13. The whole hierarchically structured particles exhibit netlike or gridlike surface morphology. If the reaction time further increases to 16 hours, the product still shows the similar morphology but grows bigger in size since more layers are formed, as seen in Figure 2.15f.

Further, Figure 2.16a presents a TEM examination of an isolated ZnO nanosheet from the sample shown in Figure 2.15d. This sheet exhibits a similar shape to that of the nanosheets in the bottom layer shown in Figure 2.15e. One end of the nanosheet, marked by an arrow in Figure 2.16a, seems to be an interface disconnected from the micropyramid's surface. The SAED pattern (the inset in Figure 2.16a) and HRTEM examination (Figure 2.16b) confirm its single crystalline nature. The angle between

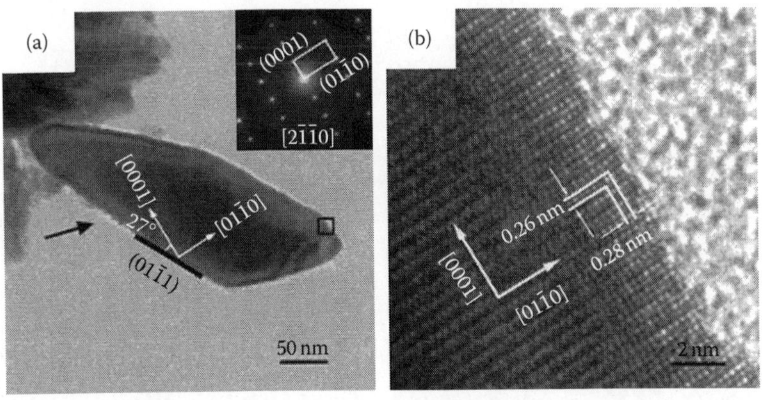

FIGURE 2.16 (a) TEM image of a typical nanosheet from the sample shown in Figure 2.14d (inset: the selected area electron diffraction pattern corresponding to the small frame area marked in the nanosheet, recorded along the [2$\bar{1}\bar{1}$0] zone), and (b) the corresponding high-resolution TEM image. The arrow marked at one end of the nanosheet indicates an interface disconnected from the micropyramid's surface. (Lu et al., *Adv. Funct. Mater.* 2008. 18. 1047–56. Copyright Wiley-VCH Verlag GmbH & Co. KGaA. Reproduced with permission.)

the end plane and the known crystal plane, $(01\bar{1}0)$, can be measured from the image to be about 27° (see Figure 2.16a), which is close to the interplanar angle between $(01\bar{1}0)$ and $(01\bar{1}1)$ planes (28.4°). Thus, the end plane of the nanosheet marked by an arrow in Figure 2.7a can be determined to be the crystal plane $\{01\bar{1}1\}$. Such results would be helpful for us to understand the growth mechanism of ZnO hierarchical structures observed here.

2.2.3.3 Influence Factors

Further experiments indicate that the concentration in the EDA aqueous solution is crucial for the formation of such complex ZnO hierarchical structures, meaning the importance of EDA as the structural director. The ZnO morphology depends on different volume ratios of DIW to EDA (V_{DIW}/V_{EDA}) when keeping other experimental conditions unchanged. In the absence of water, only nearly vertically arranged nanorod arrays are formed on the Zn foil in pure EDA (Figure 2.17a). The diameters of the nanorods fall in the range from 50 to 200 nm. When the ratio V_{DIW}/V_{EDA} increases to about 3:37, the products consist of conicle-like microparticles with

FIGURE 2.17 Field emission scanning electron microscopy (FESEM) images of the products synthesized from the reaction solution with Zn foil and different V_{DIW}/V_{EDA} values at 160°C for 12 hours: (a) 0, (b) 3:37, and (c) 1:3 (the insets 1 and 2 are a side view and a top view, respectively, for local area on the surface of a microconicle). (d) The FESEM image of the products prepared with Zn $(NO_3)_2$ instead of Zn foil at 160°C for 12 hours (V_{DIW}/V_{EDA} = 1:7). (Lu et al., *Adv. Funct. Mater.* 2008. 18. 1047–56. Copyright Wiley-VCH Verlag GmbH & Co. KGaA. Reproduced with permission.)

rough surfaces, as illustrated in Figure 2.17b. Further examination shows that there are many vertically standing nanoplatelets on the surface (see inset 1 in Figure 2.17b, a local side view). These nanoobjects do not arrange into a netlike pattern. Nevertheless, some adjacent connected nanoplatelets assume a regular angle of 60° (see inset 2 in Figure 2.17b, a local top view). By further increasing V_{DIW}/V_{EDA} to 1:7, we can obtain a hierarchical structure with a gridlike morphology for the surface, as shown in Figure 2.13. However, if V_{DIW}/V_{EDA} is too high (say, 1:3), randomly distributed microrods are observed on the Zn foil (see Figure 2.17c). These results indicate that products only with monomorphology, rather than hierarchical structures, are formed in an EDA aqueous solution with too low or too high percentages of DIW. These ZnO hierarchical architectures can be synthesized only in a binary solution with a suitable ratio of V_{DIW} to V_{EDA} (about 1:7).

In addition, the choice of zinc source in the reaction system also shows an obvious influence on the formation of ZnO hierarchical structures. We have used zinc salts, such as $Zn(NO_3)_2$, instead of zinc foil as the starting material and kept other conditions unchanged. When $Zn(NO_3)_2$ is used as the reactant, we get ZnO particles with irregular shape (~100–500 nm in size) and no hierarchical structures are formed (Figure 2.17d).

2.2.3.4 Two-Step Sequential Growth Model

When the precursor solution is heated, reactions can occur as follows [40,41]:

$$Zn + m\text{EDA} + 2H_2O \rightarrow \left[Zn\ (\text{EDA})_m\right](OH)_2 + H_2 \tag{2.9}$$

$$\left[Zn\ (\text{EDA})_m\right]^{2+} \leftrightarrow Zn^{2+} + m\text{EDA} \tag{2.10}$$

$$Zn^{2+} + 2OH^- \rightarrow Zn\ (OH)_2 \rightarrow ZnO + H_2O \tag{2.11}$$

where m is a positive integer. In the basic environment with EDA, Zn atoms at the liquid–solid (Zn foil) interface are oxidized to form soluble coordinated ions $[Zn(EDA)_m]^{2+}$, which further decompose into Zn^{2+} ions at an elevated temperature (>90°C) according to Reactions 2.9 and 2.10 [40,41]. The Zn^{2+} ions will further form $Zn(OH)_2$ with OH^-. Finally, ZnO is obtained by the decomposition of $Zn(OH)_2$ (Reaction 2.11). When the concentration of ZnO reaches supersaturation, ZnO crystal nuclei form and then grow according to the growth habit of ZnO crystal.

From the morphological evolution with reaction time, we can know that the growth of such structured ZnO could be described in two steps, as illustrated in Figure 2.18. At the first stage, ZnO hexagonal pyramid–like microcrystals were formed after nucleation. Generally, the faces perpendicular to the fast direction of growth have smaller surface areas, and the faces whose normal directions correspond to slow-growing ones thus dominate the final morphology. For ZnO crystal, the growth rates, V, along the normal directions of different low-index planes in alkaline medium are described as follows: $V_{(0001)} > V_{(10\bar{1}0)} \geq V_{(10\bar{1}\bar{1})} > V_{(10\bar{1}1)} > V_{(000\bar{1})}$ [42]. In our case, the Zn foil and the mixture solution are used to control the release rate of Zn^{2+}. As the reaction proceeds, the surfaces whose normal directions have a fast growth rate disappear, whereas the slow-growing surfaces remain. As a result, ZnO hexagonal

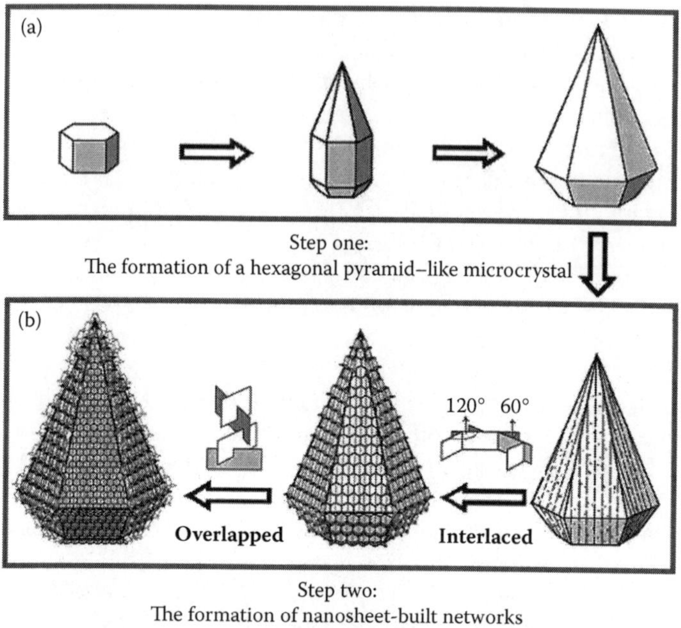

FIGURE 2.18 The schematic illustration of a two-step sequential growth model for the formation of the nanoplate-built core/shell-structured ZnO. (Lu et al., *Adv. Funct. Mater.* 2008. 18. 1047–56. Copyright Wiley-VCH Verlag GmbH & Co. KGaA. Reproduced with permission.)

pyramid–like microcrystals with the $(000\bar{1})$ basal plane and the $\{0\bar{1}11\}$ lateral planes are formed (as shown in Figure 2.15b or schematically illustrated in Figure 2.18a). This is different from the growth mechanism of tubular graphite cones by a chemical vapor deposition method reported by Zhang et al. [43,44], which consist of coaxial tubular graphite sheets. In this case, the cone-shaped structure is caused by a gradual shortening of length of the graphite sheets along the axial direction from the inner to the outer layers of the tube, and layer steps are present on the cone surface.

The second step is the formation of nanosheet networks on the facets of preformed ZnO microcrystals. After the hexagonal pyramid–like microcrystals are formed, further growth will be difficult due to their pyramid-like geometry. However, there exist inevitably defects or bulges on the lateral planes of microcrystals. Such bulges will preferentially grow along $\langle 0001 \rangle$ and $\langle 01\bar{1}0 \rangle$ (both are fast-growth directions) within the $\{2\bar{1}\bar{1}0\}$ plane and form nanosheets that are nearly vertically standing on the lateral surface of the microcrystals. In addition, there also inevitably exist some outshoots on the growing nanosheets. Such outshoots will grow and lead to the formation of secondary dendrite-like (or branched) nanosheets with $\{2\bar{1}\bar{1}0\}$ facets terminated and an interplanar angle of 60° due to the same reason. Also, third or fourth branched nanosheets could be formed on the as-grown nanosheets. Finally, more and more nanosheets with $\{2\bar{1}\bar{1}0\}$ planar surface, interlacing with and overlapping each other into a discernible multilayer and network structure, stand on the microcrystals, constituting the ZnO hierarchical micro/nanoarchitecture (see Figure 2.13, or schematically illustrated in Figure 2.18b). This growth process is similar to that of the α-MnO_2 hierarchical structures [45] and the Pt *sea urchin* [46] reported by Xia and colleagues. Finally, combining the TEM examinations shown in Figures 2.14 and 2.16, and based on the earlier discussion, the orientation relationship in this ZnO hierarchical micro/nanoarchitecture can be schematically summarized as in Figure 2.19.

Obviously, the formation of such a hierarchical structure should depend on the growth rate of ZnO crystal or the release rate of Zn^{2+} ions in the solution. Too high and too low growth rates of the crystal, corresponding to the release rate of Zn^{2+} ions in Reaction 2.9, would be unbeneficial to the aforementioned two-step growth, since a high growth rate can only lead to monomorphology (one-step growth) and a low growth rate would result in quasi-equilibrium growth. When V_{DIW}/V_{EDA} is low or no water is added to the solution, the release rate of Zn^{2+} ions in the solution, and hence the growth rate, would be slow according to Reaction 2.9. In this case, quasi-equilibrium growth of ZnO will occur, leading to the formation of nearly vertically arranged nanorod arrays on the Zn foil (see Figure 2.17a). On the contrary, a high ratio of V_{DIW} to V_{EDA}, or especially the use of $Zn(NO_3)_2$ instead of Zn foil, will induce a high release rate of Zn^{2+} ions and hence fast growth of the crystal, leading to randomly distributed particles with monomorphology (see Figure 2.17c and d). Formation of the ZnO hierarchical structures or two-step growth mode can only occur at a certain release rate of Zn^{2+} ions (here, using Zn foil as the zinc source and V_{DIW}/V_{EDA} = 3:37–1:7), which provides a kinetically favorable condition for ZnO hierarchical growth, similarly to the precursor-induced synthesis of hierarchical nanostructured ZnO [47–49].

Similarly, according to the strategy shown in Figure 2.1, we can also obtain other micro/nanostructured materials, such as TiO_2 and ZnS, using the corresponding precursors and structural directors.

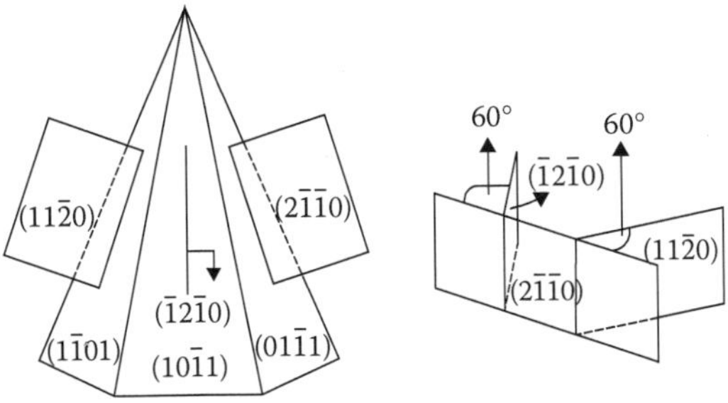

FIGURE 2.19 Schematic drawings of orientation relationship in ZnO hierarchical growth: (a) the primary hexagonal micropyramid covered with the nanosheets, and (b) some secondary or tertiary nanosheets grown from the sheet-planar surfaces of the one formed previously. (Lu et al., *Adv. Funct. Mater.* 2008. 18. 1047–56. Copyright Wiley-VCH Verlag GmbH & Co. KGaA. Reproduced with permission.)

2.3 MICRO/NANOSTRUCTURED POROUS Fe_3O_4 NANOFIBERS

Nanostructured magnetite (Fe_3O_4) is one of the most important magnetic materials and has attracted extensive attention because of its potential applications in magnetic separation [50,51], in cancer diagnostics and treatment [52], as a contrast agent in magnetic resonance imaging [53,54], and in ferrofluids [55] and other aspects [56,57]. It is well known that (quasi-) one-dimensional (1D) nanostructures are important building blocks for nanodevices [58,59]. Considerable efforts have been made to fabricate 1D nanostructures [60–66]. Also, many methods have been developed to fabricate 1D magnetite nanostructures, such as magnetic field induction [67], chemical vapor deposition [68], oxide precursor routes [69,70], and hard template methods [71]. These methods for 1D magnetite nanostructures usually need some special equipment. Furthermore, it is a tedious and limited method using an external magnetic field. Additionally, the reported 1D single crystalline magnetite nanostructures [67] and some with ordering of the magnetite ultrafine particles [72–74] usually have the lower surface-energy and magnetization, which restrict their applications in, such as adsorption as adsorbent and magnetic separation in the solution. Development of simple and effectual approaches to fabricate 1D magnetite nanostructures with high yield, large specific surface areas, and low cost still remains a challenge.

Here, we introduce a facile and green route for mass production of quasi-1D magnetite fibers with porous structure, based on protein (PR)-assisted hydrothermal method in citric aqueous solution at 200°C [75]. In a typical preparation, first, $FeCl_3 \cdot 6H_2O$ (0.01 mol) and citric acid $\left(C_6H_5O_7^{3-}\right)$ (0.02 mol) (the molar ratio of $C_6H_5O_7^{3-}$ to Fe^{3+} is 2:1) were added to a 100 mL beaker with DIW (60 mL) and continuously stirred to form a homogeneous fawn solution. In succession, glutin (1.0 g) was added to the aforementioned solution and heated at 40°C for quickly dissolving the solid glutin to form a PR colloidal solution (16.7 $g \cdot L^{-1}$). Then, the pH value of the aforementioned solution was adjusted to 7–8 by adding adequate ammonia. This precursor solution exhibited typical Tyndall effect, indicating that the PR colloidal solution or hydrosol was obtained. Second, the colloidal solution was transferred to a Teflon-lined autoclave and maintained at 200°C for 24 hours before cooling to ambient temperature naturally. The solid black products or magnetic powders in the autoclave were collected by a magnet.

2.3.1 Morphology and Structure

XRD measurement has indicated that the collected black powders or the as-prepared products correspond to the phase Fe_3O_4. The obtained Fe_3O_4 is of fiber-like morphology with nearly cylindrical shape and nanoscaled surface roughness, as illustrated in Figure 2.20a and its inset. These fibers are several tens of micrometers in length and 120 nm in mean diameter (see Figure 2.19b).

Microstructural examination was conducted for these Fe_3O_4 fibers. Under TEM fields, it was seen that the fibers consist of ultrafine nanoparticles with sizes smaller than 5 nm and are porous in structure. The surface of the fibers is thus rough, as seen in Figure 2.21a. HRTEM examination has revealed that all ultrafine nanoparticles in the fibers show nearly the same crystal orientation, as shown in Figure 2.21b. The fringes perpendicular to the axis of the fiber are separated by 4.89 Å,

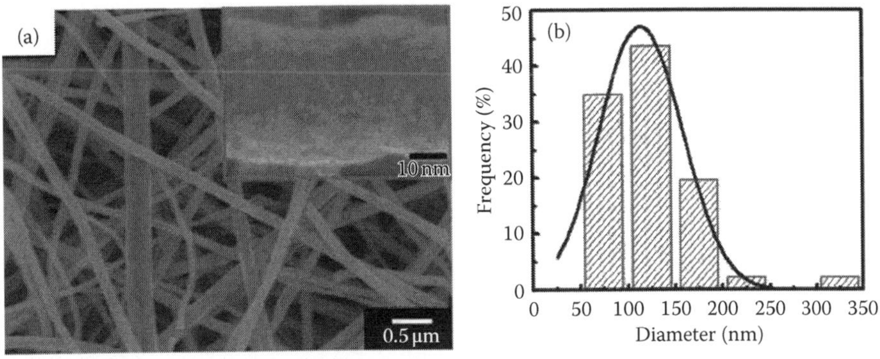

FIGURE 2.20 Morphology and size distribution of the as-prepared sample: (a) field emission scanning electron microscopy image (inset: a magnified image of a segment of a single fiber). (b) Diameter distribution of the magnetite fibers obtained from (a); solid line: Gaussian-type fitting curve. (Li et al., *J. Mater. Chem.*, 21, 11188–96, 2010. Reproduced by permission of The Royal Society of Chemistry.)

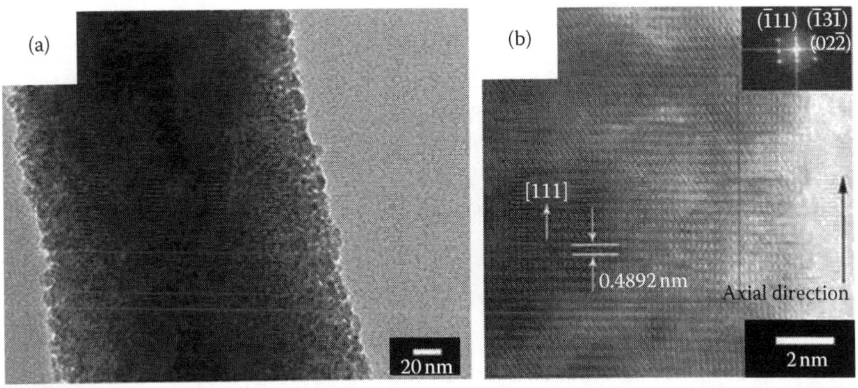

FIGURE 2.21 Microstructural examination of Fe_3O_4 nanofibers: (a) high-magnification TEM image of a single fiber, and (b) high-resolution TEM image of a partial fiber ([211] zone) (inset: numerical diffraction pattern and spot indexation with $(\bar{1}11)$, $(\bar{1}3\bar{1})$, and $(02\bar{2})$ planes of Fe_3O_4). (Li et al., *J. Mater. Chem.*, 21, 11188–96, 2010. Reproduced by permission of The Royal Society of Chemistry.)

which is in good agreement with the plane spacing of {111} of the cubic magnetite. The diffraction patterns calculated from the HRTEM image show that a majority of crystallites are of the similar orientation around the [211] zone and clearly identified by $(\bar{1}11)$, $(\bar{1}3\bar{1})$, and $(02\bar{2})$ planes, as shown in the inset of Figure 2.20b. It means that the whole single fiber is packed by ultrafine Fe_3O_4 nanoparticles with similar crystal orientation and, hence, it is porous with pore size in the nanoscale. Such 1D porous fibers could be attributed to assembling of the many ultrafine magnetite nanoparticles along the [111] direction, which is a very easy magnetization orientation of Fe_3O_4 [76]. Energy-dispersive x-ray spectroscopic microanalysis, performed during the TEM observation, has confirmed that the sample is mainly composed of Fe and oxygen (O) with only traces of carbon (C) and copper from the C-coated grid [75]. The atomic ratio of Fe to O is close to the stoichiometry of Fe_3O_4. In addition, the yield of Fe_3O_4 porous fibers can be estimated to be more than 95% based on the Fe molar numbers in the precursor solution and the final product Fe_3O_4 (by weighing). The synthesis method in this study could realize mass production of Fe_3O_4 porous fibers with high yield and low cost only if a big enough autoclave was used.

Further, nitrogen sorption measurement was conducted to evaluate the porous structure and specific surface area of such magnetite fibers. The specific surface area was thus estimated to be 123 $m^2 \cdot g^{-1}$ [75]. Since the fibers' surface area can be estimated to be about 6.4 $m^2 \cdot g^{-1}$ according to the mean diameter (120 nm), the measured surface area is mainly contributed by the walls of nanopores within the Fe_3O_4 fibers.

Figure 2.22 shows the morphological evolution of the fibers with reaction time. If the reaction time is too short (say, <4 hours), no product is found. When the reaction time is 4.5 hours, only a small amount of magnetite fibers is formed with an average diameter shorter than 100 nm, and there are many nanoparticles attached to the fibers (see Figure 2.22a). When the reaction time is in the range of 10–30 hours, more fibers are formed. The diameter distribution of the fibers narrows with increasing reaction time, whereas the mean diameter is almost unchanged (120–130 nm), as shown in Figures 2.20a and 2.22b and c. If the reaction time is longer than 30 hours, length of the fibers is shortened to below 10 μm, as shown in Figure 2.22d corresponding to the reaction time of 36 hours. A longer reaction leads to shorter fibers with nearly unchanged diameter, as illustrated in Figure 2.22e and f corresponding to the reactions for 2 and 3 days, respectively.

2.3.2 Influencing Factors

Extensive and systematic studies have revealed that many factors influence the formation and morphology of magnetite. The reaction temperature and the composition (PR content and the molar ratio of $C_6H_5O_7^{3-}/Fe^{3+}$) and pH value of the precursor solution are crucial to the formation of the magnetite fibers.

2.3.2.1 pH Value

It has been found that pH value of the precursor solution will influence the morphology of the magnetite formed [75]. If the pH is low (pH = 4–5), the products are magnetite with chain-like morphology, consisting of big particles about 500 nm in size. Only when the pH value is 7–8, high-quality magnetite fibers can be obtained, as shown in Figure 2.20a. A high pH value (9–12) cannot form magnetite fibers but cause the aggregation of ultrafine Fe_3O_4 nanoparticles.

2.3.2.2 Reaction Temperature

Temperature determines the chemical reaction rate and assembling behavior of magnetite nanoparticles and hence plays an important role in the formation of Fe_3O_4 porous fibers. It was found that no reaction took place if the temperature was

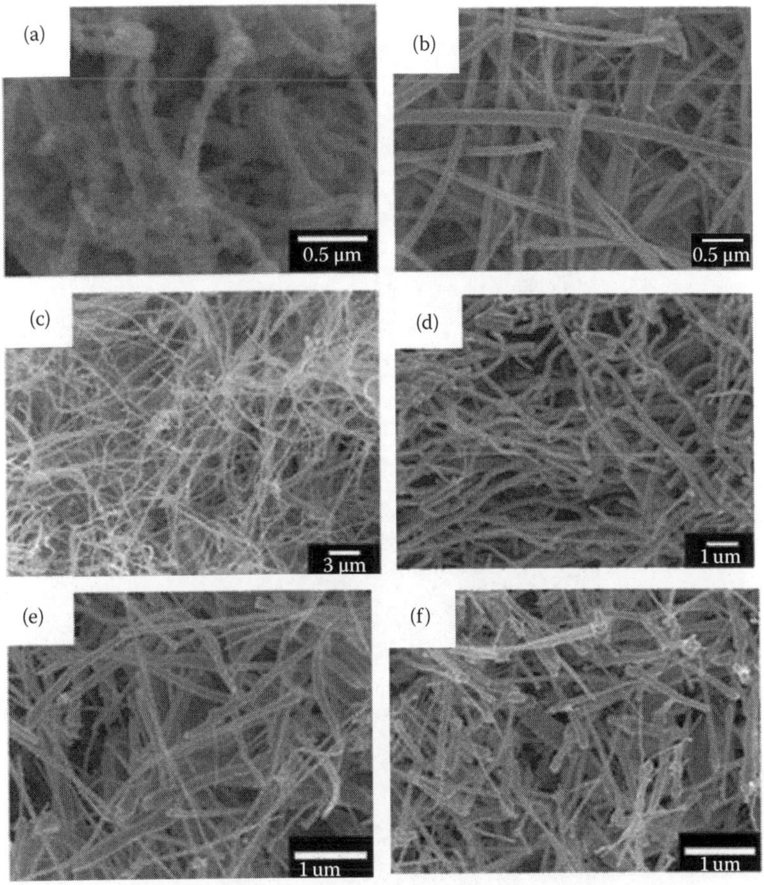

FIGURE 2.22 Morphological evolution of the Fe_3O_4 products with reaction time at 200°C: (a) 4.5, (b) 12, (c) 18, (d) 36, (e) 48, and (f) 72 hours. (Li et al., *J. Mater. Chem.*, 21, 11188–96, 2010. Reproduced by permission of The Royal Society of Chemistry.)

below 180°C and very high temperatures were not beneficial to the formation of Fe_3O_4 fibers. When the reaction temperature was about 220°C or higher, the product obtained was a mixture of thick Fe_3O_4 fibers and big spherical particles [75]. The relative content depends on the temperature, and a higher temperature leads to a larger amount of spherical particles. Only when the reaction temperature was kept around 200°C were the products homogeneous Fe_3O_4 porous fibers, as seen in Figure 2.20a.

2.3.2.3 Content of PR in the Precursor Solution

It was demonstrated that the content of PR in the solution played an irreplaceable role in the formation of the fibers. Figure 2.23 shows the morphological evolution of the final products with PR content, from 0 to 33.3 g/L, in the precursor solution.

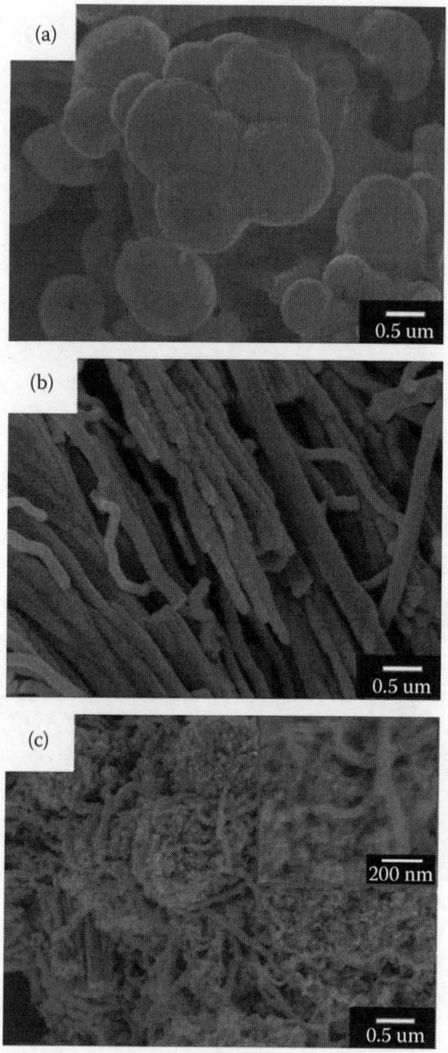

FIGURE 2.23 Morphological evolution with PR content in the precursor solution: (a) 0, (b) 13.3, and (c) 33.3 g·L^{-1} (Inset: The local magnified image). (Li et al., *J. Mater. Chem.*, 21, 11188–96, 2010. Reproduced by permission of The Royal Society of Chemistry.)

Without PR in the precursor solution, only nearly spherical microsized magnetite particles were formed (Figure 2.23a). Adding a small amount of PR (<10 g·L^{-1}) leads to the production of a mixture of magnetite fibers and spherical particles. When PR was added in a proper content (10–20 g·L^{-1}), high-quality magnetite fibers were formed, as shown in Figures 2.20a and 2.23b. But too much PR was added (say, >30 g·L^{-1}), we only obtained a mixture of magnetite fibers and aggregated nanoparticles. Figure 2.23c shows the result for the sample corresponding to a PR content of 33.3 g·L^{-1}.

2.3.2.4 Effect of $C_6H_5O_7^{3-}$

Further experiments have revealed that addition of $C_6H_5O_7^{3-}$ (Cit) is crucial to the formation of the magnetite phase. Without Cit in the precursor solution, we can only obtain α-Fe_2O_3 nanocrystals, instead of magnetite. When the molar ratio of Cit to Fe^{3+} is about 0.5:1 or higher, full magnetite phase can be formed. However, only when the molar ratio is 2:1 in the precursor solution, while maintaining the pH value at about 7–8 and the reaction time at 10–30 hours, will pure magnetite fibers be obtained, as shown in Figures 2.20a and 2.22b and c. A lower or higher molar ratio of Cit to Fe^{3+} in the precursor solution cannot lead to a high quality of fibers but a mixture of tangled nanowires and spherical magnetite particles [75]. It indicates that $C_6H_5O_7^{3-}$ in the solution not only plays a dominant role in the formation of the magnetite phase but also exhibits an important effect on magnetite morphology.

In summary, Fe_3O_4 porous fibers can be formed only under proper conditions, including moderate PR and Cit concentrations, appropriate pH value in the precursor solution, and a suitable reaction temperature.

2.3.3 Formation of Porous Oriented Magnetite Fibers

The process before the formation of magnetite nanoparticles could be described by the following reactions:

$$Fe^{3+} + 2C_6H_5O_7^{3-} + 2OH^- \xrightarrow{25°C} [Fe(C_6H_4O_7)_2]^{5-} + 2H_2O \qquad (2.12)$$

$$2[Fe(C_6H_4O_7)_2]^{5-} + 2OH^- \xrightarrow{200°C} 2Fe(OH)_2 + 4C_6H_3O_7^{3-} + 2H_2O \qquad (2.13)$$

$$6Fe(OH)_2 + O_2 \xrightarrow{200°C} 2Fe_3O_4 + 6H_2O \qquad (2.14)$$

During the preparation of the precursor solution at room temperature, when $FeCl_3$ was mixed with citric acid in DIW, soluble $[Fe(C_6H_4O_7)_2]^{5-}$ complexes were formed by the coordinate reaction between Fe^{3+} and Cit [72–74] (see Reaction 2.12). Following subsequent heating at 200°C, the $[Fe(C_6H_4O_7)_2]^{5-}$ complexes will decompose and form $Fe(OH)_2$, which is finally oxidized to Fe_3O_4 molecules (Reactions 2.13 and 2.14) [77].

2.3.3.1 Formation of Ultrafine Fe_3O_4 Nanoparticles

When the concentration of Fe_3O_4 molecules was supersaturated, ultrafine Fe_3O_4 nanoparticles were formed in the solution by nucleation and growth. Furthermore, due to the strong coordination affinity of Cit to Fe^{3+} ions [78], the citric groups will attach on the surface of Fe_3O_4 nanoparticles, which will thus be charged negatively, as seen in Figure 2.24a. The corresponding zeta potential was measured to be –35 mV, which prevents further growth and aggregation of the nanoparticles [79].

2.3.3.2 PR-Directed/Magnetic Dipole–Induced Orientation Assembling

Here, the PR is formed from the acid hydrolysis of gelatin (type A) [80], whose isoelectric point (p*I*) is about 9, and possesses long chain-like molecular structures in solution at a moderate temperature. The charged property of PR can be

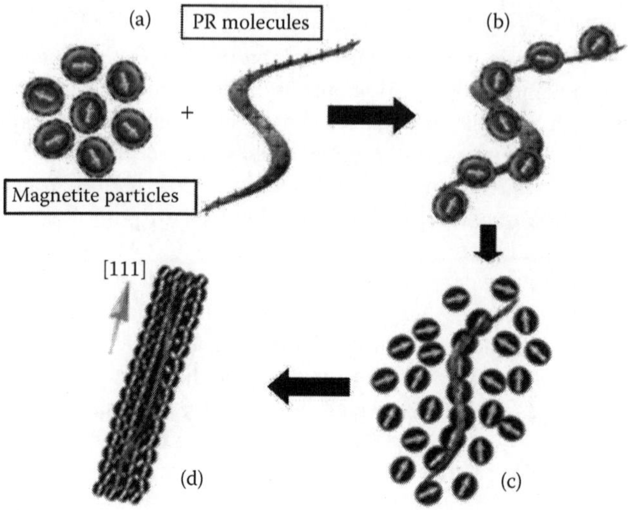

FIGURE 2.24 Schematic illustration for formation of porous oriented Fe_3O_4 fibers: (a) the negatively charged ultrafine Fe_3O_4 nanoparticles and the chain-like PR molecule with positive charges, dispersed in the solution; (b) the oriented Fe_3O_4 nanoparticle chain, formed by attachment of the nanoparticle on the PR chain; (c) the Fe_3O_4 nanoparticle chain attached to its surrounding ultrafine Fe_3O_4 nanoparticles; and (d) porous oriented Fe_3O_4 fibers. (Li et al., *J. Mater. Chem.*, 21, 11188–96, 2010. Reproduced by permission of The Royal Society of Chemistry.)

controlled by the pH value of solutions. Here, the pH value is about 7–8, lower than the PR's p*I* value, and thus PR chains will be positively charged in the solution, as seen in Figure 2.24a. Hence, the negatively charged ultrafine magnetite nanoparticles will interact with the positively charged chain-like PR molecules due to coulomb attraction, resulting in the attachment of Fe_3O_4 nanoparticles on the PR chains, as schematically shown in Figure 2.24b. Such Fe_3O_4@PR chains will continue to grow (thicker and thicker) during the reaction, by continuous attachment of the surrounding Fe_3O_4 nanoparticles in the solution on the formed Fe_3O_4@PR chains (see Figure 2.24c). On the other hand, due to the interaction of magnetic dipoles, the attached Fe_3O_4 nanoparticles will be reoriented and assemble effectively along the [111] direction, which is a very easily magnetized orientation of Fe_3O_4 [76]. Finally, if the reaction time is long enough, porous fibers packed with [111]-oriented Fe_3O_4 nanoparticles would be formed, as shown in Figure 2.24d. The PR in the core part of the fibers can be partially removed during subsequent ultrasonic rinsing in DIW due to its soluble property in water (see the following section).

2.3.3.3 Existence and Removal of PR in the Fibers

Extensive and systematic work should be done to finally clarify the formation mechanism of porous fibers. Here, we only show the existence of PR in the fibers by further experiments. In fact, Fourier transform infrared (FTIR) spectroscopic measurements have shown the presence of PR in the fibers (Figure 2.25a). TG measurement of the as-prepared sample after rinsing, in N_2, has indicated a mass loss of 4.6% in the range from 160°C to 400°C, which should be mainly attributed to the decomposition of PR, as demonstrated in Figure 2.25b.

Further, the as-prepared fibers were annealed at 600°C for 2 hours under a vacuum of 3.8×10^{-5} torr and then TEM examination was conducted, as illustrated in Figure 2.26. It has been shown that the porous fibers have been transformed into [111]-oriented single crystal fibers without pores (Figure 2.26b). In the core part, there exists a tubelike structure about 30 nm in diameter (see the arrows in Figure 2.26a). The formation of this tubelike structure should be attributed to the existence of chain-like PR in the core part of the fibers before annealing. Annealing induces the decomposition of PR and sintering of

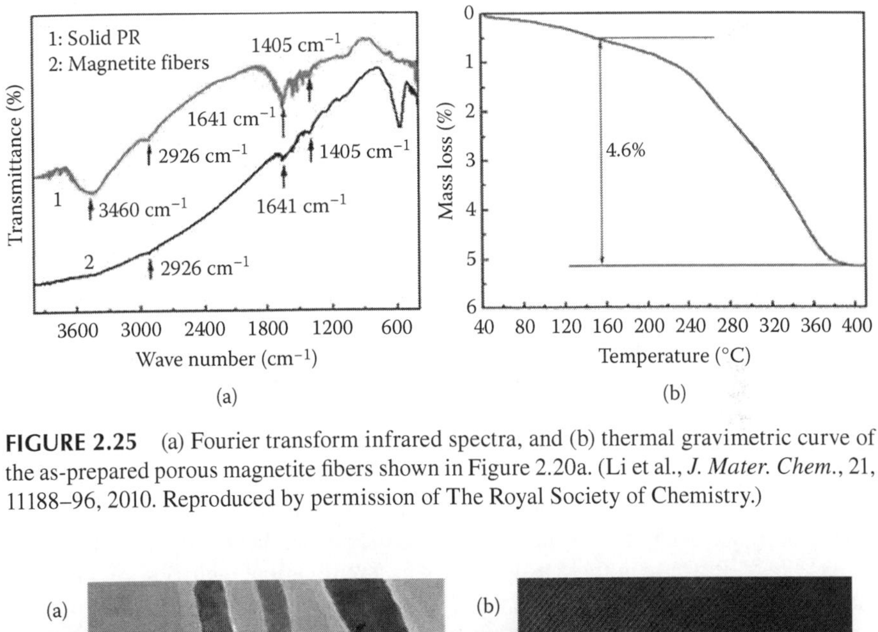

FIGURE 2.25 (a) Fourier transform infrared spectra, and (b) thermal gravimetric curve of the as-prepared porous magnetite fibers shown in Figure 2.20a. (Li et al., *J. Mater. Chem.*, 21, 11188–96, 2010. Reproduced by permission of The Royal Society of Chemistry.)

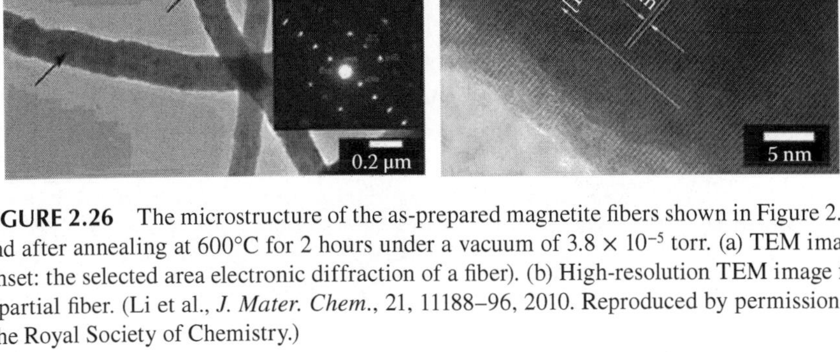

FIGURE 2.26 The microstructure of the as-prepared magnetite fibers shown in Figure 2.2a and after annealing at 600°C for 2 hours under a vacuum of 3.8×10^{-5} torr. (a) TEM image (inset: the selected area electronic diffraction of a fiber). (b) High-resolution TEM image for a partial fiber. (Li et al., *J. Mater. Chem.*, 21, 11188–96, 2010. Reproduced by permission of The Royal Society of Chemistry.)

magnetite nanoparticles, which, in turn, leads to contraction of the fibers and hence formation of the tubelike structure.

2.3.4 Effects of Reaction Parameters

Based on the discussion in Section 2.3.3.3, it is easy to understand the effects of reaction parameters on the morphology of magnetite.

Obviously, the formation of oriented porous fibers involves the attachment of Fe_3O_4 nanoparticles on PR and subsequent reorientation, which need enough time. When the reaction time is short or during the initial reaction (about 4–6 hours), we obtain only a small amount of fine fibers (Figure 2.22a). The fibers will get thicker and amount will increase till the reaction is completed (Figure 2.22c). Reaction times that are too long will induce breaking at some defective sites on the fibers, leading to the formation of shorter fibers (Figure 2.22d through f).

If the reaction temperature is too high, many Fe_3O_4 nanoparticles in the solution will aggregate before attaching on the PR due to high mobility and collision-induced connection, leading to a mixture of the fibers and big spherical aggregation [75]. At lower temperatures, Reactions 2.13 and 2.14 will not occur or Fe_3O_4 will not be formed.

When the pH value in the solution is too high (>9), the PR will not be positively charged. It means that coulomb attraction will not occur between the PR and the surrounding Fe_3O_4 nanoparticles and, hence, the long chains consisting of Fe_3O_4 nanoparticles will not be formed, only leading to the aggregation of the ultrafine nanoparticles.

Since PR plays a similar role to templates in the formation of the fibers, without PR or with too small amounts of PR in the precursor solution the fibers cannot be formed (see Figure 2.23a). However, if the PR in the solution is high, too many long chains or nanowires, consisting of the Fe_3O_4 nanoparticles, will be formed during the reaction. Many of the long chains cannot get thicker but become aggregations of Fe_3O_4 nanoparticles. Only partial nanowires can get thicker and thicker, forming fibers. Finally, the mixture of both was obtained (see Figure 2.23c).

The presence of Cit in the solution is vital to the formation of both the Fe_3O_4 phase and its fibers. It works in three aspects such as coordination, reduction, and surface modification. According to Reaction 2.12, magnetite will not be formed without Cit. Additionally, the citric groups can prevent the ultrafine Fe_3O_4 nanoparticles from aggregation [79] by surface attachment [78]. Therefore, too small amounts of Cit in the solution will not effectively prevent the aggregation of ultrafine Fe_3O_4 nanoparticles, leading to a mixture of the fibers and big spherical magnetite. On the other hand, too much Cit in the solution will neutralize positively charged PR and prevent the nanowires of PR/Fe_3O_4 from getting thicker. Some ultrafine nanowires would become spherical aggregations of Fe_3O_4 nanoparticles after reaction. Finally, a mixture of the aggregations and thin nanowires was obtained.

On this basis, we conclude that only under proper conditions, including moderate PR and Cit concentrations, appropriate pH value in the precursor solution, and suitable reaction temperature, can porous Fe_3O_4 fibers be obtained.

2.4 TREMELLA-LIKE MICRO/NANOSTRUCTURED Fe_3S_4/C COMPOSITES

The adsorption based on adsorbent has been an effective route to remove the pollutants completely without any harmful by-products remaining in water. Here, the adsorbent is a key factor in water remediation. Among normal adsorbents, carbonaceous materials have extensively been used because of their low cost and strong adsorption performance [81–83]. Generally, the carbonaceous materials are obtained by pyrolysis of C-rich precursors at a high temperature (>600°C), such as polymers [84], woods [85], and so on. However, the C produced in this way has less active adsorption sites in the adsorption process since the high temperature can diminish the functional groups on the surface. Recently, carbonaceous materials, containing functional groups (C=C, C=O, –OH, etc.) on the surface, were produced by hydrothermal treatment of glucose at a low temperature [86]. Such materials, especially the nanostructured ones, could be good adsorbents, owing to their high surface area. For instance, Yu and colleagues reported many carbonaceous nanomaterials synthesized by hydrothermal or solvothermal methods [87–90]. More works have also demonstrated that carbonaceous nanomaterials are efficient adsorbents for pollutant removal [87,91,92]. However, the carbonaceous adsorbents of nanosize are easily aggregated due to their small size, which results in reduced adsorption capacity. In addition, they are suspended in water during the adsorption process and hence difficult to separate from the water. Based on the discussion in Section 2.2.2, micro/nanostructured adsorbents could be more efficient in the adsorption process. For separation of adsorbents from water, magnetic materials could be good candidates due to their ease of collection at low magnetic fields [93]. There have been many reports about magnetic C-containing materials [93–98], such as porous C sphere composites [93] and C composites containing nickel [96], iron [94], and iron oxide [97,98]. But most of these materials were prepared by two-step methods. The facile preparation of micro/nanostructured magnetic C-containing adsorbents at low temperatures with excellent regenerative properties is still under study.

Fe_3S_4 is a useful material. It has excellent magnetic properties [99–101], electrochemical hydrogen storage properties [100], and so on. Importantly, Fe_3S_4 is an environment-friendly material and could be used as an adsorbent for the removal of heavy metals from the environment [101]. However, the lack of surface functional groups has restricted Fe_3S_4 in pollutant removal in practical environmental remediation.

Here, we introduce the one-step preparation of micro/nanostructured magnetic Fe_3S_4/C composites with tremella-like morphology by solvothermal method using EG as structural director and glucose as C source at a low temperature [102]. In a typical procedure, 0.24 g $FeCl_3 \cdot 6H_2O$ and 0.526 g thioacetamide (TAA) (CH_3CSNH_2) were dissolved in 25 mL EG and ($HOCH_2CH_2OH$), respectively. Then, the solutions were mixed together by stirring. Finally, 1 g glucose ($C_6H_{12}O_6$) was added to the mixed solution and dissolved completely (the final concentration was 20 $g \cdot L^{-1}$). The whole solution was transferred into a Teflon-lined autoclave, which was sealed and maintained at 160°C for 12 hours before being cooled to room temperature naturally. Choice of the preparation temperature at 160°C is based on glucose carbonization, which occurs at temperatures higher than 150°C [86].

2.4.1 Structure and Morphology

After the reaction of precursor solution with glucose (20 g·L^{-1}) at 160°C, dark brown products were obtained. Corresponding XRD measurements showed that all the diffraction peaks fit well with the *face-centered cubic* (fcc) Fe_3S_4 phase, as illustrated in curve I of Figure 2.27a. If the precursor solution contains no glucose, the product's phase is identical except for a higher intensity and sharper peaks of the diffraction, as seen in curve II of Figure 2.27a. The addition of glucose in the precursor does not change the crystal phase of the products.

Raman spectral measurements were carried out to examine the existence of carbonaceous materials in the products, as shown in Figure 2.27b. For the product from the glucose-contained precursor, there are two peaks around 1352 cm^{-1} (named D band) and 1574 cm^{-1} (named G band) in the Raman spectrum (see curve I in Figure 2.27b), which should be attributed to amorphous C and graphitized C, respectively [84]. The strongest peak around 1490 cm^{-1}, which can be ascribed to δCH_2 or δCH_3 [103], indicates the existence of organic functional groups in the products. In contrast, there are no such peaks for the products from the precursor without glucose, as can be seen in curve II of Figure 2.27b. As a result, the Fe_3S_4/C composites can be obtained from the precursor with glucose after solvothermal reaction. Without glucose, pure Fe_3S_4 was produced. Further, FTIR, x-ray photoelectron spectroscopy, and TG measurements have confirmed the existence of hydrophobic groups (C–H, C=C, and C–C, etc.) and functional groups on the surface of the Fe_3S_4/C composites [102].

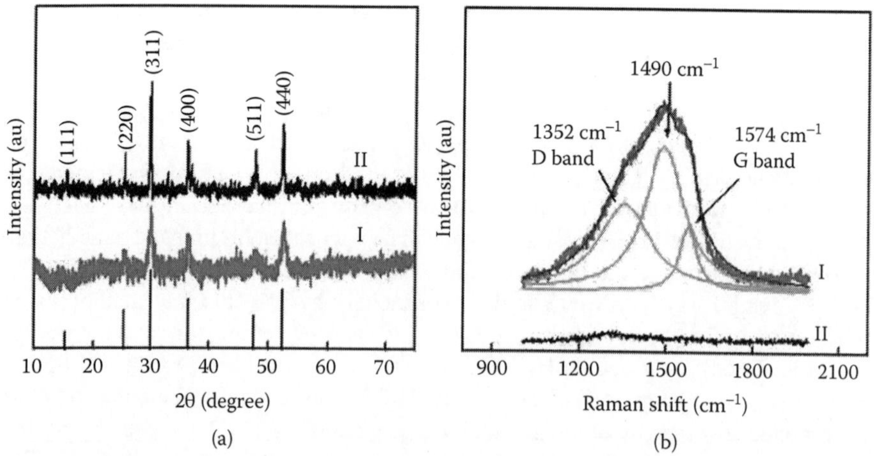

FIGURE 2.27 (a) X-ray diffraction patterns and (b) Raman spectra of the as-prepared products: curve I corresponds to the product prepared from the precursor with 20 g·L^{-1} glucose and curve II to the product prepared from the precursor without glucose. The line spectra and the marked indexes in (a) correspond to the standard diffraction and crystal planes of the face-centered cubic Fe_3S_4 phase. (Wang et al., *CrystEngComm*, 15, 2956–65, 2013. Reproduced by permission of The Royal Society of Chemistry.)

The morphology of the Fe_3S_4/C composites is typically shown in Figure 2.28. The products consist of tremella-like objects with a hierarchical micro/nanostructure. The tremellas are around 10 μm in size and built by bent nanoplates (see Figure 2.28a). The nanoplates are several micrometers in length and approximately 50 nm in thickness, as illustrated in Figure 2.28b and its inset. Energy-dispersive x-ray spectroscopy (EDS) measurements have revealed that there are Fe, sulfur (S), C, and O elements in the products with a mole ratio of 3:4:2.6:27.5, except Si element, which comes from the silicon substrate of the sample, as shown in the inset of Figure 2.28a. The EDS results demonstrate that the Fe_3S_4/C composites contain C- and O-rich functional groups, which is in good agreement with the results of the Raman spectrum.

Figure 2.29a and b shows the TEM images of a tremella-like object. SAED has revealed the single-crystal nanoplate of Fe_3S_4 with (111) planar surface (see the inset of Figure 2.29b), which is produced by an electronic beam parallel to [111]. The EDS mapping for a single nanoplate is illustrated in Figure 2.29c, indicating the existence and uniform distribution of Fe, S, C, and O elements. The Fe and S elements come from Fe_3S_4. The C and O elements can be attributed to the existence of functional groups, such as C=C, C=O, and so on. Further, HRTEM observation of the nanoplates shows that there exist resolved fringes separated by 0.20 and 0.35 nm corresponding to the lattice planes of $(\bar{4}22)$ and $(2\bar{2}0)$ in cubic Fe_3S_4, respectively, and curved fringes, which have a spacing of 0.34 nm, belonging to the lattice plane (002) of graphitized C, as shown in Figure 2.29d. The graphitized C almost covers the whole nanoplates' surface, as seen in Figure 2.29c and d. It should be mentioned that there are some dark contrasts greater than 5 nm in size in the HRTEM image of the single nanoplate (see Figure 2.29d). The dark contrasts are not from nanoparticles. In fact, the Fe_3S_4 nanoplate is covered with the partially graphitized C. The contrasts could come from different graphitized extents in different areas on the nanoplate. Such carbonaceous material, uniformly distributed on the surface of tremella-like Fe_3S_4, would serve as active sites for the adsorption [102].

FIGURE 2.28 Field emission scanning electron microscopy images of Fe_3S_4/C composites: (a) general image (inset: energy-dispersive x-ray spectroscopy results), and (b) the enlarged image of a single tremella (inset: further enlarged local image). (Wang et al., *CrystEngComm*, 15, 2956–65, 2013. Reproduced by permission of The Royal Society of Chemistry.)

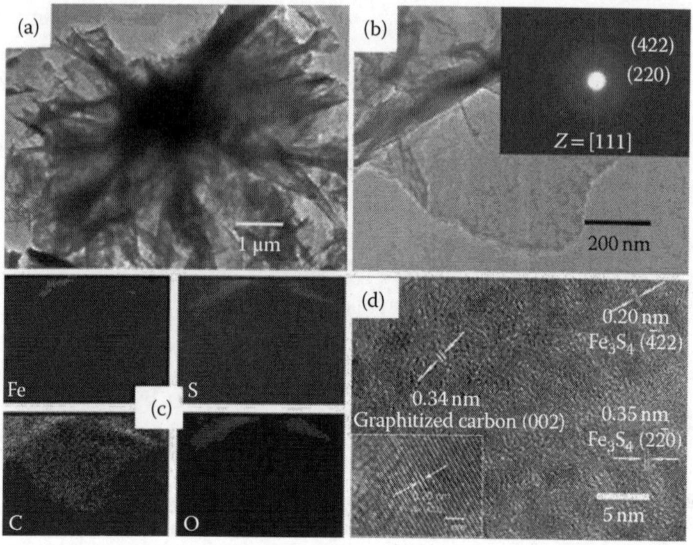

FIGURE 2.29 TEM results and composition mapping of Fe_3S_4/C: (a) TEM morphology of a single object, (b) enlarged TEM image of a single nanoplate (Inset: Selected area electronic diffraction [SAED] pattern), (c) energy-dispersive x-ray spectroscopy mapping (Fe, sulfur[S], carbon [C], and oxygen [O] of the single nanoplate), (d) high-resolution TEM image of the single nanoplate (Inset: A local magnification). (Wang et al., *CrystEngComm*, 15, 2956–65, 2013. Reproduced by permission of The Royal Society of Chemistry.)

The specific surface area for the Fe_3S_4/C composites was about 12 $m^2 \cdot g^{-1}$ based on isothermal nitrogen sorption measurement [102]. Magnetic property measurement confirmed the typical ferromagnetic behavior, as shown in Figure 2.30. The saturation magnetization (M_s), remanent magnetization (M_r), and coercivity field (H_c) at room temperature are 8.48 emu·g^{-1}, 3.04 emu·g^{-1}, and 300 Oe, respectively, and indicate that the composites are ferromagnetic, which is in good agreement with previously reported results [100,104].

2.4.2 Influence of Glucose

Further experiments have revealed that the glucose in the precursor solution not only is the C source but also determines the morphology of the products. Figure 2.31 shows the morphologies of products from precursor solutions with different amounts of glucose. Without glucose in the precursor solution, only Fe_3S_4 particles were obtained, which were aggregated into particles 3–5 μm in size (Figure 2.31a). When the glucose amount is small (say, 10 g·L^{-1}), some aggregations consisting of small sheets with around 170 nm thickness were formed as shown in Figure 2.31b. Increasing the glucose amount is beneficial to form products with tremella-like morphology (see Figure 2.28). However, if the glucose amount is too high (say, 50 g·L^{-1} or higher) the tremella-like morphology tends to disappear, and aggregated particles are produced when the amount is 100 g·L^{-1}, as typically shown in Figure 2.31c and d.

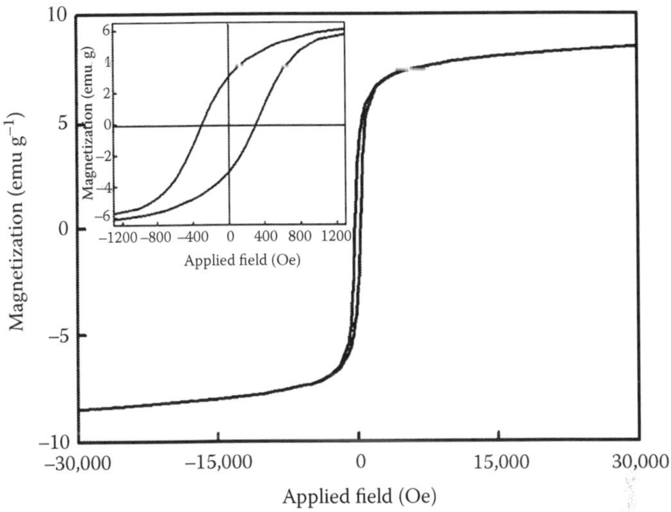

FIGURE 2.30 The hysteresis loop of Fe_3S_4/C composites measured at room temperature. The inset is the corresponding magnified hysteresis loop at a low applied field. (Wang et al., *CrystEngComm*, 15, 2956–65, 2013. Reproduced by permission of The Royal Society of Chemistry.)

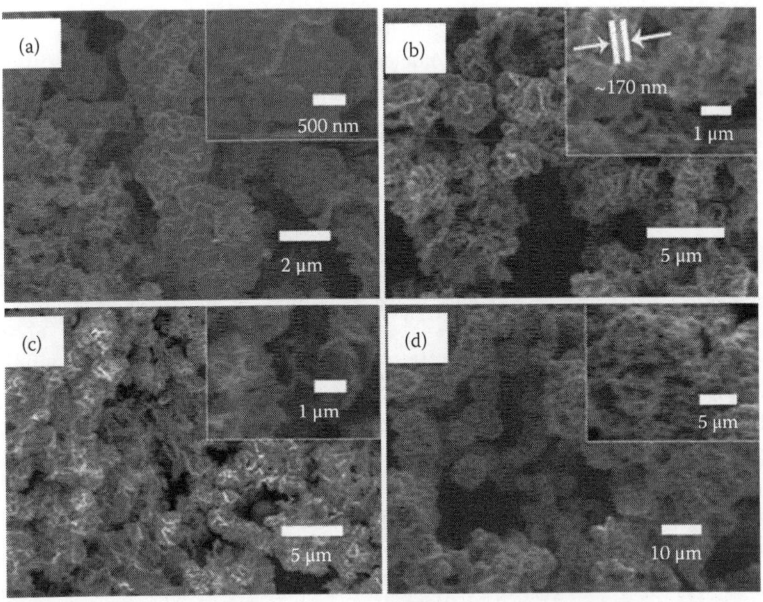

FIGURE 2.31 Scanning electron microscopy images of the products prepared in the precursor solution with (a) 0, (b) 10, (c) 50, and (d) 100 g·L^{-1} glucose (insets: the corresponding local magnified images). (Wang et al., *CrystEngComm*, 15, 2956–65, 2013. Reproduced by permission of The Royal Society of Chemistry.)

Only the appropriate concentration of glucose in the precursor solution (20 g·L^{-1}) induces tremella-like micro/nanostructured Fe_3S_4/C composites, as shown in Figure 2.28.

2.4.3 Formation of Tremella-Like Fe_3S_4/C Composites

There are $FeCl_3 \cdot 6H_2O$ and TAA dissolved in EG solution. The following reactions take place on heating at 160°C:

$$HOCH_2CH_2OH \rightarrow CH_3CHO + H_2O \quad (2.15)$$

$$2Fe^{3+} + 2CH_3CHO \rightarrow CH_3COCOCH_3 + 2Fe^{2+} + 2H^+ \quad (2.16)$$

$$CH_3CSNH_2 + H_2O \rightarrow CH_3CONH_2 + H_2S \quad (2.17)$$

$$H_2S \rightarrow 2H^+ + S^{2-} \quad (2.18)$$

$$2Fe^{3+} + Fe^{2+} + 4S^{2-} \rightarrow Fe_3S_4 \quad (2.19)$$

A portion of Fe^{3+} ions released from the dissolved $FeCl_3 \cdot 6H_2O$ would first be reduced by EG and converted into Fe^{2+} ions (Reactions 2.15 and 2.16) [105]. At the same time, the TAA in the solution would hydrolyze and release S^{2-} ions (Reactions 2.17 and 2.18). Fe_3S_4 would be formed by the reaction of Fe^{3+}, Fe^{2+}, and S^{2-} ions (Reaction 2.19).

As the reaction proceeds, the concentration of Fe_3S_4 in the solution would be saturated and, hence, Fe_3S_4 crystal nuclei are formed in the solution. Without the addition of glucose, the nuclei would subsequently grow into particles, as shown in Figure 2.31a. For the precursor solution with glucose, however, the situation is different. The glucose in the precursor solution can be oxidized by Fe^{3+}, forming gluconic acid, and thus release $CH_2OH(CHOH)_4COO^-$ anions, or the reaction

$$CH_2OH(CHOH)_4CHO \xrightarrow{[O]} CH_2OH(CHOH)_4COOH \quad (2.20)$$

would take place and release $CH_2OH(CHOH)_4COO^-$ anions. The Fe_3S_4 is a ferrous–ferric inverse thiospinel, and its crystal structure is an fcc S sublattice [106]. The plane (111) of Fe_3S_4 is polar and positively charged as more iron ions (Fe^{2+} and Fe^{3+}) than S^{2-} ions are present in this plane, as illustrated in Figure 2.32. The negatively charged $CH_2OH(CHOH)_4COO^-$ anions could be selectively adsorbed on the positively charged polar plane (111), where the charges could be compensated due to their electrostatic forces. Selective adsorption of $CH_2OH(CHOH)_4COO^-$ anions on Fe_3S_4 crystal planes can alter its surface termination and atomic arrangement, which may have a great effect on the anisotropic growth process or induce the termination of growth in the [111] direction, as previously reported [106]. In this case, preferential growth can take place within the plane (111), forming Fe_3S_4 nanoplates. From growth fundamentals, however, it could be along the close-packing directions $[1\bar{1}0]$ and/or $[11\bar{2}]$ within the (111) plane.

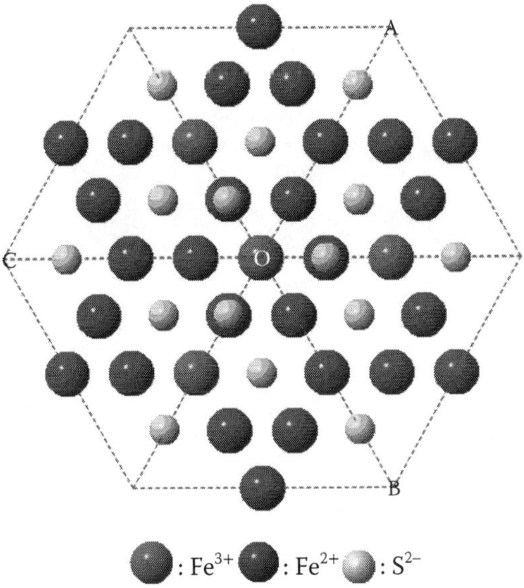

FIGURE 2.32 The schematic atomic structure of the crystal plane (111) (view along 111]). (Wang et al., *CrystEngComm*, 15, 2956–65, 2013. Reproduced by permission of The Royal Society of Chemistry.)

At the same time, the adsorbed glucose goes through polymerization and carbonization processes during solvothermal treatment. First, the glucose polymerizes to form aromatic compounds [86]. Subsequently, when the concentration reaches supersaturation, carbonaceous materials are obtained. Finally, the carbonaceous materials would further be partially graphitized during the solvothermal process, resulting in Fe_3S_4/C composites with nanoplate morphology. The formation of curved Fe_3S_4/C nanoplates could be attributed to the surface C since carbonization process of the adsorbed glucose would lead to a change in surface tension and hence influence the nanoplates' shape. As the process goes on, EG molecules will further adsorb on the surface of the Fe_3S_4/C nanoplates, resulting in a charge difference between the surfaces and the edges of the nanoplates [107], which leads to self-assembly of the curved nanoplates; thus, the tremella-like morphology is formed until the final size is attained [27,107].

On the basis of this discussion, the tremella-like objects should have been formed in the initial stage of the growth, which has been confirmed by further experiments. When the solvothermal reaction is for 1 hour, a small amount of product can be obtained. In this stage, the tremella-like objects have indeed been formed with a size of about 850 nm, as shown in Figure 2.33a. With increasing reaction time, more and more tremella-like objects are formed with continuously increasing size (see Figures 2.28 and 2.33b). Finally, carbonaceous spheres are produced by hydrothermal procedure at high glucose concentrations (>100 $g{\cdot}L^{-1}$) [86]. Here, too much glucose (50 $g{\cdot}L^{-1}$ or higher) results in the aggregation of products (Figure 2.31c and d) due to the polymerization and carbonization of the glucose in solution and that adsorbed on the Fe_3S_4, which will wrap up the Fe_3S_4 and hinder the growth of Fe_3S_4 nanoplates.

FIGURE 2.33 Field emission scanning electron microscopy images of the as-prepared products after solvothermal reaction for (a) 1 hour and (b) 4 hours. (Wang et al., *CrystEngComm*, 15, 2956–65, 2013. Reproduced by permission of The Royal Society of Chemistry.)

2.5 BRIEF SUMMARY

In summary, according to the solvothermal/hydrothermal method shown in Figure 2.1, various micro/nanostructured materials can be produced. In this chapter, we have introduced the fabrication of the micro/nanostructured materials zinc oxide, Fe_3O_4, and Fe_3S_4/C composites. These materials with novel structure can be obtained after reaction of the corresponding precursor solutions at 150°C–250°C for a certain time and/or subsequent heat treatment. The morphology of the products obtained depends on the structural directors (surfactants or organic solvents) in the precursor solutions. We can thus control the morphology and structure by choosing different precursors and structural directors. Such micro/nanostructured materials should be stable against aggregation or keep a high active surface area during adsorption and/or photocatalytic treatment in a solution, should be easy to be separated from the solution, and hence could be good candidates for environments.

REFERENCES

1. Lu, F.; Cai, W. P.; and Zhang, Y. G. 2008. ZnO hierarchical micro/nanoarchitectures: Solvothermal synthesis and structurally enhanced photocatalytic performance. *Adv. Funct. Mater.* 18: 1047–56.
2. Zeng, H. B.; Cai, W. P.; Liu, P. S.; Xu, X. X.; Zhou, H. J.; Klingshirn, C.; and Kalt, H. 2008. ZnO-based hollow nanoparticles by selective etching: Elimination and reconstruction of metal-semiconductor interface, improvement of blue emission and photocatalysis. *ACS Nano.* 2: 1661–70.
3. Jing, Z. H. and Zhan, J. H. 2008. Fabrication and gas-sensing properties of porous ZnO nanoplates. *Adv. Mater.* 20: 4547–51.
4. Han, N.; Hu, P.; Zuo, A.; Zhang, D. W.; Tian, Y. J.; and Chen, Y. F. 2010. Photoluminescence investigation on the gas sensing property of ZnO nanorods prepared by plasma-enhanced CVD method. *Sens. Actuators B Chem.* 145: 114–9.
5. Chou, T. P.; Zhang, Q. F.; Fryxell, G. E.; and Cao, G. Z. 2007. Hierarchically structured ZnO film for dye-sensitized solar cells with enhanced energy conversion efficiency. *Adv. Mater.* 19: 2588–92.

6. Zhang, Q. F.; Dandeneau, C. S.; Candelaria, S.; Liu, D. W.; Garcia, B. B.; Zhou, X. Y.; Jeong, Y. H.; and Cao, G. Z. 2010. Effects of lithium ions on dye-sensitized ZnO aggregate solar cells. *Chem. Mater.* 22: 2427–33.
7. Wang, Z. L. 2009. ZnO nanowire and nanobelt platform for nanotechnology. *Mater. Sci. Eng. R Rep.* 64: 33–71.
8. Meyer, B.; Rabaab, H.; and Marx, D. 2006. Water adsorption on ZnO(1010): From single molecules to partially dissociated monolayers. *Phys. Chem. Chem. Phys.* 8: 1513–20.
9. Kikuchi, Y.; Qian, Q. R.; Machida, M.; and Tatsumoto, H. 2006. Effect of ZnO loading to activated carbon on Pb(II) adsorption from aqueous solution. *Carbon* 44: 195–202.
10. Song, R. Q.; Xu, A. W.; Deng, B.; Li, Q.; and Chen, G. Y. 2007. From layered basic zinc acetate nanobelts to hierarchical zinc oxide nanostructures and porous zinc oxide nanobelts. *Adv. Funct. Mater.* 17: 296–306.
11. Zhou, X. F.; Hu, Z. L.; Fan, Y. Q.; Chen, S.; Ding, W. P.; and Xu, N. P. 2008. Microspheric organization of multilayered ZnO nanosheets with hierarchically porous structures. *J. Phys. Chem. C* 112: 11722–8.
12. Qiu, Y. C.; Chen, W.; and Yang, S. H. 2010. Facile hydrothermal preparation of hierarchically assembled, porous single-crystalline ZnO nanoplates and their application in dye-sensitized solar cells. *J. Mater. Chem.* 20: 1001–6.
13. Shang, T. M.; Sun, J. H.; Zhou, Q. F.; and Guan, M. Y. 2007. Controlled synthesis of various morphologies of nanostructured zinc oxide: Flower, nanoplate, and urchin. *Cryst. Res. Technol.* 42: 1002–6.
14. Xu, F.; Yuan, Z. Y.; Du, G. H.; Halasa, M.; and Su, B. L. 2007. High-yield synthesis of single-crystalline ZnO hexagonal nanoplates and accounts of their optical and photocatalytic properties. *Appl. Phys. A* 86: 181–5.
15. Cao, B. Q. and Cai, W. P. 2008. From ZnO nanorods to nanoplates: Chemical bath deposition growth and surface-related emissions. *J. Phys. Chem. C* 112: 680–5.
16. Cao, X. L.; Zeng, H.; Wang, B. M.; Xu, X.; Fang, J. M.; Ji, S. L.; and Zhang, L. D. 2008. Large scale fabrication of quasi aligned ZnO stacking nanoplates. *J. Phys. Chem. C* 112: 5267–70.
17. Cheng, J. P.; Liao, Z. M.; Shi, D.; Liu, F.; and Zhang, X. B. 2009. Oriented ZnO nanoplates on Al substrate by solution growth technique. *J. Alloys Compd.* 480: 741–6.
18. Ghoshal, T.; Kar, S.; and Chaudhuri, S. 2007. ZnO doughnuts: Controlled synthesis, growth mechanism, and optical properties. *Cryst. Growth Des.* 7: 136–41.
19. Zhang, N.; Yi, R.; Shi, R. R.; Gao, G. H.; Chen, G.; and Liu, X. H. 2009. Novel rose-like ZnO nanoflowers synthesized by chemical vapor deposition. *Mater. Lett.* 63: 496–9.
20. Xu, C.; Kim, D.; Chun, J.; Rho, K.; Chun, B.; Hong, S.; and Joo, T. 2006. Temperature-controlled growth of ZnO nanowires and nanoplates in the temperature range 250–300 degrees C. *J. Phys. Chem. B* 110: 21741–6.
21. Cao, B. Q.; Teng, X. M.; Heo, S. H.; Li, Y.; Cho, S. O.; Li, G. H.; and Cai, W. P. 2007. Different ZnO nanostructures fabricated by a seed-layer assisted electrochemical route and their photoluminescence and field emission properties. *J. Phys. Chem. C* 111: 2470–6.
22. Cao, B. Q.; Cai, W. P.; Zeng, H. B.; and Duan, G. T. 2006. Morphology evolution and photoluminescence properties of ZnO films electrochemically deposited on conductive glass substrates. *J. Appl. Phys.* 99: 073516-1–6.
23. Illy, B.; Shollock, B. A.; MacManus-Driscoll, J. L.; and Ryan, M. P. 2005. Electrochemical growth of ZnO nanoplates. *Nanotechnology* 16: 320–4.
24. Wang, X. B.; Cai, W. P.; Lin, Y. X.; Wang, G. Z.; and Liang, C. H. 2010. Mass production of micro/nanostructured porous ZnO plates and their strong structurally enhanced and selective adsorption performance for environmental remediation. *J. Mater. Chem.* 20: 8582–90.

25. Brunauer, S.; Deming, L. S.; Deming, W. E.; and Teller, E. 1940. On a theory of the van der Waals adsorption of gases. *J. Am Chem. Soc.* 62: 1723–32.
26. Brunauer, S.; Emmett, P. H.; and Teller, E. 1938. Adsorption of gases in multimolecular layers. *J. Am. Chem. Soc.* 60: 309–19.
27. Zhong, L. S.; Hu, J. S.; Liang, H. P.; Cao, A. M.; Song, W. G.; and Wan, L. 2006. Self-assembled 3D flowerlike iron oxide nanostructures and their application in water treatment. *Adv. Mater.* 18: 2426–631.
28. Yan, C. L. and Xue, D. F. 2006. Morphosynthesis of hierarchical hydrozincite with tunable surface architectures and hollow zinc oxide. *J. Phys. Chem. B* 110: 11076–80.
29. Li, C. C.; Cai, W. P.; Cao, B. Q.; Sun, F. Q.; Li, Y.; Kan, C. X.; and Zhang, L. D. 2006. Mass synthesis of large, single-crystal Au nanosheets based on a polyol process. *Adv. Funct. Mater.* 16: 83–90.
30. Wang, X. B.; Cai, W. P.; Wang, G. Z.; and Liang, C. H. 2012. Standing porous ZnO nanoplate-built hollow microspheres and kinetically controlled dissolution/crystal growth mechanism. *J. Mater. Res.* 27: 951–8.
31. Zeng, X. Y.; Yuan, J. L.; and Zhang, L. D. 2008. Synthesis and photoluminescent properties of rare earth doped ZnO hierarchical microspheres. *J. Phys. Chem. C* 112: 3503–8.
32. Zhao, W.; Song, X. Y.; Chen, G. Z.; and Sun, S. X. 2009. One-step template-free synthesis of $ZnWO_4$ hollow clusters. *J. Mater. Sci.* 44: 3082–7.
33. Verhoeven, J. D. 1975. *Fundamentals of Physical Metallurgy*. John Wiley & Sons, New York, Chapter 8.
34. Tian, Z. R.; Voigt, J. A.; Liu, J.; Mckenziei, B.; Mcdermott, M. J.; Rodriguez, M. A.; Konishi, H.; and Xu, H. F. 2003. Complex and oriented ZnO nanostructures. *Nat. Mater.* 2: 821–6.
35. Kuo, C. L.; Kuo, T. J.; and Huang, M. H. 2005. Hydrothermal synthesis of ZnO microspheres and hexagonal microrods with sheetlike and platelike nanostructures. *J. Phys. Chem. B* 109: 20115–21.
36. Verhoeven, J. D. 1975. *Fundamentals of Physical Metallurgy*. John Wiley & Sons, New York, Chapters 11 and 12.
37. Hao, Q.; Xu, L. Q.; Li, G. D.; and Qian, Y. T. 2009. Hydrothermal synthesis of microscaled Cu@C polyhedral composites and their sensitivity to convergent electron beam. *Langmuir* 25: 6363–7.
38. Li, X.; Tang, C. J.; Ai, M.; Dong, L.; and Xu, Z. 2010. Controllable synthesis of pure-phase rare-earth orthoferrites hollow spheres with a porous shell and their catalytic performance for the CO+NO reaction. *Chem. Mater.* 22: 4879–89.
39. Sui, Y. M.; Fu, W. Y.; Yang, H. B.; Zeng, Y.; Zhang, Y. Y.; Zhao, Q.; Li, Y. G. et al. 2010. Low temperature synthesis of Cu_2O crystals: Shape evolution and growth mechanism. *Cryst. Growth Des.* 10: 99–108.
40. Ma, S. C. 2003. *Basal Chemical Reactions*. Shaan'Xi Technology & Science Press, Xi'an, People's Republic of China, p. 377.
41. Gao, X. D.; Li, X. M.; and Yu, W. D. 2005. Flowerlike ZnO nanostructures via hexamethylenetetramine-assisted thermolysis of zinc-ethylenediamine complex. *J. Phys. Chem. B* 109: 1155–61.
42. Li, W. J.; Shi, E. W.; Zhong, W. Z.; and Yin, Z. W. 1999. Growth mechanism and growth habit of oxide crystals. *J. Cryst. Growth.* 203: 186–96.
43. Zhang, G. Y.; Jiang, X.; and Wang, E. G. 2003. Tubular graphite cones. *Science* 300: 472–4.
44. Zhang, G. Y.; Bai, X. D.; and Wang, E. G. 2005. Monochiral tubular graphite cones formed by radial layer-by-layer growth. *Phys. Rev. B* 71: 113411–4.
45. Li, Z. Q.; Ding, Y.; Xiong, Y. J.; Yang, Q.; and Xie, Y. 2005. One-step solution-based catalytic route to fabricate novel α-MnO_2 hierarchical structures on a large scale. *Chem. Commun.* 7: 918–20.

46. Chen, J. Y.; Herricks, T.; Geissler, M.; and Xia, Y. N. J. 2004. Single-crystal nanowires of platinum can be synthesized by controlling the reaction rate of a polyol process. *J. Am. Chem. Soc.* 126: 10854–5.
47. Liu, B.; Yu, S. H.; Zhang, F.; Li, L. J.; Zhang, Q.; Ren, L.; and Jiang, K. 2004. Ring-like nanosheets standing on spindle-like rods: Unusual ZnO superstructures synthesized from a flakelike precursor $Zn_5(OH)_8Cl_2{\cdot}H_2O$. *J. Phys. Chem. B* 108: 4338–41.
48. Jiang, C. L.; Zhang, W. Q.; Zou, G. F.; Yu, W. C.; and Qian, Y. T. 2005. Precursor-induced hydrothermal synthesis of flowerlike cupped-end microrod bundles of ZnO. *J. Phys. Chem. B* 109: 1361–3.
49. Yu, S. Y.; Wang, C.; Yu, J. B.; Shi, W. D.; Deng, R. P.; and Zhang, H. J. 2006. Precursor induced synthesis of hierarchical nanostructured ZnO. *Nanotechnology* 17: 3607–12.
50. Yavuz, C. T.; Mayo, J. T.; Yu, W. W.; Prakash, A.; Falkner, J. C.; Yean, S.; Cong, L. et al. 2006. DNA binding to protein-gold nanocrystal conjugates. *Science* 314: 964–7.
51. Thera, H.; Soumyadas, W.; Kadapu, R.; Ilton, E.; and Nathanyee, T. 2009. Reduction of Hg(II) to Hg(0) by magnetite. *Environ. Sci. Technol.* 43: 5307–13.
52. Hu, F. Q.; Wei, L.; Zhou, Z.; Ran, Y. L.; Li, Z.; and Gao, M. Y. 2006. Preparation of biocompatible magnetite nanocrystals for in vivo magnetic resonance detection of cancer. *Adv. Mater.* 18: 2553–6.
53. Brahler, M.; Georgieva, R.; Buske, N.; Muller, A.; Muller, S.; Pinkernelle, J.; Teichgraber, U.; Voigt, A.; and Bäumler, H. 2006. Magnetite-loaded carrier erythrocytes as contrast agents for magnetic resonance imaging. *Nano Lett.* 11: 2505–9.
54. Jang, J.-T.; Nah, H.; Lee, J.-H.; Moon, S. H.; Kim, M. G.; and Cheon, J. 2009. Critical enhancements of MRI contrast and hyperthermic effects by dopant-controlled magnetic nanoparticles. *Angew. Chem. Int. Ed. Engl.* 48: 1234–8.
55. Klokkenburg, M.; Erné, B. H.; Meeldijk, J. D.; Wiedenmann, A.; Petukhov, A. V.; Dullens, R. P. A.; and Philipse, A. P. 2006. In situ imaging of field-induced hexagonal columns in magnetite ferro fluid. *Phys. Rev. Lett.* 97: 185702.
56. Ceylan, S.; Friese, C.; Lammel, C.; Mazac, K.; and Kirschning, A. 2008. Inductive heating for organic synthesis by using functionalized magnetic nanoparticles inside microreactors. *Angew. Chem. Int. Ed. Engl.* 47: 8950–3.
57. Kopcanský, P.; Tomašovicová, N.; Koneracká, M.; Závišová, V.; Timko, M.; Darová, A.; and Šprincová, A. 2008. Structural changes in the 6CHBT liquid crystal doped with spherical, rodlike, and chainlike magnetic particles. *Phys. Rev. E Stat. Nonlin. Soft Matter. Phys.* 78: 011702.
58. Wang, Z. L. and Song, J. H. 2006. Piezoelectric nanogenerators based on zinc oxide nanowire arrays. *Science* 312: 242–6.
59. Law, M.; Greene, L. E.; Johnson, J. C.; Saykelly, R.; and Yang, P. 2005. Nanowire dye-sensitized solar cells. *Nat. Mater.* 4: 455–9.
60. Martin, C. R. 1994. Nanomaterials: A membrane-based synthetic approach. *Science* 266: 1961–6.
61. Dai, H.; Wong, E. W.; Lu, Y.; Fan, Z. S.; and Lieber, C. M. 1995. Synthesis and characterization of carbide nanorods. *Nature* 375: 769–72.
62. Han, W.; Fan, S.; Li, W.; and Hu, Y. 1997. Synthesis of gallium nitride nanorods through a carbon nanotube-confined reaction. *Science* 277: 1287–9.
63. Zhang, Y.; Suenaga, K.; Colliex, C.; and Iijima, S. 1998. Coaxial nanocable: Silicon carbide and silicon oxide sheathed with boron nitride and carbon. *Science* 281: 973–5.
64. Thess, A.; Lee, R.; Nikolaev, P.; Dai, H.; Petit, P.; Robert, J.; Xu, C. H. et al. 1996. Crystalline ropes of metallic carbon nanotubes. *Science* 273: 483–7.
65. Li, Y.; Ding, Y.; and Wang, Z. 1999. A novel chemical route to ZnTe semiconductor nanorods. *Adv. Mater.* 11: 847–50.

66. Prieto, A. L.; Martín-González, M.; Keyani, J.; Gronsky, R.; Sands, T.; and Stacy, A. M. 2003. The electrodeposition of high-density, ordered arrays of Bi1-xSbx nanowires. *J. Am. Chem. Soc.* 125: 2388–9.
67. Wang, J.; Chen, Q. W.; Zeng, C.; and Hou, B. Y. 2004. Magnetic-field-induced growth of single-crystalline Fe_3O_4 nanowires. *Adv. Mater.* 16: 137–40.
68. Mathur, S.; Barth, S.; Werner, U.; Ramirez, F. H.; and Rodriguez, A. R. 2008. Chemical vapor growth of one-dimensional magnetite nanostructures. *Adv. Mater.* 20: 1550–4.
69. Ding, Y.; Morber, J. R.; Snyder, R. L.; and Wang, Z. L. 2007. Nanowire structural evolution from Fe_3O_4 to ε-Fe_2O_3. *Adv. Funct. Mater.* 17: 1172–8.
70. Jia, C. J.; Sun, L.D.; Luo, F.; and Han, X. D. 2008. Large-scale synthesis of single-crystalline iron oxide magnetic nanorings. *J. Am. Chem. Soc.* 130: 16968–77.
71. Jiao, F.; Jumas, J. C.; Womes, M.; Chadwick, A. V.; Harrison, A.; and Bruce, P. G. 2006. Synthesis of ordered mesoporous Fe_3O_4 and γ-Fe_2O_3 with crystalline walls using post-template reduction/oxidation. *J. Am. Chem. Soc.* 128: 12905–9.
72. Ge, J. P.; Hu, Y.; Biasini, X. M.; Beyermann, W. P.; and Yin, Y. D. 2007. Superparamagnetic magnetite colloidal nanocrystal clusters. *Angew. Chem. Int. Ed. Engl.* 46: 4342–5.
73. Corr, S. A.; Byrne, S. J.; Tekoriute, R.; Meledandri, C. J.; Brougham, D. F.; Lynch, M.; Kerskens, C.; O'Dwyer, L.; and Gun'ko, Y. K. 2008. Linear assemblies of magnetic nanoparticles as MRI contrast. *J. Am. Chem. Soc.* 130: 4214–5.
74. Li, L.; Yang, Y.; Ding, J.; and Xue, 2010. Nanooctahedra and their magnetic field-induced two-/three-dimensional superstructure. *Chem. Mater.* 22: 3183–91.
75. Han, C. L.; Cai, W. P.; Tang, W.; Wang, G. Z.; and Liang, C. H. 2011. Protein assisted hydrothermal synthesis of ultrafine magnetite nanoparticle built-porous oriented fibers and their structurally enhanced adsorption to toxic chemicals in solution. *J. Mater. Chem.* 21: 11188–96.
76. Tabernal, P. L.; Mitra, S.; Poizot, P.; Simon, P.; and Tarascon, J. M. 2006. High rate capabilities Fe_3O_4-based Cu nano-architectured electrodes for lithium-ion battery applications. *Nat. Mater.* 5: 567–73.
77. Sugimoto, T. and Matijevic, E. J. 1980. Formation of uniform spherical magnetite particles by crystallization from ferrous hydroxide gels. *J. Colloid Interface Sci.* 74: 227–243.
78. Liu, J.; Sun, Z. K.; Deng, Y. H.; Zou, Y.; Li, C. Y.; Guo, X. H.; Xiong, L. Q.; Gao, Y.; Li, F. Y.; and Zhao, D. Y. 2009. Highly water-dispersible biocompatible magnetite particles with low cytotoxicity stabilized by citrate groups. *Angew. Chem. Int. Ed. Engl.* 48: 5875–9.
79. Deng, H.; Li, X.; Peng, Q.; Wang, X.; Chen, J.; and Li, Y. D. 2005. Monodisperse magnetic single-crystal ferrite microspheres. *Angew. Chem. Int. Ed. Engl.* 44: 2782–5.
80. Panouillé, M. and Garde, V. L. 2009. Gelation behaviour of gelatin and alginate mixtures. *Food Hydrocoll.* 23: 1074–80.
81. Méndez, A.; Fernández, F.; and Gascó, G. 2007. Removal of malachite green using carbon-based adsorbents. *Desalination* 206: 147–53.
82. Edwin Vasu, A. 2008. Removal of phenol and o-cresol by adsorption onto activated carbon. *E-J. Chem.* 5: 224–32.
83. Wang, X. B.; Liu, J.; and Xu, W. Z. 2012. One-step hydrothermal preparation of amino-functionalized carbon spheres at low temperature and their enhanced adsorption performance towards Cr(VI) for water purification. *Colloids Surf. A Physicochem. Eng. Asp.* 415: 288–94.
84. Wang, X. B.; Liu, J.; and Li, Z. J. 2009. The graphite phase derived from polyimide at low temperature. *J. Non-Cryst. Solids* 355: 72–75.
85. Figueiredo, J. L.; Valenzuela, C.; Bernalte, A.; and Encinar, J. M. 1989. Pyrolysis of holm-oak wood: Influence of temperature and particle size. *Fuel* 68: 1012–6.

86. Sun, X. M. and Li, Y. D. 2004. Colloidal carbon spheres and their core/shell structures with noble-metal nanoparticles. *Angew. Chem. Int. Ed. Engl.* 43: 597–601.
87. Chen, L. F.; Liang, H. W.; Lu, Y.; Cui, C. H.; and Yu, S. H. 2011. Synthesis of an attapulgite clay@carbon nanocomposite adsorbent by a hydrothermal carbonization process and their application in the removal of toxic metal ions from water. *Langmuir* 27: 8998–9004.
88. Liang, H. W.; Cao, X.; Zhang, W. J.; Lin, H. T.; Zhou, F.; Chen, L. F.; and Yu, S. H. 2011. Robust and highly efficient free-standing carbonaceous nanofiber membranes for water purification. *Adv. Funct. Mater.* 21: 3851–8.
89. Hu, B.; Wang, K.; Wu, L. H.; Yu, S. H.; Antonietti, M.; and Titirici, M. M. 2010. Engineering carbon materials from the hydrothermal carbonization process of biomass. *Adv. Mater.* 22: 813–28.
90. Liang, H. W.; Guan, Q. F.; Chen, L. F.; Zhu, Z.; Zhang, W. J.; and Yu, S. H. 2012. Macroscopic-scale template synthesis of robust carbonaceous nanofiber hydrogels and aerogels and their applications. *Angew. Chem. Int. Ed. Engl.* 51: 5101–5.
91. Mauter, M. S. and Elimelech, M. 2008. Environmental applications of carbon-based nanomaterials. *Environ. Sci. Technol.* 42: 5843–59.
92. Demir-Cakan, R.; Baccile, N.; Antonietti, M.; and Titirici, M.-M. 2009. Carboxylate-rich carbonaceous materials via one-step hydrothermal carbonization of glucose in the presence of acrylic acid. *Chem. Mater.* 21: 484–90.
93. Gao, M. R.; Zhang, S. R.; Jiang, J.; Zheng, Y. R.; Tao, D. Q.; and Yu, S. H. 2011. One-pot synthesis of hierarchical magnetite nanochain assemblies with complex building units and their application for water treatment. *J. Mater. Chem.* 21: 16888–92.
94. Rudge, S. R.; Kurtz, T. L.; Vessely, C. R.; Catterall, L. G.; and Williamson, D. L. 2000. Preparation, characterization, and performance of magnetic iron–carbon composite microparticles for chemotherapy. *Biomaterials* 21: 1411–20.
95. Oliveira, L. C. A.; Rios, R. V. R. A.; Fabris, J. D.; Garg, V.; Sapag, K.; and Lago, R. M. 2002. Activated carbon/iron oxide magnetic composites for the adsorption of contaminants in water. *Carbon* 40: 2177–83.
96. Gorria, P.; Sevilla, M.; Blanco, J. A.; and Fuertes, A. B. 2006. Synthesis of magnetically separable adsorbents through the incorporation of protected nickel nanoparticles in an activated carbon. *Carbon* 44: 1954–7.
97. Gupta, V. K.; Agarwal, S.; and Saleh, T. A. 2011. Chromium removal by combining the magnetic properties of iron oxide with adsorption properties of carbon nanotubes. *Water Res.* 45: 2207–12.
98. Zhang, Z. Y. and Kong, J. L. 2011. Novel magnetic Fe_3O_4@C nanoparticles as adsorbents for removal of organic dyes from aqueous solution. *J. Hazard. Mater.* 193: 325–9.
99. Dekkers, M. J. and Schoonen, M. A. A. 1996. Magnetic properties of hydrothermally synthesized greigite (Fe_3S_4)—I. Rock magnetic parameters at room temperature. *Geophys. J. Int.* 126: 360–8.
100. Cao, F.; Hu, W.; Zhou, L.; Shi, W. D.; Song, S. Y.; Lei, Y. Q.; Wang, S.; and Zhang, H. J. 2009. 3D Fe_3S_4 flower-like microspheres: High-yield synthesis via a biomolecule-assisted solution approach, their electrical, magnetic and electrochemical hydrogen storage properties. *Dalton Trans.* 42: 9246–52.
101. Watson, J. H. P.; Cressey, B.A.; Roberts, A. P.; Ellwood, D. C.; Charnock, J. M.; and Soper, A. K. 2000. Structural and magnetic studies on heavy-metal-sulphide nanoparticles produced by sulphate-reducing bacteria. *J. Magn. Magn. Mater.* 214: 13–30.
102. Wang, X. B.; Cai, W. P.; Wang, G. Z.; Wu, Z. K.; and Zhao, H. J. 2013. One-step fabrication of high performance tremella-like Fe_3S_4/C magnetic adsorbent with easy recovery and regeneration properties. *CrystEngComm* 15: 2956–65.

103. Taddei, P.; Balducci, F.; Simoni, R.; and Monti, P. 2005. Raman, IR and thermal study of a new highly biocompatible phosphorylcholine-based contact lens. *J. Mol. Struct.* 744–7: 507–14.
104. He, Z. B.; Yu, S. H.; Zhou, X. Y.; Li, X. G.; and Qu, J. F. 2006. Magnetic-field-induced phase-selective synthesis of ferrosulfide microrods by a hydrothermal process: Microstructure control and magnetic properties. *Adv. Funct. Mater.* 16: 1105–11.
105. Skrabalak, S. E.; Wiley, B. J.; Kim, M.; Formo, E. V.; and Xia, Y. N. 2008. On the polyol synthesis of silver nanostructures: Glycolaldehyde as a reducing agent. *Nano Lett.* 8: 2077–81.
106. Ma, J. M.; Chang, L.; Lian, J. B.; Huang, Z.; Duan, X. C.; Liu, X. D.; Peng, P.; Kim, T.; Liu, Z. F.; and Zheng, W. J. 2010. Ionic liquid-modulated synthesis of ferrimagnetic Fe_3S_4 hierarchical superstructures. *Chem. Commun.* 46: 5006–8.
107. Yang, J.; Li, C. X.; Quan, Z. W.; Zhang, C. M.; Yang, P. P.; Li, Y. Y.; Yu, C. C.; and Lin, J. 2008. Self-assembled 3D flowerlike Lu_2O_3 and $Lu_2O_3{:}Ln^{3+}$ (Ln = Eu, Tb, Dy, Pr, Sm, Er, Ho, Tm) microarchitectures: Ethylene glycol-mediated hydrothermal synthesis and luminescent properties. *J. Phys. Chem. C* 112: 12777–85.

3 Template-Etching Strategies

3.1 INTRODUCTION

Micro/nanostructured hollow spheres represent a promising type of nanomaterials because of their special structure as well as properties such as low density, high specific surface areas, and void properties. These spheres have proven to be excellent in widespread applications in lithium ion batteries, catalysis, sensors, drug delivery, and controlled drug release [1–4]. They have also exhibited good potential for application in environmental remediation as adsorbents [5–6], photocatalysts [7,8], and so on due to their large surface area. Soft template and hard template are the template methods commonly used in the fabrication of these micro/nanostructured hollow spheres [9]. Soft template strategy generally produces micro/nanostructured hollow spheres by using organic molecules such as ethylene glycol [10] and polyethylene glycol [11]. In addition, surfactants such as hexadecyltrimethylammonium bromide [12] are also used. The formation mechanism is generally associated with Ostwald ripening effect [13], Kirkendall effect [14], kinetically controlled amorphous spheres' dissolution process [5], and so on. Using soft template, the diameter of the obtained hollow spheres is not uniform but synthesis process is easily controlled. Hard template is another strategy to produce micro/nanostructured hollow spheres. Generally, polymer (e.g., polystyrene) or silica spheres are used as hard templates to produce the hollow spheres. In a typical layer-by-layer (LbL) process, precursors of the hollow spheres are adsorbed on the surface of the template (e.g., colloidal spheres) layer by layer to form multilayer-coated spheres [15]. The colloidal particles are then removed by calcination or exposure to solvents to obtain the hollow structures [16]. Advantages of the hard template method are low cost and easy control of the size of the hollow spheres, which depends on the diameter of the template. In this chapter, a chemical-template-synthesis strategy is introduced to demonstrate fabrication of the micro/nanostructured hollow spheres. In this strategy, the colloidal spheres not only act as templates to determine the diameter of the hollow structures but also are used as reactants.

3.2 MICRO/NANOSTRUCTURED POROUS SILICATE HOLLOW SPHERES

The basic unit of silicates is a tetrahedron-shaped anionic group with a negative four charge, which can be linked to each other in different modes and form as single units, double units, chains, sheets, rings, and framework structures. The structure has attractive chemical and physical properties that find applications as catalyst support, molecular sieve, gas adsorbent, separator, and so on [17–20]. The traditional silicate ores cost less and because of their charged surface and ion-exchange performance and also because they do not cause secondary contamination are used as adsorbents to remove pollutants in the environment. However, the adsorption capacity is low because of their low specific surface area. Their adsorption performance cannot be significantly enhanced even by milling. However, micro/nanostructured and/or porous structured silicate hollow spheres overcome the preceding shortcomings because of their high specific surface area and stable structure. They can be produced by using silica colloidal template-etching strategy, as illustrated in Figure 3.1. Initially, silica colloidal spheres considered as hard templates are immersed in metal ion–containing alkaline solution (e.g., $NH_3{\cdot}H_2O$) at a temperature for surface reaction (step I in Figure 3.1), resulting in the formation of porous silicate on the surface layer of the silica sphere; followed by etching to remove the core part silica results in the formation of porous silicate hollow spheres (step II in Figure 3.1). The formed silicate hollow spheres are then treated in an acid solution to obtain silica hollow spheres (step IV in Figure 3.1). If the silicate hollow spheres are reduced in a hydrogen environment, metal–silica hybrid hollow spheres are obtained (step III in Figure 3.1). Using various metal ion–containing alkaline solutions results in different kinds of metal–silica hybrid hollow spheres. These micro/nanostructured silicate materials find application as efficient adsorbents in pollutant removal. Here, we introduce some typical metal silicate hollow spheres, silica–metal hybrid silicate hollow spheres based on the strategy shown in Figure 3.1.

3.2.1 Copper Silicate Micro/Nanostructured Hollow Spheres

Based on the strategy shown in Figure 3.1, to obtain a novel hierarchical structure of copper silicate hollow spheres with nanotube assembled shells, silica colloidal spheres are used in copper ion–containing alkaline solution for surface reaction [21]. In a typical synthesis, monodispersed silica colloidal spheres were synthesized according to Stöber method [22]. At room temperature, analytical grade $Cu(NO3)_2{\cdot}3.5H_2O$ (0.7 mmol) and $NH_3{\cdot}H_2O$ (1 mL) were mixed in 30 mL deionized water and silica colloidal spheres (0.13 g) were homogeneously dispersed in deionized water (20 mL)

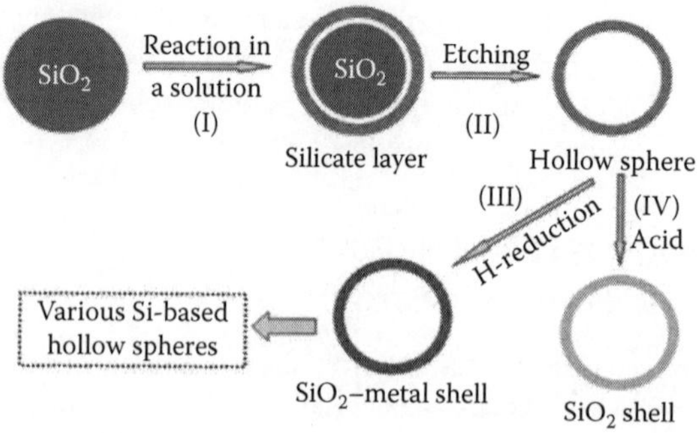

FIGURE 3.1 Schematic illustration of template-etching strategy based on silica colloidal. Step (I): immersion in metal ions–containing alkaline solution for reaction; step (II): removal of the core part silica by etching; step (III): formed silicate hollow spheres are reduced by hydrogen; step (IV): formed silicate hollow spheres are treated in an acid solution.

and were mixed by magnetic stirring. The homogeneous solution was then transferred into an autoclave at temperature 140°C; after 10 hours, blue products were collected by centrifugation and rinsed with distilled water several times until the pH was 7 and then dried in vacuum oven at 60°C for several hours.

3.2.1.1 Structure and Morphology

The X-ray diffraction (XRD) measurement indicates that the products are copper silicate [21], and morphological observation reveals they are made up of uniform microspheres with rough surface and 600 nm in diameter, as shown in Figure 3.2a. Furthermore, transmission electron microscopy (TEM) examination demonstrated that they are hollow in structure with uniform diameter and shell thickness (Figure 3.2b).

Detailed surface morphological examination confirms that a single hollow sphere is composed of a lot of nanotubes, as illustrated in Figure 3.3. Most of the nanotubes are positioned vertically on the surface of the spheres with open ends, and the diameter of all the nanotubes has a narrow size distribution. Many holes were seen on the shell surface due to the open ends of the nanotubes parallel to the electron beam, and a magnified image of the holes with the black edge and white center is shown in the top-right inset of Figure 3.3, in which three holes have an average outer diameter and inner diameter about 8 and 3.5 nm, respectively. One single nanotube on the shell exhibits two parallel dark lines and a darkish center and an open-ended structure as labeled by the arrow (see the down-right inset of Figure 3.3), which clearly indicates the unit of the shell as open-ended nanotubes. Furthermore, nitrogen isothermal adsorption measurements have shown that the specific surface area of the nanotube-based hierarchical copper silicate hollow spheres are 270 m^2/g and the pore size of the nanotubes is centered at 3.2 nm, which agrees well with the inner diameter of the nanotubes observed by TEM [21].

It should be mentioned that if the amount of $Cu(NO_3)_2{\cdot}3.5H_2O$ in the solution is increased, while other conditions remain the same, copper silicate hollow spheres with thicker shells are obtained [21].

3.2.1.2 Formation Mechanism

The growth process of the nanotube-based copper silicate micro/nanostructured hollow spheres was investigated by time-dependent experiments; Figure 3.4 shows TEM

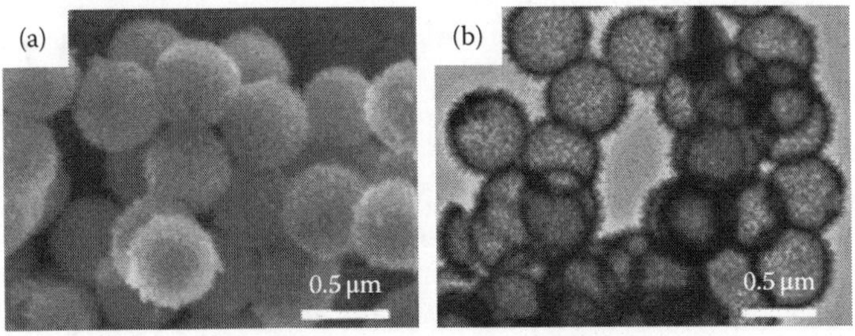

FIGURE 3.2 Field emission scanning electron microscopy (SEM) (a) and transmission electron microscopy (TEM) (b) images of the as-prepared copper silicate hollow spheres. (Wang et al., *Chem. Commun.*, December 28, 6555–7, 2008. Reproduced by permission of The Royal Society of Chemistry.)

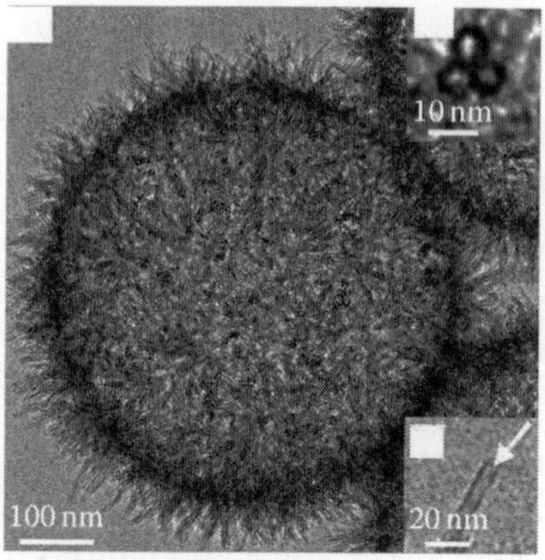

FIGURE 3.3 TEM image of a single copper silicate hollow sphere. Inset at top-right shows three holes on the surface of the shell; the inset at bottom-right shows a single nanotube on the shell. (Wang et al., *Chem. Commun.*, December 28, 6555–7, 2008. Reproduced by permission of The Royal Society of Chemistry.)

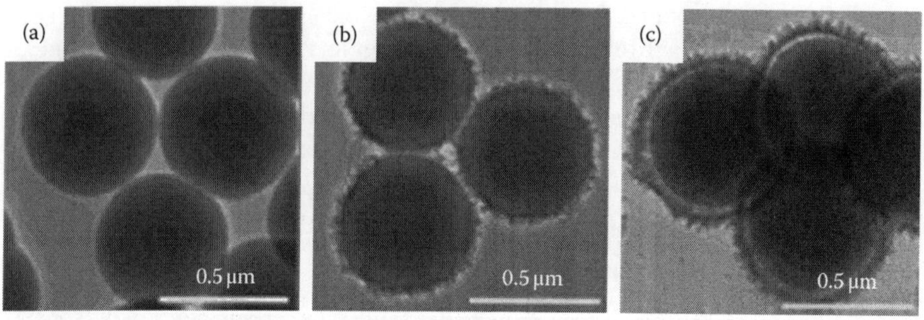

FIGURE 3.4 TEM images of the products after reaction at 140°C at different reaction times. (a) 0 hour, (b) 1 hour, (c) 2 hours. (Wang et al., *Chem. Commun.*, December 28, 6555–7, 2008. Reproduced by permission of The Royal Society of Chemistry.)

images of the corresponding products. The silica colloids in the present experiment are 500 nm in diameter as shown in Figure 3.4a, and the smooth surface of silica colloidal sphere can be seen clearly. When the reaction lasted for 1 hour, the surface of colloidal spheres became rough and bestrewed with particles (Figure 3.4b). After reaction for 2 hours, a blank boundary appeared, which indicates the formation of a core–shell structure, and a complete thin shell with an obvious rough surface was observed around the silica colloidal spheres (Figure 3.4c). When the reaction was prolonged for 10 hours, obvious voids appeared, indicating that copper silicate hollow spheres were obtained (Figure 3.2b), and the thickness of the shell was found to be uniform and thicker than that of the sample shown in Figure 3.4c.

A growth mechanism of the hollow spheres could be proposed according to the preceding experimental results. The template, which is common silica colloidal spheres, is often used as a physical template in the fabrication of the core–shell structure, and hollow spheres are usually obtained after the silica cores are dissolved in an alkaline condition [22,23]. The dissolving property implies that silicon–oxygen bonds are to form silicate ions, where ammonia solution is used as the alkaline agent. The chemical process is described as follows. Ammonia was used as the source of hydroxide ions based on ionization, and it also can be coupled with copper ions in the form of complex ions that are dispersed in the solution homogeneously. When silica colloidal spheres are heated at high temperature in ammonia solution, silicate ions are generated and react with the copper ions around the silica colloidal spheres, which results in the formation of copper silicate and preferentially deposited on the surface of silica colloidal spheres. As the reaction proceeds, silica colloidal spheres are consumed and more copper silicates are generated from the copper–ammonia complex ions and then the shell becomes complete gradually. The boundary shown in Figure 3.4c could be ascribed to the consumption of silica cores during the reaction. The copper silicate shells grew thicker and thicker until all the copper ions were transformed into copper silicate according to the preceding reaction process. Finally, the leftover silica cores were dissolved by surplus hydroxide ions at high temperature, which led to the formation of copper silicate hollow spheres. It should be pointed out that 0.7 mL of ammonia is required to dissolve silica completely in the present system. In the whole process, the silica colloidal spheres acted as a template and as a source of silicate ions, and the reagent ammonia provided hydroxyl ions and coupled with copper ions; all these put together resulted in the formation of copper silicate hollow spheres. In addition, the thickness of the shell can be controlled by adjusting the reaction parameters, for instance, thicker shells can be obtained by increasing the initial concentration of copper ions [21].

The building block of the shell layer is uniform nanotubes, as seen in Figure 3.3, and the formation process is as follows. The copper silicate generated in the reaction has a clay-type structure and is composed of asymmetric alternating sheets of silica tetrahedra and copper oxide octahedra [17]. According to the theory of L. Pauling who has pioneered studies on crystal structure of layered silicates [24], if two crystal faces of a constituent layer in a layer crystal are not equivalent, there arises a tendency for the layer to curve, one face becoming concave and one convex, and this tendency would in general not be overcome by the relatively weak force between adjacent layers. However, layer structure tends to curl at an elevated temperature and pressure according to the scrolling mechanism, which was presented to explain the formation of nanowires and nanotubes previously by many researchers [25–27]. In our experiment, from further high-magnification observation of the product formed at the initial stage (Figure 3.4b), it has been found that the particles around the silica colloidal spheres were lamellar without nanotubes, while the final product exhibited a nanotube structure. On this basis, it could be proposed that the combined effect of elevated temperature and pressure in a closed reaction system and the intrinsic asymmetric structure formed during the reaction process make the layer-structured copper silicate scroll in situ along a certain axis of the layered copper silicates. It is worth noting here that the nanotubes almost have a uniform diameter as shown in Figure 3.3, which implies that the lamellae have a uniform width before scrolling.

3.2.2 Magnesium Silicate Micro/Nanostructured Hollow Spheres

As mentioned in the preceding discussion, silica colloidal spheres release silicate ions in ammonia solution and can be used as chemical templates [21]. Similarly, magnesium silicate micro/nanostructured hollow spheres can also be fabricated by using such chemical template. The chemical-template-synthesized magnesium silicate hollow sphere presents a porous surface with a morphology and structure similar to sepiolite, which could be used as an environmentally benign and efficient adsorbent [16].

Using magnesium chloride as the source of metal ions and considering silica colloidal spheres as template, with ammonia and ammonia chloride as agents, micro/nanostructured magnesium silicate hollow spheres were obtained based on the strategy shown in Figure 3.1 [16]. Typically, silica colloidal spheres (0.1 g) are dispersed first in an alkaline solution (30 mL) containing magnesium chloride (0.75 mmol), ammonia (1 mL, 28%), and ammonia chloride (10 mmol) before reaction at 140°C for 12 hours. The silica chains are broken by hydroxide ions and so silicate ion groups are generated in the alkaline solution at high temperature [28], which react with magnesium ions to form magnesium silicate particles in situ around the silica cores (Figure 3.1, step I). In the following process, with gradual release of the silicate ions from the silica colloidal spheres, all of the magnesium ions are transformed into the magnesium silicate shell and silica/magnesium silicate core–shell structure is formed. Finally, after the remaining silica core has dissolved completely under alkaline conditions at high temperature, the well-structured magnesium silicate hollow sphere is obtained (Figure 3.1, step II).

3.2.2.1 Structure and Morphology

XRD measurement reveals that the corresponding products are magnesium silicate hydroxide hydrates with a talc structure and the ideal chemical formula is $Mg_3Si_4O_{10}(OH)_2$ [16]. The morphology of the as-prepared product is shown in Figure 3.5a. The final products are composed of spherical particles with size about 500 nm. The magnified image of a single sphere in the inset of Figure 3.5a reveals

FIGURE 3.5 The morphology of the magnesium silicate hollow microspheres prepared at 140°C for 12 hours. (a) Scanning electron microscopy image. (b) Transmission electron microscopy results. The insets are the corresponding local magnification. (Wang et al., *Chem. Eur. J.* 2010. 16. 3497–503. Copyright Wiley-VCH Verlag GmbH & Co. KGaA. Reproduced with permission.)

that the surface of the sphere is rough, porous, and bestrewn with a large amount of thin lamellae. The TEM examination demonstrates that the magnesium silicate spheres are hollow in structure, with an obviously black edge and darkish center; the shell is uniform and about 50 nm in thickness. The surface presents needlelike structures at first glance. Scanning electron microscopy (SEM) and TEM observations indicate that these hollow spheres are composed of large, narrow, and nanoscaled lamellae (Figure 3.5b). The observed needlelike structures were proposed to arise from the black edges of tilted nanoscale lamellae, which were partially or completely parallel to the electron beam.

Furthermore, time-dependent experiments were carried out as shown in Figure 3.6. The silica colloidal spheres are uniform with a diameter of 500 nm and have a smooth surface at the initial stage. After reaction for 0.5 hour, the silica colloidal spheres with smooth surface (Figure 3.4a) turned rough with some particles (Figure 3.6a), indicating that the silica colloidal spheres were partially dissolved by the hydroxyl ions. When the reaction time was prolonged to 3 hours, the hydroxyl ions persistently broke the silicon–oxygen chains of the silica colloidal spheres and the silicate ions generated continuously reacted with magnesium ions to form magnesium silicate particles, thin and complete silicate shells around the remaining silica cores were formed, as shown in Figure 3.6b. The inner silica colloidal spheres present irregular morphology, which indicates that they were gradually etched. The shell kept growing until all of the magnesium ions had been transformed into magnesium silicate particles. After 12 hours, the remaining silica cores had completely dissolved to leave the void structure (Figure 3.5). In addition, when the amount of magnesium ions in the initial precursor solution was increased, magnesium silicate hollow spheres with thicker shells were formed, as shown in Figure 3.7.

3.2.2.2 Formation of Magnesium Silicate Hollow Spheres

In this experiment, ammonia was chosen to provide hydroxide ions based on the ionization balance of ammonia. For existence of magnesium ions in the solution, ammonia chloride was a necessary additive in the precursor solution; otherwise, magnesium hydroxide was formed and deposited in the precursor ammonia solution, and large irregular particles with few spherical particles were obtained [16]. The

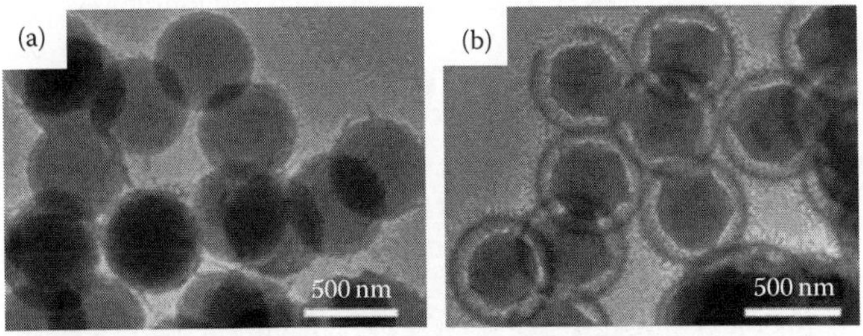

FIGURE 3.6 TEM images of magnesium silicate hollow spheres after reaction at 140°C at different reaction times. (a) 0.5 hour and (b) 3 hours. (Wang et al., *Chem. Eur. J.* 2010. 16. 3497–503. Copyright Wiley-VCH Verlag GmbH & Co. KGaA. Reproduced with permission.)

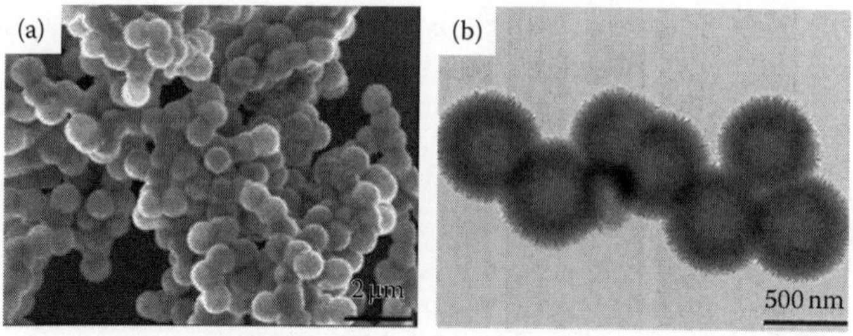

FIGURE 3.7 The morphology of the magnesium silicate micro/nanostructured hollow spheres prepared from the precursor solution with 1.5 mmol magnesium chloride and 1.2 mL ammonia (28%) at 140°C for 12 hours. (a) Scanning electron microscopy (SEM) image and (b) TEM image. (Wang et al., *Chem. Eur. J.* 2010. 16. 3497–503. Copyright Wiley-VCH Verlag GmbH & Co. KGaA. Reproduced with permission.)

reaction process is as follows. The magnesium ions react with the hydroxide ions to form magnesium hydroxide in ammonia solution; thus, the concentration of magnesium ions decreases in the solution, especially around the silica colloidal spheres, based on the deposition and dissolution balance or the reaction

$$Mg^{2+} + 2NH_3 \cdot H_2O \Leftrightarrow Mg(OH)_2 + 2NH_4^+ \tag{3.1}$$

When a large amount of silicate ions were generated from the silica colloidal spheres under alkaline conditions at high temperature and there were few magnesium ions around the silica colloidal spheres to react with the silicate ions, superfluous silicate ions generated from the silica colloidal spheres diffused into the solution and gradually led to the nucleation of magnesium silicate; thus, irregular magnesium silicate particles were formed in the ammonia solution without addition of ammonia chloride. However, when the amount of ammonia chloride was increased, in accordance with reaction (3.1), more magnesium ions reacted with the silicate ions and then grew into the magnesium silicate shell around the silica colloidal spheres, which decreased the amount of irregular particles [16]. When the amount of ammonia chloride was increased to 10 mmol in the present experiment, uniform and pure spherical particles were formed. All these results proved that the introduction of ammonium ions restrained the formation of magnesium hydroxide, in accordance with reaction (3.1). Thus, it is concluded that ammonia chloride would prevent the magnesium ions from forming magnesium hydroxide and would promise the generation of magnesium silicate particles on the surface of the silica colloidal spheres.

The formation of nanoscaled magnesium silicate lamellae in the final hollow spheres can be easily understood. According to the XRD measurements, the synthesized magnesium silicate takes the talc structure. Talc is composed of three "infinite" layers formed by sharing oxygen ions at three corners of the silica tetrahedron, as shown in Figure 3.8. A layer of octahedral-coordinated magnesium/hydroxide ions holds two layers of tetrahedral-coordinated silicon/oxygen ions together as a three-layer sheet. This "sandwich-like" three-layer unit is electrically

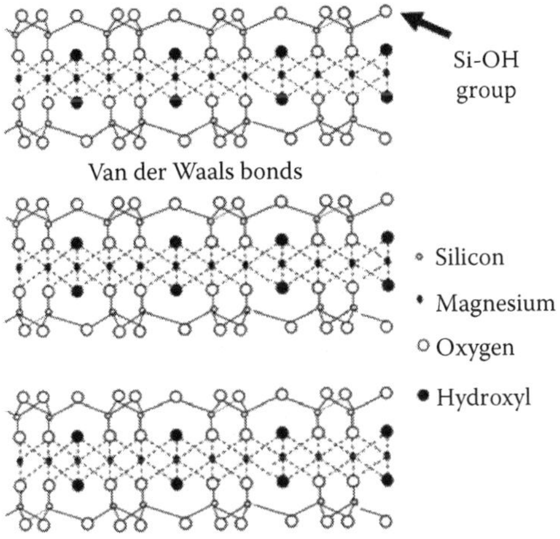

FIGURE 3.8 Schematic illustration of talc structure. (Wang et al., *Chem. Eur. J.* 2010. 16. 3497–503. Copyright Wiley-VCH Verlag GmbH & Co. KGaA. Reproduced with permission.)

neutral on the basal plane, so the crystal is held together by Van der Waals forces acting across adjacent tetrahedral layers [29]. The generated magnesium silicate thus tends to grow in the form of a lamellar structure, leading to magnesium silicate hollow spheres composed of narrow nanoscaled lamellae with a large amount of edges and faces.

Finally, it should be mentioned that the edges of talc consisting of broken bonds (Figure 3.8) exist in the form of Si–OH groups [30]. When the magnesium silicate hollow spheres were put in water, the surface Si–OH groups adsorb/dissociate hydrogen ions, which leads to the positive, negative, or uncharged sites, depending on the pH value of the solution. The magnesium silicate hollow spheres show a negative charge in deionized water.

3.2.3 Nickel Silicate and Silica–Nickel Composite Hollow Spheres

As mentioned in Section 3.1, silica colloidal spheres could be used as a physical template for the preparation of hollow spheres through coating and subsequently removing themselves in alkaline conditions, and ammonia can provide hydroxide ions continuously based on the ionization of ammonia, which can dissolve the surface of the silica spheres by breaking the silicon–oxygen bonds and forming the silicate ions. Obviously, the silicate ions react with the nickel ions to form nickel silicate particles in situ around the residual silica cores. When the silica cores are removed by etching, nickel silicate hollow spheres are obtained. Here, the nickel ions are slowly released from the nickel ammonia complex ions in ammonia solution, which could avoid the deposition of nickel ions in alkaline conditions. In this section, transformation of nickel silicate hollow spheres to silica and silica–nickel composite hollow spheres by using different treatments, such as selective leaching of nickel ions in acid solution and in situ reduction in a hydrogen atmosphere, is demonstrated as seen in steps (III) and (IV) of Figure 3.1, respectively.

3.2.3.1 Nickel Silicate Hollow Spheres

Nickel silicate hollow spheres can be prepared according to steps (I) and (II) given in Figure 3.1. Typically, $NiSO_4{\leftrightarrow}6H_2O$ (7.5 mmol) was mixed with 40 mL aqueous solution of ammonia ($NH_3{\leftrightarrow}3H_2O$, 28%, 10 mL) to form a homogeneous solution under stirring. A 1.6 g portion of silica colloidal spheres was added to 20 mL of deionized water before it was mixed with the prepared mixture. The final mixture was put into a Teflon container and heated at 90°C in a water bath under stirring. After reaction for 12 hours, a green precipitate (silica–nickel silicate core–shell spheres) is formed, which is dispersed in 50 mL of sodium hydroxide solution (1 M) and stirred for 16 hours to remove the silica core from the precipitate, resulting in nickel silicate hollow spheres.

After reaction with nickel ion–containing ammonia solution, a silica core coated with nickel silicate shell is formed. Each dark disk is embedded within a dark circle at the center as seen in Figure 3.9a. Figure 3.9b is the SEM image for some broken spheres and clearly shows the core–shell structure of the silica–nickel silicate spheres. When the silica core is dissolved in alkaline solution, the spherical morphology is undisturbed as shown in Figure 3.9c; however, the surface of the individual spheres becomes coarse and porous (Figure 3.9d and the inset). In addition, the diameter of the obtained spheres is about 700 nm, which is bigger than that of the original silica sphere templates (500 nm). XRD measurement confirms the phase structure of nickel silicate dehydrate [31].

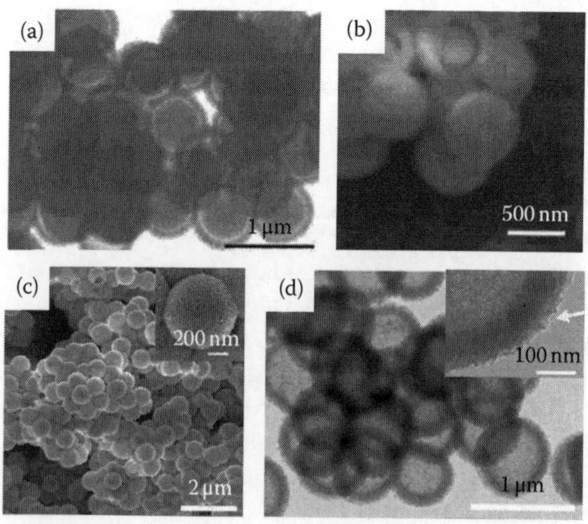

FIGURE 3.9 Morphology of the products. (a) and (b), TEM and SEM images of the silica–nickel silicate core–shell spheres, respectively; (c) and (d), SEM and TEM images of nickel silicate hollow spheres, respectively. The insets: local magnified images. (Reprinted with permission from Wang et al., 2010, 14830–4. Copyright 2010 American Chemical Society.)

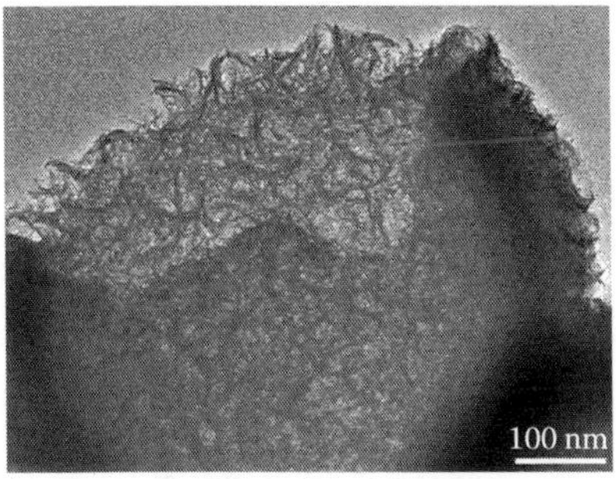

FIGURE 3.10 TEM image of a broken nickel silicate hollow sphere. (Reprinted with permission from Wang et al., 2010, 14830–4. Copyright 2010 American Chemical Society.)

From Figure 3.9d, we can see that nickel silicate spheres are hollow in structure with a shell thickness of about 100 nm. The porous shells consist of nanoscaled lamellae, as pointed out by the arrow in the inset of Figure 3.9d. Formation of the lamellae could be attributed to the layered clay structure of nickel silicate. A close examination has revealed that the lamellae interweave with each other during the growth process, which leads to a porous surface structure, as more clearly shown in Figure 3.10, corresponding to a broken nickel silicate sphere. In addition, nitrogen isothermal adsorption measurement has shown that these porous nickel silicate hollow spheres possess of quite high specific surface area up to about 350 m^2/g [31]. It should also be mentioned that the size and shell thickness of the nickel silicate hollow spheres can be controlled by the silica colloidal spheres and the amount of nickel ions in the solution.

3.2.3.2 Silica Hollow Spheres

According to step IV of the strategy shown in Figure 3.1, the obtained nickel silicate hollow spheres could be transformed into porous silica hollow spheres by a selective leaching method, which has been extensively reported [32–34]. When the metal silicate hollow spheres are immersed in an acidic solution, leaching takes place or the metal ions are extracted in the solution and a large amount of Si–OH groups are generated. The Si–OH groups are unstable and rapidly polymerize by forming Si–O–Si bonds, leading to a silica skeleton and the final porous silica spheres. In this experiment, typically, 0.1 g of nickel silicate hollow spheres was dispersed in hydrochloric acid solutions with different concentrations (4, 2, and 1 M) and stirred at 80°C. After leaching for 6 hours, silica hollow spheres were obtained.

Figure 3.11 shows the typical results, indicating large-scaled uniform porous silica spheres. From some broken spheres in Figure 3.11, we can see that the silica spheres are hollow in structure. The TEM examination further confirms the hollow structure and porous shell layer for such silica spheres, as shown in Figure 3.12. Their shell structure is different from that of the thick lamellar corresponding to the porous nickel silicate hollow spheres shown in Figures 3.9 and 3.10. The shell thickness of the porous silica spheres is about 60 nm, which is slightly thinner than that

FIGURE 3.11 Field emission SEM images of porous silica hollow spheres. (a) and (b) are low and high magnifications, respectively. (Reprinted with permission from Wang et al., 2010, 14830–4. Copyright 2010 American Chemical Society.)

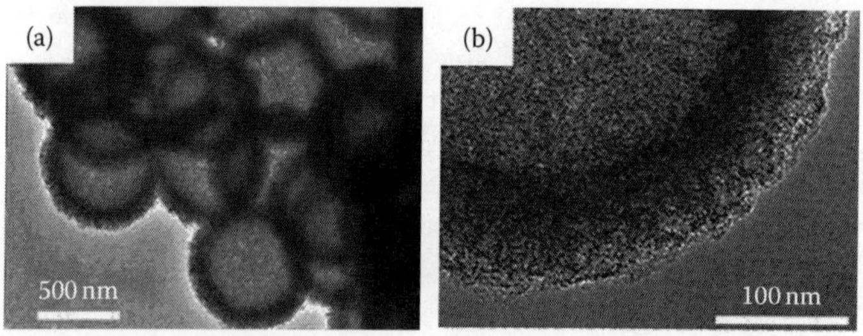

FIGURE 3.12 (a) TEM image of porous silica hollow spheres. (b) A local magnified image of (a). (Reprinted with permission from Wang et al., 2010, 14830–4. Copyright 2010 American Chemical Society.)

of the nickel silicate hollow spheres (Figure 3.9). The shrinkage of the shell layer is attributed to the removal of nickel ions from the nickel silicate skeleton. No nickel was detected in the products due to the treatment in hydrochloric acid solution [31]. This means that the silica hollow spheres can be produced by selective leaching method. Furthermore, since the size and shell thickness of nickel silicate hollow spheres can be controlled easily, porous silica hollow spheres with tunable size and shell thickness can also be obtained. The above-mentioned leaching strategy provides an alternative route to fabricate porous silica hollow spheres with controlled structure.

Further experiments have revealed that the hydrochloric acid concentration in the leach solution has a significant influence on the porous shell structure of the final silica hollow spheres. The nitrogen adsorption isotherms indicate that all of silica hollow spheres, obtained from leach solution with different hydrochloric acid concentrations, exhibit a type IV isotherm with a clear hysteresis loop, demonstrating these products with mesoporous characteristics [31]. The specific surface areas were estimated to be about 800, 760, and 643 cm^2/g, corresponding to hydrochloric acid concentrations of 4, 2, and 1 M, respectively, or the higher concentration corresponds to the higher specific surface area. For the pore size in the shell layer, however, the opposite is true, and the pore sizes are 3.4, 3.8, and 5.5 nm, respectively, corresponding to the concentrations 4, 2, and 1 M.

This phenomenon can be easily understood. When nickel silicate is immersed in the acid solution, the Si–OH groups are formed, and the polymerization of Si–OH groups results in the formation of the silica hollow spheres. Obviously, the high acid concentration induces more Si–OH groups, which results in higher the amount of silica nuclei in high acid concentration than that in low acid concentration. So, in solution with high acid concentration, more silica nanoparticles are formed with smaller size, and the silica nanoparticle-built hollow spheres are of higher specific surface area. Thus the pore size in the shell layer and the specific surface area can be controlled just by treatment of the nickel silicate hollow spheres in solution with different acid concentrations.

3.2.3.3 Silica–Nickel Composite Hollow Spheres

Silica hollow spheres with magnetic properties have attracted much attention because of their potential applications in drug delivery. The conventional method is employing magnetic nanoparticles as precursors, but the attraction between magnetic nanoparticles makes the production of magnetic silica composite hollow spheres rather difficult. Combining the strategy shown in Figure 3.1, nickel silicate hollow spheres can be used as precursors and be reduced in a hydrogen atmosphere; nickel ions would in situ be reduced to form nickel nanoparticles (step III in Figure 3.1) and the silica–nickel composite hollow spheres could be obtained easily. In the experiment, typically, 0.1 g of the nickel silicate hollow spheres was placed in a ceramic boat in the middle of a horizontal tube furnace. Hydrogen was allowed to flow (30 mL/min) in the horizontal tube oven at 450°C for 6 hours. The black powders were finally collected in the ceramic boat at room temperature.

The XRD measurement of the products confirms that nickel silicate was in situ transformed into silica–nickel composite after the hydrogen reduction for 6 hours. It was found that the intensity of the broad peak around 23°C and the peaks of nickel particles increased with the reduction duration, while the characteristic peaks of nickel silicate disappeared, indicating nickel ions reduced in a hydrogen atmosphere [31]. It should be mentioned that the peaks of nickel silicate could still be observed after reduction for 6 hours. These peaks originate from the interior core of the nickel silicate particles, since it is difficult for hydrogen molecules to enter inside and hence the remains of nickel silicate particles.

Figure 3.13a shows the morphology of the reduced products, indicating spherical shape in microscale. The broken spheres in Figure 3.6a demonstrate the hollow structure. TEM examination reveals that the spheres are hollow in structure with a uniform shell as seen in Figure 3.13b. The high-magnification TEM observation reveals that the silica–nickel composite shell bestrewed homogeneously nickel nanoparticles about 10 nm in size, which are marked by circles in Figure 3.13c. The corresponding selected area electron diffraction (SAED) pattern further confirmed the formation of nickel nanoparticles as shown in Figure 3.13d. Three ring patterns with dots, corresponding to the planar spacings of 0.203, 0.176, and 0.124 nm, can be observed, which is consistent with the plane indices (111), (200), and (222) of *fcc* nickel, respectively [35,36].

Furthermore, the effects of the reduction duration on the magnetic properties of silica–nickel composite hollow spheres are studied. Figure 3.14a shows the room

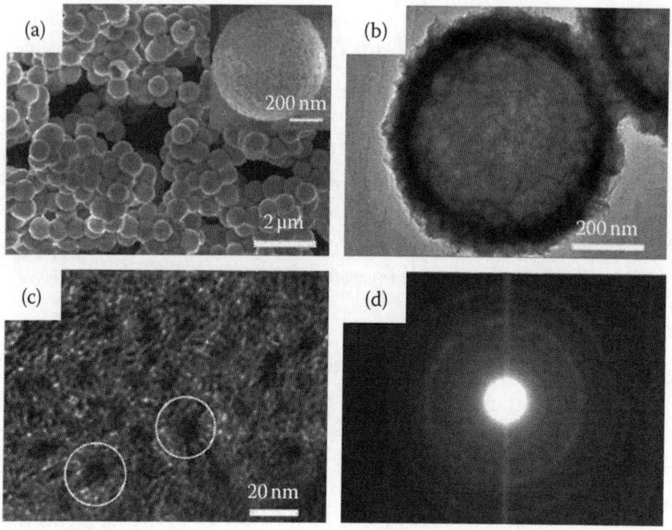

FIGURE 3.13 Morphology and structure of the silica–nickel composite hollow spheres obtained after hydrogen reduction for 6 hours. (a) Field emission SEM image (the inset is a local magnified image). (b) TEM image. (c) The high-magnification TEM image of the shell layer. (d) The selected area electron diffraction pattern corresponding to the area in (c). (Reprinted with permission from Wang et al., 2010, 14830–4. Copyright 2010 American Chemical Society.)

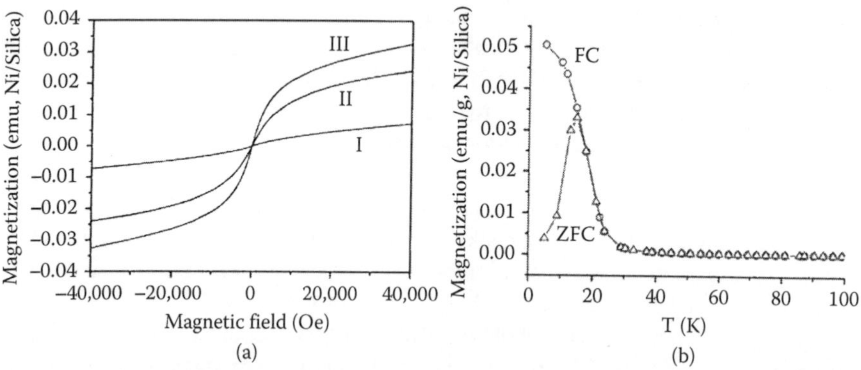

FIGURE 3.14 The magnetic properties of the silica–nickel composite hollow spheres obtained at 450°C. (a) The room temperature field-dependent magnetization curves of the silica–nickel composite after reduction for (I) 0.5 hour, (II) 1.5 hours, and (III) 6 hours. (b) Zero-field cooling (ZFC)/FC-specific magnetization in an applied field of 100 Oe for the silica–nickel composite after reduction for 6 hours. (Reprinted with permission from Wang et al., 2010, 14830–4. Copyright 2010 American Chemical Society.)

temperature field-dependent magnetization curves of the samples obtained after reduction treatment for different durations at 450°C. The magnetic saturation of the silica–nickel composite hollow spheres increases with increasing reduction time, which could be attributed to the increase in amount of nickel particles in the nickel–silica composite hollow spheres. The presence of superparamagnetic behavior in these nickel–silica composite hollows spheres was observed by using zero-field cooling (ZFC) in an applied field of 100 Oe. The corresponding ZFC susceptibility shows a blocking temperature T_b at 16 K, as illustrated in Figure 3.14b. For nickel particles, the critical diameter to form a single-domain state is about 55 nm [37]. Here, the average size of nickel nanoparticles embedded in the silica shell is estimated to be about 10 nm, from Figure 3.13c, or below the critical diameter, and hence the silica–nickel hollow spheres are superparamagnetic. These magnetic properties make possible for the hollow spheres to have applications in drug delivery and magnetic separations. In addition, since the nickel nanoparticles in the shell layer are surrounded by silica matrix, the silica–nickel composite hollow spheres are expected to be stable in air for a long time period [38].

3.3 HIERARCHICAL SiO_2@γ-AlOOH MICROSPHERES

It is well known that boehmite (γ-AlOOH) is the precursor of γ-Al_2O_3 and is one of the most important industrial catalyst supports. Recently, considerable efforts had been made to fabricate the γ-AlOOH nanostructures due to their new applications as gas sensors [39] and anion-specific adsorbents [40–42]. Many synthesis strategies have been developed for production of the nanostructured γ-AlOOH with different morphologies, including hollow nanospheres [43], nanococoons [39], nanofibers, nanowires, nanotubes [44–48], and mesoporous structures [49,50]. However, report on the preparation of hierarchically micro/nanostructured γ-AlOOH is limited, which is expected to exhibit novel performances.

As mentioned in Section 3.1, colloidal spheres can be used as templates for production of hierarchically structured microspheres. In this section, we introduce fabrication of hierarchically micro/nanostructured SiO_2@γ-AlOOH spheres in one step based on silica colloidal spherical template. In the experiment, hydrolysis of precursor sodium aluminate was promoted on basis of the dissolution property of silica colloidal spheres in alkaline solution [51,52]. Hydrolysis induces formation of lamellar γ-AlOOH and preferential deposition around the silica colloidal spheres [53]. Briefly, in a typical experiment, 0.13 g silica colloidal spheres was homogeneously dispersed in 20 mL deionized water; 0.17 g sodium aluminate and 0.48 g urea were dissolved in 30 mL deionized water under stirring. These solutions were mixed homogeneously and transferred into a Teflon autoclave and heated at a temperature of 160°C for 12 hours before cooling to room temperature. The final white products were obtained by centrifugation and washing and drying [53].

3.3.1 MORPHOLOGY AND STRUCTURE

Figure 3.15 shows the XRD results of the as-prepared products. All typical diffraction peaks can be identified as orthorhombic γ-AlOOH (curve a in Figure 3.15), indicating that the products are crystallized γ-AlOOH. The broad diffraction peaks around 23° should be ascribed to the amorphous silica colloidal spheres (curve b in Figure 3.15) [54]. SEM observation reveals that the obtained γ-AlOOH powders consist of uniform-sized spherelike objects, as shown in Figure 3.16a. The surface morphology of a single sphere can be clearly observed in the SEM image with a high

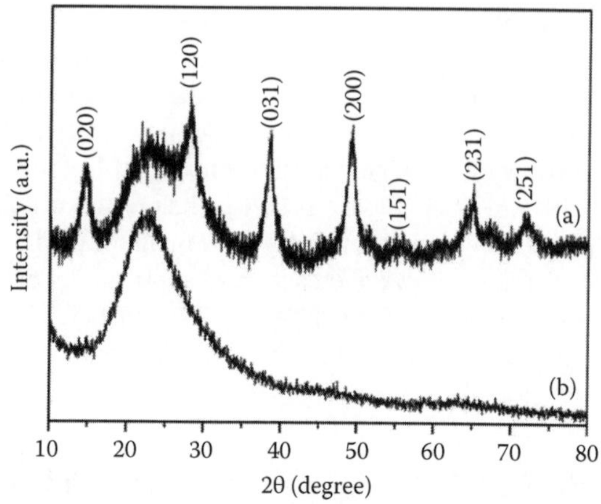

FIGURE 3.15 X-ray diffraction patterns of (a) as-prepared hierarchical SiO_2@γ-AlOOH spheres and (b) silica colloidal spheres. The indexes correspond to the crystal planes of γ-AlOOH. (The source of the material including Wang et al., *Nanotechnology*, 2009 and Institute of Physics is acknowledged.)

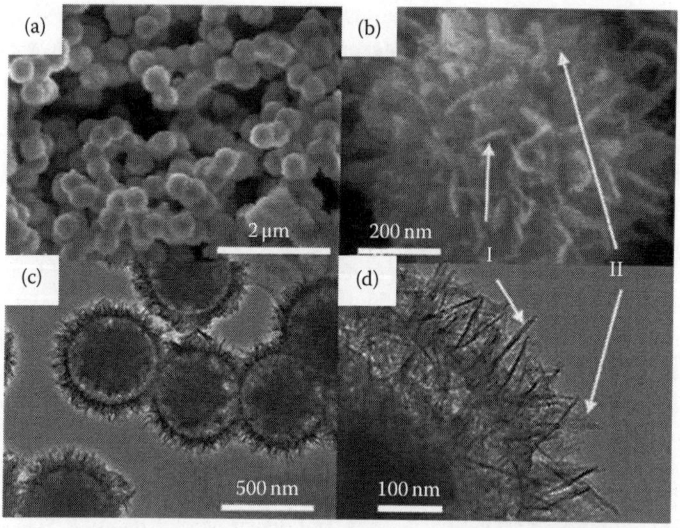

FIGURE 3.16 The morphology of hierarchical SiO_2@γ-AlOOH spheres. (a) and (b) SEM images of the general view and single sphere. (c) and (d) TEM images. (The source of the material including Wang et al., *Nanotechnology*, 2009 and Institute of Physics is acknowledged.)

magnification, as seen in Figure 3.16b. The spherical particles are of about 700 nm in diameter, possess rough surface, and consist of large amount of thin lamellae that are nearly vertically standing on the microspheres. TEM examination has demonstrated that the as-prepared spherical γ-AlOOH particles are core–shell structures with about 100-nm shell thickness. At first glance, the shell layer seems to be composed of many nanowires standing on the surface, as illustrated in Figure 3.16c and d. By a close observation, however, they are thin lamellae parallel to the electron beam (see arrow I in Figure 3.16), and many lamellae with about 5 nm in thickness can be observed (see arrow II in Figure 3.16). Furthermore, the specific surface area was estimated to be about 152 m^2/g based on the isothermal nitrogen adsorption measurements, which is larger than that of the flowerlike γ-AlOOH consisting of nanowires, reported previously [27], which causes the hierarchical surface structure of the SiO_2@γ-AlOOH spheres. Furthermore, after annealing at 400°C, the specific surface area is still as high as 140 m^2/g for the SiO_2@γ-AlOOH spheres, which indicates high structural stability [53].

3.3.2 Formation of Hierarchical SiO_2@γ-AlOOH Microspheres

3.3.2.1 Morphological Evolution with Reaction

To reveal formation of the hierarchical structure, morphological evolution with the reaction duration was observed. Figure 3.17 shows the corresponding results. The silica colloidal spheres in this experiment are of 500 nm in diameter with a smooth surface, as shown in Figure 3.17a. After reaction for 1 hour, a hazy layer consisting of thin lamellae, with a thickness of 30 nm, was formed on the surface of the silica colloidal sphere, as shown in Figure 3.17b. When the reaction time was prolonged up to 5 hours, the lamellae became thicker and denser and a blank boundary was formed between the core part and the shell, as illustrated in Figure 3.17c. When the reaction was further prolonged to 12 hours, the core–shell structures remained as such, as that shown in Figure 3.17c, except for the more thicker lamellae (Figure 3.17d).

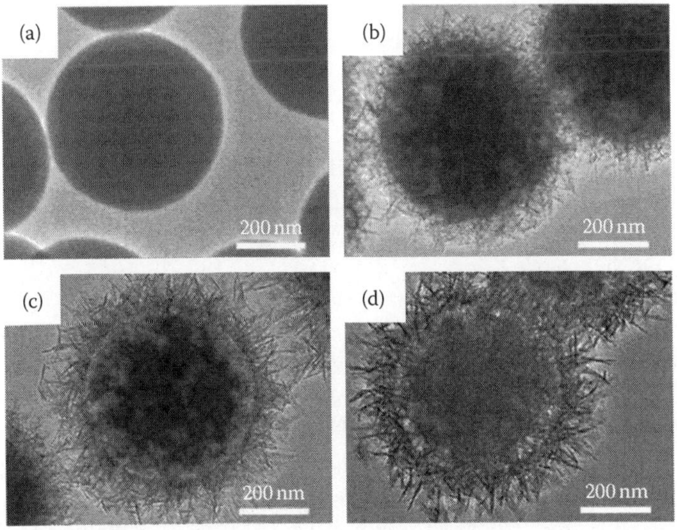

FIGURE 3.17 TEM images of the as-obtained products under different reaction times: (a) 0 hour, (b) 1 hour, (c) 5 hours, and (d) 12 hours. (The source of the material including Wang et al., *Nanotechnology* 2009 and Institute of Physics is acknowledged.)

3.3.2.2 Template-Induced Deposition Mechanism

Based on the preceding results, formation of hierarchical SiO_2@γ-AlOOH microspheres could be explained as follows. When heated at 160°C, aluminate anions in the precursor solution undergo the following reactions:

$$AlO_2^- + 2H_2O \Leftrightarrow Al(OH)_3 + OH^- \tag{3.2}$$

$$Al(OH)_3 \rightarrow \gamma AlOOH + H_2O \tag{3.3}$$

Firstly, AlO_2^- anions released from sodium aluminate hydrolyze to produce $Al(OH)_3$ and hydroxide ions (reaction 3.2); the unstable $Al(OH)_3$ further pyrolyzes to form γ-AlOOH precipitate, which occurs at a temperature >150°C [39], as per reaction 3.3. The growth of the γ-AlOOH shell is therefore described as follows. It is well known that silica colloidal spheres can be etched in alkaline environment and the silicon–oxygen bonds of silica can be broken by hydroxide ions, and alkaline solutions like sodium hydroxide are often used to dissolve the silica core in the template synthesis of hollow structures [51,55]. Here, the surface silicon–oxygen bonds of silica colloidal spheres would be broken by hydroxide ions, inducing the preferential hydrolysis of aluminate anions around the silica colloidal template. The formed $Al(OH)_3$ colloids were unstable and transformed gradually into γ-AlOOH as per reaction 3.3 at 160°C, which then deposited on the silica colloidal template. With the reaction proceeding, hydroxide ions were consumed by silica colloidal spheres, resulting in the formation of γ-AlOOH, and the gradual deposition of γ-AlOOH made the shell grow thicker. The reaction continued until all the aluminate anions were used up. Although the silica colloidal spheres were used without prior surface modification in this experiment, no irregular particles in the final products were observed (see Figure 3.16a), indicating that almost all the generated γ-AlOOH was deposited on the silica colloidal spheres. As for the deformation of the silica colloidal spheres shown in Figure 3.17c and d, it should be attributed to their partial dissolution by hydroxide ions. In other words, the silica colloidal spheres not only function as a template but also induce the hydrolysis of aluminate anions. The generated γ-AlOOH were located around the silica template and preferentially deposited on the surface of silica colloids. The formation of the hierarchical SiO_2@γ-AlOOH core–shell structured microspheres could thus be attributed to the template-induced deposition mechanism.

As mentioned in Section 3.3.1, the shells consist of lamellar γ-AlOOH in the final products (Figure 3.16), which did not scroll into nanowires or nanotubes like other layer materials under elevated temperature and pressure [17,25,56]. The γ-AlOOH crystal is layered in structure with octahedral arrangement within the lamellae, and hydroxyl ions hold the lamellae together through hydrogen bonding [44]. In acidic condition, the solution contains protons that combine with the hydroxyl oxygen lone pairs, which form aqua ligands and destroy the γ-AlOOH layers [57]. The separated layers curl into one-dimensional nanostructures via the scrolling-growth route. However, the 2D lamellar nanostructures are retained in the basic solution [58]. Here, the precursor solution was weakly basic with a pH above 8. Thus, only lamellar γ-AlOOH is produced, which interweaves with each other, forming the hierarchical surface structure.

3.3.2.3 Effect of Urea

It should be noticed that urea in the precursor solution played an important role in the deposition process, since it can release hydroxide ions through hydrolysis in the solution. When urea was not added, the final products were composed of uniform silica colloidal spheres with smooth surface and irregular-shaped γ-AlOOH particle aggregates, which indicates that γ-AlOOH nucleated homogeneously in the solution. When a small amount of urea was added, the products were a mixture of hierarchically structured spherical particles and irregular-shaped particle aggregates [53]. Only when suitable amount of urea was used, uniform hierarchically structured spherical particles were observed, as shown in Figure 3.16a. These results demonstrate that hydroxide ions released from the hydrolysis of urea restrain formation of γ-AlOOH in the solution and induce preferential hydrolysis of aluminate anions around the silica cores to form hierarchical structure. In addition, when the silica colloidal spheres were not added with the same conditions, only irregular-shaped particles were obtained. So, silica colloidal spheres played as templates, which not only induced the preferential formation and deposition of the γ-AlOOH on the silica colloidal template but also avoided agglomeration and growth of γ-AlOOH particles in the solution.

In sum, hierarchical SiO_2@γ-AlOOH spheres can be fabricated by using silica colloidal spheres in one step. Their formation could be attributed to the template-induced heterogeneous deposition, in which the dissolution properties of silica colloidal spheres in alkaline conditions played the main role under existence of urea in the solution.

3.4 STRUCTURE AND COMPONENT CONTROLLABLE HOLLOW NANOSPHERES: THE CASE OF ZnO AND NOBLE METAL CLUSTER–EMBEDDED ZnO

"Nanochambers" within nanostructured materials are of much concern [59]. Nanoobjects with hollow interior voids, including nanotubes [60,61], nanoporous materials [62], and hollow nanospheres, have also been attracting attention due to their broad and promising performances in chemical/biologic sensing, ion exchange, and so on. As a typical boundary material, hollow spheres, especially small sized ones (<100 nm in diameter), recently have inspired more and more interests, because of their "nanovoid," high surface-volume ratio, and low density, hence potential applications in many areas such as drug delivery [63], sensors [64], batteries [65], optics [66], catalysis [4], and biologic imaging [67,68]. For instance, remarkable biologic imaging performances have been obtained using gold nanocages [67] and Au_3Cu hollow nanospheres [68] as contrast agents, in which the hollow structure and large surface area were considered as important improving factors.

Up to now, several methods had been developed to synthesize hollow sphere materials. Among them are the methods based on the nanoscale Kirkendall effect, which leads to the net mass flux and formation of voids at the central parts of primal particles [14,61,69–71] and has been well demonstrated in cobalt oxide and sulfides [14] and Ni_2P [71]. Also, noble metal hollow nanoobjects, such as Au nanocages [72] and nanoshells [73], have been obtained by ion exchange. Besides, other methods were also adopted, such as template method [15,74], Ostwald ripening [75], and self-assembly [76,77]. Remarkably, micron or submicron ZnO spherelike hollow structures, such as dandelions [70], microspheres [77], mesoporous polyhedral cages [78], and urchins [79], have been fabricated, which builds a base for the potential applications of ZnO hollow spheres. However, reports on fabrication of small and thin oxide hollow nanospheres and control of their microstructure and composition (or components), which are of importance in improvement of their performances, are very limited. In this section, we introduce a simple but effective strategy to demonstrate the fabrication of the structure- and composition-controlled hollow nanospheres based on selective etching of metal/oxide core–shell structured nanoparticles [80].

3.4.1 Selective Etching Strategy for Hollow Nanospheres

The corresponding fabrication strategy is illustrated in Figure 3.18. As known already, metal and its oxide have different chemical potentials in their reactions with acid, particularly certain active metals and amphoteric oxides. In weak acid environment, the etching rate of active metal is obviously higher than that of amphoteric oxide [81]. So, when the active metal/oxide core–shell nanoparticles are immersed in a weak acid solution (see route I in Figure 3.18), the H^+ ions in the solution diffuse along lattice defects, especially, grain boundaries in the shell layer, and enter the core

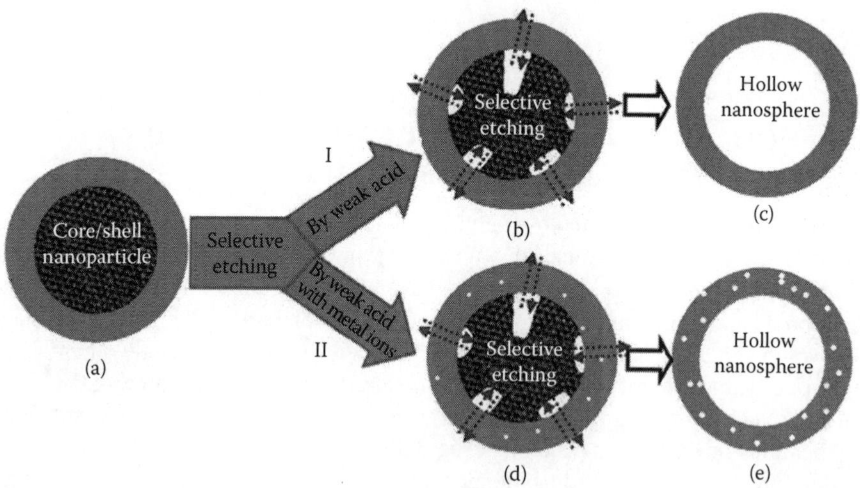

FIGURE 3.18 Schematic illustration of the selective etching strategy for hollow nanospheres. (a) A metal/oxide core–shell nanoparticle. Route I: Selective etching by weak acid. (b) Weak acid–induced preferential etching in the core part. (c) Complete consumption of the core part and formation of hollow nanosphere. Route II: Selective etching by weak acid with metal ions. (d) Weak acid–induced preferential etching in the core part and formation of few metal clusters in the shell due to diffusion of metal ions. (e) Complete consumption of the core part and formation of the hollow nanosphere embedded with metal nanoparticles (or clusters). (Reprinted with permission from Zeng et al., 2008, 1661–702. Copyright 2008 American Chemical Society.)

parts, leading to preferential etching of the core parts in addition to weak etching of the shell layer (see Figure 3.18b). A good weak acid can be used to adjust the relative etching rates of active metal in core parts and oxide in shell parts so that the core parts can be preferentially consumed. In this case, with the etching process ongoing, the core parts can be completely exhausted but the shell parts remain, which results in the formation of hollow nanospheres (see Figure 3.18c). Alternatively, if weak acids containing noble metal ions as etching agents are used (see route II in Figure 3.18), noble metal doping in the nanoshell layers can occur during the diffusion-etching process and noble metal ultrafine particles (or clusters) can be formed or embedded in the shell layers of final hollow nanospheres (see Figure 3.18d and e). We can thus control the component of the nanoshell layers.

Here, we consider ZnO as a typical example to demonstrate the effectiveness of the selective etching strategy shown in Figure 3.18, based on Zn/ZnO core–shell nanoparticles, prepared by laser ablation in liquid (LAL) [81]. A series of Zn/ZnO core–shell nanoparticle-containing colloidal solutions with average diameter <40 nm, depending on the laser powers, were firstly obtained by LAL as the sacrificial templates, as previously reported [82–84]. Figure 3.19a shows a typical result corresponding to the laser power 70 mJ/pulse. The particles are of 20 nm in average diameter and nearly spherical shaped. High-resolution transmission electron microscopy (HRTEM) examination shows that the particles are of Zn/ZnO core–shell structure and that there exist some ultrafine ZnO nanocrystals in the shell layers and many disordered areas surrounding them, as shown in Figure 3.19b and c.

3.4.2 Etching by Tartaric Acid

Based on route I in Figure 3.18, the Zn/ZnO core–shell nanoparticles shown in Figure 3.19 were taken as the sacrificial templates. When weak acid tartaric acid (TA) was gradually and continuously added to the Zn/ZnO core–shell nanoparticle-containing colloidal solution, the color of the solution changed gradually from bright

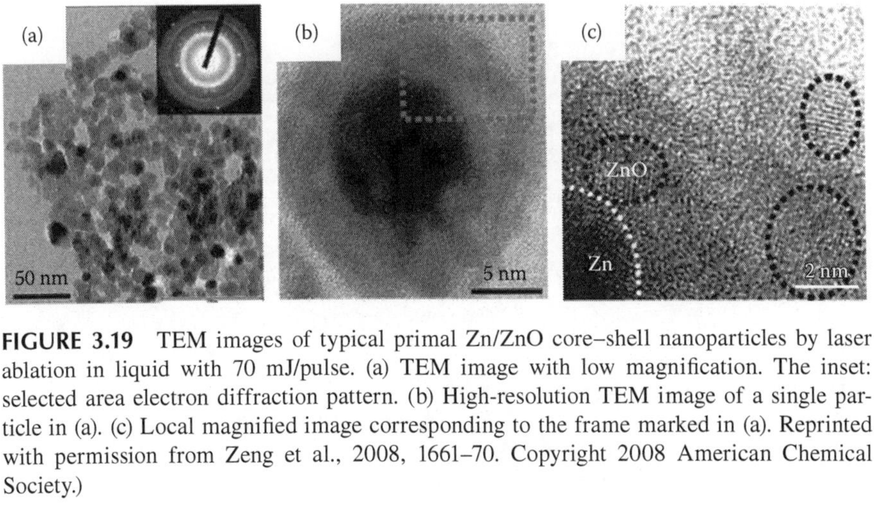

FIGURE 3.19 TEM images of typical primal Zn/ZnO core–shell nanoparticles by laser ablation in liquid with 70 mJ/pulse. (a) TEM image with low magnification. The inset: selected area electron diffraction pattern. (b) High-resolution TEM image of a single particle in (a). (c) Local magnified image corresponding to the frame marked in (a). Reprinted with permission from Zeng et al., 2008, 1661–70. Copyright 2008 American Chemical Society.)

yellow to ivory-white. Correspondingly, XRD evolution of the nanoparticles clearly demonstrates that the Zn/ZnO biphase components gradually transform to single-phase ZnO during the TA etching, as shown in Figure 3.20a. The diffraction peaks of both Zn and ZnO decrease with addition of TA, with Zn diffraction decaying much faster than ZnO and vanishing with addition of TA for 30 minutes, while the ZnO diffraction exists, as demonstrated in curve 3 in Figure 3.20a, indicating the disappearance of metal Zn after complete etching. Furthermore, the optical absorption spectral evolution of the colloidal solution shows similar results, as illustrated in Figure 3.20b. The colloidal solution before etching shows an optical absorption peak around 250 nm together with a shoulder at 357 nm (see curve 1 in Figure 3.20b), and the deep ultraviolet absorption peak is due to local surface plasmon resonance

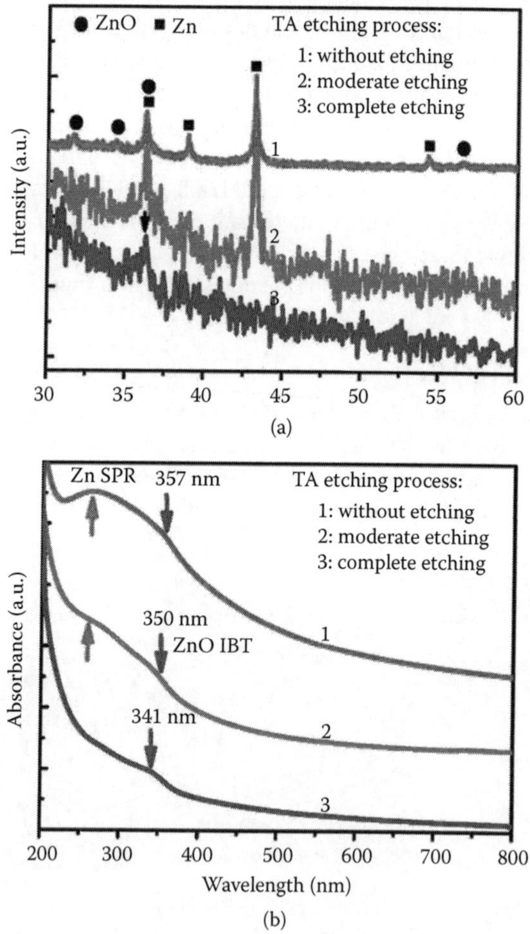

FIGURE 3.20 X-ray diffraction (a) and optical absorption (b) evolutions with tartaric acid (TA) etching process for the Zn/ZnO core–shell nanoparticles with 20-nm average original size. (Reprinted with permission from Zeng et al., 2008, 1661–70. Copyright 2008 American Chemical Society.)

(SPR) of Zn nanocores [85], while the absorption shoulder corresponds to the exciton or interband transition (IBT) absorption of ZnO [86]. SPR decreases and finally disappears with etching while the IBT absorption blueshifts down to 341 nm and becomes more dominant in the final spectrum (see curves 2 and 3 in Figure 3.20b), corresponding to a remarkable quantum size effect, indicating thinning of the ZnO shells or reduction in size of the ultrafine ZnO grains in the shell layers. All these evolutions during etching indicate occurrence of selective or preferential etching of the core–shell nanoparticles and demonstrate that the Zn/ZnO core–shell biphase nanoparticles were gradually transformed to single-phase ZnO material.

Figure 3.21 shows the optical absorption spectra for the Zn/ZnO core–shell nanoparticles with different original particle sizes (or shell thicknesses) after complete etching. With reduction in the size (also the shell thickness) of the original core–shell nanoparticles, obvious blueshift of the optical absorption edges is observed from the etched products, implying that the dimension of the etched products has been reduced into the quantum size range.

Furthermore, microstructures have been examined for the TA-etched products. Figure 3.22 shows the corresponding TEM images for the products with different original particle sizes, which clearly show hollow nanospheres that are nearly spherical shaped Moreover, etching reduces of the number of particles in the outside diameter and also shell thickness, in addition to the preferential exhaustion of the active metal Zn cores, as indicated in Figure 3.23 (estimated from TEM images).

SAED patterns of these hollow nanospheres are weak and diffused, compared with those of the original nanoparticles (see the inset of Figure 3.19a), implying decrease of crystalline degree after etching. Moreover, with reduction of hollow nanosphere diameter or shell thickness, the SAED patterns evolve from the coexistence of diffraction ring (including several diffraction spots, corresponding to ZnO) and amorphous-halation loop (typically illustrated in the inset of Figure 3.22d) to the almost completely amorphous loop (typically illustrated in the inset of Figure 3.22f),

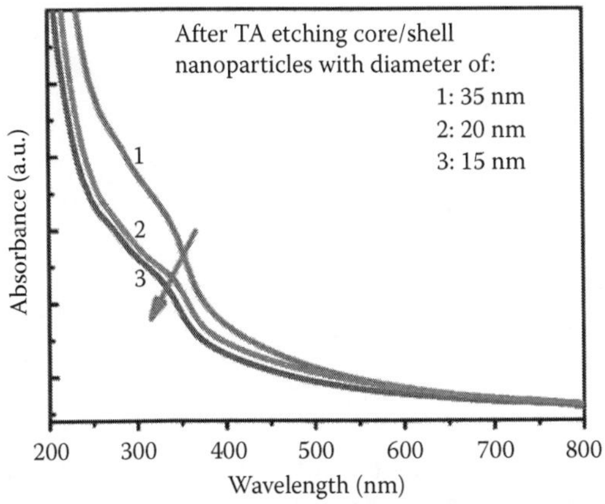

FIGURE 3.21 Optical absorption spectra of the etched products using Zn/ZnO core–shell nanoparticles with different diameters. (Reprinted with permission from Zeng et al., 2008, 1661–70. Copyright 2008 American Chemical Society.)

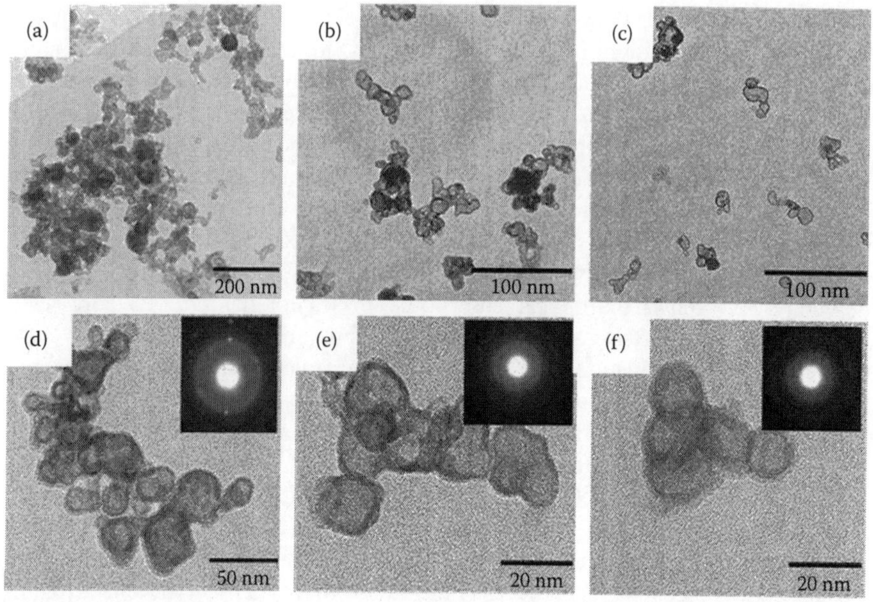

FIGURE 3.22 TEM images for different sized templates after complete etching by tartaric acid. Images (a) through (c) correspond to the Zn/ZnO core–shell nanoparticles with 35-, 20-, and 15-nm average original particle sizes, respectively. Images (d) through (f) are the local magnified images of (a) through (c), respectively, and the insets are the corresponding selected area electron diffractions. (Reprinted with permission from Zeng et al., 2008, 1661–70. Copyright 2008 American Chemical Society.)

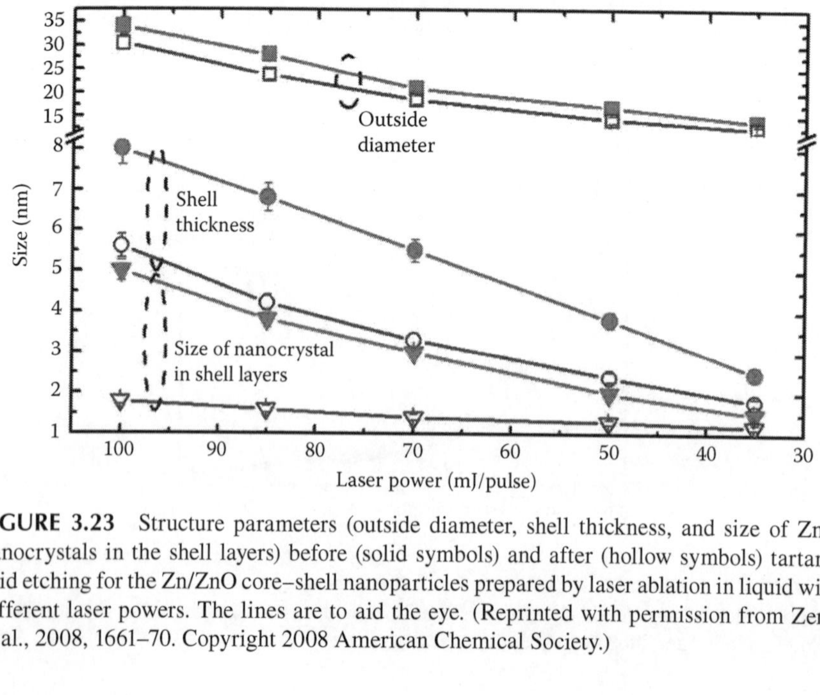

FIGURE 3.23 Structure parameters (outside diameter, shell thickness, and size of ZnO nanocrystals in the shell layers) before (solid symbols) and after (hollow symbols) tartaric acid etching for the Zn/ZnO core–shell nanoparticles prepared by laser ablation in liquid with different laser powers. The lines are to aid the eye. (Reprinted with permission from Zeng et al., 2008, 1661–70. Copyright 2008 American Chemical Society.)

indicating further decrease of crystallization degree in the ZnO shell layer with size reduction and weak acid etching-induced microstructural change in the shell layer.

HRTEM examination reveals that some ultrafine ZnO nanocrystals exist in the shell layers of hollow nanospheres, as shown in Figure 3.24 (see the dashed circles). The clear lattice fringes with 0.26-nm spacing corresponds to (002) planes of wurtzite ZnO. These ultrafine ZnO nanocrystals are surrounded by disordered areas in the shell layers. Interestingly, such weak acid etching causes a decrease in the size and number of the ZnO nanocrystals embedded in the shell layers, in addition to shell thickness or outer diameter of particles. Especially, for very small hollow nanospheres, the shell layer contains very few ultrafine nanocrystals and is nearly amorphous (see Figure 3.24d). Such microstructure and its evolution are in good agreement with XRD (see Figure 3.20a) and SAED results (see insets in Figure 3.22d through f).

The detailed comparisons of the structural parameters, including the diameter, shell thickness, and nanocrystal size, between the original core–shell nanoparticles and etching-formed hollow nanospheres are presented in Figure 3.23, clearly demonstrating the etching-induced reduction of these parameter values. Selective etching provides a simple way not only to obtain oxide hollow nanospheres but also to control their structure (diameter and shell thickness), even microstructure (embedded ZnO nanocrystals and crystallinity) by adopting proper primal nanoparticles.

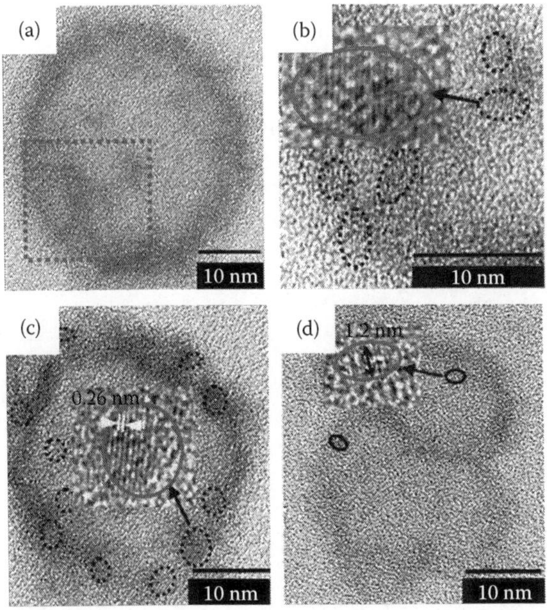

FIGURE 3.24 High-resolution transmission electron microscopy (HRTEM) images of typical hollow particles from different sized templates after etching by tartaric acid. Images (a), (c), and (d) correspond to the samples shown in images (d) through (f) of Figure 3.22, respectively. Image (b) is the magnified image corresponding to the frame marked in (a). The insets in (b) through (d) show the typical ZnO ultrafine nanocrystals embedded in the shell layers. (Reprinted with permission from Zeng et al., 2008, 1661–70. Copyright 2008 American Chemical Society.)

3.4.3 Etching by Weak Acids with Noble Metal Ions

Alternatively, according to route II in Figure 3.18, if the Zn/ZnO colloids are etched by weak acid containing noble metal ions, ZnO hollow nanospheres embedded with ultrafine noble metal nanoparticles (clusters) in the shell layer are obtained. Typically, etching by $HAuCl_4$ causes Zn SPR absorption peak to decrease and finally disappear, accompanied with blueshifting of ZnO IBT absorption shoulder, as illustrated in Figure 3.25, similarly to those by TA etching. However, an absorption peak around 550 nm appears after $HAuCl_4$ etching, which is because of the SPR of Au nanoparticles [87]. These results imply that the final products are the composites containing quantum-sized ZnO and Au nanoparticles.

TEM examination reveals formation of ZnO hollow nanospheres dispersed with Au ultrafine nanoparticles in the shell layer, as shown in Figure 3.26a. HRTEM observation reveals that the embedded ultrafine nanoparticles with high contrast are Au, which are embedded in the ZnO nanoshell layers (see Figure 3.26b and c). The Au ultrafine particles are nearly spherically shaped and about 3 nm in average size (see Figure 3.26d). Similarly, when H_2PtCl_6 is used as the etching agent, Pt–ZnO composite hollow nanospheres are formed, as shown in Figure 3.26e through g). The Pt ultrafine nanoparticles are about 2 nm in average diameter (see Figure 3.26h). These results demonstrate that the proposed selective etching strategy is suitable to fabricate not only pure oxide hollow nanospheres but also noble metal ultrafine nanoparticles (clusters)–embedded oxide hollow nanospheres, exhibiting good universality and controllability in components.

Furthermore, we can also fabricate multicomponent noble metal ultrafine particle–containing ZnO nanoshells by multistep etching process with weak acid solutions containing different noble metal ions. Typically, Au/Pt-containing ZnO hollow nanospheres were fabricated by firstly adding $HAuCl_4$ and then by adding

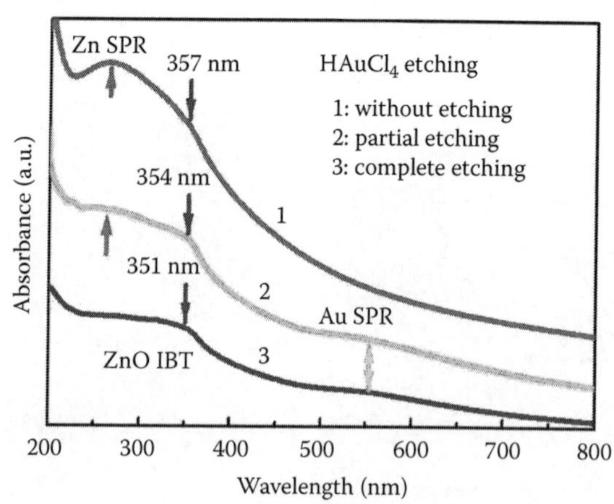

FIGURE 3.25 Optical absorption spectral evolution with the $HAuCl_4$ etching process for the Zn/ZnO core–shell nanoparticles with 20-nm average original size. (Reprinted with permission from Zeng et al., 2008, 1661–70. Copyright 2008 American Chemical Society.)

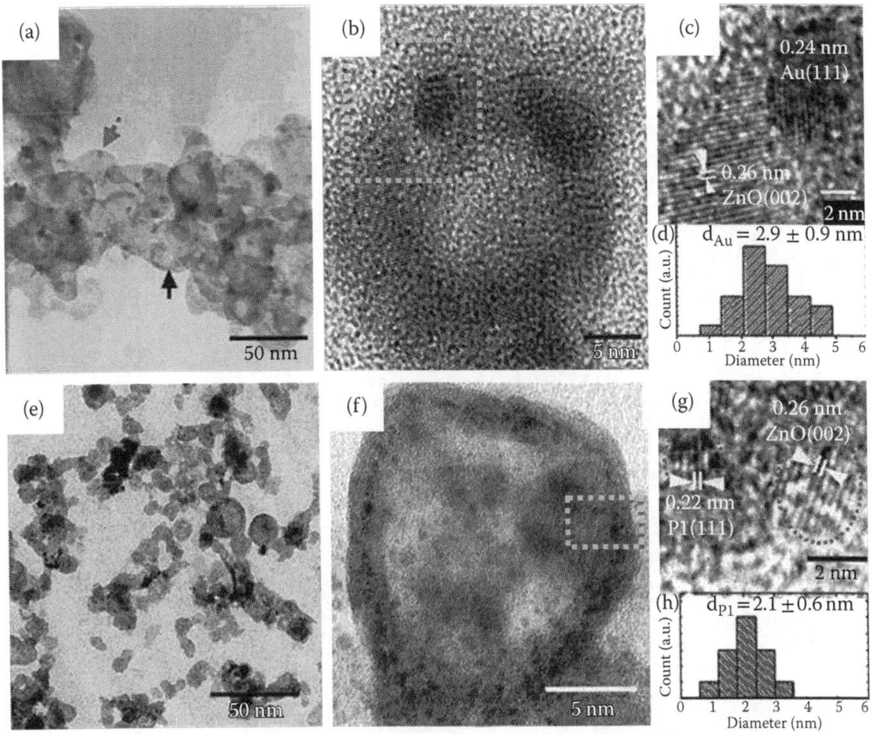

FIGURE 3.26 TEM images for the particle template with 20-nm average original size after etching by the weak acid $HAuCl_4$ (a through d) and H_2PtCl_6 (e through f). (a) and (e) Low-magnification TEM images; (b) and (f) high-resolution TEM images of a single hollow nanosphere; (c) and (g) the local magnification corresponding to the frame areas marked in (b) and (f), respectively; (d) and (h) size distribution of embedded ultrafine noble metal nanoparticles Au and Pt, respectively. (Reprinted with permission from Zeng et al., 2008, 1661–70. Copyright 2008 American Chemical Society.)

H_2PtCl_6 solution to the Zn/ZnO colloidal solution, as shown in Figure 3.27a and b. The hollow nanospheres are embedded with ultrafine particles with high contrast in the shell layers, which are similar to those in the Au–ZnO and Pt–ZnO hollow nanospheres. HRTEM examination and energy-dispersive spectroscopy (EDS) reveals the coexistence of Au and Pt ultrafine nanoparticles in the shell layers (see Figure 3.27d). The ultrafine nanoparticles are about 1.5 nm in mean size (see Figure 3.27c). In addition, it should be mentioned that the size and relative content of the noble metal nanoparticles can be controlled by the concentration of the metal ions in weak acid solution. So, this extended selective etching strategy by the weak acid with noble metal ions provides us a simple method to produce the oxide hollow nanospheres with controllable species, size, and relative content of embedded ultrafine noble metal clusters.

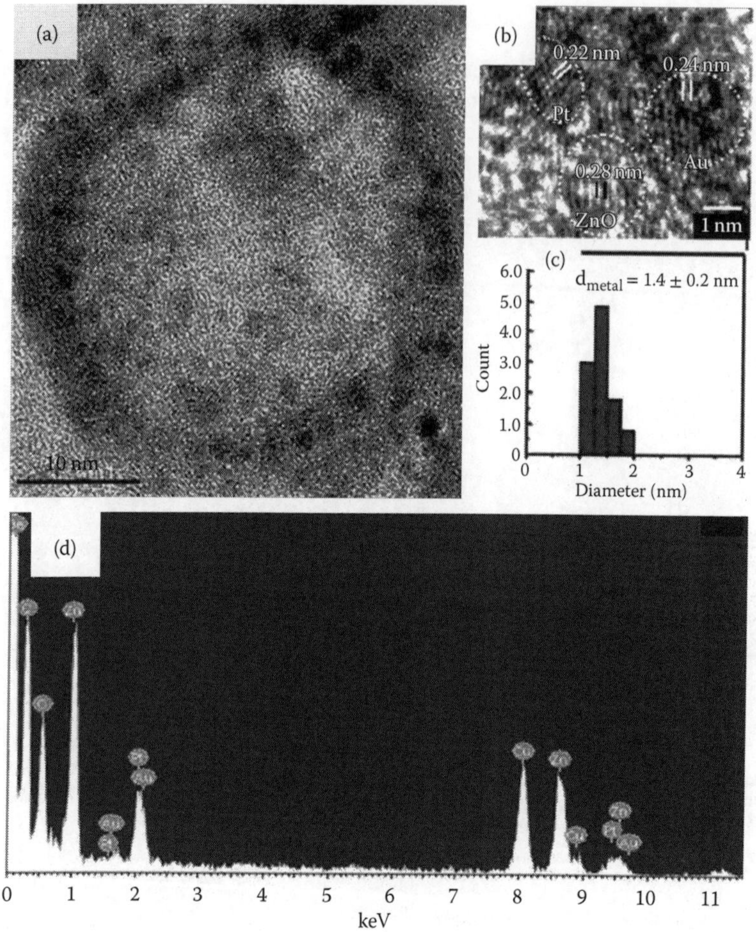

FIGURE 3.27 Microstructure and size distribution of the products with 20-nm average original size after two-step etching by $HAuCl_4$ and H_2PtCl_6. (a) HRTEM image of single hollow nanosphere. (b) The local magnified image of (a). (c) The size distribution of metal ultrafine particles. (d) Energy-dispersive spectroscopy of the etched hollow nanospheres. (Reprinted with permission from Zeng et al., 2008, 1661–70. Copyright 2008 American Chemical Society.)

3.4.4 Formation of ZnO Hollow Nanospheres

The formation of the hollow nanospheres is easily understood and attributed to the weak acid–induced preferential exhaustion of the metal Zn core parts in the core–shell nanoparticles during etching. When TA was added to the LAL-formed Zn/ZnO core–shell nanoparticle colloidal solution, the following reactions take place:

$$C_4H_6O_6 \leftrightarrow H^+ + C_4H_5O_6^- \tag{3.4}$$

$$Zn + 2H^+ \rightarrow Zn^{2+} + H_2 \tag{3.5}$$

$$ZnO + 2H^+ \leftrightarrow Zn^{2+} + H_2O \quad (3.6)$$

Firstly, H^+ ions rapidly form in the colloidal solution from $C_4H_6O_6$ molecules (reaction 3.4) and then diffuse through the ZnO shell layer and encounter and oxidize the Zn cores (reaction 3.5). The resulting Zn^{2+} ions diffuse out from the cores to the solution due to the small ion radius and hence voids are gradually produced in the core parts, as shown in Figure 3.18b. Finally, the Zn cores are completely exhausted and the hollow nanospheres are formed. At the same time, due to ZnO's amphoteric character, H^+ ions can also weakly react with the shell layers according to reaction 3.6, which leads to changes in the microstructure of the remaining ZnO nanoshell layers, including the thinning of shell layers, reduction of ZnO nanocrystal size, and increase of disordered degree. Such selective etching results in formation of the very small sized and ultrathin hollow nanospheres. In addition, the shell layers in the primal core–shell nanoparticles consist of both ultrafine nanocrystals and disordered areas [82–84], as illustrated in Figure 3.19c, which provides a great deal of flexible diffusion channels for the ions along the boundaries. On the other hand, etching leads to thinning of the shell layers and possible formation of very small gaps (or cracks) in the shell layers, which further accelerates the diffusion-redox process.

To further examine the selective etching formation of the hollow nanospheres, partial etching was carried out by adding deficient TA etching agent to the laser-induced colloidal solution. Figure 3.28a shows a single partially etched nanoparticle. The incompletely etched Zn remains in the core part. The corresponding EDS spectrum of the core part clearly exhibits the obvious Zn richness (Figure 3.28b). Such typical transitional state well coincides with the strategy illustrated in Figure 3.19b.

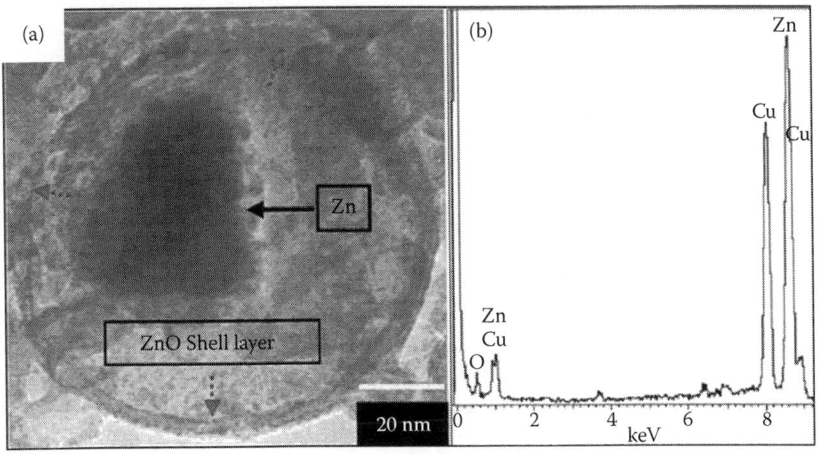

FIGURE 3.28 TEM image of one single-core–shell nanoparticle after partial etching by tartaric acid (a) and energy-dispersive spectroscopy in its core part (b). (Reprinted with permission from Zeng et al., 2008, 1661–70. Copyright 2008 American Chemical Society.)

3.4.5 Noble Metal Doping in ZnO Nanoshells

Similarly, formation of noble metal–containing ZnO hollow nanospheres can be attributed to the diffusion-redox-precipitation process during selective etching. If etching by $HAuCl_4$ solution, in addition to reactions 5 and 6), the following reaction exists:

$$3Zn + 2Au^{3+} \rightarrow 3Zn^{2+} + 2Au^{0} \tag{3.7}$$

Au^{3+} ions diffuse through the ZnO shell layers to react with Zn in the core parts, leading to reduction of gold ions and oxidization (or dissolution) of Zn according to reaction 3.7. The reduced gold enters the ZnO shell layer and precipitates in the form of ultrafine Au clusters (see Figure 3.26a). So, addition of $HAuCl_4$ solution induces the doped ZnO shells, similar etching effect as TA solution. As for addition of H_2PtCl_6, Pt^{4+} ions, similar to the effect of Au^{3+} ions, react with Zn in the core parts and are reduced by Zn according to the following reaction:

$$2Zn + Pt^{4+} \rightarrow 2Zn^{2+} + 2Pt^{0} \tag{3.8}$$

Finally, the reduced Pt^0 form ultrafine clusters in the shell layers.

3.4.6 Some Remarks on the Strategy

Altogether, a series of ZnO hollow nanospheres with diameter <40 nm and shell thickness <6 nm, including pure ZnO and ZnO embedded with Au or/and Pt ultrafine particles (clusters), can be obtained by corresponding etching processes of weak acid aqueous solutions, such as TA or chloroauric acid and/or chloroplatinic acid. Importantly, the microstructure of these hollow nanospheres can be well controlled in two aspects: (1) outside diameter, shell thickness, and ultrafine ZnO nanocrystals in the shell layers can be well controlled just by choice of original core–shell nanoparticles and subsequent weak acid etching and (2) the species, size, and content of the noble metal ultrafine nanocrystals embedded in the shells can be flexibly adjusted by etching condition of the suitable weak acid containing desired metal ions.

These results demonstrate that the selective etching strategy is universal for fabrication of oxide, metal-doped oxide, and even hybrid oxide hollow nanospheres. This strategy is suitable to obtain hollow nanospheres with controllable composition, smaller size, and higher specific surface area (compared with previous micron or submicron ZnO hollow spheres [72,83–85]). These selective etching–formed hollow nanospheres have applications in many fields such as chemical or biologic sensing, catalysts, photocatalysts, drug delivery, and biologic imaging, due to the quantum effect, high specific surface area, and the ultrafine nanocrystals in the nanoshell layers. Especially, the coupling between the different phases in thin nanoshell layers formed by doping could further enhance some of their performances. For example, noble metal doping, such as Pt, Pd, Au, Ag in TiO_2, SnO_2, and ZnO [88–91], and combination of different oxides, such as ZnO, TiO_2, and SnO_2 [92,93], are expected to greatly improve their catalytic activity, the separation of photo-generated holes and electrons, and the response to the visible light. So, this strategy may be a convenient, effective, and universal bridge toward the monocomponent or multicomponent hollow nanospheres with significant performances and applications.

3.5 BRIEF SUMMARY

The chemical-template-synthesis strategy has been introduced to demonstrate fabrication of the micro/nanostructured hollow spheres. In this strategy, the colloidal spheres not only function as template to determine the diameter of hollow structure but also are used as reactants for production of hollow spheres, which are tunable or controllable structural parameters and components, depending on the template and solution composition. Various micro/nanostructured porous silicate hollow spheres, including metal (copper, nickel, and magnesium) silicates, silica, silica–nickel composite hollow spheres, and $SiO_2@\gamma$-AlOOH microspheres, could be controllably produced based on the silica colloidal template-etching strategy shown in Figure 3.1. Also, oxide and metal cluster–embedded oxide hollow nanospheres could be produced by a simple, convenient, but effective and universal strategy, shown in Figure 3.18, based on selective etching of core–shell structured nanoparticles. The hollow nanospheres are controllable in structure and components, depending on the etching conditions and core–shell structured parameters.

REFERENCES

1. Wang, J. G.; Xiao, Q.; Zhou, H. J.; Sun, P. C.; Yuan, Z. Y.; Li, B. H.; Ding, D. T.; Shi, A. C.; and Chen, T. H. 2006. Budded, mesoporous silica hollow spheres: Hierarchical structure controlled by kinetic self-assembly. *Adv. Mater.* 18: 3284–8.
2. Wu, C. Z.; Xie, Y.; Lei, L. Y.; Hu, S. Q.; and Ou Yang, C. Z. 2006. Synthesis of new-phased VOOH hollow "dandelions" and their application in lithium-ion batteries. *Adv. Mater.* 18: 172732.
3. Choi, W. S.; Koo, H. Y.; Zhongbin, Z.; Li, Y. D.; and Kim, D. Y. 2007. Templated synthesis of porous capsules with a controllable surface morphology and their application as gas sensors. *Adv. Funct. Mater.* 17: 1743–9.
4. Kim, S. W.; Kim, M.; Lee, W. Y.; and Hyeon, T. 2002. Fabrication of hollow palladium spheres and their successful application to the recyclable heterogeneous catalyst for Suzuki coupling reactions. *J. Am. Chem. Soc.* 124: 7642–3.
5. Wang. X. B.; Cai, W. P.; Wang, G. Z.; and Liang, C. H. 2012. Standing porous ZnO nanoplate-built hollow microspheres and kinetically controlled dissolution/crystal growth mechanism. *J. Mater. Res.* 27: 951–8.
6. Zhao, X. W. and Qi, L. M. 2012. Rapid microwave-assisted synthesis of hierarchical ZnO hollow spheres and their application in Cr(VI) removal. *Nanotechnology* 23: 235604 (7 pp).
7. Zeng, Y.; Wang, X.; Wang, H.; Dong, Y.; Ma, Y.; and Yao, J. N. 2010. Multi-shelled titania hollow spheres fabricated by a hard template strategy: Enhanced photocatalytic activity. *Chem. Commun.* 46: 4312–4.
8. Xi, G. C.; Yan, Y.; Ma, Q.; Li, J. F.; Yang, H. F.; Lu, X. J.; and Wang, C. 2012. Synthesis of multiple-shell WO_3 hollow spheres by a binary carbonaceous template route and their applications in visible light photocatalysis. *Chem. Eur. J.* 18: 13949–53.
9. Lou, X. W.; (David) Archer, L. A.; and Yang, Z. C. 2008. Hollow micro-/nanostructures: Synthesis and applications. *Adv. Mater.* 20: 1–33.

10. Huang, Z. B. and Tang, F. Q. 2005. Preparation, structure, and magnetic properties of mesoporous magnetite hollow spheres. *J. Colloid Interf. Sci.* 281: 432–6.
11. Xu, Y. Y.; Chen, D. R.; Jiao, X. L.; and Xue, K. Y. 2007. Nanosized Cu_2O/PEG400 composite hollow spheres with mesoporous shells. *J. Phys. Chem. C* 111: 16284–9.
12. Xu, H. L. and Wang, W. Z. 2007. Template synthesis of multishelled Cu_2O hollow spheres with a single-crystalline shell wall. *Angew. Chem. Int. Ed. Engl.* 46: 1489–92.
13. Zeng, H. C. 2007. Ostwald ripening: A synthetic approach for hollow nanomaterials. *Curr. Nanosci.* 3: 177–81.
14. Yin, Y. D.; Rioux, R. M.; Erdonmez, C. K.; Hughes, S.; Somorjai, G. A.; and Alivisatos, A. P. 2004. Formation of hollow nanocrystals through the nanoscale Kirkendall effect. *Science* 304: 711–4.
15. Caruso, F.; Caruso, R. A.; and Möhwald, H. 1998. Nanoengineering of inorganic and hybrid hollow spheres by colloidal templating. *Science* 282: 1111–4.
16. Wang, Y. Q.; Wang, G. Z.; Wang, H. Q.; Liang, C. H.; Cai, W. P.; and Zhang, L. D. 2010. Chemical-template synthesis of micro/nanoscale magnesium silicate hollow spheres for waste-water treatment. *Chem. Eur. J.* 16: 3497–503.
17. Wang, X.; Zhuang, J.; Chen, J.; Zhou, K. B.; and Li, Y. D. 2004. Thermally stable silicate nanotubes. *Angew. Chem. Int. Ed. Engl.* 43: 2017–20.
18. Zhang, Z. T.; Han, Y.; Zhu, L.; Wang, R. W.; Yu, Y.; Qiu, S. L.; Zhao, D. Y.; and Xiao, F. S. 2001. Strongly acidic and high-temperature hydrothermally stable mesoporous aluminosilicates with ordered hexagonal structure. *Angew. Chem. Int. Ed. Engl.* 40: 1258–62.
19. Xiao, F. S.; Han, Y.; Yu, Y.; Meng, X.; Yang, J. M.; and Wu, S. 2002. Hydrothermally stable ordered mesoporous titanosilicates with highly active catalytic sites. *J. Am. Chem. Soc.* 124: 888–9.
20. Bouizi, Y.; Diaz, I.; Rouleau, L.; and Valtchev, V. P. 2005. Core-shell zeolite microcomposites. *Adv. Funct. Mater.* 15: 1955–60.
21. Wang, Y. Q.; Wang, G. Z.; Wang, H. Q.; Cai, W. P.; and Zhang, L. D. 2008. One-pot synthesis of nanotube-based hierarchical copper silicate hollow spheres. *Chem. Commun.* December 28: 6555–7.
22. Stöber, W. and Fink, A. 1968. Controlled growth of monodisperse silica spheres in micron size range. *J. Colloid Interf. Sci.* 26: 62–9.
23. Ren, N.; Wang, B.; Yang, Y. H.; Zhang, Y. H.; Yang, W. L.; Yue, Y.; Gao, H. Z.; and Tang, Y. 2005. General method for the fabrication of hollow microcapsules with adjustable shell compositions. *Chem. Mater.* 17: 2582–7.
24. Pauling, L. 1930. The structure of mechlorites. *Proc. Natl. Acad. Sci. U. S. A.* 16: 578–82.
25. Li, Y. D.; Li, X. L.; Deng, Z. X.; Zhou, B. C.; Fan, S. S.; Wang, J. W.; and Sun, X. M. 2002. From surfactant–inorganic mesostructures to tungsten nanowires. *Angew. Chem. Int. Ed. Engl.* 41: 333–5.
26. Wang, X. and Li, Y. D. 2003. Synthesis and formation mechanism of manganese dioxide nanowires/nanorods. *Chem-Eur. J.* 9: 300–6.
27. Zhang, J.; Liu, S. J.; Lin, J.; Song, H. S.; Luo, J. J.; Elssfah, E. M.; Ammar, E. et al. 2006. Self-assembly of flowerlike AlOOH (boehmite) 3D nanoarchitectures. *J. Phys. Chem. B* 110: 14249–52.
28. Iler, R. K. 1955. *The Colloid Chemistry of Silica and Silicates*, Cornell University, Ithaca, NY, pp. 6–16.
29. Huang, P. C. and Fuerstenau, D. W. 2000. The effect of the adsorption of lead and cadmium ions on the interfacial behavior of quartz and talc. *Colloid. Surface. A* 177: 147–56.
30. Casanova, H.; Orrego, J. A.; and Zapata, J. 2007. Oil absorption of talc minerals and dispersant demand of talc mineral non-aqueous dispersions as a function of talc content: A surface chemistry approach. *Colloid. Surface. A* 299: 38–44.

31. Wang, Y. Q.; Tang, C. J.; Deng, Q.; Liang, C. H.; Ng, D. H.; Kwong, F.-L.; Wang, H. Q.; Cai, W. P.; Zhang, L. D.; and Wang, G. Z. 2010. A versatile method for controlled synthesis of porous hollow spheres. *Langmuir* 26: 14830–4.
32. Aznar, A. J.; Gutierrez, E.; Diaz, P.; Alvarez, A.; and Poncelet, G. 1996. Silica from sepiolite: Preparation, textural properties, and use as support to catalysts. *Microporous Mater.* 6(2): 105–14.
33. Temuujin, J.; Okada, K.; and MacKenzie, K. 2003. Preparation of porous silica from vermiculite by selective leaching. *Appl. Clay Sci.* 22(4): 187–95.
34. Wypych, F.; Adad, L.; Mattoso, B. N.; Marangon, A. A.; and Schreiner, W. H. 2005. Synthesis and characterization of disordered layered silica obtained by selective leaching of octahedral sheets from chrysotile and phlogopite structures. *J. Colloid Interf. Sci.* 283(1): 107–12.
35. Bao, J. C.; Liang, Y. Y.; Xu, Z.; and Si, L. 2003. Facile synthesis of hollow nickel submicrometer spheres. *Adv. Mater.* 15(21): 1832–5.
36. Liu, Q.; Liu, H. J.; Han, M.; Zhu, J. M.; Liang, Y. Y.; Xu, Z.; and Song, Y. 2005. Nanometer-sized nickel hollow sphere. *Adv. Mater.* 17(16): 1995–9.
37. Lu, A. H.; Salabas, E. L.; and Schuth, F. 2007. *Halomonas shengliensis* sp. nov., a moderately halophilic, denitrifying, crude-oil-utilizing bacterium. *Angew. Chem. Int. Ed. Engl.* 46(8): 1222–44.
38. Suh, W. H. and Suslick, K. S. 2005. Magnetic and porous nanospheres from ultrasonic spray pyrolysis. *J. Am. Chem. Soc.* 127(34): 12007–10.
39. Cao, H. Q.; Zhang, L.; Liu, X. W.; Zhang, S. C.; Liang, Y.; and Zhang, X. R. 2007. Catalytic chemiluminescence properties of boehmite "nanococoons". *Appl. Phys. Lett.* 90: 193105.
40. Tanada, S.; Kabayama, M.; Kawasaki, N.; Sakiyama, T.; Nakamura, T.; Araki, M.; and Tamura, T. 2003. Removal of phosphate by aluminum oxide hydroxide. *J. Colloid Interf. Sci.* 257: 135–40.
41. Ogata, F.; Kawasaki, N.; Nakamura, T.; and Tanada, S. 2006. Removal of arsenious ion by calcined aluminum oxyhydroxide (boehmite). *J. Colloid Interf. Sci.* 300: 88–93.
42. Horanyi, G. and Kalman, E. 2004. Anion specific adsorption on Fe_2O_3 and AlOOH nanoparticles in aqueous solutions: Comparison with hematite and gamma-Al_2O_3. *J. Colloid Interf. Sci.* 269: 315–9.
43. Buchold, D. H. M. and Feldmann, C. 2007. Nanoscale gamma-AlO(OH) hollow spheres: Synthesis and container-type functionality. *Nano Lett.* 7: 3489–92.
44. Hou, H. W.; Xie, Y.; Yang, Q.; Guo, Q. X.; and Tan, C. R. 2005. Preparation and characterization of gamma-AlOOH nanotubes and nanorods. *Nanotechnology* 16: 741–5.
45. Zhao, Y. Y.; Frost, R. L.; Martens, W. N.; and Zhu, H. Y. 2007. Growth and surface properties of boehmite nanofibers and nanotubes at low temperatures using a hydrothermal synthesis route. *Langmuir* 23: 9850–9.
46. Zhu, H. Y.; Gao, X. P. D.; Song, Y.; Bai, Y. Q.; Ringer, S. P.; Gao, Z.; Xi, Y. X.; Martens, W.; Riches, J. D.; and Frost, R. L. 2004. Growth of boehmite nanoribers by assembling nanoparticles with surfactant micelles. *J. Phys. Chem. B* 108: 4245–7.
47. Zhang, J.; Wei, S. Y.; Lin, J.; Luo, J. J.; Liu, S. J.; Song, H. S.; Elawad, E. et al. 2006. Template-free preparation of bunches of aligned boehmite nanowires. *J. Phys. Chem. B* 110: 21680–3.
48. Kuiry, S. C.; Megen, E.; Patil, S. A.; Deshpande, S. A.; and Seal, S. 2005. Solution-based chemical synthesis of boehmite nanofibers and alumina nanorods. *J. Phys. Chem. B* 109: 3868–72.
49. Ren, T. Z.; Yuan, Z. Y.; and Su, B. L. 2004. Microwave-assisted preparation of hierarchical mesoporous-macroporous boehmite AlOOH and gamma-Al_2O_3. *Langmuir* 20: 1531–4.
50. Hicks, R. W. and Pinnavaia, T. 2003. Nanoparticle assembly of mesoporous AlOOH (boelamite). *Chem. Mater.* 15: 78–82.

51. Arnal, P. M.; Weidenthaler, C.; and Schuth, F. 2006. Highly monodisperse zirconia-coated silica spheres and zirconia/silica hollow spheres with remarkable textural properties. *Chem. Mater.* 18: 2733–9.
52. Skoufadis, C.; Panias, D.; and Paspaliaris, I. 2003. Kinetics of boehmite precipitation from supersaturated sodium aluminate solutions. *Hydrometallurgy* 68: 57–68.
53. Wang, Y. Q.; Wang, G. Z.; Wang, H. Q.; Liang, C. H.; Cai, W. P.; and Zhang, L. D. 2009. Template-induced synthesis of hierarchical SiO_2@gamma-AlOOH spheres and their application in Cr(VI) removal. *Nanotechnology* 20: 155604.
54. Lin, C. K.; Li, Y.; Yu, Y. M.; Yang, P. P.; and Lin, J. 2007. A facile synthesis and characterization of monodisperse spherical pigment particles with a core/shell structure. *Adv. Funct. Mater.* 17: 1459–65.
55. Fei, J. B.; Cui, Y.; Yan, X. H.; Qi, W.; Yang, Y.; Wang, K. W.; He, Q.; and Li, J. B. 2008. Controlled preparation of MnO_2 hierarchical hollow nanostructures and their application in water treatment. *Adv. Mater.* 20: 452–6.
56. Zhang, W. X.; Wen, X. G.; Yang, S. H.; Berta, Y.; and Wang, Z. L. 2003. Single-crystalline scroll-type nanotube arrays of copper hydroxide synthesized at room temperature. *Adv. Mater.* 15: 822–5.
57. Bokhimi, X.; Toledo-Antonio, J. A.; Guzman-Castillo, M. L.; and Hernandez-Beltran, F. 2001. Relationship between crystallite size and bond lengths in boehmite. *J. Solid State Chem.* 159: 32–40.
58. Chen, X. Y. and Lee, S. W. 2007. pH-dependent formation of boehmite (gamma-AlOOH) nanorods and nanoflakes. *Chem. Phys. Lett.* 438: 279–84.
59. Zeng, H. C. 2006. Synthetic architecture of interior space for inorganic nanostructures. *J. Mater. Chem.* 16: 649–62.
60. Gao, Y. H. and Bando, Y. 2002. Carbon nanothermometer containing gallium: Gallium's macroscopic properties are retained on a miniature scale in this nanodevice. *Nature* 415: 599.
61. Fan, H. J.; Kenz, M.; Scholz, R.; Nielsch, K.; Pippel, E.; Hesse, D.; Zacharias, M.; and Gösele, U. 2006. Monocrystalline spinel nanotube fabrication based on the Kirkendall effect. *Nat. Mater.* 5: 627–31.
62. Li, D.; Zhou, H.; and Honma, I. 2004. Design and synthesis of self-ordered mesoporous nanocomposite through controlled in-situ crystallization. *Nat. Mater.* 3: 65–72.
63. Suh, W. H.; Jang, A. R.; Suh, Y. H.; and Suslik, K. S. 2006. Porous, hollow, and ball-in-ball metal oxide microspheres: Preparation, endocytosis, and cytotoxicity. *Adv. Mater.* 18: 1832–7.
64. Sun, F. Q.; Cai, W. P.; Li, Y.; Jia, L. C.; and Lu, F. 2005. Direct growth of mono- and multilayer nanostructured porous films on curved surfaces and their application as gas sensors. *Adv. Mater.* 17: 2872–7.
65. Lytle, J. C.; Yan, H. W.; Ergang, N. S.; Smyrl, W. H.; and Stein, A. 2004. Structural and electrochemical properties of three-dimensionally ordered macroporous tin(IV) oxide films. *J. Mater. Chem.* 14: 1616–22.
66. Tessier, P.; Velev, O. D.; and Kalambur, A. T. 2001. Structured metallic films for optical and spectroscopic applications via colloidal crystal templating. *Adv. Mater.* 13: 396–400.
67. Chen, J.; Saeki, F.; Wiley, B. J.; Cang, H.; Cobb, M. J.; Li, Z. Y.; Au, L. et al. 2005. Gold nanocages: Bioconjugation and their potential use as optical imaging contrast agents. *Nano Lett.* 5: 473–7.
68. Su, C. H.; Sheu, H. S.; Lin, C. Y.; Huang, C. C.; Lo, Y. W.; Pu, Y. C.; Weng, J. C.; Shieh, D. B.; Chen, J. H.; and Yeh, C. S. 2007. Nanoshell magnetic resonance imaging contrast agents. *J. Am. Chem. Soc.* 129: 2139–46.
69. Peng, S. and Sun, S. 2007. Synthesis and characterization of monodisperse hollow Fe_3O_4 nanoparticles. *Angew. Chem. Int. Ed. Engl.* 46: 4155–8.

70. Liu, B. and Zeng, H. C. 2004. Fabrication of ZnO "dandelions" via a modified Kirkendall process. *J. Am. Chem. Soc.* 126: 16744–6.
71. Chiang, R. K. and Chiang, R. T. 2007. Formation of hollow Ni2p nanoparticles based on the nanoscale Kirkendall effect. *Inorg. Chem.* 46: 369–71.
72. Chen, J.; McLellan, J. M.; Siekkinen, A.; Xiong, Y.; Li, Z. Y.; and Xia, Y. 2006. Facile synthesis of gold-silver nanocages with controllable pores on the surface. *J. Am. Chem. Soc.* 128: 14776–7.
73. Schwartzberg, A. M.; Olson, T. Y.; Telley, C. E.; and Zhang, J. Z. 2006. Synthesis, characterization, and tunable optical properties of hollow gold nanospheres. *J. Phys. Chem. B* 110: 19935–44.
74. Sun, F. and Yu, J. C. 2007. Photochemical preparation of two-dimensional gold spherical pore and hollow sphere arrays on a solution surface. *Angew. Chem. Int. Ed. Engl.* 46: 773–7.
75. Yu, J.; Guo, H.; Davis, S. A.; and Mann, S. 2006. Fabrication of hollow inorganic microspheres by chemically induced self-transformation. *Adv. Funct. Mater.* 16: 2035–41.
76. Yang, H. G. and Zeng, H. C. 2004. Self-construction of hollow SnO_2 octahedra based on two-dimensional aggregation of nanocrystallites. *Angew. Chem. Int. Ed. Engl.* 43: 5930–3.
77. Gao, S.; Zhang, H.; Wang, X.; Deng, R.; Sun, D.; and Zheng, G. 2006. ZnO-based hollow microspheres: Biopolymer-assisted assemblies from ZnO nanorods. *J. Phys. Chem. B* 110: 15847–52.
78. Gao, P. X. and Wang, Z. L. 2003. Mesoporous polyhedral cages and shells formed by textured self-assembly of ZnO nanocrystals. *J. Am. Chem. Soc.* 125: 11299–305.
79. Shen, G.; Bando, Y.; and Lee, C. J. 2005. Synthesis and evolution of novel hollow ZnO urchins by a simple thermal evaporation process. *J. Phys. Chem. B* 109: 10578–83.
80. Zeng, H. B.; Cai, W. P.; Liu, P. S.; Xu, X. X.; Zhou, H. J.; Klingshirn, C.; and Kalt, H. 2008. ZnO-based hollow nanoparticles by selective etching: Elimination and reconstruction of metal-semiconductor interface, improvement of blue emission and photocatalysis. *ACS Nano* 2(8): 1661–70.
81. Jiang, Q.; Wu, Z. Y.; Wang, Y. M.; Cao, Y.; Zhou, C. F.; and Zhu, J. H. 2006. Fabrication of photoluminescent ZnO/SBA-15 through directly dispersing zinc nitrate into the as-prepared mesoporous silica occluded with template. *J. Mater. Chem.* 16: 1536–42.
82. Zeng, H. B.; Cai, W. P.; Li, Y.; Hu, J. L.; and Liu, P. S. 2005. Composition/structural evolution and optical properties of ZnO/Zn nanoparticles by laser ablation in liquid media. *J. Phys. Chem. B* 109: 18260–6.
83. Zeng, H. B.; Cai, W. P.; Hu, J. L.; Duan, G. T.; Liu, P. S.; and Li, Y. 2006. Violet photoluminescence from shell layer of Zn/ZnO core-shell nanoparticles induced by laser ablation. *Appl. Phys. Lett.* 88: 171910.
84. Zeng, H. B.; Li, Z. G.; Cai, W. P.; Liu, B. Q.; Cao, P. S.; and Yang, S. K. 2007. Microstructure control of Zn/ZnO core/shell nanoparticles and their temperature-dependent blue emissions. *J. Phys. Chem. B* 111: 14311–7.
85. Lee, J. K.; Tewell, C. R.; Schulze, R. K.; Nastasi, M.; Hamby, D. W.; Lucca, D. A.; Jung, H. S.; and Hong, K. S. 2005. Synthesis of ZnO nanocrystals by subsequent implantation of Zn and O species. *Appl. Phys. Lett.* 86: 183111.
86. Bahnemann, D. W.; Kormann, C.; and Hoffmann, M. R. 1987. Preparation and characterization of quantum-size zinc oxide: A detailed spectroscopic study. *J. Phys. Chem.* 91: 3789–98.
87. Fu, G. H.; Cai, W. P.; Kan, C. X.; Li, C. C.; and Zhang, L. D. 2003. Controllable optical properties of Au/SiO_2 nanocomposite induced by ultrasonic irradiation and thermal annealing. *Appl. Phys. Lett.* 83: 36–8.
88. Zhong, L.; Hu, J.; Cui, Z.; Wan, L.; and Song, W. 2007. In-situ loading of noble metal nanoparticles on hydroxyl-group-rich titania precursor and their catalytic applications. *Chem. Mater.* 19: 4557–62.

89. Zheng, Y.; Zheng, L.; Zhan, Y.; Lin, X.; Zheng, Q.; and Wei, K. 2007. Ag/ZnO heterostructure nanocrystals: Synthesis, characterization, and photocatalysis. *Inorg. Chem.* 46: 6980–6.
90. Reddy, E. P.; Sun, B.; and Smirniotis, P. G. 2004. Transition metal modified TiO_2-loaded MCM-41 catalysts for visible-and UV-light driven photodegradation of aqueous organic pollutants. *J. Phys. Chem. B* 108: 17198–205.
91. Zhao, W.; Chen, C.; Li, X.; and Zhao, J. 2002. Photodegradation of sulforhodamine-B dye in platinized titania dispersions under visible light irradiation: Influence of platinum as a functional co-catalyst. *J. Phys. Chem. B* 106: 5022–8.
92. Wang, W.; Zhu, Y.; and Yang, L. 2007. ZnO–SnO_2 hollow spheres and hierarchical nanosheets: Hydrothermal preparation, formation mechanism, and photocatalytic properties. *Adv. Funct. Mater.* 17: 59–64.
93. Giuseppe, M.; Vincenzo, A.; Maria, L.; Cristina, M.; Leonardo, P.; Vicente, R.; Mario, S.; Richard, J.; and Anna, M. 2001. Preparation characterization and photocatalytic activity of polycrystalline ZnO/TiO_2 systems. 1. Surface and bulk characterization. *J. Phys. Chem. B* 105: 1026–32.

4 Electrospinning and In Situ Conversion

4.1 INTRODUCTION

Fabrication of fibers by electrospinning was first reported by J. F. Cooley, in 1902 [1]. However, there have been only a few publications about its application in the fabrication of thin fibers until 1993 [2–5]. It is since the research done by Dosh and others in 1995 [6] that the electrospinning technique has attracted much attention, because they demonstrated the fabrication of thin fibers from a broad range of organic polymers. The electrospinning technology is a simple and versatile method, or applying the electrical drawing force directly on a jet body to generate the fibers with nano- to submicroscaled diameters [7–9]. Figure 4.1 is the schematic illustration of an electrospinning setup, which consists of a high-voltage electrical source, an electrospinning-jet syringe, and a collector (a grounded conductor [or metal plate]). During electrospinning, the high-voltage electrical source is an electrode contacting with the spinning solution. The other electrode is linked with the collector. Under a high voltage (15–30 kV), the solution in the electrospinning-jet syringe will be charged. The induced charges are distributed on the surface of the syringe nozzle. The surface charges are acted on by the repellent force from the same charges and the coulombic attraction from the applied electric field. Because of the viscousness or surface tension of the solution, liquid droplets will not effuse from the syringe nozzle. With increasing electric field, the liquid droplets in the nozzle will be changed into a conic shape or Taylor cone due to the repellent force from the same charges and the coulombic attraction from the out electric field [10,11]. When the applied electric field is high enough (higher than a critical value), the electric field force will overcome the surface tension, and the syringe nozzle will jet out the charged stream. The stream is in the electric field and drawn into thin and long fibers [12,13], as shown in Figure 4.1. With drawing of the stream and evaporation of the solvent, the jet stream will be solidified, forming fibers on the metal plate. The final diameter ranges from hundreds of microns to tens of nanometers depending on the solution properties (surface tension, composition, and concentration) and the electrospinning parameters, including nozzle size, electric field, and distance from the collector. Obviously, this method has the advantages of easy operation, low cost, and not needing the use of coagulation chemistry or high temperatures to obtain solid micro/nanostructured fibers from solution.

There have been extensive reports on the fabrication of nanofibers by the electrospinning technique in the past decade [14]. The prepared products include inorganic oxide-nanofibers (oxides [15,16], composite oxides [17,18], and metal-doped oxides [19–22]), metal oxides (or sulfides)/macromolecule (polymer) composite fibers

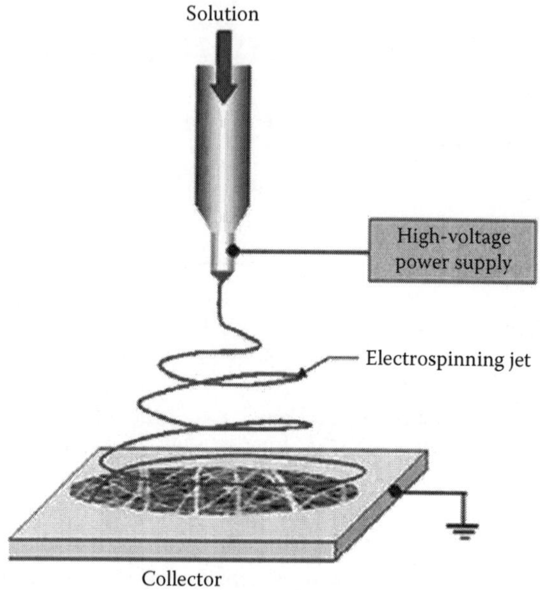

FIGURE 4.1 Schematic illustration of the electrospinning setup.

[23–25], and metal nanoparticles/polymer nanofibers [26–28]. In this chapter, we introduce some works to demonstrate how the electrospinning technique is used for fabrication of various micro/nanostructured fibers, including mainly (1) polymer/inorganic compound composite porous nanofibers by direct electrospinning, (2) composite porous nanofibers based on in situ conversion by basification of electrospun nanofibers, (3) hierarchically micro/nanostructured polymer-oxide fibers based on the electrospun nanofiber–templated hydrothermal route, and (4) metal porous nanotube films by plasma etching of metal-coated electrospun polymer nanofibers.

4.2 POLYACRYLONITRILE/FERROUS CHLORIDE COMPOSITE POROUS NANOFIBERS

Chromates ($Cr_2O_7^{2-}$, CrO_4^{2-}, and $HCrO_4^-$) are highly soluble and mobile in aquatic systems [29]. They are carcinogenic and highly toxic to humans, animals, and plants. Reactions that reduce hexavalent chromium [Cr(VI)] to Cr(III) are environmentally beneficial since the latter is relatively less hazardous. However, the simple and effective methods for removal of Cr are still urgently to be developed.

It is well known that some functional groups (>C=O, –CN, OC–N, etc.) of organic compounds can interact with metal ions. Especially, the nitrile group of polyacrylonitrile (PAN), a common, cheap, and nontoxic raw material used in the textile industry, can form a weak complex with cations, or $CH_2CH(CN)^- \cdots M$ (M = Zn^{2+}, Fe^{3+}, Fe^{2+}, Li^+, Na^+, and K^+) [30–33]. For instance, Reicha [34] reported metal–oxygen and metal–nitrogen complexes in the polymer of acrylonitrile (AN)–ferric chloride ($FeCl_3$). Obviously, for fabricating a PAN/$FeCl_2$ composite film by polymerizing an AN monomer and ferric chloride, a large amount of the complexes $CH_2CH(CN)\cdots Fe(II/III)$ should exist in the film. Such a film would strongly adsorb Cr(VI) in a solution and, further, the Fe(II) in the composite film could reduce it to Cr(III). This means that such a composite film with Fe(II) could be used for Cr(VI) reduction and removal. On this basis, in this section we introduce the fabrication of PAN/$FeCl_2$ composite nanofiber films by the aforementioned electrospinning technology [35].

In a typical experiment, 6 mL of dimethylformamide (DMF) solution of PAN (10 %wt, molecular weight (MW) = ~60,000) containing 0.12 g $FeCl_2 \cdot 4H_2O$ was put into a syringe with an internal hole diameter of 0.7 mm. The injection rate of the solution was controlled to be about 0.3 mL/h. The composite nanofibers (containing 10.6 %wt $FeCl_2$) were collected on a rotatable drum to form a netlike film, by an applied voltage of 22 kV, while keeping the spinneret tip to collector distance at 25 cm. The final fiber film was about 250 mm × 250 mm in dimension. Similarly, pure PAN nanofiber without iron element was also prepared under the same conditions for reference.

4.2.1 Morphology and Structure

Figure 4.2a shows the obtained uniform PAN/$FeCl_2$ composite film by electrospinning, which corresponds to a photograph of the film with about 10.6 %wt in $FeCl_2$ content and 15 μm in thickness. This film is composed of fibers. The fibers are randomly arranged, forming a netlike porous film, as seen in Figure 4.2b. For the products electrospun from the pure PAN without iron element, the morphology is similar. All fibers fall into the range from 100 to 300 nm in diameter and center around 200 nm, as illustrated in Figure 4.2c. Further, the N_2 sorption measurement of such a nanofiber film has exhibited a type II isotherm [36], indicating monolayer adsorption at low pressure and multilayer adsorption at high pressure, and has also shown a hysteresis loop between adsorption and desorption, which means a porous structure, as illustrated in Figure 4.2d. The specific surface area was thus evaluated to be about 10 $m^2 \cdot g^{-1}$ according to the Brunauer–Emmett–Teller (BET) equation [37]. From the average diameter of the nanofibers (200 nm), however, their specific surface area was estimated to be only about 2 $m^2 \cdot g^{-1}$. It means that the measured surface area of the nanofiber film is not mainly a contribution of the nanofibers' surface. Further, a pore-diameter analysis confirmed that there existed many nanosized pores (about 2.5 nm in diameter) within the fibers, or that the fibers were porous, as indicated in the inset of Figure 4.2d. Obviously, the nanosized pores in the nanofibers mainly contribute to the surface area of the nanofiber film.

X-ray diffraction (XRD) measurements exhibited no diffraction peak for the composite nanofiber film, indicating an amorphous structure, in contrast to $FeCl_2$ powders and pure PAN nanofibers, both of which have crystal structures, as illustrated in Figure 4.3a. Transmission electron microscopy (TEM) observations have also revealed that the composite nanofibers are a homogeneous noncrystalline phase (see Figure 4.3b). Further, it was found, by high-resolution TEM examination, that channel-like and nanosized pores were randomly distributed in the nanofibers, as shown in Figure 4.4a.

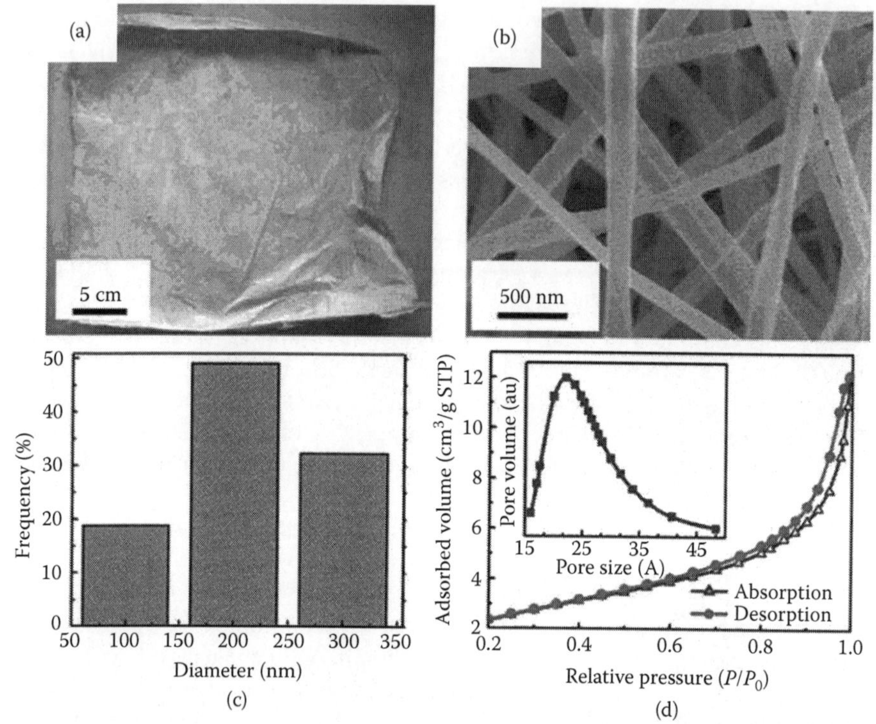

FIGURE 4.2 Structure and morphology of electrospun products: (a) a photograph of the filmlike electrospun product, (b) field emission scanning electron microscopy (FESEM) image of the products, (c) a histogram of the fibers' diameter from (b), and (d) N_2 sorption isotherm for the sample shown in (a). Inset shows the corresponding pore diameter distribution. (Lin et al., *J. Mater. Chem.*, 21, 991–7, 2011. Reproduced by permission of The Royal Society of Chemistry.)

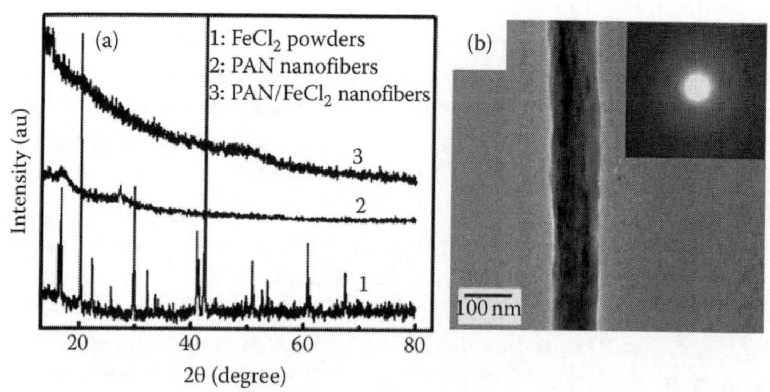

FIGURE 4.3 Phase structure analysis of nanofibers: (a) x-ray diffraction patterns and (b) transmission electron microscopy (TEM) image of a single polyacrylonitrile/$FeCl_2$ nanofiber. The inset shows the selected area electron diffraction pattern. (Lin et al., *J. Mater. Chem.*, 21, 991–7, 2011. Reproduced by permission of The Royal Society of Chemistry.)

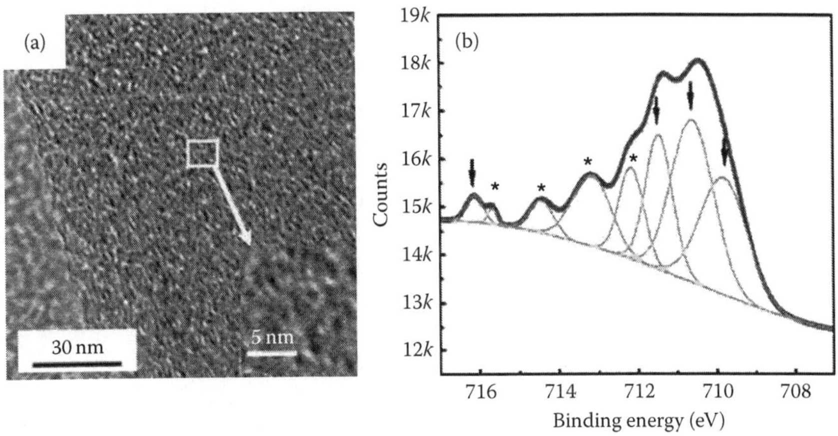

FIGURE 4.4 (a) High-resolution transmission electron microscopy image of a single polyacrylonitrile (PAN)/$FeCl_2$ nanofiber: the inset shows the locally magnified image corresponding to the area of the rectangular mark. (b) X-ray photoelectron spectroscopy of Fe2p for the PAN/$FeCl_2$ nanofibers. Arrows correspond to the Fe2p in Fe(II); dots correspond to the Fe2p in Fe(III). (Lin et al., *J. Mater. Chem.*, 21, 991–7, 2011. Reproduced by permission of The Royal Society of Chemistry.)

The formation of porous and amorphous nanofibers is easily understood and can be attributed to the rapid evaporation of the solvent during electrospinning, which blocks the crystallization of macromolecular chains and decreases the density of the fibers. Thus, the fibers become porous in structure in contrast to the bulk material [38]. Such porous structures provide a convenient route for substance exchange between the inner fiber and the surroundings.

Similarly, PAN/$FeCl_2$ composite nanofiber films with different amounts of iron salt could be obtained, with similar morphology and structure to the aforementioned one. Here, it should be pointed out that composite nanofibers with too much iron salt (say, >50 %wt $FeCl_2$) would become very brittle and lack practicability. Also, the diameter of the nanofibers could be controlled simply by the concentration of the electrospinning solution. A lower concentration induced thinner composite nanofibers.

4.2.2 Existence of $FeCl_2$ within Fibers

The valence state of iron in composite fibers was studied by x-ray photoelectron spectroscopy (XPS) measurement, as shown in Figure 4.4b. The main peaks at 709.8, 710.6, 711.4, and 716.1 eV correspond to the $Fe2p_{3/2}$ of Fe(II) (see the arrows marked in Figure 4.4b) [39]. It means that Fe(II) was still kept well in the nanofibers, which could be attributed to Fe(II) complexing with the nitrile group of PAN. This is also in agreement with previous reports [40]. The existence of Fe(II) would strongly suppress the crystallization of polymer molecules of PAN [41]. Here, it should be mentioned that there also exists a small amount of Fe(III) in the nanofibers, probably due to oxidization during preservation. The relatively small peaks at 712.2, 713.2, 714.4, and 715.6 eV, shown as asterisks in Figure 4.4b, correspond to the signals from Fe(III) [39].

Importantly, the obtained PAN/$FeCl_2$ nanofibers have exhibited excellent performance of Cr removal from a $Cr_2O_7^{2-}$-containing solution in one step [35]. The Cr removal capability is more than 110 mg Cr/g $FeCl_2$, which was much higher than the previously reported values of the other nanomaterials as Cr(VI) removal adsorbents. In contrast, the pure PAN nanofibers cannot remove Cr, the mixture of PAN nanofibers with $FeCl_2$ powders, and the PAN/$FeCl_2$-cast films can only remove Cr insignificantly. These results provide an effective route for the development of environmental remediation materials. The PAN/$FeCl_2$ composite porous nanofibers could be good candidates or new materials used for efficient Cr removal from wastewater and deep purification of pollutant water (details are given in Chapter 6).

4.2.3 Extension of the Electrospinning Strategy: In Situ Conversion

4.2.3.1 Polyacrylonitrile/FeOOH Composite Nanofibers

It should be mentioned that the Fe(II) in PAN/$FeCl_2$ composite nanofibers could be partially dissolved in water, which is not beneficial to the treatment of contaminated water. If PAN/$FeCl_2$ composite nanofibers are further treated by basification, as illustrated in Figure 4.5, insoluble nanoparticle-containing composite nanofibers will be formed.

As an example, the composite nanofibers shown in Figure 4.2 were immersed in 10% ammonia aqueous solution for 24 hours before washing and drying. A fiber film was finally obtained, as shown in Figure 4.6. The fiber film still kept the original configuration. XRD and TEM examination confirmed that the fibers were still amorphous in structure (see the inset of Figure 4.6b). Further, the valence states of elements in the composite fibers were examined by XPS measurements, as demonstrated in Figure 4.7, corresponding to the binding energy spectra of oxygen and Fe. The Ols spectrum is of Gaussian–Lorenzian peak type and can be decomposed, by

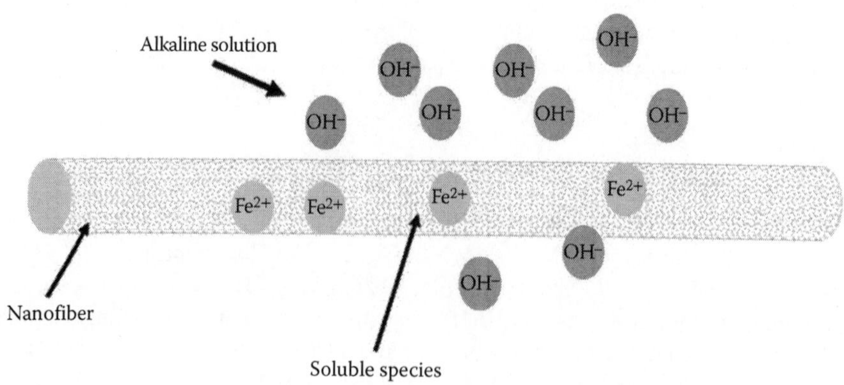

FIGURE 4.5 Schematic illustration of the polyacrylonitrile/$FeCl_2$ composite nanofiber in a basic solution.

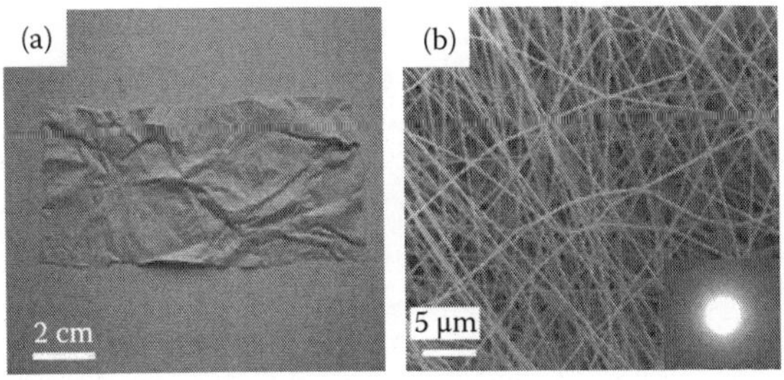

FIGURE 4.6 Basified nanofibers: (a) a photograph of the basified nanofiber film and (b) the corresponding FESEM image. The inset shows the selected area electron diffraction pattern.

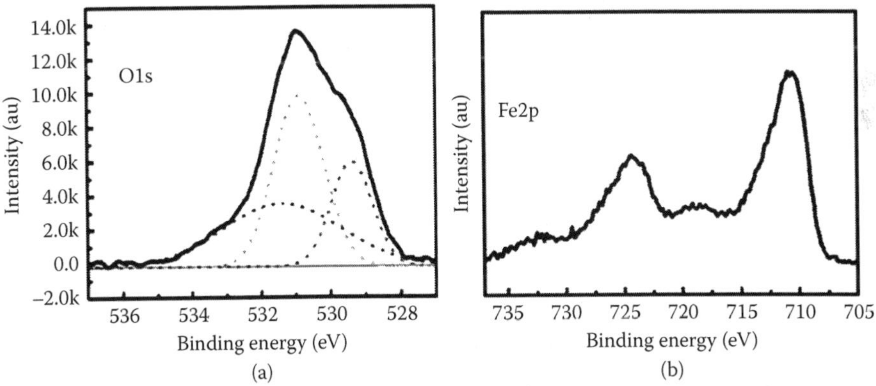

FIGURE 4.7 X-ray photoelectron spectroscopy results of polyacrylonitrile/$FeCl_2$ composite nanofibers after reaction with an ammonia aqueous solution: (a) O1s spectra. Dotted lines are fitting results. (b) Fe2p spectrum.

fitting, into three peaks, as illustrated in Figure 4.7a. The peaks at 530.0, 531.4, and 532.6 eV should correspond to the binding energies of the oxygen in the iron-related oxide, hydroxy, and adsorbed water, respectively [42,43]. Also, the peaks at 711.1 eV ($Fe2p_{3/2}$) and 724.5 eV ($Fe2p_{1/2}$) with a spacing of 13.4 eV confirmed the existence of Fe(III) (see Figure 4.7b). On the basis of the aforementioned analysis, the Fe and the oxygen in the fibers exist in the forms of Fe(III) and $(OH)^-$. In addition, the treated fiber film is light yellow in color, in agreement with that of $Fe(OH)_3$. So, the reacted products should be FeOOH. The soluble salt was converted into an insoluble hydrate, and the iron was fixed in the fibers.

4.2.3.2 Polyacrylonitrile/$Mg(OH)_2$ Composite Nanofibers

Magnesium hydrate [$Mg(OH)_2$] is of high activity, has strong adsorption performance and innocuity, and is hence an environment-friendly material. It has been extensively used for treating wastewater [44–46]. Recently, nanoscaled magnesium hydrate for remediation of heavy metal pollutant has been concerned [47]. However, how to load it on a suitable carrier is an important issue.

Here, we introduce the fabrication of $Mg(OH)_2$ nanoparticle–loaded PAN fibers based on the electrospinning and in situ conversion strategy, as schematically illustrated in Figure 4.8. Briefly, by using $MgCl_2$, instead of $FeCl_2$, as a component of the precursor solution we can obtain PAN/$MgCl_2$ composite nanofibers by the electrospinning technique, similar to that mentioned at the beginning of this section [see step (I) in Figure 4.8]. Then, the electrospun PAN/$MgCl_2$ composite nanofibers were immersed in a basic (ammonia) solution for 12 hours [see step (II) in Figure 4.8]. The PAN/$Mg(OH)_2$ composite nanofibers were obtained after washing and drying the basified fibers. Figure 4.9 shows the typical results. Obviously, after basification

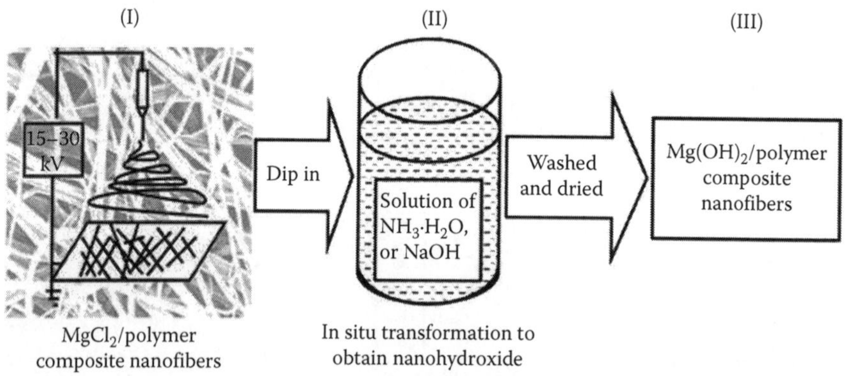

FIGURE 4.8 Schematic illustration for fabrication of polyacrylonitrile/$Mg(OH)_2$ composite nanofibers based on basification of preformed nanofibers.

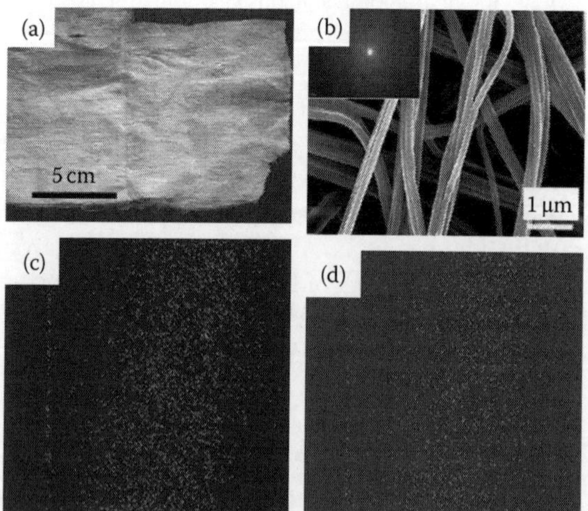

FIGURE 4.9 Morphology and composition distribution of polyacrylonitrile/$Mg(OH)_2$ composite nanofibers: (a) the photograph of the fiber film after basification and (b) FESEM image. The inset shows the selected area electron diffraction pattern. The mapping of (c) Mg and (d) oxygen elements in a single fiber is also shown.

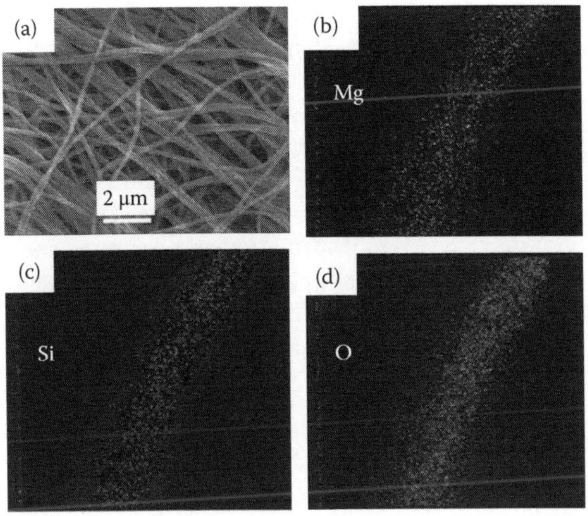

FIGURE 4.10 Morphology and composition distribution of polyacrylonitrile/$MgSiO_3$ composite nanofibers: (a) FESEM image. The mapping of (b) Mg, (c) Si, and (d) oxygen (O) elements in a single fiber is also shown.

the morphology of the composite nanofibers did not change; but the $MgCl_2$ in the fibers was converted to $Mg(OH)_2$, as clearly demonstrated in Figure 4.9c and d, corresponding to the element mapping of Mg and oxygen in the fibers.

Similarly, according to the strategy shown in Figure 4.8, by putting the PAN/$MgCl_2$ composite nanofibers into Na_2SiO_3 aqueous solution, instead of ammonia, for 12 hours, we could finally obtain PAN/$MgSiO_3$ composite nanofibers with a nearly unchanged morphology, as shown in Figure 4.10. Such an in situ conversion strategy based on the basification of electrospinning-induced nanofibers is of universal acceptability and can produce various composite nanofibers. These new materials can be good candidates for environmental remediation.

4.3 HIERARCHICALLY MICRO/NANOSTRUCTURED PAN@γ-AlOOH FIBERS

Although some nanosized powders could be used as effective adsorbents of contaminants [48–50], some vital defects hinder their practical applications due to their ease of agglomeration and difficulty in collection after use. As an alternative approach to avoid agglomeration, inorganic hierarchically micro/nanostructured materials or microsized objects with nanostructures, such as dandelion-like CuO [51] and ZnO [52], flowerlike iron oxide [53] and γ-AlOOH [54], and hierarchical MnO [55] (see Chapters 2 and 3), have been successfully prepared. These materials have steady activity due to their structural geometry. However, their separation after use in water is still a challenge due to their existence in the form of powders.

It is well known that polymers are excellent recyclable materials with good elasticity and pliability. Recently, the electrospun polymer fibers have exhibited great potential in environment remediation because it is freestanding and recyclable easily and can be integrated with the other matrixes conveniently [56–58]. In addition, the fiber-structured materials have some attractive advantages, such as being comparatively low cost, applicability to various materials, and the ability to generate relatively large-scale integrated films, which are important in some practical applications [35]. Obviously, if micro/nanostructured inorganic materials grow on the surface of electrospun polymer fibers, a flexible composite fiber film would be obtained with micro/nanostructures. Such a composite material could possess the advantages of both polymer fibers and micro/nanostructured inorganic materials and, hence, overcome the shortcomings of powdered nanomaterials. In this section, the electrospun fiber–templated hydrothermal strategy is introduced for the fabrication of hierarchically micro/nanostructured PAN@γ-AlOOH composite nanofibers.

4.3.1 Electrospun Nanofiber–Templated Hydrothermal Route

Figure 4.11 shows the corresponding strategy. Briefly, electrospun polymer fibers are used as the flexible template. The fibers are immersed in a precursor solution and hydrothermally treated. Nanostructured γ-AlOOH grows on the fibers' surface, forming hierarchically micro/nanostructured composite fibers during the hydrothermal reaction. Further, tubular hollow fibers will be obtained after subsequent calcination of the composite fibers. As a typical example, here electrospun PAN fibers, as mentioned in Section 4.2, are directly used as the substrate for γ-AlOOH growth and can also be used as the sacrificial template to produce the micro/nanostructured hollow or tubular inorganic fibers (see Figure 4.11c).

In experiments [59], typically 0.01 g electrospinning-induced PAN fiber film was mixed with 1.5 g aluminum powder and put in hexamethylenetetramine (HMTA) ($C_6H_{12}N_4$) aqueous solution (1 %wt). Then, the mixture was sealed and heated to 110°C for hydrothermal reaction for 10 hours. The final products were obtained by taking out the fibers, washing with deionized water, and drying in a vacuum oven.

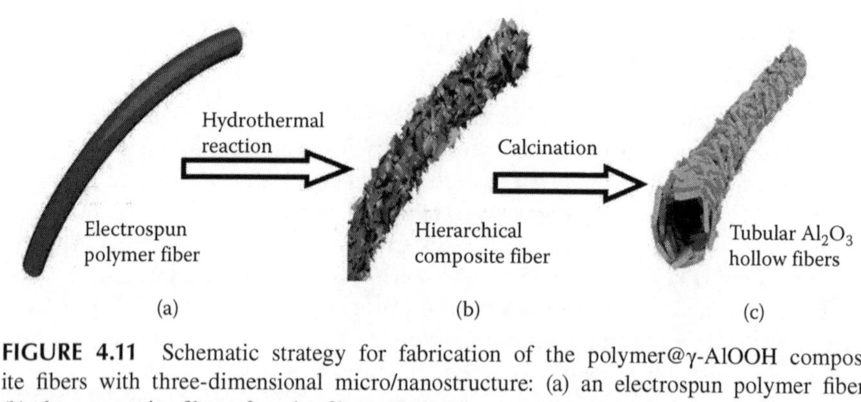

FIGURE 4.11 Schematic strategy for fabrication of the polymer@γ-AlOOH composite fibers with three-dimensional micro/nanostructure: (a) an electrospun polymer fiber, (b) the composite fiber after the fiber-templated hydrothermal reaction, and (c) a tubular hollow fiber after calcination of the composite fiber. (Lin et al., *RSC Adv.*, 2, 1769–73, 2012. Reproduced by permission of The Royal Society of Chemistry.)

4.3.2 Morphology and Structure

Figure 4.12a shows the freestanding electrospun PAN fiber film, which is white in color and consists of randomly arranged fibers, similar to that shown in Figure 4.2a and b. Most of the fibers fall in the range from 200 to 400 nm in diameter. After subsequent hydrothermal reaction, the film's color is unchanged. XRD measurement revealed that the phase constitution before and that after the reaction are different, as shown in Figure 4.12b. The peaks at 17° and 27° correspond to the diffraction of the PAN fibers [60]. After reaction, the new peaks can be identified as orthorhombic γ-AlOOH, indicating the formation of PAN@γ-AlOOH.

Further, SEM observation indicated that the fibers became much thicker than the pure PAN fibers (about 1 μm in diameter) after the reaction, as shown in Figure 4.12c, meaning the formation of the core-shell structured PAN@γ-AlOOH fibers. The fibers exhibit a micro/nanostructure several tens of microns in length.

FIGURE 4.12 (a) FESEM image of electrospun polyacrylonitrile (PAN) fibers, and the inset is a photograph of the fiber film; (b) x-ray diffraction patterns of the electrospun PAN fibers (1) and the PAN@γ-AlOOH composite fibers (2); (c) FESEM image of the PAN@γ-AlOOH composite fibers; (d) a local magnified image of (c), and the inset is a further magnified image. (Lin et al., *RSC Adv.*, 2, 1769–73, 2012. Reproduced by permission of The Royal Society of Chemistry.)

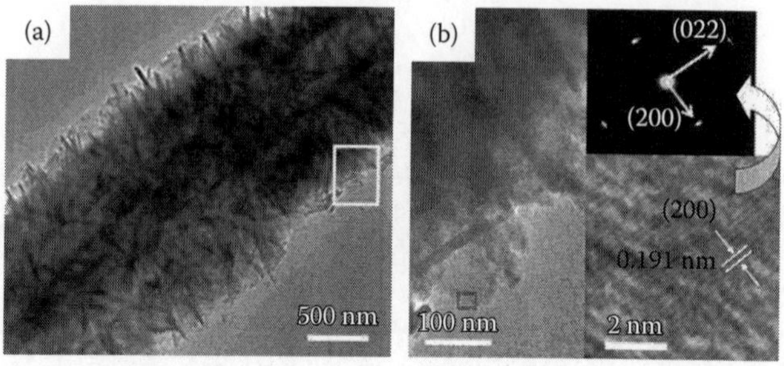

FIGURE 4.13 Microstructural examination of the prepared composite fibers: (a) TEM image of a single as-prepared PAN@γ-AlOOH fiber and (b) a local magnified image corresponding to the rectangular area marked in (a); the down-right inset shows a high-resolution TEM image corresponding to the rectangular area marked in (b) and the up-right inset shows the selected area electron diffraction pattern of a nanoplate vertical to the electron beam. (Lin et al., *RSC Adv.*, 2, 1769–73, 2012. Reproduced by permission of The Royal Society of Chemistry.)

The products are homogeneously distributed on the fibers. Local magnification has revealed that the products are cross-linked γ-AlOOH nanoplates, standing nearly vertically on the fibers, exhibiting a nest-like architecture, as shown in Figure 4.12d. The nanoplates are 80–150 nm in edge size and 15–40 nm in thickness (see the inset in Figure 4.12d).

A TEM examination was conducted. Figure 4.13a shows the morphology corresponding to a segment of the composite fiber. The selected area electron diffraction (SAED) pattern, for a nanoplate perpendicular to the electron beam, confirmed that the nanoplate is a single crystal with planar orientation $[0\bar{1}1]$, as illustrated in Figure 4.13b and its insets. In addition, the specific surface area of the composite fibers was determined to be about 19 $m^2 \cdot g^{-1}$ by the corresponding isothermal nitrogen sorption measurement and the BET equation [37]. Further, the AlOOH content in the composite fibers was estimated to be 47.4 %wt, by measuring the difference in film weight before and after the hydrothermal reaction. So, the real specific surface area of the cross-linked γ-AlOOH nanoplates in the composite fibers should be about 40 $m^2 \cdot g^{-1}$. Here, it should be mentioned that such composite fiber film is still flexible.

4.3.3 Influence Factors

4.3.3.1 Annealing Treatment

According to the strategy shown in Figure 4.11, if the composite fibers were subsequently calcined at a high temperature the PAN fibers would be burned away and the hollow or tubular hierarchical micro/nanostructured fibers should be formed. Figure 4.14 shows the SEM and TEM examinations corresponding to calcination at 550°C. The nanoplates are still vertically standing and cross-linked, but the PAN fibers have been removed and the tubular hollow structure

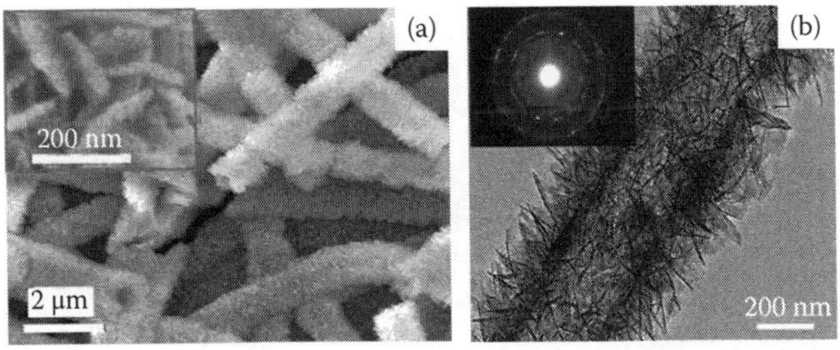

FIGURE 4.14 Microstructural examination of the composite fibers after calcination at 550°C: (a) FESEM image of calcinated fibers, showing a tubular hollow structure. The inset is a local magnified image. (b) TEM image of a single calcinated fiber. The inset shows the selected area electron diffraction pattern of a nanoplate. (Lin et al., *RSC Adv.*, 2, 1769–73, 2012. Reproduced by permission of The Royal Society of Chemistry.)

has been left. According to previous reports [61–63], the phase transformation from boehmite to alumina takes place in the temperature range from 250°C to 550°C. So the obtained hollow fibers should be Al_2O_3. Thus, the γ-AlOOH nanoplate became the polycrystal Al_2O_3 (see the SAED pattern in Figure 4.14b). Such structured materials could be used as catalysts [64] or catalyst supports [65–67].

4.3.3.2 Effect of Reaction Duration

It can be expected that reaction duration is crucial to the formation of PAN@γ-AlOOH micro/nanostructured fibers. When reaction time is short (say, 3 hours or less), only very few ultrafine nanoplates are formed, standing on the PAN fibers (Figure 4.15a). For a reaction duration of 6 hours, much more nanoplates were formed on the fibers. Only when the reaction was long enough (say, 10 hours or longer) could the cross-linked and dense nanoplates be observed homogeneously, standing nearly vertically on the fibers (Figure 4.12c and d).

FIGURE 4.15 FESEM images of the composite fibers obtained at 110°C for different reaction times: (a) 3 hours and (b) 6 hours.

FIGURE 4.16 FESEM images of the composite fibers obtained from the reaction solution with different $C_6H_{12}N_4$ amounts; the reaction time was 10 hours: (a) 0.1%, (b) 0.25%, and (c) 10% in weight. (d) A local magnification of (c).

4.3.3.3 Effect of Hexamethylenetetramine Addition

Further experiments have indicated that the concentration of $C_6H_{12}N_4$ (or HMTA) in the precursor solution is important to the formation of γ-AlOOH nanoplates during the hydrothermal reaction. When its concentration is low (say, 0.1 %wt), only a few nanoplates were formed on the PAN fibers, as illustrated in Figure 4.16a. A higher concentration (say, 0.25 %wt) induced more nanoplates (see Figure 4.16b). When the concentration was very high (say, 10 %wt), however, the product after the reaction for 10 hours was a mixture of big particles (about 300 nm in size) and floc, densely surrounding the fibers, as seen in Figure 4.16c and d. Only an appropriate concentration (say, 1 %wt) can lead to the production of cross-linked γ-AlOOH nanoplates, standing nearly vertically on the fibers, forming a nest-like micro/nanoarchitecture (Figure 4.12c and d).

4.3.3.4 Effect of Reaction Temperature

Also, the reaction temperature was crucial to the morphology of the final products. Under the conditions of 1 %wt HMTA in the precursor solution and 10 hours of reaction time, a too-low reaction temperature (say, 90°C or lower) induced no products on the PAN fibers, or no reaction took place, whereas a too-high temperature (say, 140°C or higher) led to the formation of a mixture with big blocks and thin nanoplates on the PAN fibers, as shown in Figure 4.17. Only the reaction at an appropriate temperature (around 110°C) produces the cross-linked γ-AlOOH nanoplates homogeneously standing on the fibers with a nest-like micro/nanoarchitecture (Figure 4.12c and d).

FIGURE 4.17 FESEM image of the products obtained from reaction at 140°C for 10 hours (the precursor solution contains 1% hexamethylenetetramine).

4.3.4 Formation of Micro/Nanostructured PAN@γ-AlOOH Fibers

The formation of a PAN@γ-AlOOH fiber is easily understood. Briefly, when the precursor solution was heated at 110°C, the following reactions occurred [59]:

$$C_6H_{12}N_4 + 6H_2O \rightarrow 6HCHO + 4NH_3 \quad (4.1)$$

$$NH_3 + H_2O \rightarrow NH_4^+ + OH^- \quad (4.2)$$

$$Al + 2H_2O + OH^- \rightarrow Al(OH)_4^- + H_2 \quad (4.3)$$

$$Al(OH)_4^- \rightarrow AlOOH + H_2O + OH^- \quad (4.4)$$

At the beginning of the hydrothermal reaction, HTMA (or $C_6H_{12}N_4$) first reacts with water to produce ammonia (Reaction 4.1) [68,69]. The ammonia in turn hydrolyzes and OH^- ions are generated, leading to an alkaline solution or an increase in pH value (Reaction 4.2). In such a basic solution, with continuing $C_6H_{12}N_4$ hydrolysis, aluminum atoms would react with excessive hydroxyl ions to form $Al(OH)_4^-$ (Reaction 4.3) [70–73]. The $Al(OH)_4^-$ is unstable in the hydrothermal condition [74,75] and decomposes to form AlOOH (Reaction 4.4). When the concentration of AlOOH reaches supersaturation, γ-AlOOH nuclei would be formed on the PAN fibers.

It is well known that γ-AlOOH has a layered structure with octahedral arrangement within the lamellae, and the hydroxyl ions hold the lamellae together through hydrogen bonding [76,77]. Also, there are reports that appropriate basic conditions favor the formation of two-dimensional (2D) γ-AlOOH nanosheets [70,74,78] and the 2D lamellar structure would be retained in basic solutions [79]. Here, the initial reaction solution was weakly basic with a pH above 8. The platelike γ-AlOOH nuclei would preferentially grow along the [100] and [011] directions within the plane $(0\bar{1}1)$. Because of the geometric limitation, only the (200)-oriented γ-AlOOH nuclei can grow continuously; the other plane-oriented nuclei cannot grow sufficiently. Finally, cross-linked γ-AlOOH nanoplates are formed, standing nearly vertically on the fibers, exhibiting a nest-like micro/nanoarchitecture (see Figure 4.12d).

In addition, based on Reactions 4.1 through 4.3, the effect of HTMA in the precursor solution on the products' morphology should be understood easily. A low concentration of HTMA corresponds to a low OH^- concentration and hence production of only a few $Al(OH)_4^-$ ions, which induce slow formation of the products (Reactions 4.3 and 4.4), and only a few γ-AlOOH nanoplates within the same reaction time (see Figure 4.16a and b), whereas a too-high HTMA concentration, corresponding to a high OH^- concentration, would induce fast formation and growth of γ-AlOOH, resulting in densely distributed products with irregular morphology. The effect of the reaction temperature is also similar to that of HTMA. At a low temperature Reaction 4.1 cannot take place and no product is formed, whereas a too-high temperature gives rise to quick formation and growth of γ-AlOOH, leading to products with dispersed sizes (Figure 4.17).

All in all, the strategy shown in Figure 4.11 is simple, controllable, and hence suitable for the mass production of PAN@γ-AlOOH micro/nanostructured fibers. Such fibers can be converted into the tubular Al_2O_3 hollow fibers with the nearly unchanged morphology after calcination in air. These materials are high in specific surface area, are convenient to collect and separate from solution, and hence can be used as efficient adsorbents for the removal of contaminants from solutions. It has been demonstrated that the as-prepared PAN@γ-AlOOH fibers had a much higher adsorption capacity than the normal γ-AlOOH nanopowders and exhibited very good recycling performance as an adsorbent to remove the heavy metal ions Cr(VI) from model wastewater [59]. Also, it is expected to have the other potential applications in such as sensitive material or catalyst.

4.4 ELECTROSPUN NANOFIBER–BASED AG POROUS NANOTUBE FILMS

Ag nanostructured porous films have outstanding properties and potential applications in catalysis [80], superhydrophobicity [81], lithium ion batteries [82], and so on. Especially, they can be used for the substrates of surface-enhanced Raman scattering (SERS) sensors owing to their high density of surface nanogaps, and hence significant local electromagnetic field enhancement ability, and large surface area to adsorb the detected molecules [83–85]. Many methods have been developed for their fabrication, such as electrodeposition [86], electrophoresis [87], chemical reduction [88], selective etching [89], thermal decomposition [90], and template routes (polyethyleneimine hydrogel [91] and colloidal crystal [81]). Here we will introduce a new template strategy or the plasma etching Ag-coated electrospun-nanofiber template for Ag porous nanotube-built films with the contaminant-free surface and tunable structure [92].

4.4.1 Ag-Coated Nanofiber Template–Plasma Etching Strategy

Figure 4.18 schematically shows the nanofiber template–plasma etching strategy. Briefly, polystyrene nanofibers (PNs) are first prepared by electrospinning, as shown in Figure 4.18a. The electrospun PNs are then coated with Ag by ion beam sputtering and subsequently removed by dissolution. Thus, tubelike (or hollow) Ag nanofibers are formed (see Figure 4.18b and c). The Ag tube–packed film was etched (or bombarded) by plasma to form micro/nanostructured Ag porous films with a clean surface. Such films should be contaminant free, built of Ag porous nanotubes, and homogeneous in macrosize but rough and porous in nanoscale. Each nanotube block is micro/nanostructured with evenly distributed nanopores on the tube walls (see Figure 4.18d).

In a typical experiment, polystyrene (homopolymer, MW = 220,000) was dissolved in DMF to form a 25 wt% transparent solution. Such a solution was placed in a syringe equipped with a blunt metal needle 0.7 mm in hole diameter. The PN-packed film was thus fabricated by electrospinning on a setup with a needle tip–collector distance of 25 cm and voltage of 20 kV. Figure 4.19a shows a typical morphology of PNs by electrospinning, which are 300 nm in mean diameter and randomly arranged, forming a loose fiber film of white color several centimeters in size and about 20 μm in thickness.

Further, the PNs were used as templates. According to the strategy shown in Figure 4.18, a PN film was coated with 40 nm Ag shell by ion beam sputtering at a deposition rate of about 0.1 nm·s^{-1}. The inner PNs were then removed by dissolution in CH_2Cl_2 solution. The fiber film was thus obtained, as shown in the inset of Figure 4.19a (a photograph of the film). The corresponding morphology is illustrated in Figure 4.19b. Obviously, compared with the original PNs these fibers, which are composed

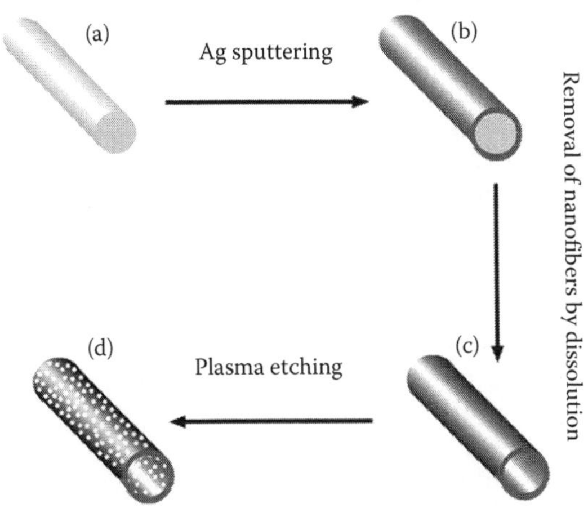

FIGURE 4.18 Schematic illustration of the strategy for fabricating an Ag porous nanotube: (a) the electrospun polystyrene nanofiber (PN), (b) the Ag-coated PN, (c) the Ag hollow (tubelike) nanofiber, and (d) the porous Ag tube (or hollow nanofiber). (Reprinted with permission from He et al., 2011, 1551–5. Copyright 2011 American Chemical Society.)

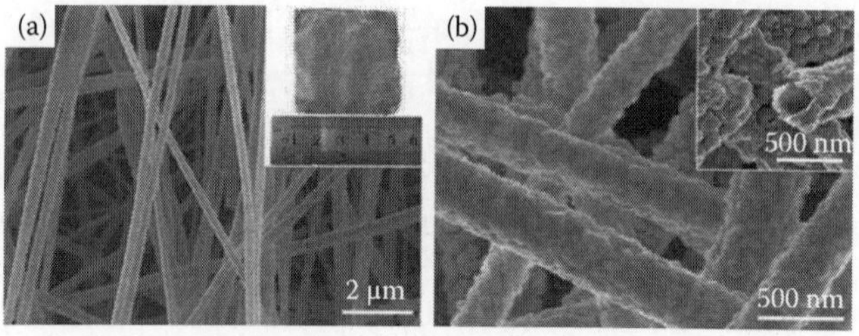

FIGURE 4.19 FESEM images of electrospun polystyrene nanofiber (PN) film (a) before and (b) after Ag coating and subsequent removal of the PNs. Inset in (a) shows a photograph corresponding to (b). Inset in (b) shows the cross-sectional view of a single tube (about 40 nm in thickness). (Reprinted with permission from He et al., 2011, 1551–5. Copyright 2011 American Chemical Society.)

FIGURE 4.20 A low-magnified FESEM image in the edge region of the film shown in the inset of Figure 4.19 (a). The circle marks correspond to the end parts of the fibers, showing hollow structure.

of nanoparticles tens of nanometers in size, are bigger in diameter due to the coated Ag shells. However, they are still isolated well from each other and keep a fibrous morphology. Cross-sectional view has revealed that the Ag fibers are tubular structured or hollow, as shown in the inset of Figure 4.19b, due to ion sputtering–induced nonshadow deposition [93] and removal of PNs. Figure 4.20 presents a low-magnified field emission scanning electron microscopy (FESEM) image in the edge region of the film, clearly showing more fibrous end parts with tubular or hollow structure.

4.4.2 Morphology and Structure of Ag Porous Nanotube Films

The Ag nanotube–packed film shown in the inset of Figure 4.19a was finally bombarded by Ar plasma for 40 minutes under the conditions of 0.15 torr pressure, 100 W input power, and 8 standard cubic feet per hour (SCFH) flow rate of Ar gas [92]. After etching, numerous nanopores are formed on tube walls, as shown in Figure 4.21. All the nanotubes still keep the fibrous profile without distortion. The nanopores are rough and irregular and are nanometers to tens of nanometers in scale. In other words, such Ag porous nanotubes are micro/nanostructured, and the porous tube-built film is homogeneous in macroscale but hierarchically rough and porous in micro/nanoscale. Moreover, this film should be contaminant free due to the plasma bombardment during fabrication. Importantly, the film is freestanding and flexible and can be tailored freely or transferred to other substrates [92]. We could thus obtain a homogeneous Ag porous nanotube–built film, with a thickness much higher than 20 μm, by putting two or more pieces of thin homogeneous films together.

4.4.3 Structural Tunability of Ag Porous Nanotube Films

Further, based on the strategy shown in Figure 4.18 the film architecture should be tunable by the PN template's configuration, Ag sputtering time, and plasma etching conditions, which have been confirmed by further experiments [92]. The arrangement and distribution density of the porous nanotubes in the film can be easily controlled by the electrospun PN template's configuration. Typically, Figure 4.22 shows the film consisting of aligned Ag porous nanotubes, which was obtained from the aligned PN film. Also, from the electrospun PN films with different diameters and number densities of fibers, the corresponding Ag porous nanotube films could be obtained by subsequent coating and plasma etching, as demonstrated in Figure 4.23, which shows tunable nanotubes' size and number density. This controllability of the structural polymorphism is important in practical applications.

FIGURE 4.21 FESEM images of the sample shown in Figure 4.19b after 40 minutes of plasma bombardment (at different magnifications). (Reprinted with permission from He et al., 2011, 1551–5. Copyright 2011 American Chemical Society.)

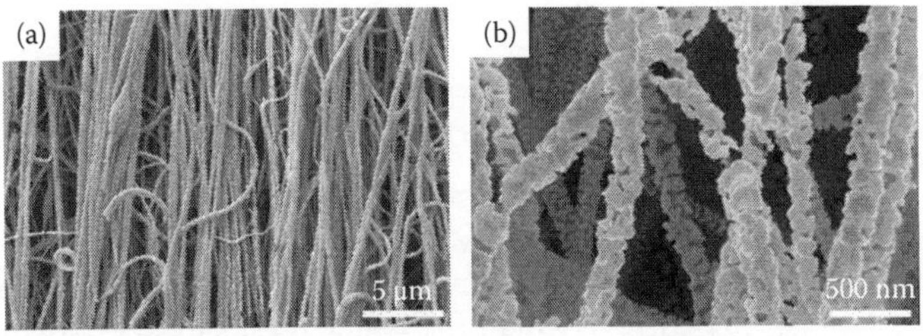

FIGURE 4.22 FESEM images of Ag porous nanotube films by plasma etching the Ag-coated aligned electrospun polystyrene nanofibers.

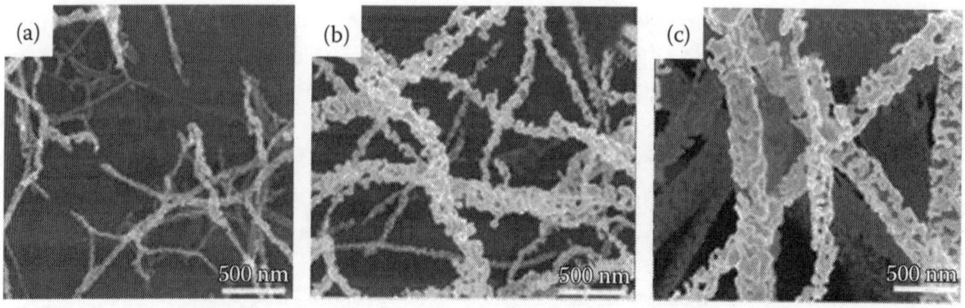

FIGURE 4.23 FESEM images of the porous nanotube-built films obtained from electrospun polystyrene nanofiber templates with different fibers' diameters and number densities.

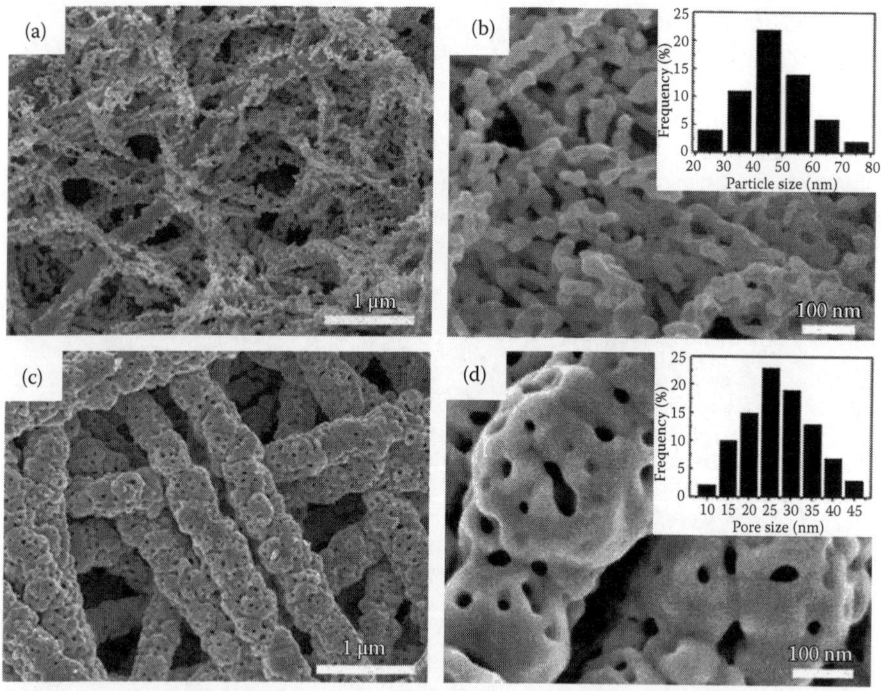

FIGURE 4.24 FESEM images of the porous nanotube-built films obtained by plasma etching the polystyrene nanofiber templates coated with Ag shells of different thicknesses: (a) and (b) 30 nm in shell thickness (at different magnifications), and the inset in (b) shows particle size distribution; (c) and (d) 180 nm shell thickness (at different magnifications), and the inset in (d) shows pore size distribution. (Reprinted with permission from He et al., 2011, 1551–5. Copyright 2011 American Chemical Society.)

Further, the pore shape in the porous nanotubes was strongly associated with the Ag coating's thickness. Representatively, Figure 4.24 shows the porous nanotube-built films obtained by plasma etching PN templates coated with Ag shells 30 and 180 nm in thickness. For the former, the film is packed by broken Ag nanotubes (Figure 4.24a), which consist of smaller nanoparticles (20–80 nm in size), and exhibits a nanoparticle-aggregated structure, as shown in Figure 4.24b and its inset. For the latter, the film is built by integrated nanotubes, with a scraggy and smooth surface as well as homogeneously distributed nanoholes (10–45 nm in size), and the nanoholes are formed at the cupped sites on the tubes' surface (Figure 4.24c and d and the inset). Moreover, number and size of nanopores in the tube vary with the plasma etching duration. Shorter etching corresponds to lesser and smaller nanopores, or vice versa. After etching for enough time (say, >20 minutes), a large number of nanopits together with nanopores were formed [92].

4.4.4 Formation of Nanopores on Tube Walls

The formation of nanopores on tube walls by plasma bombardment can be described as follows: Ag, unlike other metals, has a high surface mobility, even at room temperature [94]. During plasma etching, the Ar^+ ion bombardment will lead to the densification and even surface melting (or sputtering) of the deposited Ag shells, due to energy transfer. For the PNs with thick Ag shells, such densification will induce the contraction of Ag shells and hence the formation of nanoholes or nanopits on the cupped sites, as shown in Figure 4.24c. Also, the surface melting would result in a smooth surface (see Figure 4.24d). For the PNs with thin Ag coatings, which are of lower heat capacity, the tubelike morphology will be destroyed due to the ion sputtering–induced etching (see Figure 4.24a).

To sum up, the film architecture can be easily controlled by the template configuration, Ag coating thickness, and the plasma etching condition. Such an Ag porous nanotube-built film is flexible and can be transferred to any other substrates and hence has great potential for practical applications in many fields, such as catalysis, sensors, and nanodevices. Due to its unique structure and clean surface, this film exhibits strong SERS activity with good stability and reproducibility and shows the possibility of molecule-level detection [92]. Also, the strategy shown in Figure 4.18 is universal for fabricating other metal porous nanotube-built films.

4.5 BRIEF SUMMARY

In summary, the electrospinning technique is simple, flexible, controllable, and hence suitable for the mass production of various micro/nanostructured fiber materials, including polymer nanofibers, inorganic oxide nanofibers (oxides, composite oxides, and metal-doped oxides), metal oxides (or sulfides)/macromolecule (polymer) composite fibers, and metal nanoparticles/polymer nanofibers. Importantly, the electrospun nanofibers could be used as templates for many other composite nanofiber materials by the in situ conversion strategy. In this chapter, in addition to the fabrication of PAN/ferrous chloride composite porous nanofibers by one-step electrospinning, we have mainly introduced fabrication of PAN/inorganic [FeOOH, $Mg(OH)_2$, and $MgSiO_3$] composite porous nanofibers based on in situ conversion by basification of electrospun PAN fibers, preparation of hierarchically micro/nanostructured PAN@γ-AlOOH fibers based on the electrospun PAN fiber–templated hydrothermal route, and production of Ag porous nanotube-built films by plasma etching of Ag-coated electrospun PNs.

REFERENCES

1. Cooley, J. F. 1902. Apparatus for electrically dispersing fluids. US Patent 692631.
2. Baumgarten, P. K. 1971. Electrostatic spinning of acrylic microfibers. *J. Colloid. Interface Sci.* 36: 71.
3. Larrondo, L. and Manley, R. S. J. 1981. Electrostatic fiber spinning from polymer melts.1. Experimental observations on fiber formation and properties. *J. Polym. Sci. Pol. Phys.* 19: 909–20.
4. Larrondo, L. and Manley, R. S. J. 1981. Electrostatic fiber spinning from polymer melts.2. Examination of the flow field in an electrically driven jet. *J. Polym. Sci. Pol. Phys.* 19: 921–32.
5. Larrondo, L. and Manley, R. S. J. 1981. Electrostatic fiber spinning from polymer melts.3. Electrostatic deformation of a pendant drop of polymer melt. *J. Polym. Sci. Pol. Phys.* 19: 933–40.
6. Dosh, I. J.; Srinivasan, G.; and Reneker, D. H. 1995. A novel electrospinning process. *Polym. News* 20: 206–07.
7. Dzenis, Y. 2004. Spinning continuous fibers for nanotechnology. *Science* 304: 1917–19.
8. Zhang, D.; Karki, A. B.; Rutman, D.; Young, D. P.; Wang, A.; Cocke, D.; Ho, T. H.; and Guo, Z. 2009. Electrospun polyacrylonitrile nanocomposite fibers reinforced with Fe_3O_4 nanoparticles: Fabrication and property analysis. *Polymer.* 50: 4189–98.
9. Zhu, J.; Wei, S.; Chen, X.; Karki, A. B.; Rutman, D.; Young, D. P.; and Guo, Z. 2010. Electrospun polyimide nanocomposite fibers reinforced with core-shell Fe-FeO nanoparticles. *J. Phys. Chem. C.* 114: 8844–50.
10. Taylor, G. 1964. Disintegration of water drops in electric field. *P. Roy. Soc. A-Math. Phy.* 280: 383–97.
11. Reznik, S. N.; Yarin, A. L.; and Theron, A. 2004. Coaxial liquid-liquid flows in tubes with limited length. *J. Fluid. Mech.* 516: 349–77.
12. Deitzel, J. M.; Kleinmeyer, J. D.; and Hirvonen, J. K. 2001. Controlled deposition of electrospun poly(ethylene oxide) fibers. *Polymer* 42: 8163–70.
13. Yarin, A. L.; Koombhongse, S.; and Reneker, D. H. 2001. Taylor cone and jetting from liquid droplets in electrospinning of nanofibers. *J. Appl. Phys.* 90: 4836–46.
14. Li, D. and Xia, Y. N. 2004. Electrospinning of nanofibers: Reinventing the wheel? *Adv. Mater.* 16: 1151–70.
15. Macias, M.; Chacko, A.; and Ferraris, J. P. 2005. Electrospun mesoporous metal oxide fibers. *Micropor. Mesopor. Mat.* 86: 1–13.
16. Viswanathamurthi, P.; Kim, H. Y.; and Lee, D. R. 2003. Vanadium pentoxide nanofibers by electrospinning. *Scripta Mater.* 49: 577–81.
17. Yang, X. H. and Liu, Y. C. 2007. Fabrication of Cr_2O_3/Al_2O_3 composite nanofibers by electrospinning. *J. Mater. Sci.* 42: 8470–72.
18. Zhan, S. H.; Jiao, X. L.; and Song, Y. 2007. Mesoporous TiO_2/SiO_2 composite nanofibers with selective photocatalytic properties. *Chem. Commun.* 20: 2043–45.
19. Liu, L.; Li, S. C.; and Wang, L. Y. 2009. Preparation, characterization, and gas-sensing properties of Pd-doped In_2O_3 nanofibers. *Mater. Lett.* 63: 1975–77.
20. Ji, T. H.; Yang, F.; and Lu, Y. Y. 2009. Synthesis and visible-light photocatalytic activity of Bi-doped TiO_2 nanobelts. *Mater. Lett.* 63: 2044–46.
21. Li, L. 2009. Synthesis and ethanol sensing properties of Fe-doped SnO_2 nanofibers. *Mater. Lett.* 63: 917–19.
22. Liu, L.; Wang, L. Y.; and Li, S. C. 2009. Improved ethanol sensing properties of Cu-doped SnO_2 nanofibers. *Mater. Lett.* 63: 2041–43.
23. Im, J. S.; Kim, M.; and Lee, Y. S. 2008. Preparation of PAN-based electrospun nanofiber webs containing TiO_2 for photocatalytic degradation. *Mater. Lett.* 62: 3652–55.

24. Faridi-Majidi, R. and Sharifi-Anjani, N. 2007. In situ synthesis of iron oxide nanoparticles on poly(ethylene oxide) nanofibers through an electrospinning process. *J. Appl. Polym. Sci.* 105: 1351–55.
25. Wang, C.; Tong, Y. B.; and Sun, Z. Y. 2007. Preparation of one-dimensional TiO_2 nanoparticles within polymer fiber matrices by electrospinning. *Mater. Lett.* 61: 5125–28.
26. Formo, E.; Campbell, D.; and Xia, Y. 2008. Functionalization of electrospun TiO_2 nanofibers with Pt nanoparticles and nanowires for catalytic applications. *Nano Lett.* 8: 668–72.
27. Patel, A. C.; Li, S. X.; and Wang, C. 2007. Electrospinning of porous silica nanofibers containing silver nanofibers for catalytic applications. *Chem. Mater.* 19: 1231–38.
28. Jin, W. J.; Jeon, H. J.; and Kim, J. 2007. A study on the preparation of poly(vinyl alcohol) nanofibers containing silver nanoparticles. *Syn. Met.* 157: 454–59.
29. Hug, S. J.; Laubscher, H. U.; and James, B. R. 1997. Iron (III) catalyzed photochemical reduction of chromium (VI) by oxalate and citrate in aqueous solutions. *Environ. Sci. Technol.* 31: 160–70.
30. Padhye, M. R. and Karandikar, A. V. 1985. The effect of alkali salt on solvent polyacrylonitrile interaction. *J. Appl. Poly. Sci.* 30: 667–73.
31. Du, J. and Zhang, X. 2008. Role of polymer-salt-solvent interactions in the electrospinning of polyacrylonitrile/iron acetylacetonate. *J. Appl. Poly. Sci.* 109: 2935–41.
32. Phadke, M. A.; Musale, D. A.; Kulkarni, S. S.; and Karode, S. K. 2005. Poly (acrylonitrile) ultrafiltration membranes. I. Polymer-salt-solvent interactions. *J. Polym. Sci. Pol. Phys.* 43: 2061–73.
33. Melby, L. R.; Mahler, W.; Mochel, W. E.; Harder, R. J.; Hertler, W. R.; and Benson, R. E. 1962. Substituted quinodimethans.2. Anion-radical derivatives and complexes of 7,7,8,8-tetracyanoquinodimethan. *J. Am. Chem. Soc.* 84: 3374–87.
34. Reicha, F. M. 1995. Spectral, thermogravimetric, and electrical studies on a novel prepared acrylonitrile-ferric chloride polymer complex.2. *New Polym. Mat.* 4: 265–76.
35. Lin, Y.; Cai, W.; Tian, X.; Liu, X.; Wang, G.; and Liang, C. 2011. Polyacrylonitrile/ferrous chloride composite porous nanofibers and their strong Cr-removal performance. *J. Mater. Chem.* 21: 991–97.
36. Brunauer, S.; Deming, L. S.; Deming, W. E.; and Teller, E. 1940. On a theory of the van der Waals adsorption of gases. *J. Am. Chem. Soc.* 62: 1723–32.
37. Brunauer, S.; Jefferson, M. E.; Emmett, P. H.; and Hendricks, S. B. 1931. Equilibria in the iron-nitrogen system. *J. Am. Chem. Soc.* 53: 1778–86.
38. Luoh, R. and Hahn, H. T. 2006. Electrospun nanocomposite fiber mats as gas sensors. *Compos. Sci. Technol.* 66: 2436–41.
39. Grosvenor, A. P.; Kobe, B. A.; Biesinger, M. C.; and McIntyre, N. S. 2004. Investigation of multiple splitting of Fe2p XPS spectra and bonding in iron compounds. *Surf. Interface Anal.* 36: 1564–74.
40. Zhang, D.; Wei, S.; Kaila, C.; Su, X.; Wu, J.; Karki, A. B.; Young, D. P.; and Guo, Z. 2010. Carbon-stabilized iron nanoparticles for environmental remediation. *Nanoscale* 2: 917–19.
41. Kato, K.; Eika, Y.; and Ikada, Y. 1997. In situ hydroxyapatite crystallization for the formation of hydroxyapatite/polymer composites. *J. Mater. Sci.* 32: 5533–43.
42. Biesinger, M. C.; Brown, C.; and Mycroft, J. R. 2004. X-ray photoelectron spectroscopy studies of chromium compounds. *Surf. Interface Anal.* 36: 1550–63.
43. Astrup, T.; Stipp, S. L. S.; and Christensen, T. H. 2000. Immobilization of chromate from coal fly ash leachate using an attenuating barrier containing zero-valent iron. *Environ. Sci. Technol.* 34: 4163–68.
44. Tan, B. H.; Teng, T. T.; and Omar, A. K. M. 2000. Removal of dyes and industrial dye wastes by magnesium chloride. *Water Res.* 34: 597–601.

45. Tralful, I. and Lester, J. N. 2001. Conditions influencing the precipitation of magnesium ammonium phosphate. *Water Res.* 35: 4191–99.
46. Musvoto, E. V.; Wentzeli, M. C.; and Ekama, G. A. 2000. Integrated chemical-physical processes modeling-II simulating aeration treatment of anaerobic digester supernatants. *Water Res.* 34: 1869–80.
47. Zhang, S.; Cheng, F. Y.; and Tao, Z. L. 2006. Removal of nickel ions from wastewater by $Mg(OH)_2$/MgO nanostructures embedded in Al_2O_3 membranes. *J. Alloy. Compound.* 426: 281–85.
48. Theron, J.; Walker, J. A.; and Cloete, T. E. 2008. Nanotechnology and water treatment: Applications and emerging opportunities. *Crit. Rev. Microbiol.* 34: 43–69.
49. Savage, N. and Diallo, M. S. 2005. Nanomaterials and water purification: Opportunities and challenges. *J. Nanopart. Res.* 7: 331–42.
50. Bottero, J.-Y.; Rose, J.; and Wiesner, M. R. 2006. Nanotechnologies: Tools for sustainability in a new wave of water treatment processes. *Integr. Environ. Assess. Manag.* 2: 391–5.
51. Liu, B. and Zeng, H. C. 2004. Mesoscale organization of CuO nanoribbons: Formation of "dandelions." *J. Am. Chem. Soc.* 126: 8124–25.
52. Liu, B. and Zeng, H. C. 2004. Fabrication of ZnO "dandelions" via a modified Kirkendall process. *J. Am. Chem. Soc.* 126: 16744–46.
53. Zhong, L.-S.; Hu, J.-S.; Liang, H.-P.; Cao, A.-M.; Song, W.-G.; and Wan, L.-J. 2006. Self-assembled 3D flowerlike iron oxide nanostructures and their application in water treatment. *Adv. Mater.* 18: 2426–31.
54. Zhang, J.; Liu, S.; Lin, J.; Song, H.; Luo, J.; Elssfah, E. M.; Ammar, E. et al. 2006. Self-assembly of flowerlike AlOOH (boehmite) 3D nanoarchitectures. *J. Phys. Chem. B.* 110: 14249–52.
55. Li, Z. Q.; Ding, Y.; Xiong, Y. J.; Yang, Q.; and Xie, Y. 2005. One-step solution-based catalytic route to fabricate novel alpha-MnO_2 hierarchical structures on a large scale. *Chem. Commun.* 7: 918–20.
56. Ma, H.; Yoon, K.; Rong, L.; Mao, Y.; Mo, Z.; Fang, D.; Hollander, Z.; Gaiteri, J.; Hsiao, B.S.; and Chu, B. 2010. High-flux thin-film nanofibrous composite ultrafiltration membranes containing cellulose barrier layer. *J. Mater. Chem.* 20: 4692–704.
57. Yoon, K; Hsiao, B. S.; and Chu, B. 2009. High flux ultrafiltration nanofibrous membranes based on polyacrylonitrile electrospun scaffolds and crosslinked polyvinyl alcohol coating. *J. Membrane Sci.* 338: 145–52.
58. Yoon, K.; Kim, K.; Wang, X. F.; Fang, D. F.; Hsiao, B. S.; and Chu, B. 2006. High flux ultrafiltration membranes based on electrospun nanofibrous PAN scaffolds and chitosan coating. *Polymer* 47: 2434–41.
59. Lin, Y.; Cai, W.; He, H.; Wang, X.; and Wang, G. 2012. Three-dimensional hierarchically structured PAN@gamma-AlOOH fiber films based on a fiber templated hydrothermal route and their recyclable strong Cr(VI)-removal performance. *RSC Adv.* 2: 1769–73.
60. Mathur, R. B.; Bahl, O. P.; Mittal, J.; and Nagpal, K. C. 1991. Structure of thermally stabilized PAN fibers. *Carbon* 29: 1059–61.
61. Kloprogge, J. T.; Ruan, H. D.; and Frost, R. L. 2002. Thermal decomposition of bauxite minerals: Infrared emission spectroscopy of gibbsite, boehmite and diaspore. *J. Mater. Sci.* 37: 1121–29.
62. Lin, C. P. and Wen, S. B. 2002. Variations in a boehmite gel and oleic acid emulsion under calcination. *J. Am. Ceram. Soc.* 85: 1467–72.
63. Suvaci, E.; Simkovich, G.; and Messing, G. L. 2000. The reaction-bonded aluminum oxide process: I, the effect of attrition milling on the solid-state oxidation of aluminum powder. *J. Am. Ceram. Soc.* 83: 299–305.
64. Ma, C.; Chang, Y.; Ye, W.; Shang, W.; and Wang, C. 2008. Supercritical preparation of hexagonal gamma-alumina nanosheets and its electrocatalytic properties. *J. Colloid Interf. Sci.* 317: 148–54.

65. Pattamakomsan, K.; Ehret, E.; Morfin, F.; Gelin, P.; Jugnet, Y.; Prakash, S.; Bertolini, J.C.; Panpranot, J.; and Aires, F. J. C. S. 2011. Selective hydrogenation of 1,3-butadiene over Pd and Pd-Sn catalysts supported on different phases of alumina. *Catal. Today*, 164: 28–33.
66. Gong, M.-S.; Kim, J.-U.; and Kim, J.-G. 2010. Preparation of water-durable humidity sensor by attachment of polyelectrolyte membrane to electrode substrate by photochemical crosslinking reaction. *Sensor. Actuat. B-Chem.* 147: 539–47.
67. Venkatesan, B. M.; Dorvel, B.; Yemenicioglu, S.; Watkins, N.; Petrov, I.; and Bashir, R. 2009. Highly sensitive, mechanically stable nanopore sensors for DNA analysis. *Adv. Mater.* 21: 2771–76.
68. Xia, Z.; Sha, J.; Fang, Y.; Wan, Y.; Wang, Z.; and Wang, Y. 2010. Purpose built ZnO/ $Zn(5)(OH)_8Ac_2$center dot $2H_2O$ architectures by hydrothermal synthesis. *Cryst. Growth Des.* 10: 2759–65.
69. Govender, K.; Boyle, D. S.; Kenway, P. B.; and O'Brien, P. 2004. Understanding the factors that govern the deposition and morphology of thin films of ZnO from aqueous solution. *J. Mater. Chem.* 14: 2575–91.
70. Chen, X. Y.; Huh, H. S.; and Lee, S. W. 2007. Hydrothermal synthesis of boehmite (gamma-AlOOH) nanoplatelets and nanowires: pH-controlled morphologies. *Nanotechnology* 18: 285608.
71. Liang, H.; Liu, L.; Yang, Z.; and Yang, Y. 2010. Facile hydrothermal synthesis of uniform 3D gamma-AlOOH architectures assembled by nanosheets. *Cryst. Res. Technol.* 45: 195–98.
72. Xu, Z. P. and Lu, G. Q. 2005. Hydrothermal synthesis of layered double hydroxides (LDHs) from mixed MgO and Al_2O_3: LDH formation mechanism. *Chem. Mater.* 17: 1055–62.
73. Feng, C.; Wei, Q.; Wang, S.; Shi, B.; and Tang, H. 2007. Speciation of hydroxyl-Al polymers formed through simultaneous hydrolysis of aluminum salts and urea. *Colloid. Surf. A.* 303: 241–48.
74. Zanganeh, S.; Kajbafvala, A.; Zanganeh, N.; Mohajerani, M. S.; Lak, A.; Bayati, M. R.; Zargar, H. R.; and Sadrnezhaad, S. K. 2010. Self-assembly of boehmite nanopetals to form 3D high surface area nanoarchitectures. *Appl. Phys. A-Mater.* 99: 317–21.
75. Benezeth, P.; Palmer, D. A.; and Wesolowski, D. J. 2008. Dissolution/precipitation kinetics of boehmite and gibbsite: Application of a pH-relaxation technique to study near-equilibrium rates. *Geochim. Cosmochim. Ac.* 72: 2429–53.
76. Hou, H. W.; Xie, Y.; Yang, Q.; Guo, Q. X.; and Tan, C. R. 2005. Preparation and characterization of gamma-AlOOH nanotubes and nanorods. *Nanotechnology* 16: 741–45.
77. Bokhimi, X.; Toledo-Antonio, J. A.; Guzman-Castillo, M. L.; and Hernandez-Beltran, F. 2001. Relationship between crystallite size and bond lengths in boehmite. *J. Solid State Chem.* 159: 32–40.
78. Cai, W.; Yu, J.; and Jaroniec, M. 2010. Template-free synthesis of hierarchical spindle-like gamma-Al_2O_3 materials and their adsorption affinity towards organic and inorganic pollutants in water. *J. Mater. Chem.* 20: 4587–94.
79. Chen, X. Y. and Lee, S. W. 2007. pH-dependent formation of boehmite (gamma-AlOOH) nanorods and nanoflakes. *Chem. Phys. Lett.* 438: 279–84.
80. Huang, J.; Vongehr, S.; Tang, S.; Lu, H.; Shen, J.; and Meng, X. 2009. Ag dendrite-based Au/Ag bimetallic nanostructures with strongly enhanced catalytic activity. *Langmuir.* 25: 11890–96.
81. Li, Y.; Li, C.; Cho, S. O.; Duan, G.; and Cai, W. 2007. Silver hierarchical bowl-like array: Synthesis, superhydrophobicity, and optical properties. *Langmuir* 23: 9802–07.
82. Lee, Y. J.; Lee, Y.; Oh, D.; Chen, T.; Ceder, G.; and Belcher, A. M. 2010. Biologically activated noble metal alloys at the nanoscale: For lithium ion battery anodes. *Nano Lett.* 10: 2433–40.

83. Wang, C.-C. 2008. Surfaced-enhanced Raman scattering-active substrates prepared through a combination of argon plasma and electrochemical techniques. *J. Phys. Chem. C.* 112: 5573–78.
84. Hirai, Y.; Yabu, H.; Matsuo, Y.; Ijiro, K.; and Shimomura, M. 2010. Arrays of triangular shaped pincushions for SERS substrates prepared by using self-organization and vapor deposition. *Chem. Commun.* 46: 2298–300.
85. Banholzer, M. J.; Millstone, J. E.; Qin, L.; and Mirkin, C. A. 2008. Rationally designed nanostructures for surface-enhanced Raman spectroscopy. *Chem. Soc. Rev.* 37: 885–97.
86. Cherevko, S.; Xing, X.; and Chung, C.-H. 2010. Electrodeposition of three-dimensional porous silver foams. *Electrochem. Commun.* 12: 467–70.
87. He, H.; Cai, W.; Lin, Y.; and Chen, B. 2010. Au nanochain-built 3D netlike porous films based on laser ablation in water and electrophoretic deposition. *Chem. Commun.* 46: 7223–25.
88. Gutes, A.; Carraro, C.; and Maboudian, R. 2010. Silver dendrites from galvanic displacement on commercial aluminum foil as an effective SERS substrate. *J. Am. Chem. Soc.* 132: 1476–77.
89. Zhang, Q. and Zhang, Z. 2010. On the electrochemical dealloying of Al-based alloys in a NaCl aqueous solution. *Phys. Chem. Chem. Phys.* 12: 1453–72.
90. Lee, S. J. and Kim, K. 2003. Development of silver film via thermal decomposition of layered silver alkanecarboxylates for surface-enhanced Raman spectroscopy. *Chem. Commun.* 2: 212–3.
91. Jin, R. H. and Yuan, J. J. 2005. Fabrication of silver porous frameworks using poly(ethyleneimine) hydrogel as a soft sacrificial template. *J. Mater. Chem.* 15: 4513–17.
92. He, H.; Cai, W.; Lin, Y.; and Dai, Z. 2011. Silver porous nanotube built three-dimensional films with structural tunability based on the nanofiber template-plasma etching strategy. *Langmuir* 27: 1551–55.
93. Duan, G.; Lv, F.; Cai, W.; Luo, Y.; Li, Y.; and Liu, G. 2010. General synthesis of 2D ordered hollow sphere arrays based on nonshadow deposition dominated colloidal lithography. *Langmuir* 26: 6295–302.
94. Li, Z.; Tong, W. M.; Stickle, W. F.; Neiman, D. L.; Williams, R. S.; Hunter, L. L.; Talin, A. A.; Li, D.; and Brueck, S. R. J. 2007. Plasma-induced formation of Ag nanodots for ultra-high-enhancement surface-enhanced Raman scattering substrates. *Langmuir* 23: 5135–38.

5 Structurally Enhanced Photocatalysis Properties

5.1 INTRODUCTION

In Chapters 2 through 4, we have introduced several methods for the fabrication of micro/nanostructured materials. These methods are simple, flexible, controllable, and hence suitable for the mass production of micro-/nanostructured materials with good reproducibility, under benign conditions and a friendly environment. These micro/nanostructured materials are composed of microsized objects with nanostructures. Because of their special architectures, they possess the advantages of both nanosized and microscaled materials, that is, they have not only the high surface activity and high specific surface area of nanomaterials but also the antiaggregation property, stable structure, and easy separation property of microsized materials. More importantly, these materials exhibit strong structurally enhanced performances, such as significantly enhanced adsorption and catalysis performances especially for semiconducting oxides with micro/nanostructures, and hence can be used for the highly efficient removal of contaminants in water. Here, the phrase "structurally enhanced properties" means that material performance is significantly enhanced by structural tuning, or some performances of micro/nanostructured materials, due to their special structures, are much better than those of the same materials with normal structures. In this chapter, we introduce the photocatalytic properties of some typical oxide materials with special micro/nanostructures to demonstrate their structurally enhanced performances.

5.2 ZnO-BASED MICRO/NANOARCHITECTURES

An important application of ZnO is as a photocatalyst in environmental protection [1–5]. Through the photocatalytic generation of hydrogen peroxide [6,7], ZnO can be utilized for the degradation of organic pollutants in nearly neutral solutions and the sterilization of bacteria and viruses [8,9]. Recently, semiconductor photocatalysts of nanometer scale have become very attractive due to their different physical and chemical properties from bulk materials [10,11]. Since photocatalytic reaction occurs at surfaces, the nanosized semiconductor will increase the decomposition rate due to the increased surface area. However, it should be mentioned that nanoscaled objects (such as nanoparticles, nanorods, and nanosheets) tend to aggregate during aging, which results in an unwanted reduction in active surface area. A practical method to prevent the particles from aggregating is to immobilize the photocatalysts in the form of a thin film on a substrate [5,6]. In this case, however, the efficiency is significantly lower than that of the corresponding suspensions. Therefore, nanostructured ZnO, which is stable against aggregation and possesses a higher surface-to-volume ratio, is required in its environmental remediation applications. It has been confirmed that the ZnO with micro/nanostructures can overcome the aforementioned shortcomings, showing both high surface-to-volume ratio and stability against aggregation. Such structured ZnO has shown strong structure-induced enhancement of photocatalytic performance and exhibited much better photocatalytic property and durability than those exhibited by other nanostructured ZnO powders, such as the powders of nanoparticles, nanosheets, and nanoneedles.

5.2.1 Nanoplate-Built Core–Shell-Structured ZnO Objects

Nanoplate-built core–shell-structured ZnO objects are shown in Figure 2.12. Such a micro/nanoarchitecture is composed of microsized conic-like particles, which are built by many alternating nanosheets with a thickness of 10 nm as a highly branched nanoarchitecture standing on hexagonal pyramid–like core microcrystals, showing high surface-to-volume ratio and stability against aggregation [12].

Photocatalytic property was measured for the reaction solution, in a flask, containing 60 mL methyl orange (MO) aqueous solution, with an initial concentration of 5.0×10^{-5} M and a pH of 6.4, and 30 mg as-synthesized micro/nanostructured ZnO as catalyst. The solution was magnetically stirred thoroughly in the dark to reach the adsorption equilibrium of the MO on the catalyst before exposure to ultraviolet (UV) light irradiation from a 60 W low-pressure mercury lamp (254 nm, 5 cm in distance from the flask). Optical absorption spectra for the UV-irradiated solution were recorded to evaluate the evolution of the MO content in the solution.

5.2.1.1 Structural Dependence of Photocatalytic Activity

The corresponding photocatalytic activity in the degradation of the well-known organic azo dye MO, a typical pollutant in the textile industry, was evaluated. Figure 5.1 shows the optical absorption spectra of the MO aqueous solution with the as-prepared micro/nanostructured ZnO powders shown in Figure 2.12 after exposure to UV light for different periods. The main absorption peak at 464 nm, corresponding to the MO molecules, decreases rapidly with extension of the exposure time and completely disappears after about 80 minutes. If there is further exposure, no absorption peak was observed in the whole spectrum, indicating the complete decomposition of MO. It is clear that the intense orange color of the starting solution gradually disappears with increasing exposure time to the UV light.

To demonstrate the structure-induced enhancement of photocatalytic performance of micro/nanostructured ZnO, corresponding experiments were performed for some other nanostructured ZnO powders (nanoneedles, nanosheets, and commercial nanoparticles) and the commercial photocatalyst Degussa P25 titania, under the same conditions, for reference. Figure 5.2 shows the results of the MO normalization concentration in the solution versus the exposure time from the optical absorbance measurements at 464 nm for all samples. Without any catalyst, only an insignificant decrease in the concentration of MO was detected after UV irradiation (curve a in Figure 5.2). Addition of catalysts leads to obvious degradation of MO,

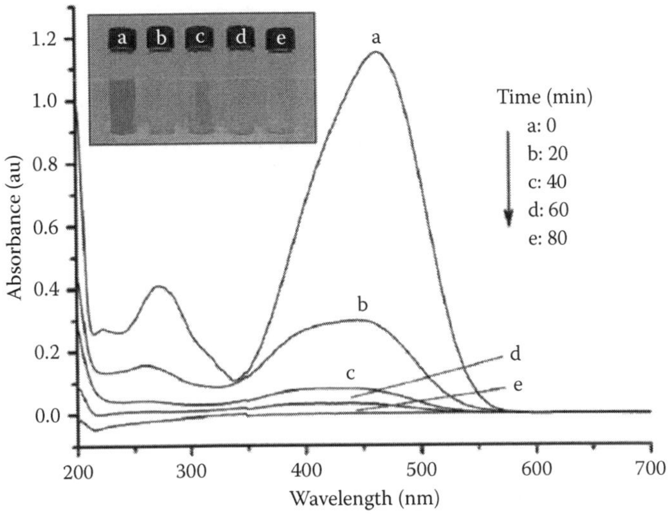

FIGURE 5.1 Time-dependent optical absorbance spectra for methyl orange solution with a starting concentration of 5.0×10^{-5} M (60 mL) in the presence of micro/nanoarchitectured ZnO (30 mg) after its exposure to ultraviolet light for different periods. Inset: corresponding time-dependent color change. (Lu et al., *Adv. Funct. Mater.* 2008. 18. 1047–56. Copyright Wiley-VCH Verlag GmbH & Co. KGaA. Reproduced with permission.)

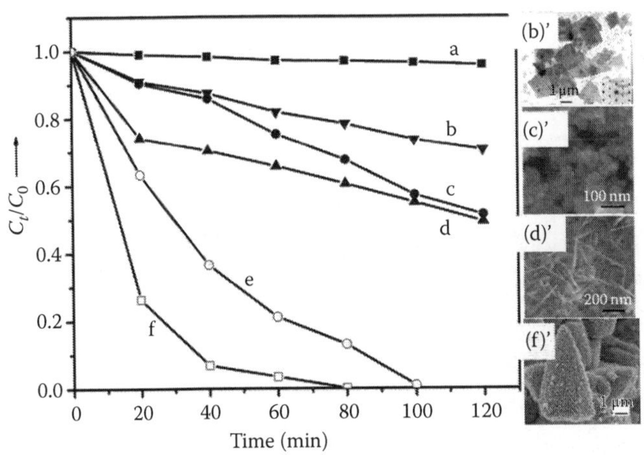

FIGURE 5.2 Methyl orange (MO) normalization concentrations (from optical absorbance measurements at 464 nm) in the solution (60 mL) with different catalysts (30 mg) versus exposure time to ultraviolet light. Starting MO concentration is $C_0 = 5.0 \times 10^{-5}$ M. (a) Without any catalyst, (b) with ZnO nanosheet powders, (c) with ZnO nanoparticle powders, (d) with ZnO nanoneedle powders, (e) with Degussa P25 titania powders, and (f) with micro/nanoarchitectured ZnO. The images on the right, (b)', (c)', (d)', and (f)', are the morphologies corresponding to the catalysts in (b), (c), (d), and (f). (Lu et al., *Adv. Funct. Mater.* 2008. 18. 1047–56. Copyright Wiley-VCH Verlag GmbH & Co. KGaA. Reproduced with permission.)

and the photocatalytic activity depends on their morphology. The activity increases in turn for the nanostructured ZnO powders, nanosheets (curve b), nanoparticles (curve c), and nanoneedles (curve d), but it is still lower than that of Degussa P25 titania (curve e). For the ZnO with the micro/nanostructure, however, the activity is much higher than that of the reference ZnO and even Degussa P25 (curve f). The MO solution was completely decolorized by using the micro/nanostructured ZnO after UV irradiation for 80 minutes.

5.2.1.2 Durability of Photocatalytic Activity

In addition, durability of photocatalytic activity was inspected by the reuse of catalysts in fresh MO solution under UV light irradiation. Figure 5.3 demonstrates the photodegradation results within three cycles for micro/nanostructured ZnO, Degussa P25 titania, and ZnO nanoneedles (80-minute irradiation for each cycle). For Degussa P25 titania, there is no significant change in the activity even after reuse for three times, exhibiting the best durability. Comparatively, the micro/nanostructured ZnO shows slight decrease in photocatalytic activity but more stability than ZnO nanoneedles under UV irradiation. The decoloration percentage of MO solution decreased only from 100% to 95% after three cycles.

5.2.1.3 Structure-Induced Enhancement of Photocatalytic Activity and Durability

The photocatalytic superiority of micro/nanostructured ZnO over other nanostructured ZnO powders is easily understood. It can be attributed to their special structural features. The micro/nanostructured ZnO possesses a higher specific surface area

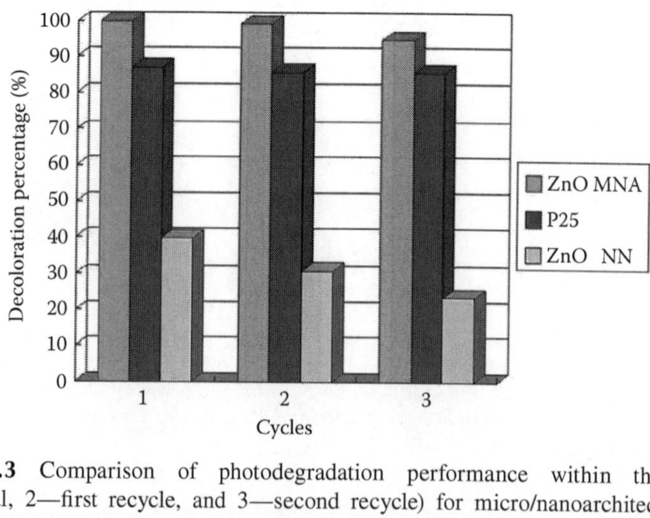

FIGURE 5.3 Comparison of photodegradation performance within three cycles (1—original, 2—first recycle, and 3—second recycle) for micro/nanoarchitectured ZnO (ZnO MNA, green), Degussa P25 titania (P25, blue), and ZnO nanoneedles (ZnO NN, pearl blue). (Lu et al., *Adv. Funct. Mater.* 2008. 18. 1047–56. Copyright Wiley-VCH Verlag GmbH & Co. KGaA. Reproduced with permission.)

(ca. 185 $m^2 \cdot g^{-1}$) than the reference samples (Degussa P25 powders: ca. 50 $m^2 \cdot g^{-1}$; the commercial ZnO nanoparticles: ca. 30.6 $m^2 \cdot g^{-1}$) [12], which is beneficial to the enhancement of photocatalytic performance. On the other hand, the vertical and net-like or grid-like arrangement of nanosheets, as well as the conic shape of the ZnO micro/nanostructured objects, prevents aggregation effectively and thus keeps a large active surface area. Comparatively, for the reference samples, such as Degussa P25 TiO_2 nanopowders, the unwanted aggregation during the reaction usually induces a significant decrease in the active surface area and thus lowers the photocatalytic performance [9,13]. Additionally, the unique structure of the micro/nanostructured ZnO may also increase the photocatalytic efficiency. The nanosheet thickness of 10 nm is close to the region where quantum size effect is prominent. The nanosheets' bandgap broadening induced by quantum size effect would not only bring higher redox potentials but also promote the transfer of electrons from the conductive bands of nanosheets with high electric potential to those of the core-part micropyramid with low electric potential. Then the probability of photogenerated electron–hole pair recombination could be reduced, which in turn enhances charge-transfer rates in the materials [9]. Thus, good photocatalytic performance is obtained.

As for photocatalytic durability, the decrease in photodegradation efficiency of the recycled ZnO catalysts can be attributed to the photocorrosion rather than pH-dependent dissolution (chemical corrosion) of ZnO, since the experiments were conducted in a nearly neutral solution (pH = 6.4). Photocorrosion is a main obstacle for the working of photocatalysts such as ZnO and CdS but not TiO_2 [14,15]. Here, the micro/nanostructured ZnO shows higher photocatalytic durability than ZnO nanoneedles, despite having lower durability than Degussa P25 titania, under UV light irradiation. The better single-crystalline nature of the synthesized micro/nanostructured ZnO would contribute to their good durability [16].

Based on the aforementioned, the advantages of such micro/nanostructured ZnO powders as photocatalysts compared with other nanostructured ZnO powders (nanoparticles, nanoneedles, and nanosheets) are as follows: (1) a high surface-to-volume ratio with effective prevention of further aggregation to keep the high-catalytic activity area due to the grid-like structure; (2) easier to separate and recycle than common nanocrystals due to the larger sizes of the ZnO particles; (3) higher redox potentials of size-quantized nanosheets standing on the microconic particles as a result of an increase in bandgap energy, which in turn enhances the charge-transfer rates in the system and reduces the nonradiation recombination of electron–hole pairs.

5.2.2 Noble Metal Cluster–Embedded ZnO Composite Hollow Spheres

ZnO hollow nanoparticles (HNPs) and noble metal cluster–embedded ZnO HNPs were fabricated by laser ablation in liquid (LAL) and subsequent selective etching, as described in Section 3.4 and shown in Figures 3.22 and 3.26. The photocatalytic experiments were performed for a 20 mL aqueous solution with 10^{-5} M MO and 10 mg catalyst powders under irradiation by a UV lamp 125 W in power and 365 nm in wavelength. The photocatalytic performance was evaluated by monitoring the optical absorption peak of MO at 464 nm, as mentioned in Section 5.2.1.

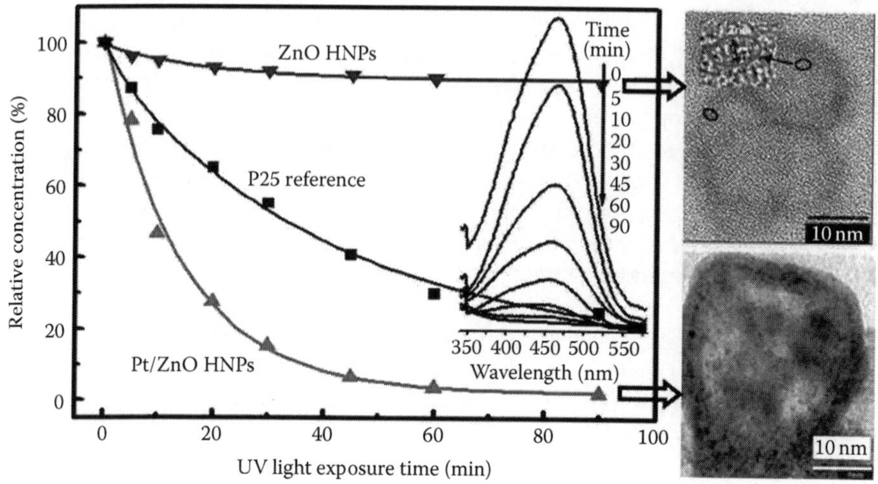

FIGURE 5.4 Methyl orange photodegradation rate of pure ZnO and Pt/ZnO hollow nanoparticles compared with reference P25 powders. (Reprinted with permission from Zeng et al., 2008, 1661–70. Copyright 2008 American Chemical Society.)

5.2.2.1 Photocatalytic Activity

Figure 5.4 shows the photocatalysis efficiencies of typical ZnO HNPs, in the degradation of MO, compared with a conventional TiO_2 reference sample (P25). For pure ZnO HNPs, relatively poor photodegradation ability can be seen. However, for Pt cluster–embedded ZnO HNPs (or Pt/ZnO HNPs for short) 90% of the MO can be degraded in 40 minutes in this experimental case (the Au/ZnO and Au–Pt/ZnO HNPs are similarly effective), which is faster than that of the P25 reference [17]. Such an improvement can be attributed to the several outstanding features of Pt cluster–embedded ZnO HNPs, including small sizes of noble metal clusters and shell thickness of ZnO, high surface-to-volume ratio, and effective electron–hole separation of Schottky barriers [18–20]. Also, the whole-surface contact of embedded noble metal ultrafine particles to the ZnO matrix plays an important role, which adequately acts as metal–semiconductor heterojunctions, facilitates charge transfer, and hence improves photocatalysis efficiency.

5.2.2.2 Photostability

One of the biggest problems in the photocatalysis of ZnO is its photostability, which is related to crystallinity, charge transfer, and so on. Therefore, to be a good photocatalyst it has to be stable and reusable. Since in the noble metal–semiconductor system, metal acts as a reservoir for electrons, and holes will move to surface of the semiconductors and can thus oxidize the organic dyes. Electrons at the surface can reduce Zn^{2+} to Zn, namely, photocorrode ZnO. But in the case of noble metal cluster–embedded ZnO, electrons are confined to the metal part; then, the photocorrosion should be depressed.

The aforementioned great improvement in photocatalysis efficiency of Pt/ZnO over pure ZnO HNPs could also be attributed to its photostability. The high disorder degree of pure ZnO HNPs greatly favors the photocorrosion effect, resulting in low photocatalysis efficiency. However, the ZnO matrix in Pt/ZnO HNPs is of high crystallinity; thus, the photocorrosion effect should be insignificant. Therefore, the sacrificial template–selective etching strategy (see Section 3.4) has opened a simple, effective, and low-cost process for the fabrication of noble metal–semiconductor interface-abundant photocatalysts and has good universality, such that it can be extended to noble metal–TiO_2 systems.

5.2.2.3 Pt/ZnO Porous Shells

It has been found that combining the selective etching strategy described in Section 3.4 the porous shell can be produced by sonication-assisted selective etching of Zn/ZnO core–shell nanoparticles, which were prepared by LAL [21]. Figure 5.5 presents the typical results of Pt/ZnO porous shells by sonication-assisted selective etching. The transmission electron microscopy (TEM) image shows that, after etching, these core–shell nanoparticles have become hollow, but their shell surfaces are obviously porous. No separated Pt nanoparticles are found, which indicates that Pt has been incorporated into the porous ZnO nanoshells during the etching process,

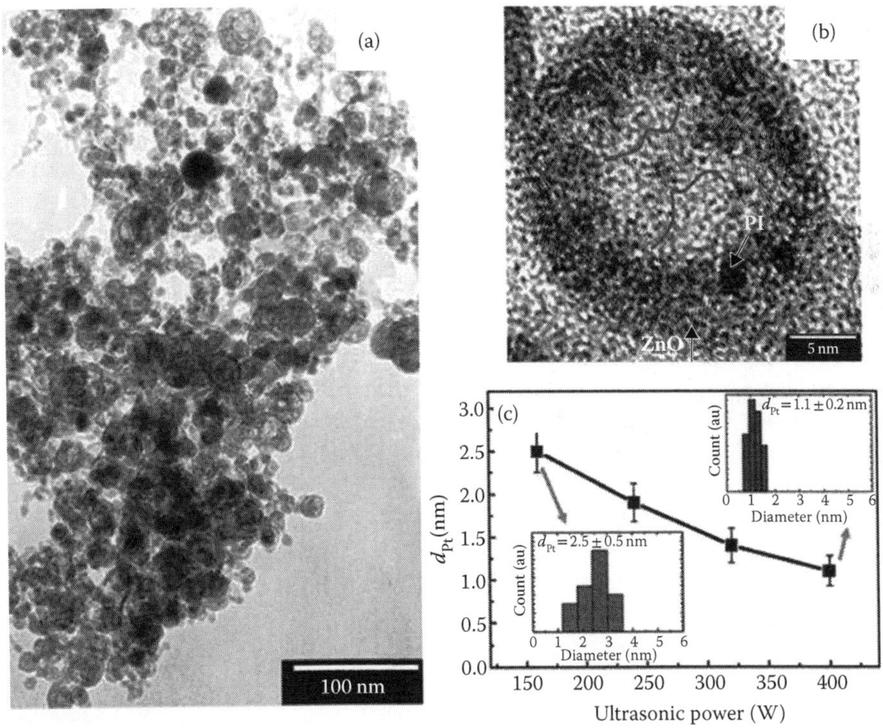

FIGURE 5.5 Pt/ZnO porous nanocages by sonication-assisted selective etching: (a) transmission electron microscopy (TEM) image with low magnification, (b) high-resolution TEM image of a single nanocage, and (c) the size dependence of Pt clusters on the power of ultrasonic irradiation. (Reprinted with permission from Zeng et al., 2008, 19620–4. Copyright 2008 American Chemical Society.)

forming Pt/ZnO porous nanocages. The microstructure of Pt/ZnO porous nanocages is revealed by the high-resolution TEM image in Figure 5.5b. The Pt clusters are embedded in the ZnO matrix (shell). On the one hand, the ZnO shell matrix was found to be crystalline according to their clear lattice fringes. In the central part of the two-dimensional projection of this particle, ZnO crystals are observed with a strap region to continuously joint the opposite shell walls (forming the hollow framework), but two blanks exist in its two side regions as the curves marked (forming the surface holes), which indicates formation of the porous cage. On the other hand, the Pt clusters (see the dashed arrow in Figure 5.5b) with higher contrast due to their higher electron density than the ZnO matrix (see the solid arrow) can be easily seen in a large quantity. Correspondingly, the lattice fringes with spacing of 0.22 and 0.28 nm can be identified for Pt (111) and ZnO (100) planes, respectively. Furthermore, it was found that the sizes of Pt clusters are dependent on the power of ultrasonic irradiation. With the increase of ultrasonic power from 160 to 400 W, the average diameter of Pt clusters decreases from 2.5 to 1.1 nm and the size distribution becomes narrower, showing high controllability.

The formation of Pt/ZnO porous nanocages is the result of the H_2PtCl_6–$C_4H_6O_6$ two-step weak acid etching process. The H_2PtCl_6 etching induces the hollow cores and Pt cluster-embedded nanoshells, and the $C_4H_6O_6$ etching forms porous nanoshells. For the former effect, composite nanoshells could be formed by a diffusion–redox–deposition process during etching. Besides the weak reaction of H^+ with the ZnO shells, the reaction between Pt^{4+} and Zn cores will take place. Pt^{4+} ions diffuse through the ZnO shells to react with Zn in the core parts, leading to the deoxidization of Pt^{4+} ions and the oxidization of Zn. The deoxidized Pt^0 is precipitated in the ZnO shells, resulting in embedded ultrafine Pt nanoparticles, whereas the oxidized Zn^{2+} diffuses out, resulting in void cores. In fact, here, the Zn/ZnO core/shell nanoparticles play the role of sacrificial templates with nanoscale, but the synchronous reaction–diffusion behavior of Pt is, to some extent, similar to the nanoscale Kirkendall effect [22], in which the two-way diffusion takes place due to existence of the material density gradient. Ultrasonic irradiation plays an important role in providing the driving force and the dispersing function. The increased ultrasonic power would increase the reaction rate and sites. The corresponding size reduction of Pt nanoparticles could be ascribed to the ultrasonic irradiation–induced more homogeneous occurrence of the diffusion–redox–deposition process in more diffusion channels, which reduces the ion number involved in each channel, and hence the size of the precipitated Pt nanoparticles. Finally, $C_4H_6O_6$-induced shell opening is the result of the weak reaction of a weak acid with ZnO due to its amphoteric feature. It was found that the stronger acid induces rapid cataclysm of ZnO shells, but the weaker acid cannot induce shell opening.

From the aforementioned characterizations and analyses, two outstanding features of these nanocages are worthy of being noticed. On the one hand, considering the small size and high density of these Pt clusters, without doubt, great masses of metal–semiconductor (Pt/ZnO) interfaces have been constructed in the final composite nanocages. In other words, abundant Schottky barriers have been formed, and these nanocages have thus strong ability to separate the photo-generated charges. Moreover, the embedding state, rather than the usual surface attachment,

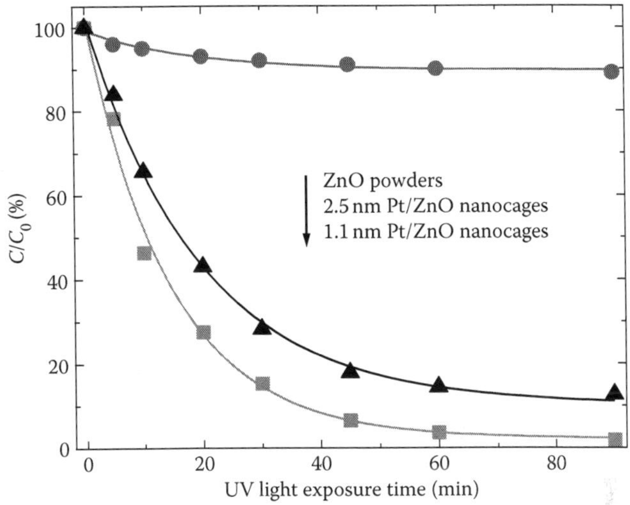

FIGURE 5.6 Methyl orange photodegradation rate of ZnO powders and Pt/ZnO porous nanocages with Pt cluster diameters of 2.5 and 1.1 nm. (Reprinted with permission from Zeng et al., 2008, 19620–4. Copyright 2008 American Chemical Society.)

would further improve the effective interfaces and favor this separating ability. On the other hand, the hollow and porous structure would greatly increase the specific surface area. Thus, these Pt/ZnO porous nanocages could have remarkable performance in charge separation–favoring fields, such as photocatalysis.

The results of photocatalytic degradation of MO using Pt/ZnO nanocages are illustrated in Figure 5.6, compared with commercial ZnO powders (with particle sizes from 500 to 2000 nm). Obviously, the Pt/ZnO nanocages have much better photocatalytic activity than the common ZnO powders. Furthermore, the photocatalytic efficiency of the nanocages slightly increases with the reduction of Pt cluster size. Especially for 1.1 nm Pt/ZnO nanoshells, MO was quickly degraded about 90% in 40 minutes under this experimental condition [21]. As mentioned earlier, such an improved photocatalytic activity can be attributed to the more effective electron–hole separation and larger specific surface area due to the special structure of these Pt/ZnO porous nanocages.

5.2.3 Ag Nanoparticle–Decorated Nanoporous ZnO Microrods

The photocatalytic efficiency of ZnO can be effectively enhanced by surface modification with noble metal nanoparticles [21,23–28]. The enhanced photocatalytic performances can be attributed to the special metal–semiconductor interface, allowing for the establishment of the Schottky barrier to facilitate the charge separation [29,30]. This may also be attributed to the enhanced adsorption capability and kinetic properties introduced by the specific interactions at the metal–semiconductor interface [31] and increased productivity of active hydroxyl radicals due to the high surface concentration of hydroxyls resulting from the interaction of metal nanoparticles with the semiconductors [18]. There are some reports of the ZnO surface being modified with different noble metals in nanoparticle forms to improve the photocatalytic activity [21,23–28]. Among them, Ag nanoparticle–modified ZnO has shown significantly improved photocatalytic degradation performance toward organic contaminants [18,27,28,32–36]. Ag nanoparticles have been successfully decorated on ZnO with different geometries/morphologies such as nanofibers [33], rods [27,37,38], and nanospheres [18,32]. All resultant Ag nanoparticle–decorated ZnO photocatalysts reported to date exhibit an enhanced photocatalytic performance toward the degradation of organic contaminants such as MO, rhodamine B, and orange G [18,27,28,35,36].

Herein, we introduce the enhanced UV and solar light photocatalytic activities and improved stability of uniquely configured Ag nanoparticle–decorated nanoporous ZnO micrometer rods (Ag-NPs/n-ZnO MRs) [39]. Nanoporous ZnO micrometer rods (n-ZnO MRs) composed of ZnO nanoparticles were fabricated via a facile solvothermal-assisted heat treatment method, as previously reported in the literature [40,41]. Ag nanoparticles were decorated on the n-ZnO MRs by solar light photoreduction of Ag ions [39]. Figure 5.7 shows the typical morphology and microstructure of Ag-NPs/n-ZnO MRs. The sample consists of rods, about 90–150 nm in diameter and about 0.5–3 μm in length (Figure 5.7a). TEM examination showed that the product had a typical nanoporous rodlike structure. The Ag nanoparticles

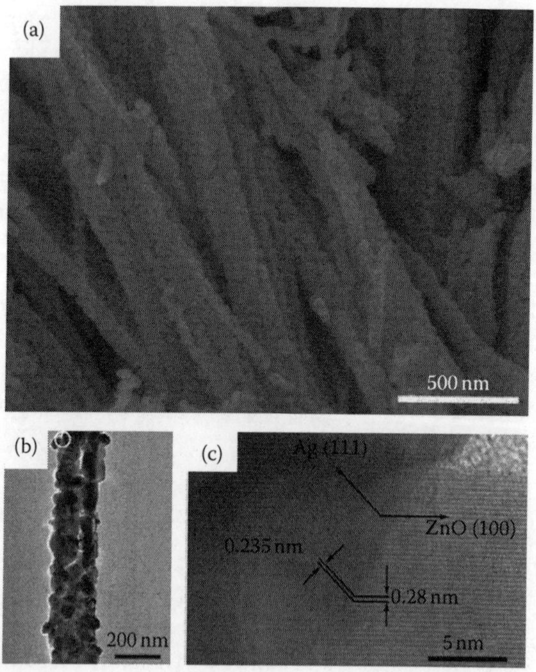

FIGURE 5.7 Morphology and microstructure of Ag nanoparticle–decorated nanoporous ZnO micrometer rods (Ag-NPs/n-ZnO MRs) with Ag loading at 1.66 %wt: (a) scanning electron microscopy image, (b) TEM image of a single microrod attached to Ag nanoparticles, and (c) high-resolution TEM image of the area marked with a circle in (b). (Reprinted with permission from Deng et al., 2012, 6030–7. Copyright 2012 American Chemical Society.)

with a diameter of 20–50 nm were found to attach to the surface of n-ZnO MRs (Figure 5.7b). A high-resolution TEM image at the Ag–ZnO interface has revealed an interplanar spacing of about 0.28 nm, corresponding to the (100) plane of ZnO, whereas an interplanar spacing of 0.235 nm could be assigned to the (101) plane of Ag (Figure 5.7c). Further, it has been revealed, by nitrogen adsorption/desorption isothermal measurements, that the samples were mesoporous with an obvious bipore size distribution mainly around 3–30 nm and a specific surface area of 21 $m^2 \cdot g^{-1}$ [39].

5.2.3.1 Photocatalytic Activity

Photocatalytic performances of Ag-NPs/n-ZnO MRs in UV and solar light regions were investigated via the degradation of methylene blue (MB), which is a typical cationic organic pollutant usually discharged by the textile industry after use. In the experiments, the micrometer rods (20 mg) were added to 80 mL of 1.25×10^{-5} $mol \cdot L^{-1}$ MB solution. A 300 W high-pressure mercury lamp with maximum emission at 365 nm was used as the UV light source. A 500 W xenon lamp with main wavelength in 365–720 nm was used as the solar light source for sunlight photocatalysis. The quantitative determination of MB was performed by monitoring the intensity change of its optical absorption peak [39].

Figure 5.8a shows the photocatalytic activity and kinetics of as-prepared Ag nanoparticles, n-ZnO MRs, and Ag-NPs/n-ZnO MRs for degradation of MB under UV and solar light irradiation, in which the result of a control solution without any catalyst is also included for comparison. The self-degradation of MB (when no catalyst was used) was less than 17% under high-pressure mercury-lamp irradiation for 25 minutes. It was observed that the degradation of MB by using only Ag nanoparticles was insignificant, and Ag-NPs/n-ZnO MRs displayed higher photocatalytic efficiency than n-ZnO MRs. It took almost 20 minutes for n-ZnO MRs to completely degrade MB, whereas only 10 minutes were needed when Ag-NPs/n-ZnO MRs, with 1.2–2.2 %wt in Ag loading, were used. The photocatalytic degradation kinetics could be described by pseudo-first-order kinetics [28,32], or

$$\ln\frac{C_0}{C} = kt \tag{5.1}$$

where k is a pseudo-first-rate kinetic constant and t is the irradiation time. C_0 and C are the MB concentrations in the solution before and after irradiation for t, respectively. Thus, the photodegradation of MB could be considered as a pseudo-first-order reaction in kinetics. The slope of the linear plot should correspond to the rate constant k. The variations in $\ln(C_0/C)$ as a function of irradiation time are given in Figure 5.8b. The k value of the Ag-NPs/n-ZnO MRs was more than double of that for the n-ZnO MRs.

Under solar light irradiation, the corresponding results are demonstrated in Figure 5.8c and d. The self-degradation of MB was less than 10% in the entire irradiation process. The Ag-NPs/n-ZnO MRs with an Ag loading of 1.7 %wt exhibited the best photocatalytic performance, and its k value was 5.6 times larger than that of n-ZnO MRs. The photocatalytic efficiency and k value was ever-increasing with rise of the Ag loading from 0 to 1.7%wt, but decreased if the Ag loading is higher than 1.7%wt. These results further demonstrate that the amounts of Ag nanoparticles had a significant impact on

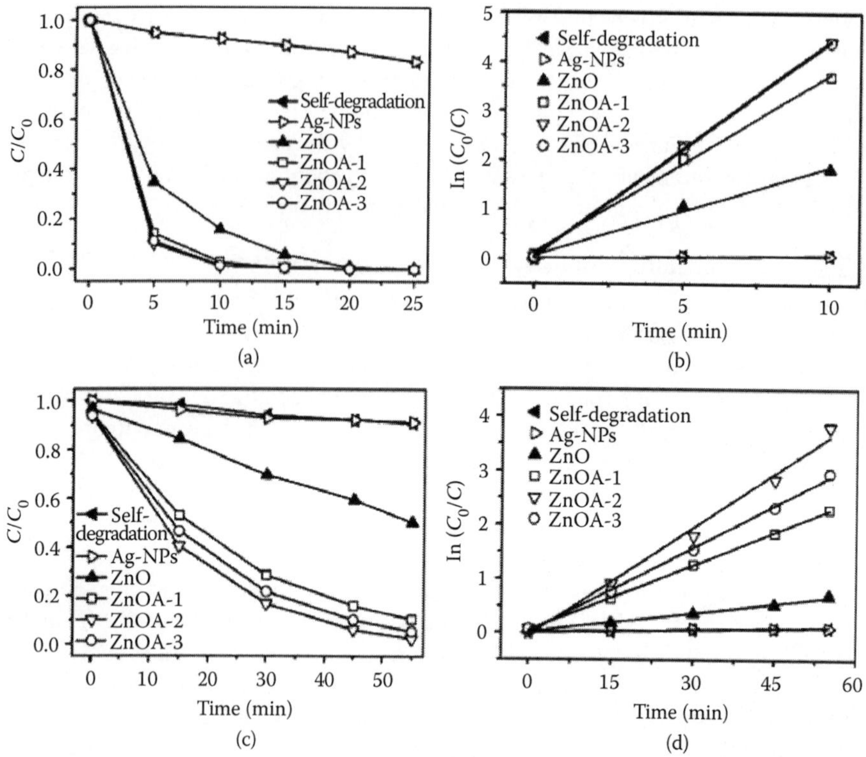

FIGURE 5.8 Photocatalytic activity and kinetics of as-prepared Ag nanoparticles (Ag-NPs), nanoporous ZnO micrometer rods (n-ZnO MRs), and Ag nanoparticle–decorated nanoporous ZnO micrometer rods (Ag-NPs/n-ZnO MRs) for the degradation of methylene blue: (a) and (b) under ultraviolet irradiation and (c) and (d) under solar light irradiation. ZnOA-1, ZnOA-2, and ZnOA-3 correspond to Ag-NPs/n-ZnO MRs with Ag loading amounts of 1.2, 1.7, and 2.2 %wt, respectively. (Reprinted with permission from Deng et al., 2012, 6030–7. Copyright 2012 American Chemical Society.)

Ag-NPs/n-ZnO MRs' photocatalytic activity. When the Ag loading was optimal, thermodynamically, the Ag nanoparticles could act as an electron well. The electrons on the surface of n-ZnO MRs could effectively move toward Ag nanoparticles. However, some Ag nanoparticles might act as recombination centers and inhibit the photocatalytic activity of Ag-NPs/n-ZnO MRs, which is similar to previous reports [42].

5.2.3.2 Photostability

Further, the stability of photocatalytic performance in UV and solar light regions was studied. The Ag-NPs/n-ZnO MRs with 1.7 %wt Ag loading were used to degrade MB dye in five repeated cycles. Figure 5.9 shows the corresponding results. The photocatalytic performance of Ag nanoparticle–modified ZnO exhibited effective photostability under UV (Figure 5.9a) and solar light irradiation (Figure 5.9c), where the photocatalytic efficiency reduced only by 1.45% and 1.91% after five cycles, respectively. Two possible reasons were responsible for the favorable photostability,

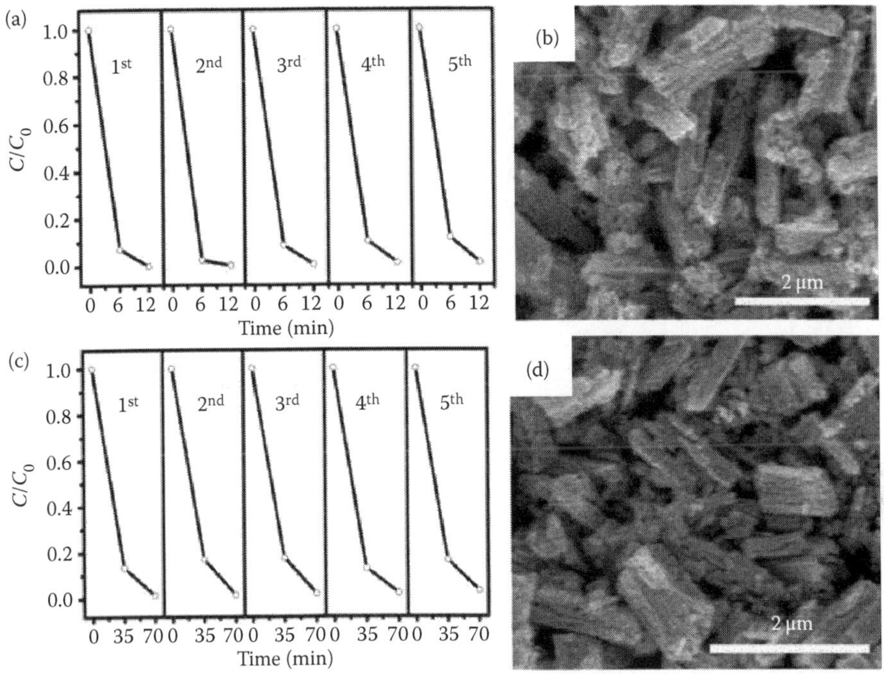

FIGURE 5.9 Five photocatalytic degradation cycles of methylene blue using Ag nanoparticle–decorated nanoporous ZnO micrometer rods (Ag-NPs/n-ZnO MRs) with 1.7 %wt in Ag loading as catalyst, and field emission scanning electron microscopy (FESEM) images of the catalyst after five cycles: (a) and (b) under ultraviolet light irradiation and (c) and (d) under solar light irradiation. (Reprinted with permission from Deng et al., 2012, 6030–7. Copyright 2012 American Chemical Society.)

as follows: first, the micro/nanostructure photocatalysts possess the advantages of both nanosized (high activity) and micrometer-sized (structural stability) materials [12,43]. It could be observed that the reused Ag-NPs/n-ZnO MRs kept the original structure with a few attached particles (Figure 5.9b and d). Second, the superior crystalline quality of the n-ZnO MRs via annealing at 450°C for 2 hours and the Ag nanoparticles deposited on the defect sites of the ZnO rods could significantly reduce the amount of surface defects in the photocatalyst and effectively inhibit photocorrosion, leading to the improvement of the catalysts in terms of photostability [44,45].

5.2.3.3 Interface-Induced Enhancement of Photocatalytic Activity

There have been extensive reports about the enhanced photocatalytic activity of the photocatalysts modified with noble metals [18,34,42,46–54]. The enhancement of photo-catalytic activity could mainly be associated with the interface between the metal and the semiconductor. The degradation of MB using Ag-NPs/n-ZnO MRs under UV and solar light irradiation could be described by such an interface, as illustrated in Figure 5.10. The enhanced photocatalytic performance in the UV region could be attributed to the formation of Schottky barriers at the interface between the Ag-NPs and the n-ZnO MRs, which facilitated the segregation of charges and

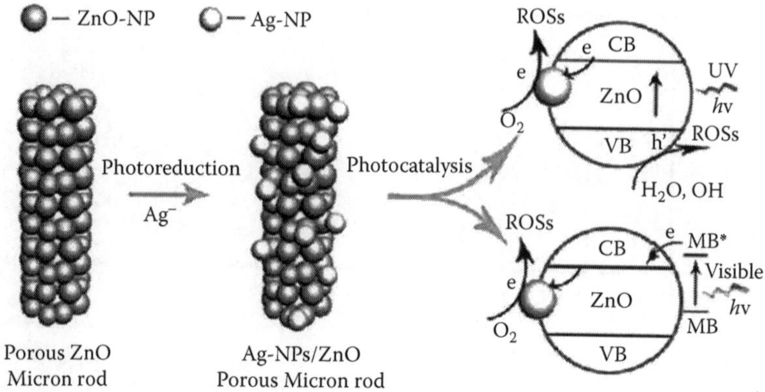

FIGURE 5.10 Schematic illustration of the photosensitized degradation process of methylene blue (MB) using Ag nanoparticle–decorated nanoporous ZnO micrometer rods (Ag-NPs/n-ZnO MRs) as catalyst under ultraviolet (UV) and solar light irradiation. CB, conductance band; ROS, reactive oxidative species; VB, valence band. (Reprinted with permission from Deng et al., 2012, 6030–7. Copyright 2012 American Chemical Society.)

prevented charge recombination [34,42,46,49]. The photocatalysis of Ag-NPs/n-ZnO MRs under UV irradiation involved multiple steps: (1) formation of Schottky barriers at metal–semiconductor interaction, (2) excitation of ZnO micrometer rods by UV light, (3) production of electron–hole pairs, (4) generation of reactive oxidative species (ROSs), and (5) mineralization of the organic compounds by ROSs [18,47]. For the photocatalysis under solar light irradiation, in addition to the enhancement in the UV region, the adsorbed MB would be photoactivated by the visible light (shown in Figure 5.10), which is followed by electron transfer from the excited MB (MB*) to the conduction band of ZnO [47,48,50,52,53]. The electrons on the surface of ZnO were subsequently trapped by the Ag nanoparticles, which separated the $MB^{\bullet +}$ and the electron, preventing the recombination process [54]. Moreover, the surface plasma resonance of Ag nanoparticles excited by the solar light improved excitation of the surface electrons and transfer of the interfacial electrons [42,46,49]. Separated electrons might then be consumed by the oxygen molecules dissolved in the solution to generate various ROSs, promoting photocatalysis [51].

Here, the unique structure of Ag-NPs/n-ZnO MRs is crucial to achieving excellent photocatalytic activity and stability. Both the high-crystalline ZnO nanoparticles of 20 nm and Ag nanoparticles of 20–50 nm are responsible for the high photocatalytic activity, whereas the microstructure of n-ZnO MRs improve the stability [12,55]. Also, the n-ZnO MRs built with ZnO nanoparticles possess larger specific surface areas than those of micrometer rods, which is favorable for the higher photocatalytic efficiency. Additionally, the good crystalline quality of the annealed ZnO rods with significantly reduced surface defect sites would also improve photocorrosion resistance and photostability.

All in all, the loading of Ag nanoparticles on ZnO microsized rods had a significant effect on the photocatalytic efficiency of MB. The results have demonstrated that nanostructured Ag-NPs/n-ZnO MRs exhibit enhanced photocatalytic performance and high stability an can be reused in the degradation of the organic dye MB, which can be a good candidate material for organic pollutants' remediation.

5.3 MICRO/NANOSTRUCTURED α-Fe_2O_3 POROUS SPHERES

Hematite (α-Fe_2O_3) has a bandgap of 1.9–2.2 eV and is a potential semiconductor catalyst of visible light driving [56]. Moreover, compared with other narrow-bandgap semiconductors (CdS, WO_3, $ZnIn_2S_4$, and Bi_2WO_6) [57–60] α-Fe_2O_3 exhibits better environmental compatibility, lower cost, and higher stability against photocorrosion in aqueous solutions [61]. In addition, it is easy to separate α-Fe_2O_3 powders from a solution by using a magnet, due to its magnetic property. In this section, we introduce the photocatalytic activity of spherical micro/nanostructured α-Fe_2O_3 under visible light for the degradation of rhodamine 6G (R6G) to demonstrate its structural enhancement performance.

The micro/nanostructured α-Fe_2O_3 spheres were synthesized by the hydrothermal reaction of a precursor solution containing $FeCl_3 \cdot 6H_2O$, $C_6H_8O_6$, and $CO(NH_2)_2$ at 160°C, and annealing at 500°C [62]. Figure 5.11 shows the typical morphology. α-Fe_2O_3 spheres are 0.5–5 μm in profile size and consist of radialized nanolamella structures. Their high magnification has revealed that the nanolamellas are composed of interlinked and elongated particles with a diameter of about 5–30 nm, which are seen more clearly in the inset of Figure 5.11a. Further, TEM examination showed that the microspheres had a porous structure (Figure 5.11b). This is a typical micro/nanostructure. The specific surface area has been estimated to be about 20 $m^2 \cdot g^{-1}$ [62]. Also, such α-Fe_2O_3 spheres exhibit ferromagnetic behaviors at room temperature, which is similar to some other reports [56,63–66]. The remanent magnetization and coercivity were found at 0.67 $emu \cdot g^{-1}$ and 65 Oe, respectively [62]. The microspheres could thus be easily separated and collected through the magnetic separation process.

5.3.1 Photocatalytic Activity

The optical absorption spectral measurements for α-Fe_2O_3 have revealed that a significant absorption occurred below a wavelength of 600 nm [62]. It means

FIGURE 5.11 The typical morphology of micro/nanostructured α-Fe_2O_3 spheres: (a) FESEM image. Inset: a local magnification. (b) transmission electron microscopy (TEM) image of a single sphere. (Reprinted with permission from Liu et al., 2012, 9704–13. Copyright 2012 American Chemical Society.)

that the micro/nanostructured α-Fe_2O_3 porous spheres under study could function using visible light.

In the study, 40 mg of the as-prepared catalysts were dispersed in 1.25 × 10^{-5} mol·L^{-1} R6G aqueous solution (40 mL, pH = 6.5) with the addition of 0.3 mL H_2O_2 solution (30 %wt) and irradiated by a fluorescent lamp 40 W in power and 400–700 nm in wavelength. The visible light photocatalytic activity of the α-Fe_2O_3 porous spheres was thus evaluated. The degradation process was monitored through the change in intensity of the optical absorption peak of R6G at 526 nm. It has been shown that in the presence of α-Fe_2O_3 porous spheres and under visible light irradiation, the absorption peak of R6G at 526 nm rapidly decreased in intensity and disappeared completely after about 60 minutes. Figure 5.12a shows the comparative photocatalytic results of the micro/nanostructured α-Fe_2O_3 porous spheres, nano- and micron-sized α-Fe_2O_3 particles. The degradation rate of R6G by the α-Fe_2O_3 porous spheres was much faster than the rates of the nano- and micron-sized particles at the same exposure time. After 60 minutes of irradiation, most of the R6G was decomposed in the solution with the α-Fe_2O_3 porous spheres, whereas 10% and 64% of R6G still remained in the solutions with the nano- and micron-sized α-Fe_2O_3 particles, respectively.

The degradation kinetic reaction can also be described by Equation 5.1. Figure 5.12b gives the corresponding results of variations in $\ln(C_0/C)$ as a function of irradiation time. The apparent rate constant k values, which reflect the reaction rate of a photocatalytic process, were obtained by linear fitting. The k value for α-Fe_2O_3 porous spheres was twice that of α-Fe_2O_3 nanoparticles and nearly 12 times larger than that of α-Fe_2O_3 micron-sized particles, indicating that the k value was enhanced by using the micro/nanostructured α-Fe_2O_3 porous spheres.

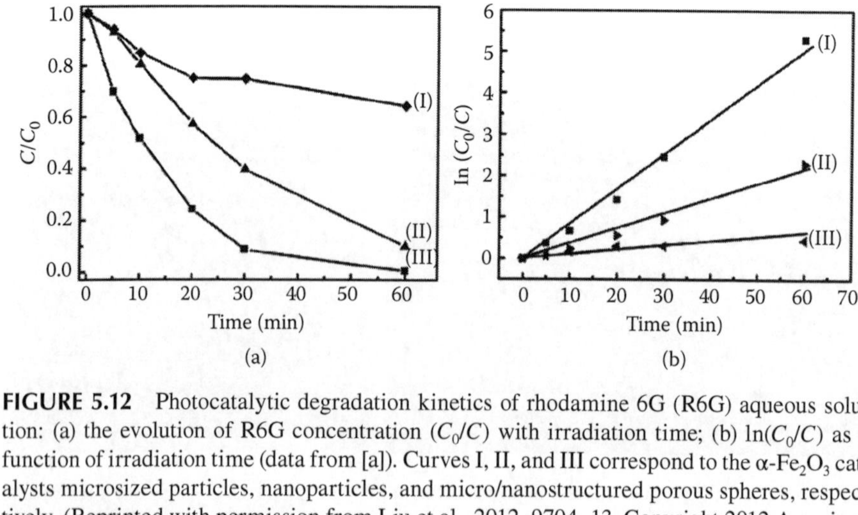

FIGURE 5.12 Photocatalytic degradation kinetics of rhodamine 6G (R6G) aqueous solution: (a) the evolution of R6G concentration (C_0/C) with irradiation time; (b) $\ln(C_0/C)$ as a function of irradiation time (data from [a]). Curves I, II, and III correspond to the α-Fe_2O_3 catalysts microsized particles, nanoparticles, and micro/nanostructured porous spheres, respectively. (Reprinted with permission from Liu et al., 2012, 9704–13. Copyright 2012 American Chemical Society.)

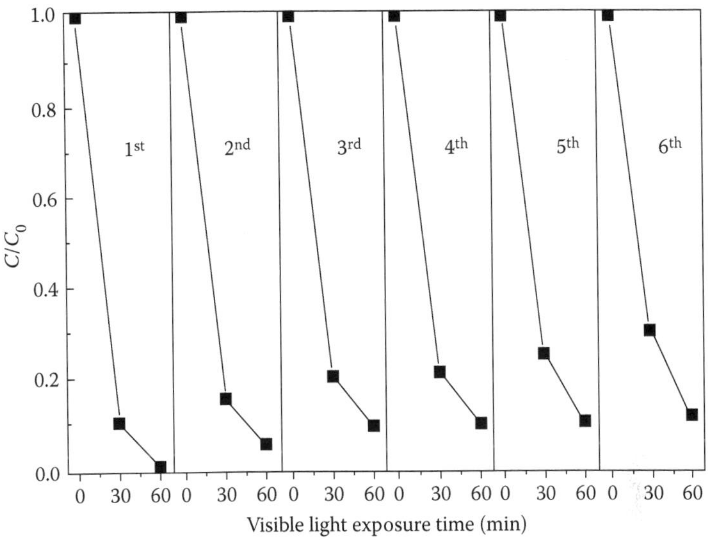

FIGURE 5.13 Photodegradation performance of micro/nanostructured α-Fe_2O_3 porous spheres to 1.25×10^{-5} mol·L^{-1} rhodamine 6G aqueous solutions within six cycles under visible light irradiation. (Reprinted with permission from Liu et al., 2012, 9704–13. Copyright 2012 American Chemical Society.)

5.3.2 Photostability

The photostability of micro/nanostructured α-Fe_2O_3 porous spheres was evaluated by reusing the catalysts in a fresh R6G solution under visible light irradiation. Figure 5.13 shows the recycled experiments of the photodegradation of R6G for six cycles (60 minutes of irradiation for each cycle). It was observed that 100% of R6G was photodegraded when the catalyst was used for the first cycle. After six successive cycles, 90% of R6G was photodegraded; but the photocatalytic activity of the α-Fe_2O_3 porous spheres was almost unchanged, indicating that this catalyst was photostable during the photodegradation of R6G. Here, a special advantage of using this catalyst was that it could be easily separated and collected by an external magnet, which was much easier than the centrifugal separation of nonmagnetic ZnO or TiO_2 nanoparticles from the solution.

5.3.3 Structurally Induced Photocatalytic Enhancement

The high photocatalytic feature of micro/nanostructured α-Fe_2O_3 porous spheres was attributed to the high specific surface area together with their special porous structure. Importantly, these characteristics could control the rate of release of •OH and thus improve the utilization efficiency of •OH. It is well known that •OH plays a key role in the photocatalytic reaction of α-Fe_2O_3 and H_2O_2 [56,67]. The generated •OH exists for a short time (10^{-9} second) [68] and is generally annihilated before it reacts with pollutants. Not all •OH have high effective efficiency. In the photocatalytic reaction, it is important to control the slow release of •OH, improve the efficiency of using •OH, and avoid the annihilation phenomenon. Using micro/nanostructured materials is a feasible method to enhance the efficiency in the use of •OH. It has been indicated, by electron spin resonance spectral measurements, that •OH indeed existed in the photocatalytic degradation reaction of R6G and the micro/nanostructured α-Fe_2O_3 porous spheres with H_2O_2 under visible light irradiation [62]. Further experiments have indicated that the photocatalytic activity of R6G was significantly improved by the cooperation of micro/nanostructured α-Fe_2O_3 porous spheres with H_2O_2 compared to the use of individual micro/nanostructured α-Fe_2O_3 porous spheres or H_2O_2 [62].

There are also two unique virtues of pores in the micro/nanostructured spheres to enhance visible light photocatalytic performance: (1) a hierarchical porous superstructure provides ideal channels for easy and fast diffusion of R6G molecules into the internal surface of the porous spheres, which greatly increases the collision probability between active radicals and R6G molecules, leading to a higher degradation efficiency than other nano- and micron-sized particle powders; (2) the nanopores have a bipore size distribution in the porous spheres [62]. The bigger-pore-size nanopores along the radial are transparent to visible light, which greatly increases the utilizing efficiency of visible light, allowing multiple reflections of visible light within the interior, which facilitates more efficient use of the light source and enhances light harvesting, leading to an increased quantity of available •OH to participate in the photocatalytic degradation of R6G [69–71]. Therefore, the aforementioned results indicate that the degradation of R6G aqueous solution was mainly caused by photocatalytic degradation reaction instead of adsorption on the porous spheres under visible light irradiation, and the micro/nanostructured α-Fe_2O_3 porous spheres showed structurally enhanced visible light photocatalytic activity.

5.4 MICRO/NANOSTRUCTURED $Bi_{0.5}Na_{0.5}TiO_3$

Bismuth sodium titanate (BNT) ($Bi_{0.5}Na_{0.5}TiO_3$), a kind of perovskite (ABO_3-type) structure, is a promising alternative for green ferroelectric ceramics without lead elements [72,73]. The bulk properties of BNT have been extensively studied due to its versatile applications in modern electric devices [74–78]. Nanostructuring of artificial materials could induce novel properties. Therefore, nanoscaled BNT structures, including BNT nanoparticles, nanospheres, and nanocubes, have attracted great efforts [79–81]. There are also a few reports on hierarchical BNT micro/nanostructures [43]. Here, we introduce the photocatalytic performances of hierarchical BNT micro/nanostructures in the photodegradation of MO. In comparison with the spherical and cubic BNT powders, owing to the large surface area, BNT micro/nanostructures show a superior performance.

5.4.1 Morphology and Structure

Hierarchically micro/nanostructured BNT was fabricated based on the in situ self-assembly of a BNT nanocrystal under hydrothermal conditions by treatment of the precursor solution or the citric acid aqueous solution containing $Ti(C_4H_9O)_4$,

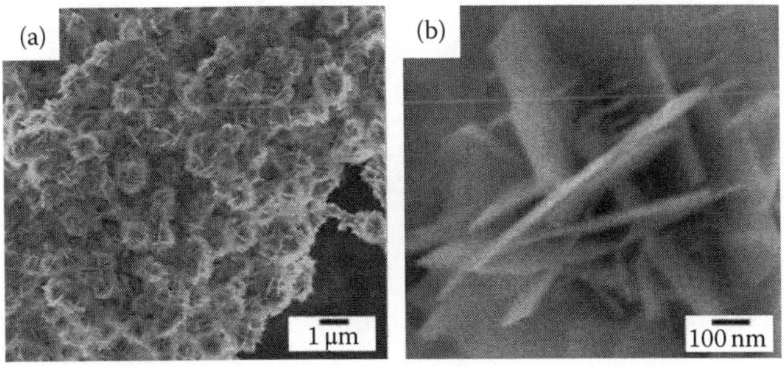

FIGURE 5.14 The typical morphology of hierarchically micro/nanostructured bismuth sodium titanate: (a) low-magnification FESEM image and (b) the local magnified image. (Li et al., *J. Mater. Chem.*, 19, 2253–8, 2009. Reproduced by permission of The Royal Society of Chemistry.)

$Bi(NO3)_3 \leftrightarrow 5H_2O$, and NaOH at 170°C for 10 hours [43]. Figure 5.14a shows its typical morphology, flowerlike spheres 200–500 nm in diameter. High-magnification examination reveals that each flowerlike sphere is composed of cross-linked and vertically standing nanosheets about 100 nm in width and 300 nm in length (Figure 5.14b). This BNT exhibits the typical hierarchical micro/nanostructure morphology.

5.4.2 Photocatalytic Activity

The hierarchical BNT micro/nanostructure when exposed to the environment could be used in the photodegradation of dye molecules. To ensure the feasibility, its bandgap via the corresponding UV–visible adsorption measurement was studied. The absorption cutoff wavelength of the as-prepared BNT sample was 450 nm. The energy bandgap of BNT hierarchical micro/nanostructures was thus estimated to be about 3.08 eV [43]. Therefore, a UV lamp 365 nm in wavelength was chosen as the light source for the photocatalytic measurements. Still the photocatalytic degradation of MO solution was examined by using the prepared BNT crystals as catalyst. The absorption spectra of an aqueous solution of MO in the presence of the hierarchically micro/nanostructured BNT after irradiation for different periods of time by a UV light were measured [43]. The photocatalytic performance was thus evaluated by monitoring the main absorption peak of MO at 464 nm, as mentioned in Section 5.2.

Figure 5.15 shows the photocatalytic results corresponding to the as-synthesized BNT catalysts with different morphologies, including spherical and cubic shapes. The degradation rate of MO using hierarchically micro/nanostructured BNT (curve 1) was much higher than that using spherical (curve 2) and cubic (curve 3) BNT. After irradiation for 120 minutes, about 92% of MO was degraded for the former, but it was only 47% for spherical BNT and 16% for cubic BNT. The addition of BNT catalysts leads to significant degradation of MO, and their photocatalytic activity depends on the morphology.

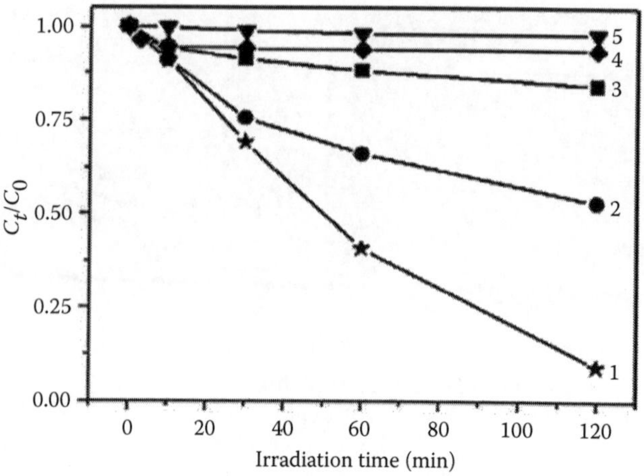

FIGURE 5.15 Methyl orange (MO) normalization concentration in the solution (80 mL) with different bismuth sodium titanate (BNT) catalysts (20 mg) versus exposure time to ultraviolet (UV) light. Starting MO concentration: $C_0 = 1 \times 10^{-5}$ M. Curve 1: hierarchically micro/nanostructured BNT; curve 2: microsized spherical BNT; curve 3: microsized cubic BNT; curve 4: hierarchically micro/nanostructured BNT without UV light irradiation; curve 5: without any catalyst under UV light irradiation. (Li et al., *J. Mater. Chem.*, 19, 2253–8, 2009. Reproduced by permission of The Royal Society of Chemistry.)

Further, two additional experiments were carried out: (1) adsorption experiment in darkness to examine whether the physisorption/chemisorption of hierarchically micro/nanostructured BNT plays important roles in the fast decrease of MO concentration and (2) UV light irradiation of the solution without any catalyst to determine the natural decrease of MO concentration. The result is shown in curves 4 and 5 of Figure 5.15. Obviously, the adsorption effect on the decrease of MO concentration is very small and only leads to a 5.4% decrease of the MO concentration after 120 minutes of stirring in darkness (curve 4 in Figure 5.15). It should be noted that it took only about 1 minute to reach maximum adsorption on the BNT nanosheets with about 4.0% decrease of the MO concentration. This fast adsorption and about 5.4% decrease of the MO concentration are considered to be associated with the peculiar hierarchical micro/nanostructure of BNT. As for the conditions without any catalyst under UV light irradiation, very little MO photodegradation could be observed (about 2.0% within 120 minutes; curve 5 in Figure 5.15). So, the room-temperature photodegradation of micro/nanostructured BNT dominates the decrease in MO concentration.

The high photocatalytic activity of micro/nanostructured BNT could be attributed to the electric field formed between the $(Bi_2O_2)^{2+}$ layer and the $(Bi_2Ti_3O_{10})^{2-}$ layer in the crystal structure of the BNT, inducing the separation of photoelectrons and photogenerated holes [82]. Further, the specific surface area of the catalysts increases, resulting in an increase in the adsorption percentages of MO molecules. The micro/nanostructured BNT has a higher surface area ($29 m^2 \cdot g^{-1}$) than microsized spherical BNT ($15\ m^2 \cdot g^{-1}$) and microsized cubic BNT ($6.0\ m^2 \cdot g^{-1}$) [43] and provides more active sites for photocatalytic reaction, exhibiting excellent photocatalytic activity.

5.5 BRIEF SUMMARY

In summary, we have introduced the photocatalytic properties of some typical oxide materials with special micro/nanostructures, such as a semiconductor oxide (ZnO), a metal/ZnO composite oxide, a magnetic oxide (α-Fe_2O_3), and bismuth sodium titanate ($Bi_{0.5}Na_{0.5}TiO_3$). Because of their high specific surface area and high antiaggregation (structural stability), which keep high active surface area during use, these micro/nanostructured materials have exhibited much more efficient and stronger photocatalytic activity and higher catalytic durability than the nanostructured materials and the microsized particle powders, demonstrated their structurally enhanced photocatalysis performances, and hence could be used as high efficient photocatalysts for removal of contaminants in water. These works have demonstrated that the photocatalytic performance of the catalysts can be improved by designing and fabricating the appropriate micro/nanostructures. Such structure-induced enhancement of photocatalytic performance could also be applicable to other catalyst materials, such as TiO_2.

REFERENCES

1. Gong, B.; Peng, Q.; Na, J.-S.; and Parsons, G. N. 2011. Highly active photocatalytic ZnO nanocrystalline rods supported on polymer fiber mats: Synthesis using atomic layer deposition and hydrothermal crystal growth. *Appl. Catal. A: Gen.* 407: 211–6.
2. Xu, T.; Zhang, L.; Cheng, H.; and Zhu, Y. 2011. Significantly enhanced photocatalytic performance of ZnO via graphene hybridization and the mechanism study. *Appl. Catal. B: Environ.* 101: 382–7.
3. Ohno, T.; Bai, L.; Hisatomi, T.; Maeda, K.; and Domen, K. 2012. Photocatalytic water splitting using modified GaN:ZnO solid solution under visible light: Long-time operation and regeneration of activity. *J. Am. Chem. Soc.* 134: 8254–9.
4. Kavitha, T.; Gopalan, A. I.; Lee, K.-P.; and Park, S.-Y. 2012. Glucose sensing, photocatalytic and antibacterial properties of graphene-ZnO nanoparticle hybrids. *Carbon* 50: 2994–3000.
5. Wang, X.; Zhang, Q.; Wan, Q.; Dai, G.; Zhou, C.; and Zou, B. 2011. Controllable ZnO architectures by ethanolamine-assisted hydrothermal reaction for enhanced photocatalytic activity. *J. Phys. Chem. C* 115: 2769–75.
6. Hoffman, A. J.; Carraway, E. R.; and Hoffmann, M. R. 1994. Photocatalytic production of H_2O_2 and organic peroxides on quantum-sized semiconductor colloids. *Environ. Sci. Technol.* 28: 776–85.
7. Carraway, E. R.; Hoffman, A. J.; and Hoffmann, M. R. 1994. Photocatalytic oxidation of organic-acids on quantum-sized semiconductor colloids. *Environ. Sci. Technol.* 28: 786–93.
8. Sato, K.; Aoki, M.; and Noyori, R. 1998. A "green" route to adipic acid: Direct oxidation of cyclohexenes with 30 percent hydrogen peroxide. *Science* 281: 1646–7.
9. Ye, C.; Bando, Y.; Shen, G.; and Golberg, D. 2006. Thickness-dependent photocatalytic performance of ZnO nanoplatelets. *J. Phys. Chem. B* 110: 15146–51.
10. Curri, M. L.; Comparelli, R.; Cozzoli, P. D.; Mascolo, G.; and Agostiano, A. 2003. Colloidal oxide nanoparticles for the photocatalytic degradation of organic dye. *Mat. Sci. Eng. C-Mater.* 23: 285–9.
11. Hariharan, C. 2006. Photocatalytic degradation of organic contaminants in water by ZnO nanoparticles: Revisited. *Appl. Catal. A-Gen.* 304: 55–61.
12. Lu, F.; Cai, W.; and Zhang, Y. 2008. ZnO hierarchical micro/nanoarchitectures: Solvothermal synthesis and structurally enhanced photocatalytic performance. *Adv. Funct. Mater.* 18: 1047–56.

13. Hu, J. S.; Ren, L. L.; Guo, Y. G.; Liang, H. P.; Cao, A. M.; Wan, L. J.; and Bai, C. L. 2005. Mass production and high photocatalytic activity of ZnS nanoporous nanoparticles. *Angew. Chem. Int. Ed.* 44: 1269–73.
14. Van Dijken, A.; Janssen, A. H.; Smitsmans, M. H. P.; Vanmaekelbergh, D.; and Meijerink, A. 1998. Size-selective photoetching of nanocrystalline semiconductor particles. *Chem. Mater.* 10: 3513–22.
15. Neppolian, B.; Choi, H. C.; Sakthivel, S.; Arabindoo, B.; and Murugesan, V. 2002. Solar/UV-induced photocatalytic degradation of three commercial textile dyes. *J. Hazard. Mater.* 89: 303–17.
16. Xu, H. and Wang, W. 2007. Template synthesis of multishelled Cu_2O hollow spheres with a single-crystalline shell wall. *Angew. Chem. Int. Ed.* 46: 1489–92.
17. Zeng, H.; Cai, W.; Liu, P.; Xu, X.; Zhou, H.; Klingshirn, C.; and Kalt, H. 2008. ZnO-based hollow nanoparticles by selective etching: Elimination and reconstruction of metal-semiconductor interface, improvement of blue emission and photocatalysis. *ACS Nano* 2: 1661–70.
18. Height, M. J.; Pratsinis, S. E.; Mekasuwandumrong, O.; and Praserthdam, P. 2006. Ag-ZnO catalysts for UV-photodegradation of methylene blue. *Appl. Catal. B: Environ.* 63: 305–12.
19. Wu, J. J. and Tseng, C. H. 2006. Photocatalytic properties of nc-Au/ZnO nanorod composites. *Appl. Catal. B: Environ.* 66: 51–7.
20. Ammari, F.; Lamotte, J.; and Touroude, R. 2004. An emergent catalytic material: Pt/ZnO catalyst for selective hydrogenation of crotonaldehyde. *J. Catal.* 221: 32–42.
21. Zeng, H.; Liu, P.; Cai, W.; Yang, S.; and Xu, X. 2008. Controllable Pt/ZnO porous nanocages with improved photocatalytic activity. *J. Phys. Chem. C* 112: 19620–4.
22. Park, W. I. and Yi, G. C. 2004. Electroluminescence in n-ZnO nanorod arrays vertically grown on p-GaN. *Adv. Mater.* 16: 87–90.
23. Jing, L. Q.; Wang, D. J.; Wang, B. Q.; Li, S. D.; Xin, B. F.; Fu, H. G.; and Sun, J. Z. 2006. Effects of noble metal modification on surface oxygen composition, charge separation and photocatalytic activity of ZnO nanoparticles. *J. Mol. Catal. A: Chem.* 244: 193–200.
24. Wang, Q.; Geng, B.; and Wang, S. 2009. ZnO/Au hybrid nanoarchitectures: Wet-chemical synthesis and structurally enhanced photocatalytic performance. *Environ. Sci. Technol.* 43: 8968–73.
25. Xu, Y.; Xu, H.; Li, H.; Xia, J.; Liu, C.; and Liu, L. 2011. Enhanced photocatalytic activity of new photocatalyst Ag/AgCl/ZnO. *J. Alloy. Compd.* 509: 3286–92.
26. Jing, L. Q.; Cai, W. M.; Sun, X. J.; Hou, H.; Xu, Z. L.; and Du, Y. G. 2002. Preparation and characterization of Pd/ZnO and Ag/ZnO composite nanoparticles and their photocatalytic activity. *Chin. J. Catal.* 23: 336–40.
27. Zheng, Y.; Zheng, L.; Zhan, Y.; Lin, X.; Zheng, Q.; and Wei, K. 2007. Ag/ZnO heterostructure nanocrystals: Synthesis, characterization, and photocatalysis. *Inorg. Chem.* 46: 6980–6.
28. Li, X. Z. and Li, F. B. 2001. Study of Au/Au^{3+}-TiO_2 photocatalysts toward visible photooxidation for water and wastewater treatment. *Environ. Sci. Technol.* 35: 2381–7.
29. Linsebigler, A. L.; Lu, G. Q.; and Yates, J. T. 1995. Photocatalysis on TiO_2 surfaces: Principles, mechanisms, and selected results. *Chem. Rev.* 95: 735–58.
30. Tada, H.; Teranishi, K.; Inubushi, Y.-i.; and Ito, S. 2000. Ag nanocluster loading effect on TiO_2 photocatalytic reduction of bis(2-dipyridyl)disulfide to 2-mercaptopyridine by H_2O. *Langmuir* 16: 3304–9.
31. Lu, W.; Gao, S.; and Wang, J. 2008. One-pot synthesis of Ag/ZnO self-assembled 3D hollow microspheres with enhanced photocatalytic performance. *J Phys. Chem. C* 112: 16792–800.

32. Lai, Y.; Meng, M.; and Yu, Y. 2010. One-step synthesis, characterizations and mechanistic study of nanosheets-constructed fluffy ZnO and Ag/ZnO spheres used for rhodamine B photodegradation. *Appl. Catal. B: Environ.* 100: 491–501.
33. Lin, D.; Wu, H.; Zhang, R.; and Pan, W. 2009. Enhanced photocatalysis of electrospun Ag–ZnO heterostructured nanofibers. *Chem. Mater.* 21: 3479–84.
34. Gu, C.; Cheng, C.; Huang, H.; Wong, T.; Wang, N.; and Zhang, T.-Y. 2009. Growth and photocatalytic activity of dendrite-like ZnO@Ag heterostructure nanocrystals. *Cryst. Growth Des.* 9: 3278–85.
35. Georgekutty, R.; Seery, M. K.; and Pillai, S. C. 2008. A highly efficient Ag-ZnO photocatalyst: Synthesis, properties, and mechanism. *J. Phys. Chem. C* 112: 13563–70.
36. Zheng, Y.; Chen, C.; Zhan, Y.; Lin, X.; Zheng, Q.; Wei, K.; and Zhu, J. 2008. Photocatalytic activity of Ag/ZnO heterostructure nanocatalyst: Correlation between structure and property. *J. Phys. Chem. C* 112: 10773–7.
37. Ren, C.; Yang, B.; Wu, M.; Xu, J.; Fu, Z.; Lv, Y.; Guo, T.; Zhao, Y.; and Zhu, C. 2010. Synthesis of Ag/ZnO nanorods array with enhanced photocatalytic performance. *J. Hazard. Mater.* 182: 123–9.
38. Chen, T.; Zheng, Y.; Lin, J.-M.; and Chen, G. 2008. Study on the photocatalytic degradation of methyl orange in water using Ag/ZnO as catalyst by liquid chromatography electrospray ionization ion-trap mass spectrometry. *J. Am. Soc. Mass. Spectrom.* 19: 997–1003.
39. Deng, Q.; Duan, X.; Ng, D. H. L.; Tang, H.; Yang, Y.; Kong, M.; Wu, Z.; Cai, W.; and Wang, G. 2012. Ag nanoparticle decorated nanoporous ZnO microrods and their enhanced photocatalytic activities. *ACS Appl. Mater. Interfaces* 4: 6030–7.
40. Duan, X.; Wang, G.; Wang, H.; Wang, Y.; Shen, C.; and Cai, W. 2010. Orientable pore-size-distribution of ZnO nanostructures and their superior photocatalytic activity. *CrystEngComm* 12: 2821–5.
41. Yang, L.; Wang, G.; Tang, C.; Wang, H.; and Zhang, L. 2005. Synthesis and photoluminescence of corn-like ZnO nanostructures under solvothermal-assisted heat treatment. *Chem. Phys. Lett.* 409: 337–41.
42. Sung-Suh, H. M.; Choi, J. R.; Hah, H. J.; Koo, S. M.; and Bae, Y. C. 2004. Comparison of Ag deposition effects on the photocatalytic activity of nanoparticulate TiO_2 under visible and UV light irradiation. *J. Photochem. Photobiol. A: Chem.* 163: 37–44.
43. Li, J.; Wang, G.; Wang, H.; Tang, C.; Wang, Y.; Liang, C.; Cai, W.; and Zhang, L. 2009. In situ self-assembly synthesis and photocatalytic performance of hierarchical $Bi_{0.5}Na_{0.5}TiO_3$ micro/nanostructures. *J. Mater. Chem.* 19: 2253–8.
44. Fu, H.; Xu, T.; Zhu, S.; and Zhu, Y. 2008. Photocorrosion inhibition and enhancement of photocatalytic activity for ZnO via hybridization with C60. *Environ. Sci. Technol.* 42: 8064–9.
45. Kislov, N.; Lahiri, J.; Verma, H.; Goswami, D. Y.; Stefanakos, E.; and Batzill, M. 2009. Photocatalytic degradation of methyl orange over single crystalline ZnO: Orientation dependence of photoactivity and photostability of ZnO. *Langmuir* 25: 3310–5.
46. Herrmann, J. M.; Tahiri, H.; Ait-Ichou, Y.; Lassaletta, G.; González-Elipe, A. R.; and Fernández, A. 1997. Characterization and photocatalytic activity in aqueous medium of TiO_2 and Ag-TiO_2 coatings on quartz. *Appl. Catal. B: Environ.* 13: 219–28.
47. Wu, T.; Liu, G.; Zhao, J.; Hidaka, H.; and Serpone, N. 1998. Photoassisted degradation of dye pollutants. V. Self-photosensitized oxidative transformation of rhodamine B under visible light irradiation in aqueous TiO_2 dispersions. *J. Phys. Chem. B* 102: 5845–51.
48. Kamat, P. V. 1993. Photochemistry on nonreactive and reactive (semiconductor) surfaces. *Chem. Rev.* 93: 267–300.
49. Zhao, G.; Kozuka, H.; and Yoko, T. 1996. Sol-gel preparation and photoelectrochemical properties of TiO_2 films containing Au and Ag metal particles. *Thin Solid Films* 277: 147–54.

50. Hagfeldt, A. and Graetzel, M. 1995. Light-induced redox reactions in nanocrystalline systems. *Chem. Rev.* 95: 49–68.
51. Ryu, J. and Choi, W. 2004. Effects of TiO_2 surface modifications on photocatalytic oxidation of arsenite: The role of superoxides. *Environ. Sci. Technol.* 38: 2928–33.
52. Zhao, D.; Chen, C.; Wang, Y.; Ma, W.; Zhao, J.; Rajh, T.; and Zang, L. 2007. Enhanced photocatalytic degradation of dye pollutants under visible irradiation on Al (III)-modified TiO_2: Structure, interaction, and interfacial electron transfer. *Environ. Sci. Technol.* 42: 308–14.
53. Rehman, S.; Ullah, R.; Butt, A.; and Gohar, N. 2009. Strategies of making TiO_2 and ZnO visible light active. *J. Hazard. Mater.* 170: 560–9.
54. Xiong, Z.; Zhang, L. L.; Ma, J.; and Zhao, X. S. 2010. Photocatalytic degradation of dyes over graphene-gold nanocomposites under visible light irradiation. *Chem. Commun.* 46: 6099–101.
55. Wang, X.; Cai, W.; Lin, Y.; Wang, G.; and Liang, C. 2010. Mass production of micro/nanostructured porous ZnO plates and their strong structurally enhanced and selective adsorption performance for environmental remediation. *J. Mater. Chem.* 20: 8582–90.
56. Yu, J.; Yu, X.; Huang, B.; Zhang, X.; and Dai, Y. 2009. Hydrothermal synthesis and visible-light photocatalytic activity of novel cage-like ferric oxide hollow spheres. *Cryst. Growth Des.* 9: 1474–80.
57. Zhang, H. and Zhu, Y. 2010. Significant visible photoactivity and antiphotocorrosion performance of CdS photocatalysts after monolayer polyaniline hybridization. *J. Phys. Chem. C* 114: 5822–6.
58. Li, L.; Krissanasaeranee, M.; Pattinson, S. W.; Stefik, M.; Wiesner, U.; Steiner, U.; and Eder, D. 2010. Enhanced photocatalytic properties in well-ordered mesoporous WO_3. *Chem. Commun.* 46: 7620–2.
59. Lei, Z.; You, W.; Liu, M.; Zhou, G.; Takata, T.; Hara, M.; Domen, K.; and Li, C. 2003. Photocatalytic water reduction under visible light on a novel $ZnIn_2S_4$ catalyst synthesized by hydrothermal method. *Chem. Commun.* 17: 2142–3.
60. Wu, J.; Duan, F.; Zheng, Y.; and Xie, Y. 2007. Synthesis of Bi_2WO_6 nanoplate-built hierarchical nest-like structures with visible-light-induced photocatalytic activity. *J. Phys. Chem. C* 111: 12866–71.
61. Van de Krol, R.; Liang, Y.; and Schoonman, J. 2008. Solar hydrogen production with nanostructured metal oxides. *J. Mater. Chem.* 18: 2311–20.
62. Liu, G.; Deng, Q.; Wang, H.; Ng, D. H.; Kong, M.; Cai, W.; and Wang, G. 2012. Micro/nanostructured α-Fe_2O_3 spheres: Synthesis, characterization, and structurally enhanced visible-light photocatalytic activity. *J. Mater. Chem.* 22: 9704–13.
63. Sun, B.; Horvat, J.; Kim, H. S.; Kim, W.-S.; Ahn, J.; and Wang, G. 2010. Synthesis of mesoporous α-Fe_2O_3 nanostructures for highly sensitive gas sensors and high capacity anode materials in lithium ion batteries. *J. Phys. Chem. C* 114: 18753–61.
64. Zhu, L.-P.; Xiao, H.-M.; Liu, X.-M.; and Fu, S.-Y. 2006. Template-free synthesis and characterization of novel 3D urchin-like α-Fe_2O_3 superstructures. *J. Mater. Chem.* 16: 1794–7.
65. Jiao, F.; Harrison, A.; Jumas, J.-C.; Chadwick, A. V.; Kockelmann, W.; and Bruce, P. G. 2006. Ordered mesoporous Fe_2O_3 with crystalline walls. *J. Am. Chem. Soc.* 128: 5468–74.
66. Zhu, L.-P.; Xiao, H.-M.; and Fu, S.-Y. 2007. Template-free synthesis of monodispersed and single-crystalline cantaloupe-like Fe_2O_3 superstructures. *Cryst. Growth Des.* 7: 177–82.
67. Xie, H.; Li, Y.; Jin, S.; Han, J.; and Zhao, X. 2010. Facile fabrication of 3D-ordered macroporous nanocrystalline iron oxide films with highly efficient visible light induced photocatalytic activity. *J. Phys. Chem. C* 114: 9706–12.

68. Valko, M.; Izakovic, M.; Mazur, M.; Rhodes, C. J.; and Telser, J. 2004. Role of oxygen radicals in DNA damage and cancer incidence. *Mol. Cell. Biochem.* 266: 37–56.
69. Wang, H.; Li, G.; Jia, L.; Wang, G.; and Tang, C. 2008. Controllable preferential-etching synthesis and photocatalytic activity of porous ZnO nanotubes. *J. Phys. Chem. C* 112: 11738–43.
70. Tian, G.; Chen, Y.; Zhou, W.; Pan, K.; Tian, C.; Huang, X.-r.; and Fu, H. 2011. 3D hierarchical flower-like TiO2 nanostructure: Morphology control and its photocatalytic property. *CrystEngComm* 13: 2994–3000.
71. Li, H.; Bian, Z.; Zhu, J.; Zhang, D.; Li, G.; Huo, Y.; Li, H.; and Lu, Y. 2007. Mesoporous titania spheres with tunable chamber structure and enhanced photocatalytic activity. *J. Am. Chem. Soc.* 129: 8406–7.
72. Pronin, I.; Syrnikov, P.; Isupov, V.; Egorov, V.; and Zaitseva, N. 1980. Peculiarities of phase transitions in sodium-bismuth titanate. *Ferroelectrics* 25: 395–7.
73. Smlenskii, G. A.; Isupv, V. A.; Afranovskaya, A. I.; and Krainik, N. N. 1961. New ferroelectrics of complex composition. *Sov. Phys. Solid State (Engl. Transl.)* 2: 2651–4.
74. Lee, H. N.; Christen, H. M.; Chisholm, M. F.; Rouleau, C. M.; and Lowndes, D. H. 2005. Strong polarization enhancement in asymmetric three-component ferroelectric superlattices. *Nature* 433: 395–9.
75. Vanderah, T. A. 2002. Talking ceramics. *Science* 298: 1182–4.
76. Suchanicz, J.; Roleder, K.; Kania, A.; and Hańaderek, J. 1988. Electrostrictive strain and pyroeffect in the region of phase coexistence in $Na_{0.5}Bi_{0.5}TiO_3$. *Ferroelectrics* 77: 107–10.
77. Roleder, K.; Suchanicz, J.; and Kania, A. 1989. Time dependence of electric permittivity in $Na_{0.5}Bi_{0.5}TiO_3$ single crystals. *Ferroelectrics* 89: 1–5.
78. Lencka, M. M.; Oledzka, M.; and Riman, R. E. 2000. Hydrothermal synthesis of sodium and potassium bismuth titanates. *Chem. Mater.* 12: 1323–30.
79. Ma, Y.; Cho, J.; Lee, Y.; and Kim, B. 2006. Hydrothermal synthesis of Bi(1/2) Na(1/2) TiO_3 piezoelectric ceramics. *Mater. Chem. Phys.* 98: 5–8.
80. Pookmanee, P.; Uriwilast, P.; and Phanichpant, S. 2004. Hydrothermal synthesis of fine bismuth titanate powders. *Ceram. Int.* 30: 1913–5.
81. Jing, X.; Li, Y.; and Yin, Q. 2003. Hydrothermal synthesis of $Na_{0.5}Bi_{0.5}TiO_3$ fine powders. *Mater. Sci. Eng. B* 99: 506–10.
82. Yao, W. F.; Wang, H.; Xu, X. H.; Shang, S. X.; Hou, Y.; Zhang, Y.; and Wang, M. 2003. Synthesis and photocatalytic property of bismuth titanate $Bi_4Ti_3O_{12}$. *Mater. Lett.*57: 1899–902.

6 Structurally Enhanced Adsorption Performances for Environment

6.1 INTRODUCTION

Environmental pollution, induced by some contaminants, such as organic pollutants and heavy metals from release of industrial production, is of much concern [1–3]. There have been many useful methods and techniques developed for environmental remediation [4–8]. Among them, removal of pollutants based on adsorption of adsorbents is an important and effective way [9]. In this case, the key issue is the high-efficient adsorbents. Fabrication of the adsorbents with high surface activity, high specific surface area, and strong selective adsorption are expected. Although the nanosized particle powders could be candidates for such absorbents, they are very easily aggregated, leading to unwanted reduction in the active surface area. In Chapter 5, we have introduced the hierarchical micro/nanostructured materials, which are composed of microsized objects with nanostructures, and exhibited the significantly structurally enhanced photocatalytic performances. Similarly, such materials also exhibit significant enhanced adsorption performance compared with the normally structured ones due to their high specific surface area and structural stability. Therefore, these materials could be the candidates of high-efficient adsorbents for removal of some pollutants, such as heavy metal ions, organic contamination in water (or solutions), due to their easy separation from solution during application in the environmental remediation. Recently, there have been some reports on the removal of pollutants using micro/nanostructured materials as absorbents [8–13]. For instance, Zhong et al. [9] prepared three-dimensional (3D) flowerlike iron oxide with micro/nanostructures and used them as absorbents to remove As(V) and Cr(VI) from water. Some new micro/nanostructured adsorbents, such as mesoporous MgO [10], flowerlike ceria [11], and hierarchical meso-/macroporous aluminum phosphonate hybrid materials, were fabricated for adsorption of toxic pollutants in water [12]. In this chapter, we discuss the adsorption properties of some special micro/nanostructured materials including the oxide semiconductors [ZnO], magnetic compounds [Fe_3O_4, FeS/C], composites, and metal silicates, to the organic and heavy metal contaminant, to demonstrate their structurally enhanced adsorption performances, and also to show the possibility and validity of the micro/nanostructured materials as the promising adsorbents for contaminant removal.

6.2 MICRO/NANOSTRUCTURED ZnO AS ADSORBENTS

It is well known that ZnO is a promising candidate for photocatalysts [14,15], sensors [16], solar cells [17], and so on [18]. There have been extensive reports in these fields. Also, ZnO is an environmentally friendly material and its surface has many functional groups, such as hydroxyl groups, which can be active sites for adsorption [19,20]. ZnO with micro/nanostructure could thus be a good candidate as adsorbent for wastewater treatment. However, reports on micro/nanostuctured ZnO as adsorbent for environmental remediation are very limited. In this section, we discuss the adsorption properties of ZnO with two kinds of special micro/nanostructures to the heavy metals (cations).

6.2.1 Adsorption of Cu(II) on Porous ZnO Nanoplates

The micro/nanostructured porous ZnO nanoplates were fabricated based on solvothermal method, using ethylene glycol as the morphology director, followed by annealing process, as shown in Figure 2.2. The as-prepared ZnO nanoplates contain two terminal nonpolar planes that are several microns in dimension and 10–15 nm in thickness. The nanoplates are porous with pore diameter of 5–20 nm and high specific surface area of 147 m^2/g. Such porous ZnO nanoplates with high specific surface area can strongly adsorb heavy metal ions in wastewater. Here, we consider Cu(II) ions as the adsorbate to demonstrate their adsorptive capacity. All the adsorption experiments were carried out at 25°C in the dark [21]. $CuCl_2{\cdot}2H_2O$ was used as the source of Cu(II) ions. The pH value of Cu(II) aqueous solutions was adjusted between 4 and 6 by adding HCl or NaOH solution (0.1 M). Then, 5 mg of the prepared adsorbent (or the porous ZnO nanoplates) was added to 10 mL solution of the adsorbate under stirring and kept for 10 hours to establish adsorption equilibrium. The commercial ZnO nanopowders (~5 m^2/g in specific surface area) were used for comparison.

6.2.1.1 Adsorption Measurements of Cu(II)

The adsorption isotherms of Cu(II) ions on the porous ZnO nanoplates and the commercial ZnO nanoparticles are shown in Figure 6.1. The adsorption amount of Cu(II) ions increased with the Cu(II) ion concentration in the solution for both the commercial ZnO nanopowders and the porous ZnO nanoplates. For the former, the adsorption saturates when the concentration of Cu(II) ions in the solution is 200 mg/L. For the latter, however, the adsorption does not saturate even when the concentration of Cu(II) ions in the solution is higher than 2200 mg/L (the corresponding $CuCl_2$ concentration is 4.66 g/L, still much lower than its aqueous solubility [757 g/L, at 25°C], [22]). The adsorptive capacity is much higher than that (~54 mg/g) of the surface functionalized activated carbon reported previously [23]. Compared with the commercial ZnO nanopowders, such micro/nanostuctured porous ZnO plates exhibit significant structurally induced enhancement of adsorption performance.

Further analysis has revealed that the porous ZnO nanoplates and commercial ZnO nanoparticles have different adsorption behaviors. The adsorption isotherm of the

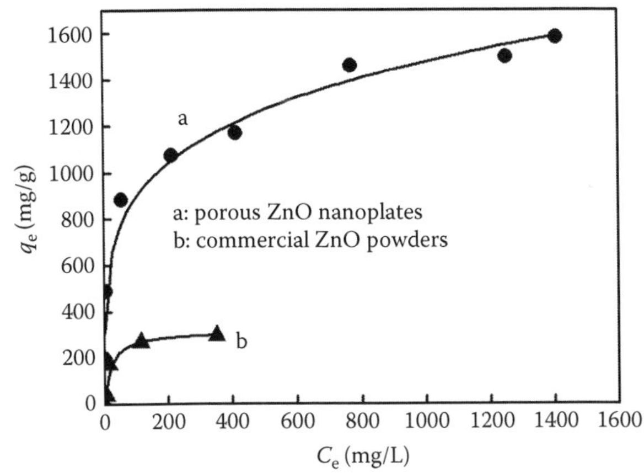

FIGURE 6.1 The adsorption isotherms of Cu(II) ions. Curves a and b correspond to adsorption on the porous ZnO nanoplates and commercial ZnO nanopowders, respectively. (Wang et al., *J. Mater. Chem.*, 20, 8582–90, 2010. Reproduced by permission of The Royal Society of Chemistry.)

former is subject to Freundlich model, which describes adsorption on heterogeneous surfacc [24], or

$$q_e = K_F \times C_e^{1/n} \tag{6.1}$$

where q_e (mg/g[1]) is the equilibrium adsorption amount and C_e is the final concentration of an adsorbate in solution (mg/L). K_F ($mg^{(1-1/n)} \cdot L^{(1/n)}/g$) and n are the parameters reflecting the adsorption capacity and the adsorption intensity, respectively. The corresponding plots of $lgq_e - lgC_e$ exhibit good linear relationship, as shown in Figure 6.2a. The parameters K_F and n take the values 324.2 and 4.6, respectively. In contrast, the isotherm of the commercial ZnO nanopowders can be described by Langmuir model, which describes monolayer adsorption on uniform surface [25], or

$$q_e = \frac{q^0 K_L C_e}{1 + K_L C_e} \tag{6.2}$$

where q^0 is the saturated adsorption amount in mg/g and K_L is the constant in L/mg, depicting the affinity in the process of adsorption. The corresponding plots of $C_e/q_e - C_e$ are illustrated in Figure 6.2b, indicating good linear relationship. The parameters q^0 and K_L took values of 315.60 mg/g and 0.05 L/mg, respectively.

6.2.1.2 Structurally Enhanced Adsorption

The strong adsorption ability of the porous ZnO nanoplates is due to their structure and their high specific surface area. Although the planar surface of the ZnO nanoplates is nonpolar $(10\bar{1}0)$ plane, there should exist a large number of polar sites on the wall of pores within the plates due to the porous geometry. During subsequent exposure to ambient air or in water, hydroxyl groups are formed on these sites [20].

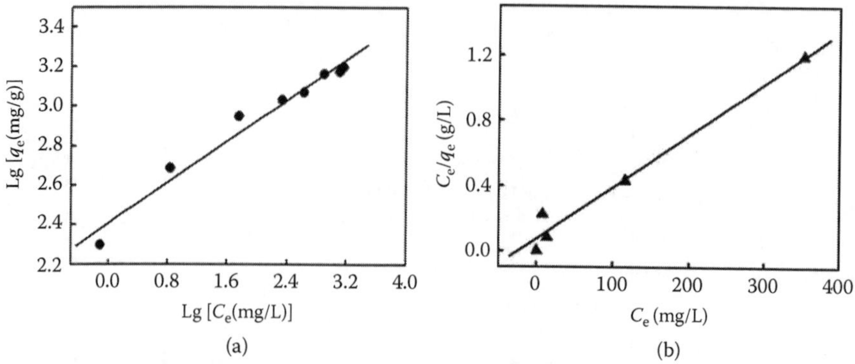

FIGURE 6.2 Plots of $lgq_e - lgC_e$ (a) and $C_e/q_e - C_e$ (b) for adsorption of Cu(II) ions on the porous ZnO nanoplates and commercial ZnO nanopowders, respectively (data from Figure 6.1). (Wang et al., *J. Mater. Chem.*, 20, 8582–90, 2010. Reproduced by permission of The Royal Society of Chemistry.)

The Fourier transform infrared (FTIR) spectroscopic measurement confirms the existence of hydroxyl groups at the surface of the porous ZnO nanoplates [21]. These hydroxyl groups act as active adsorptive sites [26]. The adsorption of Cu(II) ions onto the pore walls of porous ZnO nanoplates is illustrated in Figure 6.3. The hydrated Cu(II) ions or $Cu(H_2O)_6^{2+}$ can react with the hydroxyl groups and form Cu-O weak bond through Lewis interaction [27]. In addition, the adsorbed hydrated Cu(II) ions could partially hydrolyze and form Cu–OH, leading to the formation of Cu–O–Cu on the pore walls. Therefore, multilayer adsorption could take place on the porous ZnO nanoplates, exhibiting strong adsorptive performance and Freundlich-type adsorption behavior.

Furthermore, x-ray photoelectron spectroscopic (XPS) measurements were performed to examine the interaction between the Cu(II) ions and the porous ZnO nanoplates. Figure 6.4 shows the results before and after adsorption. The Cu 2p peaks at 932.76, 942.10, 952.65, and 961.57 eV indicate the existence of Cu(II) on the ZnO nanoplates after the adsorption, as shown in curve 2 of Figure 6.4a. Importantly, the binding energy (BE) values of Cu 2p3/2 (932.76 eV) and Cu 2p1/2 (952.65 eV) are close to those of CuO (932.70 and 952.70 eV) [28,29], confirming the interaction between Cu(II) and the surface of porous ZnO nanoplates and the formation of Cu–O weak bonds due to the interaction with hydroxyl groups, which is in good agreement with the previous report [26]. Also, the BE values of Zn 2p3/2 and Ols peaks after adsorption were lower than those before adsorption. This could also be attributed to existence of the Cu–O bonds that changed the chemical environment of Zn and O atoms, as illustrated in Figure 6.4b and c. In addition, further experiment has revealed that adsorption of Cu(II) will slightly decrease the pH value of the initial Cu(II) solution. For instance, for Cu(II) solution with initial concentration of 100 mg/L, the pH value decreases from 6 to 5.8 after adsorption by porous ZnO nanoplates, which confirms the H^+ release due to the reaction of Cu(II) ions with the hydroxyl groups on the ZnO surface.

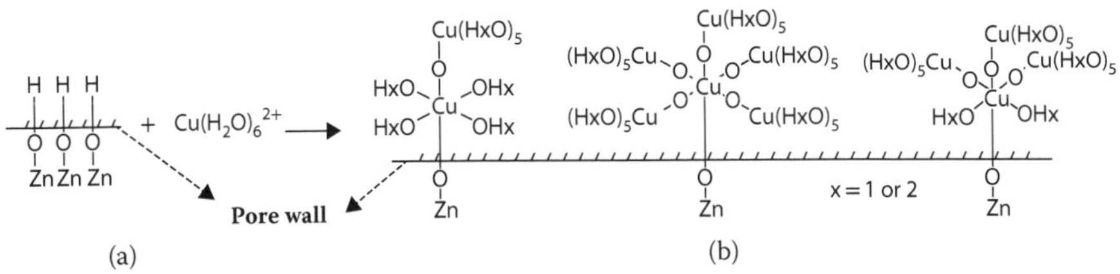

FIGURE 6.3 The adsorption illustration of Cu(II) ions on the polar sites of pore walls within the porous ZnO nanoplates. (Wang et al., *J. Mater. Chem.*, 20, 8582–90, 2010. Reproduced by permission of The Royal Society of Chemistry.)

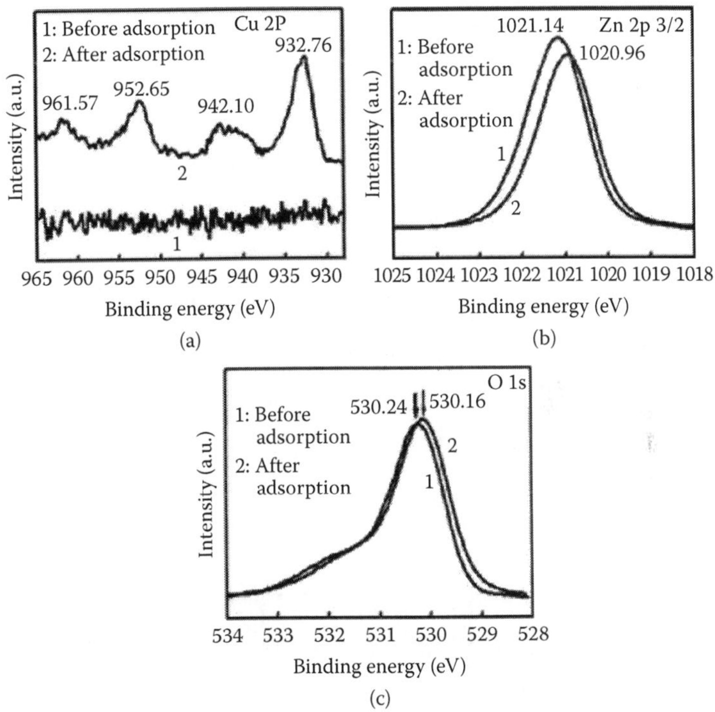

FIGURE 6.4 X-ray photoelectron spectroscopic spectra of the porous ZnO nanoplates before and after adsorption of Cu(II) ions for Cu 2p (a), Zn 2p3/2 (b), and Ols (c). (Wang et al., *J. Mater. Chem.*, 20, 8582–90, 2010. Reproduced by permission of The Royal Society of Chemistry.)

As for the commercial ZnO nanopowders, the morphology observation has revealed that the ZnO particles are dominated by polyhedrons [21]. Therefore, Cu(II) ions could adsorb on the uniform surface of the commercial ZnO particles in monolayer or Langmuir type and hence can be saturated in the solution with low Cu(II) concentration.

6.2.1.3 Extension to Anion Adsorption

As demonstrated previously, the metal cationic contaminants can be strongly adsorbed on the porous ZnO nanoplates, which exhibit structurally enhanced adsorption. For the anionic contaminants (such as chromate and methyl orange), however, the opposite is true. No or only insignificant adsorption was observed on the porous ZnO nanoplates, depending on the pH values of the solution. At pH = 9, closely to the point of zero charge (PZC) of ZnO, no adsorption was observed on the porous ZnO nanoplates for the chromate anions and methyl orange, as illustrated in Figure 6.5a corresponding to the optical absorption spectra of the aqueous solution of methyl orange before and after addition of the porous ZnO nanoplates. The concentration of methyl orange was unchanged after addition of the adsorbents (the same for chromate solution). This is normal in a pH region closely to or above the PZC of the adsorbent, which cloaks nonspecific anion sorption properties. Furthermore, under weak acidic condition (pH = 6), only insignificant adsorption of both anionic contaminants occurs on the porous ZnO nanoplates, as shown in Figure 6.5b and c. These could be attributed to their nonspecific anion sorption properties.

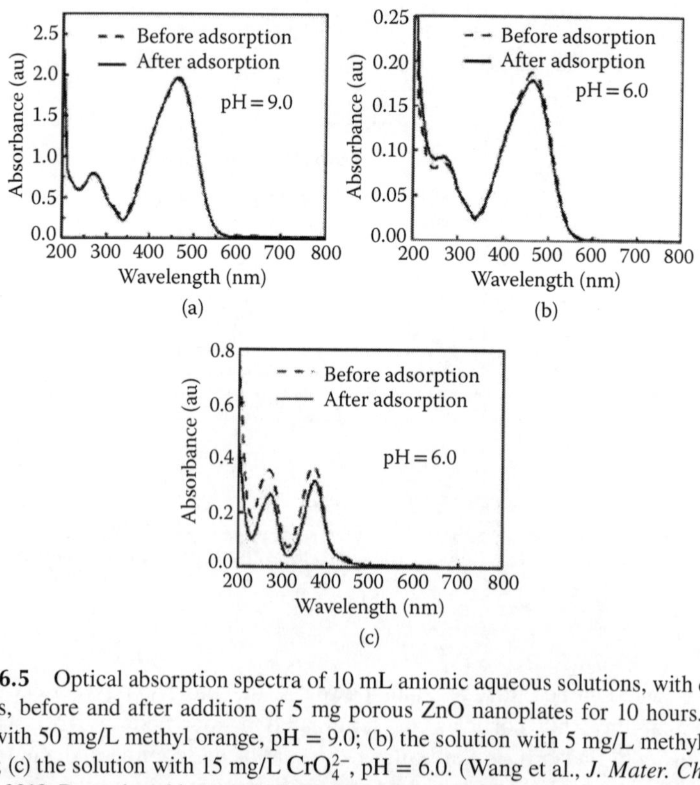

FIGURE 6.5 Optical absorption spectra of 10 mL anionic aqueous solutions, with different pH values, before and after addition of 5 mg porous ZnO nanoplates for 10 hours. (a) The solution with 50 mg/L methyl orange, pH = 9.0; (b) the solution with 5 mg/L methyl orange, pH = 6.0; (c) the solution with 15 mg/L CrO_4^{2-}, pH = 6.0. (Wang et al., *J. Mater. Chem.*, 20, 8582–90, 2010. Reproduced by permission of The Royal Society of Chemistry.)

6.2.2 Standing Porous Nanoplate–Built ZnO Hollow Microspheres

As mentioned previously, ZnO with micro/nanostructure could be a good candidate as adsorbent for wastewater treatment. The porous micro/nanostructured nanoplates with high specific surface area has shown excellent adsorption performance of Cu(II) ions in aqueous solution and demonstrated the validity of the porous ZnO nanoplates as the promising adsorbent for contaminant removal. However, these porous nanoplates were easy to stack together or overlap, due to the planar geometry, and hence decreased the surface area exposed to the solution. The structurally enhanced adsorption performance of the micro/nanostructured material was only partially exhibited. Obviously, if these porous nanoplates are assembled into a micro/nanostructure and all the porous nanoplates are vertically standing and cross-linked, as shown in Figure 2.6, the stacking or pileup of the nanoplates will not take place, or the surface area within the porous nanoplates will be sufficiently exposed to the solution. The structurally enhanced adsorption performance would thus be sufficiently exhibited. In this section, we discuss the adsorption properties of the ZnO hollow microspheres built with standing porous nanosheets, which were produced based on a modified hydrothermal route using citrate as the structural director followed by annealing treatment (see Section 2.2), to demonstrate the structurally enhanced adsorption performance of heavy metal cations, compared with the porous nanoplates or commercial ZnO nanopowders.

The standing porous nanoplate–built ZnO hollow spheres are shown in Figure 2.6. The hollow spheres are honeycomb-like and microsized (~5–20 μm in diameter) with specific surface area of 46 m^2/g. The shell layers are built by the vertically standing, cross-linked porous nanoplates with an exposed surface of nonpolar $(10\bar{1}0)$ planes. The nanoplates are porous with mean pore size of ~12 nm in and thickness of ~25 nm. It has been shown that such ZnO hollow microspheres with exposed porous nanosheet surface presented here are high-efficient adsorbents for heavy metal removal from wastewater [30]. Here, we take Cu(II), Pb(II), and Cd(II) as the adsorbates to demonstrate their adsorption capacity and behaviors on the porous ZnO hollow microspheres. The adsorption experiments were carried out at 25°C in the dark. $CuCl_2 \cdot 2H_2O$, $Cd(NO_3)_2 \cdot 4H_2O$, and $Pb(NO_3)_2$ were, respectively, used as the sources of Cu(II), Cd(II), and Pb(II) ions. A certain amount of prepared catalysts (5 mg for adsorption of Cu(II) and 10 mg for adsorption of Cd(II) and Pb(II)) was added to 10 mL solution with different initial concentrations under stirring and kept for 10 hours to establish adsorption equilibrium. The pH value was adjusted between 4 and 6 for Cu(II) solution and 6 for Cd(II) and Pb(II) solutions by adding NaOH (0.1 M) or HNO_3 (0.1 M) solution.

6.2.2.1 Adsorptive Performance of Heavy Metal Ions

6.2.2.1.1 High-Efficient Adsorption of Cu(II) Ions

The adsorption isotherm of Cu(II) ions on the porous nanoplate–built ZnO hollow microspheres is shown in curve 1 of Figure 6.6a. The adsorption amount of Cu(II) ions increased with the Cu(II) ion concentration in the solution. The adsorption amount is higher than 1400 mg Cu in each gram of the ZnO hollow microspheres. Importantly,

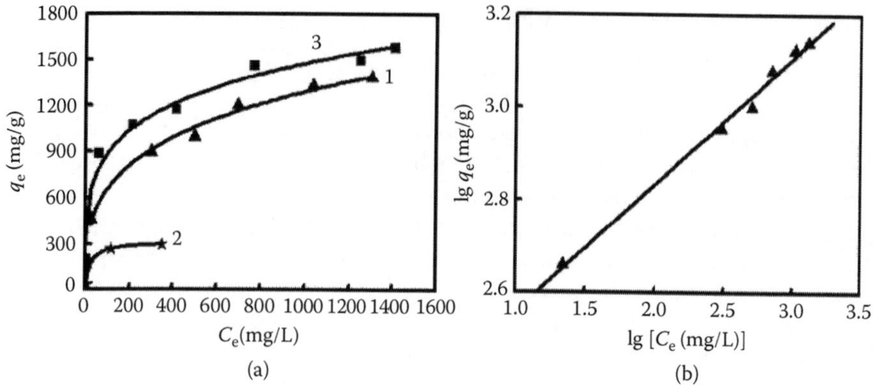

FIGURE 6.6 The adsorption performance of Cu(II) ions on different absorbents. (a) The adsorption isotherms on the porous nanosheet–built ZnO hollow microspheres (curve 1), the commercial ZnO nanopowders (curve 2), and the pure porous ZnO nanoplates (curve 3). (Wang et al., *J. Mater. Chem.*, 20, 8582–90, 2010. Reproduced by permission of The Royal Society of Chemistry.) (b) The plots of $lgq_e - lgC_e$ corresponding to curve 1 in (a). (Wang et al., 199–205, Copyright 2013, with permission from Elsevier. Reprinted from *Colloids Surf. A*, 422.)

no saturated adsorption was found even when the initial concentration of Cu(II) in the solution was higher than 2000 mg/L. The adsorption capacity is close to that of the above mentioned porous ZnO nanoplates with much higher specific surface area (147 m^2/g) [21] (curve 3 in Figure 6.6a). It means that the ZnO hollow microspheres with exposed porous nanosheet surface exhibit much higher adsorption efficiency than the pure porous nanoplates, since it can keep high exposed active surface area during use due to its special geometry. Correspondingly, for the commercial ZnO nanopowders, the adsorption is much lower and saturated when the concentration of Cu(II) ions in the solution is up to 300 mg/L (curve 2 in Figure 6.6a).

Like the porous ZnO nanoplates, the adsorption isotherm of the porous ZnO hollow microspheres (curve 1 in Figure 6.6a) also follows Freundlich model, or Equation 6.1. Figure 6.6b gives the corresponding plots of $lgq_e - lgC_{e,}$ exhibiting good linear relationship. The parameters K_F and n took values of 177 and 3.5, respectively, which are close to those of the pure porous ZnO nanoplates with much higher specific surface area. Therefore, the hollow microspheres with exposed porous nanosheet surface is superior to the porous nanoplates in terms of adsorption efficiency, since the latter is inevitably overlapped during use and decreases the exposed active surface area.

6.2.2.1.2 Adsorption of Pb(II) and Cd(II) Ions

Furthermore, the adsorption can extend to the other metal ions. Here, Pb(II) and Cd(II) were taken as typical heavy metal ions (or adsorbates), which are very toxic pollutants in wastewater and difficult to be removed, to demonstrate the structurally enhanced adsorptive performance of the ZnO hollow microspheres with exposed porous nanosheet surface. Figure 6.7 illustrates their isothermal adsorption curves of the adsorbents (the porous ZnO hollow microspheres and the commercial ZnO nanopowders). The adsorption amounts increase with increase of concentration of adsorbates for all samples. The adsorption capacities of the porous ZnO hollow microspheres are much higher than that of the commercial ZnO nanopowders and also larger than those of the reported carbon adsorbent (10.9 mg/g for adsorption of Cd(II) and 97.1 mg/g for adsorption of Pb(II) ions) [31], which follows Langmuir model. The order of adsorption capacity of these three metal ions is Cu(II) > Pb(II) > Cd(II), as listed in Table 6.1.

Similarly, for Pb(II) ions, the adsorption behavior on the porous ZnO hollow microspheres is also subject to Freundlich isotherm [24], or Equation 6.1, as shown in Figure 6.8a, indicating unsaturated adsorption. The parameters K_F and n are 32.2 ($mg^{(1-1/n)}\cdot L^{(1/n)}g$) and 4.3, respectively. For adsorption of Cd(II) ions on the porous ZnO hollow microspheres, however, there exists a saturated value, or it follows Langmuir model [25], or Equation 6.2. The corresponding plot of $C_e/q_e - C_e$ is shown in curve 1 of Figure 6.8b. The q^0 and K_L values are thus estimated to be 28.1 (mg/g) and 0.02 (L/mg), respectively. For the commercial ZnO nanopowders, the adsorptions of all three adsorbates are subject to Equation 6.2, or monolayer adsorption, as shown in Figure 6.2b and in plots 1' and 2 of Figure 6.8b. The corresponding parameter values are listed in Table 6.1. It can be seen that the adsorption capacities of Cu(II) and Pb(II) are about five times higher on the porous ZnO hollow microspheres than those on the commercial ZnO nanopowders, exhibiting significantly structurally enhanced adsorption performances, while it is only about two times for Cd(II) ion's adsorption.

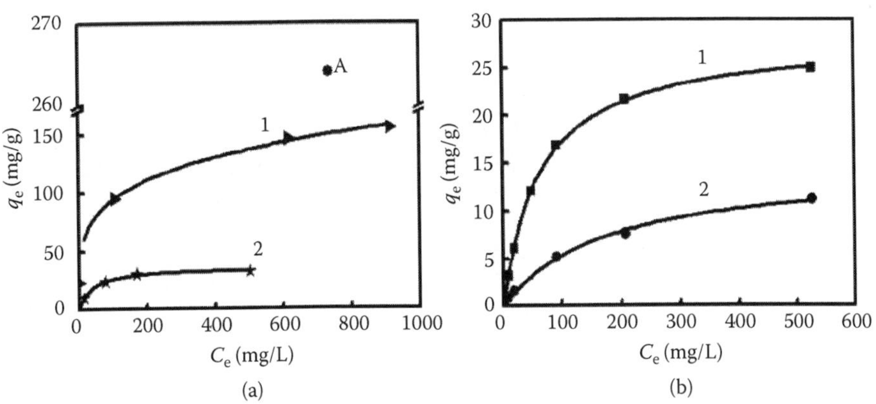

FIGURE 6.7 The adsorption isotherms of Pb(II) ions (a) and Cd(II) ions (b). Curves 1 and 2 correspond to the adsorption on the porous nanosheet–built ZnO hollow microspheres and the commercial ZnO nanopowders, respectively. Point A in (a) corresponds to the result of Ni(II) adsorption on the ZnO hollow microspheres (25°C, pH = 6.0). (Reprinted from *Colloids Surf. A*, 422, Wang et al., 199–205, Copyright 2013, with permission from Elsevier.)

TABLE 6.1
The Adsorption Parameters of Cu(II), Pb(II), and Cd(II) on Different Adsorbents

	Porous ZnO Hollow Spheres			Commercial ZnO Powders		
	Cu(II)	Pb(II)	Cd(II)	Cu(II)	Pb(II)	Cd(II)
Adsorption Type	**Freundlich**	**Freundlich**	**Langmuir**	**Langmuir**	**Langmuir**	**Langmuir**
cf_e (mg/g)	>1400	>160	/	/	/	/
cf_e (mmol/g)	>22.03	>0.77	/	/	/	/
cf^e (mg/g)	1	/	28.1	315.6	35.9	15.4
cf^e (mmoL/g)	1	/	0.3	5.0	0.17	0.14
K_L (L/mg)	/	/	0.02	0.05	0.03	0.005
K_F (mg$^{(1-1/n)}$·L$^{(1/n)}$/g)	177.0	32.2	/	/	/	/
n	3.48	4.28	/	/	/	/

Source: Reprinted from *Colloids Surf. A*, 422, Wang et al., 199–205, Copyright 2013, with permission from Elsevier.

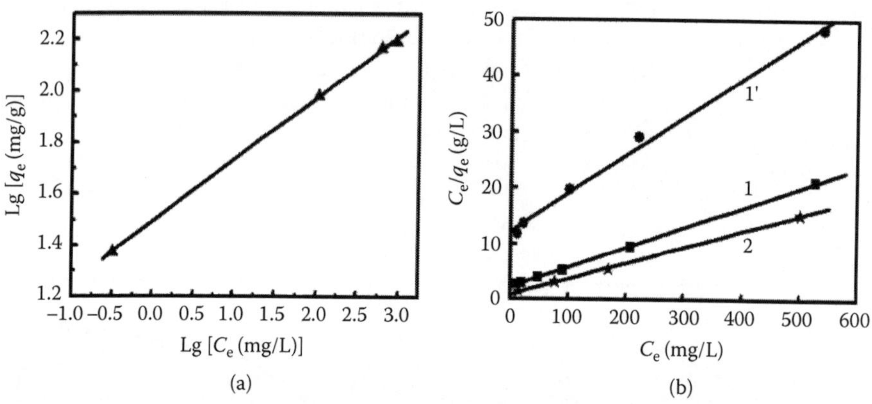

FIGURE 6.8 (a) The plot of lgq_e versus lgC_e for adsorption of Pb(II) ions on the porous nanosheet–built ZnO hollow microspheres (Freundlich model). (b) The plots of C_e/q_e versus C_e (Langmuir model). Lines 1 and 1' are for adsorption of Cd(II) ions on the porous ZnO hollow microspheres and the commercial ZnO nanopowders, respectively. Line 2 is for adsorption of Pb(II) ions on the commercial ZnO nanopowders (data from Figure 6.7). (Reprinted from *Colloids Surf. A*, 422, Wang et al., 199–205, Copyright 2013, with permission from Elsevier.)

6.2.2.2 Structurally Enhanced Adsorption and Its Electronegativity Dependence

The strong adsorption performances of the porous nanosheet–built ZnO hollow microspheres should be related to their unique structure. Such material consists of vertically standing and cross-linked porous nanosheets, which is stable against aggregation and hence keeps the surface area within the porous nanoplates sufficiently exposed to the solution. The structurally enhanced adsorption performance would thus be sufficiently exhibited, showing much higher adsorption capacity to the Cu(II), Pb(II), or Cd(II) ions than those on the commercial ZnO nanopowders and the reported carbon adsorbents.

It is well known that hydroxyl groups can be formed on the surface of ZnO during exposure to ambient air or in water, resulting in negatively charged surface [20]. These hydroxyl groups would be actively adsorptive sites [26] and interact with Cu(II), Pb(II), or Cd(II) species to form bonding of Cu–O, Pb–O, or Cd–O by Lewis interaction [27], which has been confirmed by the XPS spectral measurements. Typically, Figure 6.9 shows the BE spectra of O1s and Zn2p3/2 for the porous ZnO hollow microspheres before and after adsorption of Cd(II) ions. Before adsorption, the spectrum of the porous ZnO hollow microspheres consists of $O1s_A$ (530.23 eV) and $O1s_B$ (531.40 eV), which can be ascribed to zinc oxide [32] and hydroxide [33], respectively (Figure 6.9a). After adsorption of Cd species, $O1s_B$ is at 531.7 eV, close

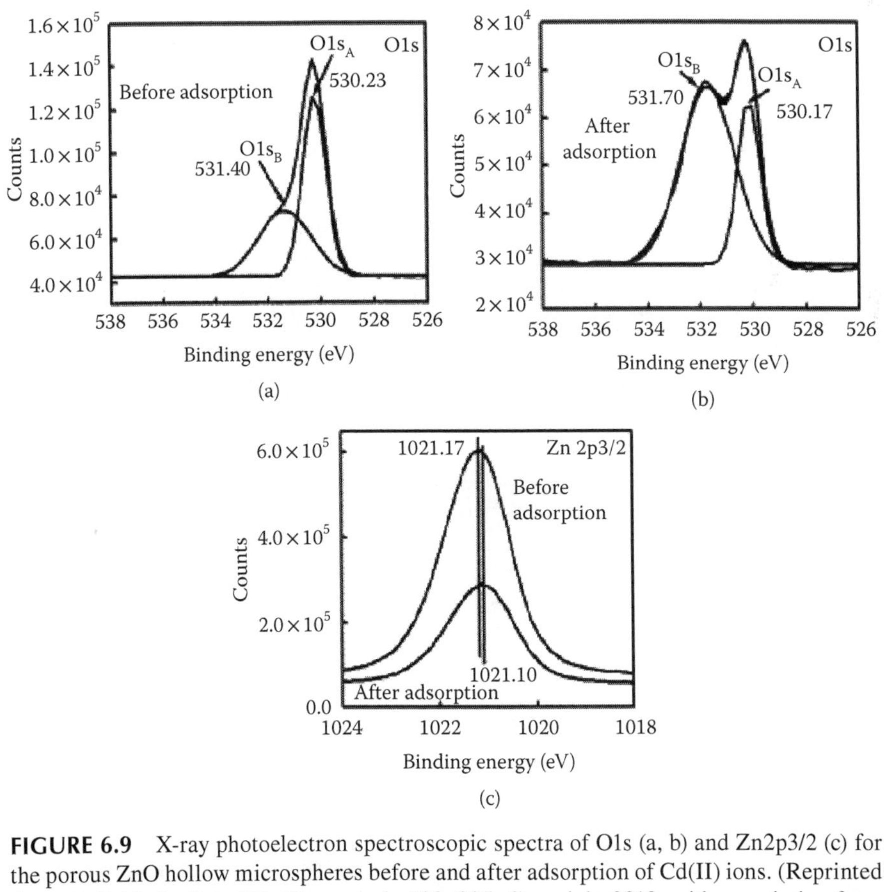

FIGURE 6.9 X-ray photoelectron spectroscopic spectra of O1s (a, b) and Zn2p3/2 (c) for the porous ZnO hollow microspheres before and after adsorption of Cd(II) ions. (Reprinted from *Colloids Surf. A*, 422, Wang et al., 199–205, Copyright 2013, with permission from Elsevier.)

to that in CdO (531.6 eV) [34], indicating that Zn–O–H is replaced by Zn–O–Cd, or formation of Cd–O weak bonding, as shown in Figure 6.9b. Correspondingly, the BE value of Zn 2p3/2 also shifts slightly after adsorption, due to the change in the chemical environment for the surface zinc atoms (see Figure 6.9c).

As mentioned in Section 2.2.2.1, the planar surface of the exposed and cross-linked ZnO nanosheets, in the porous ZnO hollow spheres, is the nonpolar $(10\bar{1}0)$ plane. However, there should still exist a large number of polar sites on the wall of pores within the plates due to the porous geometry. It is because of the porous geometry that there exist different active adsorption sites, which should correspond to the different adsorption energy levels. Generally, the interaction between the surface of ZnO (or the hydroxyl groups) and the heavy metal ions is closely related to the electronegativity of heavy metal ions [35]. The high electronegativity values would induce the strong interaction, which, in turn, corresponds to the more adsorption sites with different adsorption energy levels and even multilayer adsorption, showing Freundlich adsorption type, while the low electronegativity values will lead to less adsorptive sites, due to the weak interaction, exhibiting monolayer adsorption (or Langmuir model). Table 6.2 gives the electronegativity values for some typical heavy metal ions. Here, the electronegativity value is high for Cu(II) and Pb(II) and low for Cd(II), hence keeping the adsorption capacity order as Cu(II) > Pb(II) > Cd(II).

Furthermore, extended adsorptive measurements were conducted for confirmation of such electronegativity dependence of the adsorption performance, by using Ni(II) as an adsorbate, whose electronegativity (1.91) (Table 6.2) is between those of Cu(II) and Pb(II). For 10 mL Ni(II) aqueous solution with initial concentration of 1000 mg/L, after addition of 10 mg porous ZnO hollow microspheres, the equilibrium adsorption amount is 264.1 mg/g (see point A in Figure 6.7a). The adsorption amount of Ni(II) is also between those of Pb(II) and Cu(II).

Similarly, for Cd(II) ions, they could be mainly adsorbed on the homogeneous exposed nonpolar surface due to their low electronegativity value, leading to Langmuir-type adsorption. However, for Cu(II) and Pb(II) ions with larger electronegativity values, they can be adsorbed on the polar sites on the pore walls within the nanosheets, in addition to the exposed nonpolar planar surface, leading to Freundlich-type adsorptive performance owing to their heterogeneous surface adsorption, as demonstrated in Figures 6.6b and 6.8.

In addition, the XPS spectral measurements were performed to reveal the interaction between the metal ions and surface of the adsorbent (or the hydroxyl groups),

TABLE 6.2
The Pauling Electronegativity of Some Heavy Metal Ions

Metal Ions	Cu(II)	Ni(II)	Pb(II)	Cd(II)
Pauling electronegativity	2.00 [36]	1.91 [36]	1.87 [37]	1.69 [38]

Source: Reprinted from *Colloids Surf. A*, 422, Wang et al., 199–205, Copyright 2013, with permission from Elsevier.

as demonstrated in Figure 6.10, corresponding to the results after Cd(II) and Pb(II) adsorption on the porous ZnO hollow microspheres. The BE value of Cd3d5/2 is 405.74 eV, which is much lower than that of CdO (407.38 eV) [39] (Figure 6.10a). It means that the adsorbed Cd species are weakly connected with hydroxyl groups on the surface of ZnO, only forming Cd–O weak binding. For Pb(II) adsorption, however, the BE value of Pb4f 7/2 is 138.15 eV, which is very close to that of PbO (138.20 eV) [26] (Figure 6.10b), indicating that Pb(II) strongly interacts with the surface of ZnO through hydroxyl groups, forming Pb–O bonding.

As for adsorption on the commercial ZnO nanopowders with Langmuir type, it could be attributed to the homogeneous polyhedral surfaces of the ZnO particles [30].

6.2.3 Brief Remarks

Altogether, the possibility and validity of the micro/nanostructured ZnO (such as the porous ZnO nanoplates and the porous nanosheet–built ZnO hollow microspheres) as the adsorbents for contaminant removal have been demonstrated, which show significantly structurally enhanced adsorption of the heavy metal cations, compared with the commercial ZnO nanopowders due to their unique micro/nanostructures. The adsorption behavior can be described by Langmuir model or Freundlich model, depending on the electronegativity of the heavy metals. For metals with high electronegativity, due to the large interaction with the adsorbent, it can adsorb on the active sites, with different adsorptive energy levels, on the adsorbent, showing Freundlich-type adsorption. Otherwise, metals with low electronegativity only exhibit Langmuir-type adsorption. Here, it should be pointed out that the ZnO with micro/nanostructures as adsorbents are conditional, since there exists pH value–dependent solubility of ZnO in the aqueous solution [40]. The low pH value corresponds to high Zn dissolution equilibrium concentration when pH value is below 10 [40]. Further experiment has shown that, at pH = 4, the equilibrium Zn concentration in the aqueous solution is up to 8 ppm for the

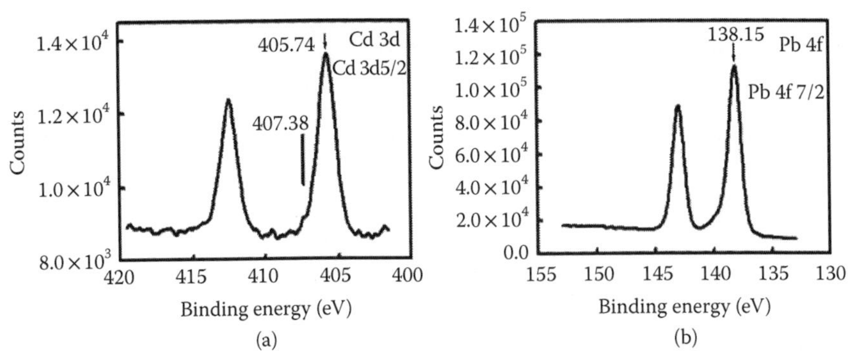

FIGURE 6.10 X-ray photoelectron spectroscopic spectra of Cd3d (a) and Pb4f (b) for the porous ZnO hollow microspheres after adsorption of Cd(II) and Pb(II) ions, respectively. Vertical line indicates the position of Cd 3d5/2 in CdO. (Reprinted from *Colloids Surf. A*, 422, Wang et al., 199–205, Copyright 2013, with permission from Elsevier.)

micro/nanostructured ZnO. It means that the micro/nanostructured ZnO should be used as the adsorbent in the neutral or weak acidic (basic) solution. In addition, the regeneration of the adsorbent after use is expected. This could be realized by surface modification of the adsorbent. Batch or column sorption experiment and in-depth research about such micro/nanostructured ZnO nanoplates as adsorbents are required to be undertaken.

6.3 COMPOSITE POROUS NANOFIBERS FOR REMOVAL OF CR(VI)

Chromium is a heavy metal and commonly used or generated by a number of industrial processes, such as electroplating, leather tanning, metal polishing, pigment manufacture, tannery facilities, mineral extraction, and also as a common ingredient for protective coatings, pigments, and stainless steel [41]. Once released into the environment from industrial processes [42,43], chromium becomes a contaminant in soil and groundwater. Chromium usually exists in the form of chromates ($Cr_2O_7^{2-}$, CrO_4^{2-}, and $HCrO_4^-$) in aquatic systems. Chromates are highly soluble and mobile [43]. They are carcinogenic and highly toxic to humans, animals, and plants. Reduction of hexavalent chromium [Cr(VI)] to Cr(III) is environmentally beneficial since the latter is relatively less hazardous. However, there still exists the possibility of Cr(III) oxidation back to Cr(VI) [44]. Generally, Cr(VI) is removed from water by reduction of Cr(VI) to Cr(III) using reducing agents such as highly soluble Fe(II) salts ($FeSO_4{\cdot}7H_2O$, $FeCl_2{\cdot}4H_2O$, etc.), sulfur dioxide, or sodium bisulfite, before precipitation of hydrated oxide through addition of NaOH [45,46]. However, such process needs costly treatment equipment and creates large amount of sludge and potential hazards to the environment due to landfill leaching and inefficient recovery of the treated metals for reusage [47]. Although some nanomaterials, such as activated carbon, carbon-encapsulated magnetic NPs, and carbon-coated Fe nanoparticles, exhibited good potential for Cr(VI) removal in wastewater [48–50], there are some problems to be resolved before being put into practice. Recently, it has been reported that the polyacrylonitrile (PAN)/$FeCl_2$ composite porous nanofibers (NFs) has a property of excellent Cr removal from a $Cr_2O_7^{2-}$-containing solution [51]. The PAN/$FeCl_2$ NFs could be fabricated by electrospinning, as previously described [51]. They are porous and amorphous in structure and of 100–300 nm in diameter and 10 m^2/g in specific surface area, as shown in Figure 4.2. The Cr-removal capability of such micro/nanostructured composite materials is more than 110 mg Cr/g $FeCl_2$, which is much higher than the previously reported values of the other nanomaterials as Cr(VI)-removal adsorbents, in addition to the easy separation from solution. The details are discussed in this section.

6.3.1 Cr-Removal Performance

Aqueous solutions with different Cr(VI) concentrations were prepared by dissolving $K_2Cr_2O_7$ in deionized water. The pH value of the solution was adjusted to be 5 by adding hydrochloric acid and sodium hydroxide solution. Cr-removal experiments were then performed by adding the as-prepared samples (or composite NF films) to the Cr(VI) solutions at room temperature, before sampling the solution at defined time intervals for analysis of Cr content in the solution [51].

6.3.1.1 High Cr-Removal Capacity

After 0.5 g of the composite NF film, shown in Figure 4.2a orb, containing 4.17 × 10^{-4} mol iron element, was put into 100 mL of $K_2Cr_2O_7$ aqueous solution (containing 1.33 × 10^{-4} mol Cr element, or 69.1 ppm) for 3 days, a nearly colorless clarifying solution was obtained, and the composite NF film changed from gray to yellow due to the Cr adsorption. Curve 1 in Figure 6.11a shows total Cr content in the solution as a function of time. The Cr content sharply decreases from 69.1 to 40 ppm in 5 minutes and to 31.1 ppm in 60 minutes. Correspondingly, pH value of the solution drops from the initial value 5 to 2.3 [51]. Subsequently, Cr content was ever-decreasing to 9.3 ppm in 3 days and pH value was nearly unchanged (<3.0). The Cr-removal capability was thus estimated to be about 11.7 mg Cr/g NFs, corresponding to 110 mg Cr/g $FeCl_2$. This is much higher than the previously reported values of the other nanomaterials as Cr(VI)-removal adsorbents, such as ceria hollow nanospheres (15.4 mg Cr/g CeO_2) [52], magnetite Fe_3O_4 nanoparticles (2.95 mg Cr/g Fe_3O_4) [53], γ-Fe_2O_3 nanoparticles or mesoporous γ-Fe_2O_3 (14.6–15.6 mg Cr/g γ-Fe_2O_3) [54], and the nanosized akaganeite (80 mg Cr/g) [45]. These nanomaterials should be centrifugalized to be separated from the contaminated solution after treatment. Therefore, such composite NFs are advantageous as adsorbents over the general nanomaterials with respect to Cr(VI) removal. Comparatively, the mixture of PAN NF film with the same amount of $FeCl_2$ as the composite NFs only removed Cr insignificantly during the initial 5 minutes (curve 3 in Figure 6.11a). The cast film of PAN/$FeCl_2$ is more

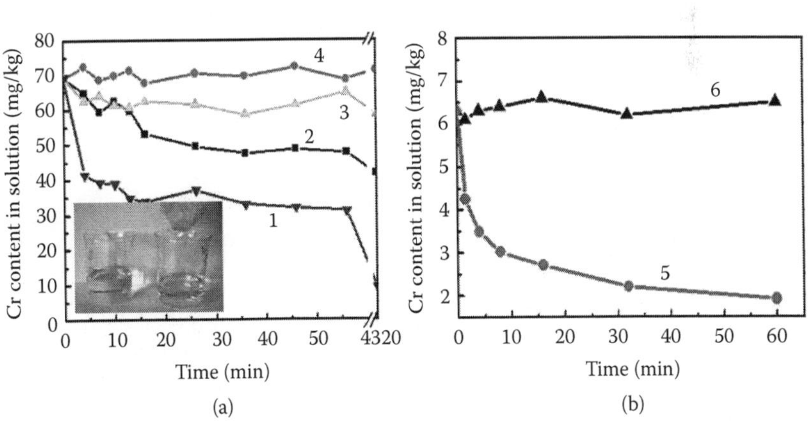

FIGURE 6.11 The Cr-removal measurements in the $K_2Cr_2O_7$ aqueous solutions with initial pH = 5, Cr content 69.1 ppm (a) and 6.3 ppm (b) after addition of different materials (see the details in the text). Curves 1–4: addition of polyacrylonitrile (PAN)/$FeCl_2$ nanofibers (NFs), cast film of PAN/$FeCl_2$, the mixture of $FeCl_2$ powders with PAN NFs, and pure PAN NFs, respectively. Curves 5 and 6: addition of PAN/$FeCl_2$ NFs, pure PAN NFs, respectively. Inset: the photo before and after addition of the PAN/$FeCl_2$ NFs to the $K_2Cr_2O_7$ solution for 3 days. (Lin et al., *J. Mater. Chem.*, 21, 991–7, 2011. Reproduced by permission of The Royal Society of Chemistry.)

effective than the mixture for longer time but much lower than the composite NFs for Cr removal (curve 2 in Figure 6.11a). As for the pure PAN NF film, it cannot remove Cr from the solution (see curve 4 in Figure 6.11a). So, only the PAN/$FeCl_2$ composite NFs exhibited good Cr-removal performance from the $Cr_2O_7^{2-}$ solution.

Also, the experiments have revealed that Cr-removal capacity for the composite NF film depends on the Cr content in the solution. For a solution with much lower Cr content, the capacity will slightly decrease. Curve 5 in Figure 6.11b shows the results corresponding to the following experimental conditions. A measure of 0.094 g composite NF film shown in Figure 4.2 (containing 8.35×10^{-5} mol Fe(II)) was soaked in 100mL Cr(VI) solution containing 1.2×10^{-5} mol or 6.3 ppm Cr. The Cr content in the solution could decrease up to 1.9 ppm in 60 minutes and finally to 1.6 ppm after 10 days. The Cr-removal capacity was estimated to be about 4.4 mg Cr/g NFs (or 41.5 mg Cr/g $FeCl_2$), partially close to the preceding results. This demonstrates that such composite NFs can be used for deepening purification of pollutant water. Correspondingly, pure PAN NFs cannot remove Cr from the solution (see curve 6 in Figure 6.11b). These results indicate that such composite NFs are a kind of effective pollutant-eliminating material, which can easily realize one-step removal of Cr from the solution and be facilely integrated into a device, such as filtration membrane [55], extending the application of electrospinning technique in environment remediation.

6.3.1.2 Influences of the pH Value and Fe(II) Content

Further experiments have indicated that the Cr-removal capacity is also associated with the pH value in the Cr-containing aqueous solution and the Fe(II) content in the composite NF films [51].

The initial pH value of the Cr aqueous solution has significant influence on the Cr-removal capacity of the composite NFs, as shown in Figure 6.12a. We can see

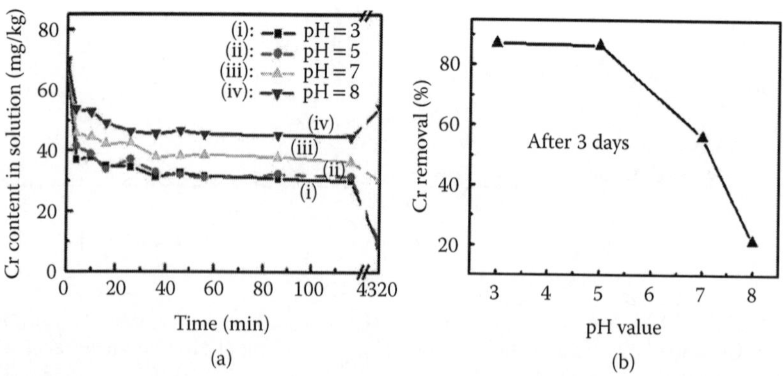

FIGURE 6.12 Initial pH value dependence of Cr-removal performances in $K_2Cr_2O_7$ aqueous solutions with initial Cr content 69.1 ppm. (a) The influence of the initial pH value on Cr-removal measurements; (b) the initial pH value–dependent Cr-removal capacity after 3 days (data from [a]). (Lin et al., *J. Mater. Chem.*, 21, 991–7, 2011. Reproduced by permission of The Royal Society of Chemistry.)

that the acidic solution is beneficial for Cr removal. Figure 6.12b illustrates the pH value–dependent Cr-removal capacity after 3 days. The Cr-removal capacity increases with the decrease of the pH value to 3. Fortunately, the practical industrial wastewater is usually the solution with neutral or weak acidity. So, the composite NF film is very much suitable for its treatment. As for the solution with strong acidity (pH < 1), the active component in NFs is unstable and dissolves [51].

Furthermore, Fe(II) content dependence of Cr-removal capacity for the composite NF film was also examined, as demonstrated in Figure 6.13a corresponding to the Cr removal as function of the soaking time in the solution for the composite NFs with different Fe(II) contents, under the same experimental conditions as those shown in curve 1 of Figure 6.11a. The significant Cr removal takes place within the initial 10 minutes for all samples, after which relatively slow removal is exhibited. The Cr-removal capacity (in unit weight of the NFs) in 60 minutes as a function of $FeCl_2$ content in the NF films could be obtained from Figure 6.3a, as shown in Figure 6.13b. When the $FeCl_2$ content in the NFs is less than 7%, the Cr removal, in unit weight of the NFs, in 60 minutes increases nearly linearly with the $FeCl_2$ content. It will reach the maximum if the content is up to about 11%wt. The higher $FeCl_2$ content decreases the Cr-removal capacity, as seen in Figure 6.13b.

Finally, it should be pointed out that few iron ions in the composite NFs would be released during soaking in the Cr-containing acidic solution, depending on the pH value of the solution. For instance, the concentration of iron ions increased by 84 ppm in the solution with initial pH 5 after 3 days. When the initial pH value of the solution is about 7, however, scarcely iron ions were detected [51].

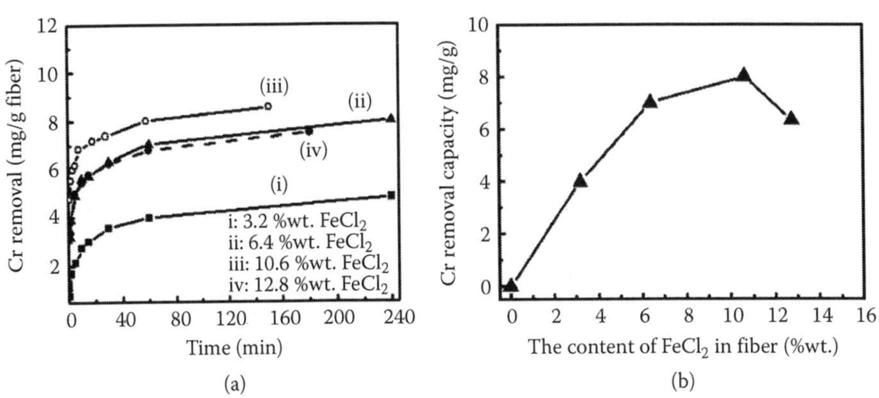

FIGURE 6.13 Dependence of Cr-removal performance on the Fe(II) content in the composite nanofibers (NFs). (a) Cr removals as a function of the soaking time in the solution for the composite NFs with different Fe(II) contents, under the same experimental conditions as those shown in curve 1 of Figure 6.11a. (b) The Cr-removal capacity in 60 minutes as a function of the Fe content in the composite NFs (data from [a]). (Lin et al., *J. Mater. Chem.*, 21, 991–7, 2011. Reproduced by permission of The Royal Society of Chemistry.)

6.3.2 PAN---Fe(II) Complexes–Induced Cr Adsorption

The removal process of the composite NFs could mainly be attributed to the complexes PAN---Fe(II) [56] in the porous NFs, which adsorb Cr from the solution. As we know, PAN is a polar polymer and partly self-associated by interchain dipole–dipole interaction at the CN bonds [56]. Here, the polymerization from dimethylformamide solution of PAN containing $FeCl_2$, during electrospinning, would form large amounts of weak coordination bonds PAN---Fe(II) in the composite polymer. The existence of Fe(II) in the composite NFs has been confirmed by XPS [51].

6.3.2.1 Hydrolyzation and Cr Adsorption

It is well known that reaction of iron ions with H_2O would form a series of products, such as $Fe(OH)^+$, $Fe(OH)_2$, $Fe(OH)_2^+$, $Fe(OH)^{2+}$, and $Fe(OH)_3$ [42]. Here, hydrolyzation reaction would initially take place, when the composite NFs with the complexes PAN---Fe(II) were put in the aqueous solution containing Cr(VI), forming $PAN\cdots Fe^{(II)}(OH)^+$, $PAN\cdots Fe^{(II)}(OH)_2$, $PAN\cdots Fe^{(III)}(OH)^{2+}$, $PAN\cdots Fe^{(III)}(OH)_2^+$, $PAN\cdots Fe^{(III)}(OH)_3$, and so on (or PANFe(II/III)OH for short) preferentially on the pore walls and the surface, or the reaction:

$$\text{PAN---Fe(II)} + H_2O \rightarrow [\text{PANFe(II/III)OH}]\downarrow + H^+ \quad (6.3)$$

took place, where PANFe(II/III)OH represents

$$\begin{cases} PAN\cdots Fe^{(II)}(OH)^+ \\ PAN\cdots Fe^{(III)}(OH)^{2+} \\ PAN\cdots Fe^{(II)}(OH)_2 \\ PAN\cdots Fe^{(III)}(OH)_2^+ \\ PAN\cdots Fe^{(III)}(OH)_3 \\ \cdots\cdots \end{cases} \quad (6.3')$$

These hydrolyzed products (PANFe(II/III)OH) will hold back most iron ions escaping from the NFs, leading to decrease of pH value of the solution due to consumption of OH^- from the solution.

Furthermore, among the hydrolyzed products, $PAN\cdots Fe^{(II)}(OH)^+$, $PAN\cdots Fe^{(III)}(OH)^{2+}$, and $PAN\cdots Fe^{(III)}(OH)_2^+$ on the pore walls within the porous composite NFs and in the surface of the NFs would adsorb the negative groups of $Cr_2O_7^{2-}$ from the solution due to their positively charged property. The adsorption on $PAN\cdots Fe^{(II)}(OH)^+$ could further induce the redox reaction. As previously reported [41,42,44], Fe(II) in the oxyhydroxides was oxidized to Fe(III) and Cr(VI) was reduced to Cr(III), forming the product $PAN\cdots Fe_{1-x}Cr_x(OH)_3$ and $PAN\cdots Fe_{1-x}Cr_x(OH)_x$ [57], or the reactions

$$\text{PANFe(II)OH}^+ + Cr_2O_7^{2-} \rightarrow PAN\cdots Fe_{1-x}^{(III)}Cr_x^{(III)}(OH)_3 + H^+ \quad (6.4)$$

and/or

$$\text{PANFe(II)OH}^+ + Cr_2O_7^{2-} \rightarrow PAN\cdots Fe_{1-x}^{(III)}Cr_x^{(III)}(OH)_x + H^+ \quad (6.4')$$

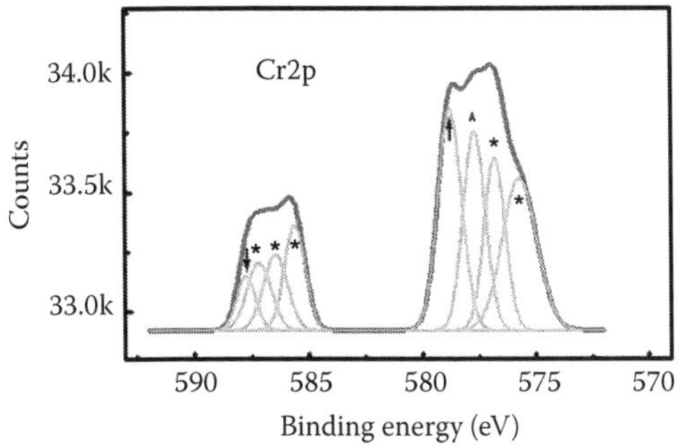

FIGURE 6.14 Narrow region x-ray photoelectron spectroscopic spectra of Cr2p for the polyacrylonitrile/$FeCl_2$ nanofibers after soaking in the $K_2Cr_2O_7$ aqueous solutions for 3 days. Star marks correspond to the Cr2p in Cr(III); arrow marks correspond to Cr2p in Cr(VI). (Lin et al., *J. Mater. Chem.*, 21, 991–7, 2011. Reproduced by permission of The Royal Society of Chemistry.)

would occur within the porous NFs, where $0 < x < 1$. The existence of Cr(III) in the composite NFs has been confirmed by XPS measurements, for the sample after soaking in $K_2Cr_2O_7$ solution, as demonstrated in Figure 6.14. In addition to the peaks at 578.7 and 587.9 eV, corresponding to the BE of adsorptive Cr(VI), the other peaks (star marks in Figure 6.14) are in good agreement with the published XPS spectra for oxides or hydroxides of Cr(III) (e.g., $Cr(III)_xO_y$ and $Cr(OH)_3$) [58–60]. It means that the composite porous NFs could not only remove Cr from wastewater but also reduce Cr(VI) to Cr(III), alleviating the Cr-induced toxicity.

Obviously, the high Cr-removal performance of the composite NFs should be attributed to their porous structure, that is, large surface area. The Fe(II) inside PAN is inactive since Cr that diffuses into the PAN should be insignificant, if any under the experimental condition (in a limited time period). After the composite porous NFs were added to the solution with Cr(VI), a quick decrease of Cr content in the initial stage (5 minutes) from the complexes on the surface of the porous NFs was found, due to their quick contact with $Cr_2O_7^{2-}$ in the solution. Subsequently, relatively slow Cr-removal takes place because the solution cannot easily enter the nanosized pores of the porous NFs, as seen by curves i–iii in Figures 6.12a and 6.13a. For the mixture of pure PAN NFs with $FeCl_2$ powders, there should exist small amount of complexes PAN---Fe(II) formed on the PAN NFs' surface during mixing, leading to the insignificant decrease of Cr content only in the initial stage (curve 3 in Figure 6.11a). Similarly, the low Cr-removal capacity for the cast film of PAN/$FeCl_2$ can be attributed to its low specific surface area (despite its porous structure). As for the sample of pure PAN, the electron-rich group (–CN) in the NFs presents negative charge on the surface [55], cannot adsorb $Cr_2O_7^{2-}$ radicals, and hence cannot remove Cr(VI) (curve 4 in Figure 6.11a).

6.3.2.2 Initial pH Value and Fe Content Dependence

As mentioned in Section 6.3.2.1, when the composite NFs with the complexes PAN---Fe(II) are exposed to aqueous solution, hydrolyzation reaction occurs according to reaction 6.3 and forms a series of products (see Equation 6.3'). However, the relative content among these hydrolyzed products depend on the initial pH value of the solution. Obviously, the lower initial pH value in the aqueous solution will induce the higher relative content of the hydrolyzed products PAN···$Fe^{(II)}(OH)^+$and PAN···$Fe^{(III)}(OH)^{2+}$, which is beneficial for the adsorption of $Cr_2O_7^{2-}$ radicals, leading to higher Cr-removal capacity, as shown in Figure 6.12. Otherwise, the high initial pH value induces formation of more PAN···$Fe^{(II)}(OH)_2$ and PAN···$Fe^{(III)}(OH)_3$ and less PAN···$Fe^{(II)}(OH)^+$ and PAN···$Fe^{(III)}(OH)^{2+}$. In this case, the Cr-removal capacity would be much lower. So, the composite NFs should be used for Cr removal in the acidic solution.

As for the dependence of the Cr removal on Fe content in the NFs, it is easily understood. Since the amount of complexes PAN---Fe(II) formed increases with rise in the Fe content in the composite NFs, the Cr-removal capacity in unit weight of the NFs increases nearly linearly with the Fe content when the content is not too high (e.g., <7% $FeCl_2$) (see Figure 6.13b). If the Fe content is high enough (about 11% $FeCl_2$ in this case), nearly all the –CN groups of PAN in the NFs are bonded with Fe(II), in which the Cr-removal capacity reaches the maximum (Figure 6.13b). When the Fe content is higher, the superfluous Fe(II) within the NFs would be dissolved into the Cr(VI)-contaminated aqueous solution during Cr-removal treatment. The Fe(II) dissolved into the solution will react with the $Cr_2O_7^{2-}$ ions in the solution, or the following reaction occurs:

$$Cr_2O_7^{2-} + 6Fe^{2+} + 14H^+ \rightarrow 2Cr^{3+} + 6Fe^{3+} + 7H_2O \qquad (6.5)$$

Obviously, such reaction would decrease the concentration of $Cr_2O_7^{2-}$ ions in the solution and hence reduce the adsorption amount of Cr in the NFs. So, when the Fe content is too high, the Cr-removal capacity decreases, as shown in Figure 6.13.

6.3.3 Micro/Nanostructured PAN@γ-AlOOH Fibers

It is well known that polymer is an excellent recyclable material with good elasticity and pliability [61]. As demonstrated in Section 6.3.1.1, the electrospun polymer fiber mats (or films) have exhibited great potential in environment remediation because it is freestanding and easily recyclable and can be integrated with the other matrixes conveniently [62–64]. Also, the fiber possesses several attractive advantages, such as comparatively low cost, applicability to various materials, and the ability to generate relatively large-scale integrated film, which are very important in practical applications [51]. Obviously, if the micro/nanostructured inorganic materials grow on the electrospun polymer fibers' surface, a flexible composite fiber film would be formed with 3D micro/nanostructures, as discussed in Section 4.3. Such structured composite material is expected to possess advantages of both the polymer fibers and micro/nanostructured inorganic materials and hence is to overcome the shortcomings of powdered nanomaterials.

In this section, we discuss the enhanced adsorption capacity and good recycling performance of the micro/nanostructured PAN@γ-AlOOH fibers as the adsorbent to remove heavy metal ions, Cr(VI), from model wastewater. The micro/nanostructured polymer (PAN) @γ-AlOOH composite fibers were fabricated based on electrospun fiber templated hydrothermal strategy [65]. PAN@γ-AlOOH micro/nanostructured material consists of cross-linked γ-AlOOH nanoplates nearly vertically standing on the PAN fibers and with thickness of 20–40 nm, as shown in Figure 4.12. This material could be used as an efficient adsorbent for removal of Cr(VI) from the solution with good reusability. Here, the adsorption capacity of γ-AlOOH (containing 47.4%wt. in the composite fibers) is employed to represent that of the composite fibers since the pure PAN cannot remove Cr from the solution, as mentioned in Section 6.3.1.

6.3.3.1 Adsorption Isotherms of Cr(VI) Ions

The adsorption isotherm of Cr(VI) ions (pH = 3) on the PAN@γ-AlOOH composite fibers, shown in Figure 4.12, is shown in curve 1 of Figure 6.15a, which demonstrates much stronger adsorption than the γ-AlOOH nanopowders (32 m^2/g in specific surface area) (curve 4 of Figure 6.15a), prepared according to the method previously described [13]. Quantitatively, the adsorption isotherms of Cr(VI) ions on the composite fibers and the powders are well subject to Freundlich model [24], or Equation 6.1, as shown in Figure 6.15b. Further experiments have revealed that the adsorption capacity of the composite fibers depends on the pH value of the Cr(VI)-containing solutions decreasing with rising pH value (see curves 1, 2, 3 in Figure 6.15a). Here, it should be pointed out that the adsorption treatment induces a slight increase of pH value of the solution.

6.3.3.2 Regeneration and Reusability of the PAN@γ-AlOOH Fibers

Furthermore, the regeneration and reusability of the PAN@γ-AlOOH fibers as an adsorbent for Cr(VI) removal was studied, and the experiments were carried out in four consecutive adsorption/desorption cycles. Firstly, 0.1 g composite fibers was added to 50 mL solution with Cr(VI) concentration 30 mg/L and at pH = 3, for 24 hours, to reach adsorption equilibrium. Subsequently, it was fished out by

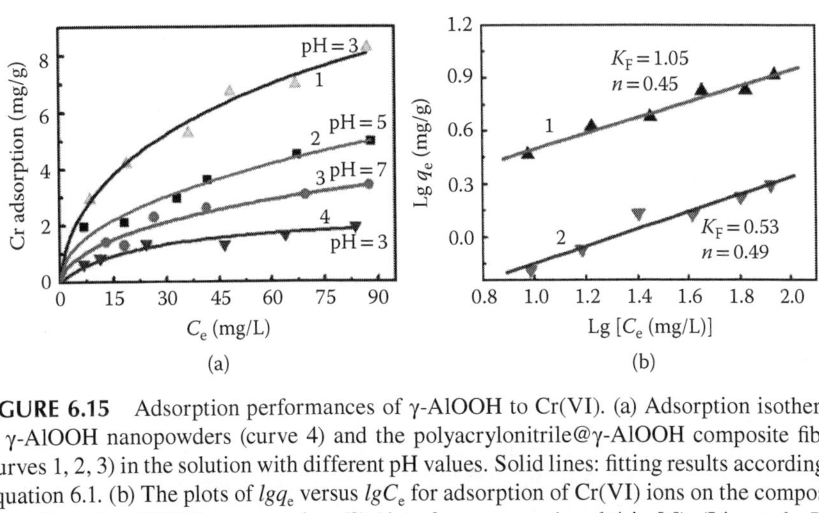

FIGURE 6.15 Adsorption performances of γ-AlOOH to Cr(VI). (a) Adsorption isotherms of γ-AlOOH nanopowders (curve 4) and the polyacrylonitrile@γ-AlOOH composite fibers (curves 1, 2, 3) in the solution with different pH values. Solid lines: fitting results according to Equation 6.1. (b) The plots of lgq_e versus lgC_e for adsorption of Cr(VI) ions on the composite fibers (1) and γ-AlOOH nanopowders (2) (data from curves 1 and 4 in [a]). (Lin et al., *RSC Adv.*, 2, 1769–73, 2012. Reproduced by permission of The Royal Society of Chemistry.)

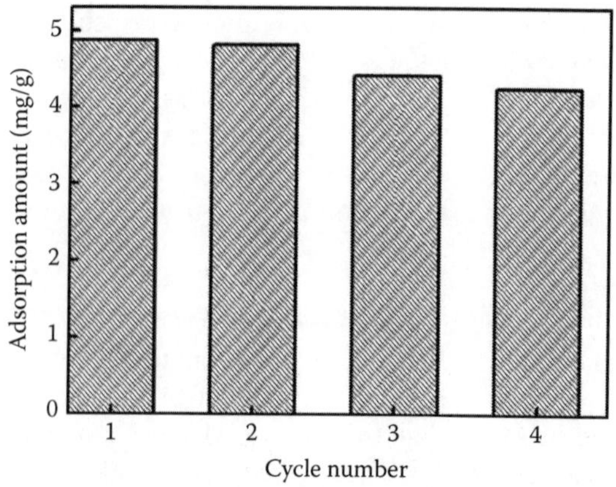

FIGURE 6.16 The cycled adsorption capacity of polyacrylonitrile@γ-AlOOH fibers to Cr(VI) in cycled adsorption/desorption experiments at pH = 3. (Lin et al., *RSC Adv.*, 2, 1769–73, 2012. Reproduced by permission of The Royal Society of Chemistry.)

tweezers and washed to neutrality by deionized water to remove the free Cr(VI) ions. After drying for 5 hours at room temperature, the washed composite fibers were then desorbed by soaking in 40 mL extraction medium (0.01 M NaOH solution) for 24 hours. Then the fibers were taken out, washed, and dried again for reuse. It has been shown that such PAN@γ-AlOOH fibers have stable recycling adsorption ability. Figure 6.16 demonstrates such recycling adsorption capacity for the composite fibers. The PAN@γ-AlOOH fibers still retained more than 85% of its original Cr(VI)-adsorption capacity after four cycles, demonstrating the stable recycling adsorption ability.

6.3.3.3 Structurally Enhanced and Protonation-Dependent Adsorption

The high adsorption capacity for the hierarchically structured PAN@γ-AlOOH fibers can mainly be attributed to their special structure. Furthermore, since the cross-linked γ-AlOOH nanoplates nearly vertically standing on the fibers are of high structural stability, they would prevent the morphology from destroying, during recycling experiments, and keep the high active surface area, showing good cycling performance. Importantly, different from the powdered adsorbent, which should be separated by centrifugation after remediation, the composite fibers can be simply collected and reclaimed by tweezers after experiment.

The pH value–dependent adsorption capacity could be attributed to the surface protonation of the adsorbent (γ-AlOOH) in acidic solution. At a low pH value, the γ-AlOOH surface is highly protonated and hence favors adsorption of Cr(VI) ions, which mainly exist in the form of $HCrO_4^-$ [66]. The surface protonation degree should decrease with increase in pH value, leading to reduction of adsorption capacity (see curves 1, 2, 3 in Figure 6.15a). In addition, the protonation is a proton-consuming reaction and would increase the pH values of the solution [67]. This has been confirmed by the experiment. Typically, the pH value increases to 4.2 after adsorption for the solution with initial pH = 3 in this study.

Altogether, due to the 3D micro/nanostructure-induced stability against aggregation, the PAN@γ-AlOOH composite fibers possess much higher adsorption capacity and good recyclability than the conventional γ-AlOOH nanopowders, as an adsorbent to remove the heavy metal ions, Cr(VI), from solution. These fibers could serve as a candidate adsorbent for environment remediation because of their effective recyclable performance and easy separation from solution.

6.4 MAGNETIC MICRO/NANOSTRUCTURED MATERIALS AS ADSORBENTS

As mentioned in Section 6.1, the adsorption based on adsorbents has been used because it can remove the pollutants completely without any harmful by-products left in water. The magnetic micro/nanostructured materials could be the ideal adsorbents. Also, for separation of adsorbents from water, magnetic materials could be a good candidate due to their easy collection at low magnetic fields [50,68]. In this section, we discuss the adsorption performances of the pollutants on two kinds of magnetic micro/nanostructured materials to demonstrate the importance of the materials with magnetic properties in pollutant removal.

6.4.1 Tremella-Like Fe_3S_4/C Magnetic Adsorbent

Among the normal adsorbents, carbonaceous materials have extensively been used due to their low cost and strong adsorption performance [69–73]. Also, there have been many reports about the magnetic carbon-containing materials [50,74–78], such as porous carbon sphere composites [50] and carbon composites containing nickel [74], iron [75], and iron oxide [76–78]. Fe_3S_4 has excellent magnetic properties [79–84], electrochemical hydrogen storage properties [80], and so on. Importantly, Fe_3S_4 is an environmental friendly material and could be used as an adsorbent for removal of heavy metals from the environment [81]. Here, we discuss the adsorption performances of the pollutant MB on the carbon-modified Fe_3S_4 or Fe_3S_4/C composite with micro/nanostructure, which has rich surface functional groups as demonstrated in Section 2.4, to demonstrate its possibility as a good adsorbent for pollutant removal in practical environmental remediation. Such micro/nanostructured magnetic Fe_3S_4/C composites, using solvothermal method with glucose as the carbon source at a low temperature, show tremella-like morphology, as shown in Figure 2.27, with the magnetic property and highly dispersed carbon on the surface, which supplies functional groups [85].

All the adsorption measurements were performed at 25°C. Herein, we take MB, which is a dye pollutant, as an example to demonstrate the enhanced adsorption performance of the tremalla-like magnetic Fe_3S_4/C composites compared with Fe_3S_4 nanoparticles without carbon. MB was initially dissolved in distilled water to obtain solutions with different initial concentrations. The initial pH value of the solutions was adjusted to 9.0 by adding HCl or NaOH solution. A measure of 10 mg adsorbent was added to each solution (10 mL) with agitation and maintained for up to 24 hours to establish adsorption equilibrium.

The adsorbent was separated from the solution quickly by a magnet after different adsorption intervals [85].

6.4.1.1 Adsorption Kinetics

Let us first examine the adsorption kinetics. Representatively, Figure 6.17a gives the optical absorbance spectra of 10 mL solution with initial MB concentration 10 mg/L and 10 mg Fe_3S_4/C composites after adsorption for different time periods. The adsorbent can be separated from the solution easily by a magnet (see the inset of Figure 6.17a). The color of MB solution became lighter within few minutes, indicating that the tremella-like magnetic Fe_3S_4/C composites can adsorb MB effectively and quickly. The optical absorption peak at 664 nm, corresponding to MB, almost disappears within 5 minutes. MB is removed completely after 20 minutes, as shown by curve 6 in Figure 6.17a.

According to the Beer–Lambert law [86], the maximal absorption value of MB corresponds to the wavelength at 664 nm. The concentration change of MB in the solution during adsorption was thus calculated. Correspondingly, Figure 6.17b demonstrates the concentrations of MB in the solution as a function of adsorption time. Furthermore, it has been found that the kinetic adsorption performance of MB on the Fe_3S_4/C composites can be well described by pseudo-second-order kinetic model [87–89], or

$$\frac{dq_t}{dt} = k(q_e - q_t)^2 \tag{6.6}$$

where q_t (mg/g) is the adsorption amount time t (minutes) and k (g/mg/min) is the rate constant. This model means that the driving force ($q_e - q_t$) is proportional to the

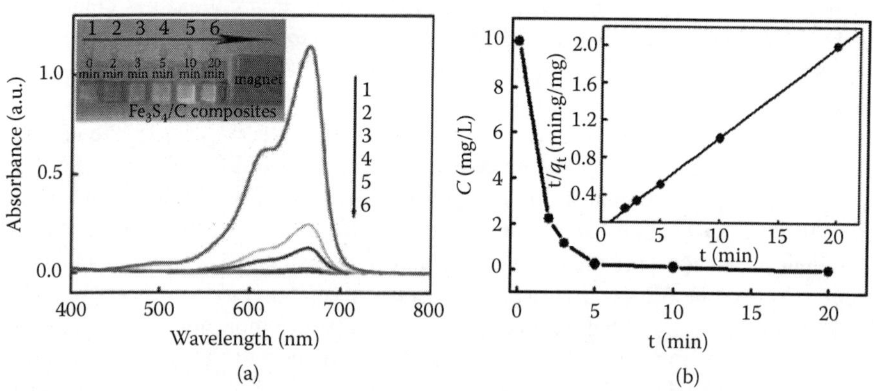

FIGURE 6.17 The adsorption kinetics of methylene blue (MB) on the Fe_3S_4/C composites. (a) Optical absorbance curves of 10 mL solution with MB 10 mg/L in initial concentration and 10 mg Fe_3S_4/C composites after different adsorption times. Curves 1–6 correspond to 0, 2, 3, 5, 10, and 20 minutes, respectively. The inset is photos of the MB solutions after adsorption for the corresponding time and separation of the composites from solution by a magnet. (b) Concentration evolution of MB in the solution with adsorption time. The inset: Plot of pseudo-second-order sorption kinetics of MB on Fe_3S_4/C composites (data from [b]). (Wang et al., *CrystEngComm,* 15, 2956–65, 2013. Reproduced by permission of The Royal Society of Chemistry.)

available fraction of active sites [90]. Furthermore, by integral and rearrangement, Equation 6.6 can be written as follows:

$$\frac{t}{q_t} = \frac{1}{kq_e^2} + \frac{1}{q_e}t \tag{6.7}$$

The plot of t / q_t versus t should be a linear relation. The constant k can thus be estimated from intercept of the plot. The inset of Figure 6.17b presents the corresponding plot, showing good linear relationship, indicating that kinetic adsorption behavior is well subject to pseudo-second-order kinetic model. The k value was determined to be 0.22 g/mg/min, which is much higher than that (0.0014 g/mg/min) of MB adsorption on the carbon adsorbent at the same initial concentration as previously reported [91]. These results indicate that MB can highly efficiently adsorb on the Fe_3S_4/C composites.

6.4.1.2 Adsorption Isotherms

By isothermal adsorption measurements, the micro/nanostructured tremella-like Fe_3S_4/C composite exhibits significantly structurally enhanced adsorption performance. Figure 6.18 gives the adsorption isotherms of MB on the Fe_3S_4/C composites and Fe_3S_4 nanoparticles. The equilibrium adsorption amount (or q_e) increases with the concentration of MB in the solution at lower concentration. At higher concentration, adsorption is saturated for both adsorbents. The saturated adsorption amount is much higher for the Fe_3S_4/C (~95 mg/g) than that for the Fe_3S_4 nanoparticles (~25 mg/g), showing significantly enhanced adsorption performance.

Furthermore, it has been revealed that the adsorption isotherms of both samples can be fitted according to Langmuir model, or Equation 6.2. The corresponding plots of $C_e/q_e - C_e$ are demonstrated in Figure 6.18b, exhibiting good linear relations. The

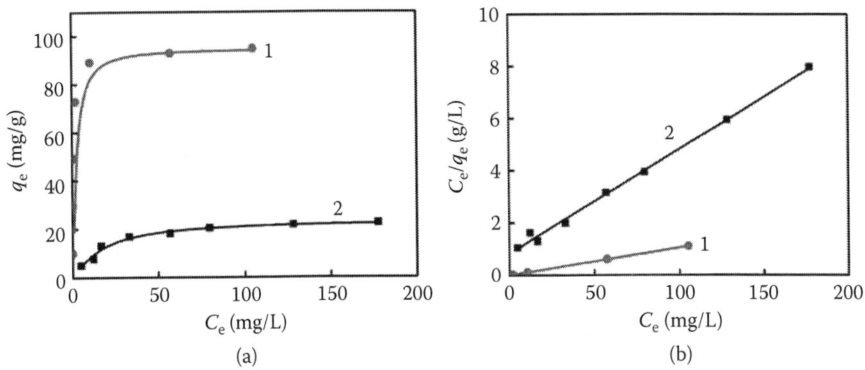

FIGURE 6.18 Adsorption behavior of methylene blue on the Fe_3S_4/C composites shown in Figure 2.27 (curve 1) and the Fe_3S_4 nanoparticles (curve 2). (a) Adsorption isotherms. The solid lines are the fitting results by Langmuir model (Equation 6.2). (b) The plots of $C_e/q_e - C_e$ for adsorption of MB (data from [a]). (Wang et al., One-step fabrication of high performance micro/nanostructured Fe_3S_4/C magnetic adsorbent with easy recovery and regeneration properties, *CrystEngComm,* 15, 2956–65, 2013. Reproduced by permission of The Royal Society of Chemistry.)

parameters were thus estimated to be $q^0 = 95$ mg/g, $K_L = 1.73$ L/mg for Fe_3S_4/C, and $q^0 = 25$ mg/g, $K_L = 0.049$ L/mg for Fe_3S_4. The constant value of K_L is much higher for the former than the latter indicating much stronger affinity to MB for the composites.

Here, it should be mentioned that morphology of the composites is almost unchanged after MB adsorption, indicating good structural stability of the tremella-like adsorbent during adsorption. Similarly, the XRD pattern after MB adsorption also shows the Fe_3S_4 phase, except the high diffraction background around 23°, which could be due to the adsorbed MB [85].

6.4.1.3 Regenerative Adsorption Performance

As we know, regeneration is an important performance for adsorbents. Some adsorbents could be regenerated by heating, which burns out the adsorbed organic pollutants, as previously reported [92,93]. However, for pollutants that cannot be pyrolyzed at high temperature, heating is not feasible. Besides, heating may not only release toxic gases but also change the structure of adsorbents, resulting in reduced adsorption capacity.

Herein, we discuss the regenerative adsorption performance of the Fe_3S_4/C composites. Recycling measurements of the adsorbent were carried out using ethanol as elution solution, which was adjusted to pH 5.0 by adding HCl or NaOH solution. The regenerated adsorbent was again immersed in 10 mL MB solution to evaluate its recycling efficiency (E), which is determined by

$$E(\%) = \frac{q_r}{q_0} \times 100\% \tag{6.8}$$

where q_0 and q_r are the adsorption amount (mg/g) of the as-prepared adsorbent and the regenerated adsorbent, respectively.

It has been found that the MB adsorbed on Fe_3S_4/C can be removed by soaking, as shown in Figure 6.19a. When the MB-adsorbed Fe_3S_4/C is immersed in the ethanol elution solution, the solution's color deepens with the soaking time, as demonstrated in the inset of Figure 6.19a. The corresponding optical absorbance spectra show that 30-minute soaking can almost completely desorb the MB from Fe_3S_4/C composites, as illustrated in Figure 6.19a. The MB concentrations in the ethanol elution solution after soaking for different time periods could be estimated by the optical absorbance spectral curve of 10 mg/L MB ethanol solution (curve 6 in Figure 6.19a), as shown in Figure 6.12b. After soaking in 10 mL ethanol for 30 minutes, ~90% adsorbed MB could be removed for 10 mg Fe_3S_4/C with 0.1 mg adsorbed MB. For comparison, if the MB-adsorbed Fe_3S_4/C is soaked in water at the same conditions, the MB removal rate is several times lower than that in ethanol, although MB is more soluble in water than in ethanol [85].

Furthermore, the adsorption capacity of the regenerated adsorbent was evaluated. Firstly, 10 mg regenerated adsorbent was added to 10 mL of 10 mg/L MB solution. After 20 minutes, the adsorbent was separated from solution by a magnet before washing by ethanol for reuse. The MB concentration of the remnant solutions could be obtained by the optical absorbance spectral measurements. The recycling efficiency of the regenerated Fe_3S_4/C composites could thus be estimated according to

Equation 6.8, as shown in Figure 6.20, corresponding to the adsorption efficiency as a function of recycling times. The adsorption efficiency is gradually decreased with the recycling times because some of the active sites are occupied by MB molecules and could not be washed out completely by ethanol solvent in each cycle. However, the adsorption efficiency is still 92% after recycling for two times.

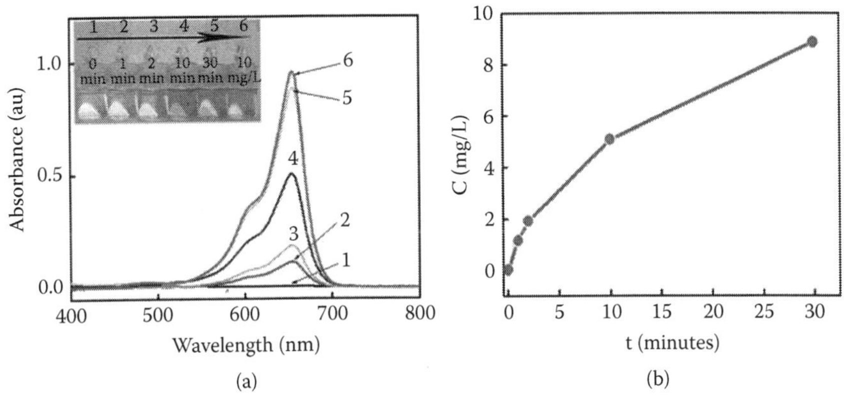

FIGURE 6.19 Desorption measurements of methylene blue (MB)-adsorbed Fe_3S_4/C composites. (a) Optical absorbance spectra of 10 mL ethanol elution solution after addition of 10 mg MB-adsorbed Fe_3S_4/C composites, which is shown in curve 6 of Figure 6.17a, for different time periods. Curves 1–5 correspond to 0, 1, 2, 10, and 30 minutes in ethanol, respectively. Curve 6 corresponds to the ethanol solution with 10 mg/L MB. The inset is the photo of the ethanol solutions after addition of the MB-adsorbed Fe_3S_4/C composite for different time periods. (b) The MB concentration in the ethanol elution solution as a function of the soaking time. (Wang et al., *CrystEngComm,* 15, 2956–65, 2013. Reproduced by permission of The Royal Society of Chemistry.)

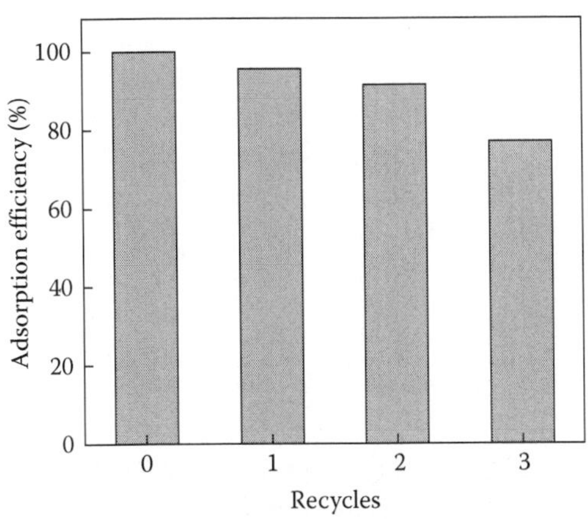

FIGURE 6.20 Adsorption efficiency of the Fe_3S_4/C composite as a function of recycling times. (Wang et al., *CrystEngComm,* 15, 2956–65, 2013. Reproduced by permission of The Royal Society of Chemistry.)

6.4.1.4 Enhanced Surface Carbon-Induced Adsorption and Desorption

The enhanced adsorption and reusable performance, for the Fe_3S_4/C adsorbent, could be easily understood, which can be attributed to its micro/nanostructure and surface carbon-induced functional groups. High surface area would lead to high adsorption capacity. However, the specific surface area (12 m^2/g) for the Fe_3S_4/C adsorbent is only twice as high as that of Fe_3S_4 nanoparticles (5 m^2/g) and the saturated adsorption amount for the former is nearly four times as high as that of the latter. Obviously, in addition to the structurally enhanced adsorption, the surface functional groups should also be responsible for the enhanced adsorption capacity.

There exist many functional groups on the surface of Fe_3S_4/C composites, such as C=C, C=O, C–O, C–H, and –OH groups [85]. Due to existence of C=O, –OH groups, the surface of Fe_3S_4/C is negatively charged [35]. MB is an organic positively charged molecules, which contains many functional groups, such as N=C and S=C. Therefore, the Fe_3S_4/C composites can adsorb the positively charged MB molecules by electrostatic interaction [92]; besides, there are some π–π interactions [94] between the adsorbent's surface and MB molecules. The FTIR spectrum for the MB-adsorbed composite has confirmed the strong interaction between MB and the Fe_3S_4/C composites, as shown in curve 2 of Figure 6.21. The peak at 1115 cm^{-1} is assigned to the C–N signal derived from MB molecules, indicating the adsorption of MB on Fe_3S_4/C. The signal of –OH shifts from 3430 to 3450 cm^{-1} after adsorption of MB (see curves 1 and 2 in Figure 6.21). This could be attributed to the increased electron cloud density of –OH after MB adsorption due to formation of hydrogen bonding between the –OH and MB molecules, such as O–H...N and O–H...S, indicating strong chemical bonding with the adsorbate.

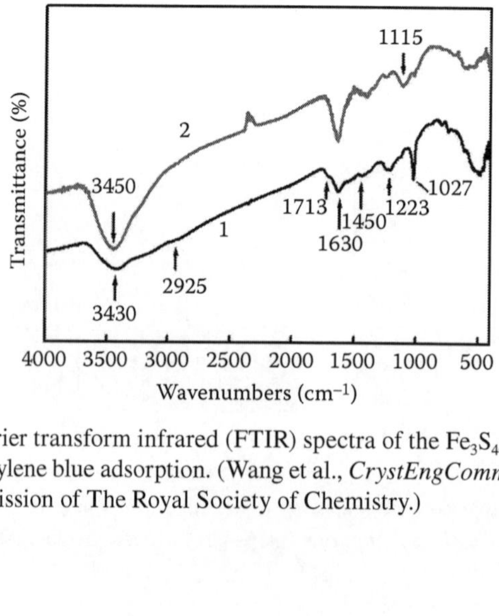

FIGURE 6.21 Fourier transform infrared (FTIR) spectra of the Fe_3S_4/C composites before (1) and after (2) methylene blue adsorption. (Wang et al., *CrystEngComm,* 15, 2956–65, 2013. Reproduced by permission of The Royal Society of Chemistry.)

For good reusable performance of Fe_3S_4/C, it can also be attributed to the carbon-induced functional groups. During soaking in the elution solution, the ethanol molecules interact with hydroxyl groups on the surface of the Fe_3S_4/C composites and form the hydrogen bonding O–H…O, which is stronger than the hydrogen bonding between MB and the adsorbent (or the O–H…N, O–H…S between MB and hydroxyl groups of the adsorbent). The adsorbed MB would thus be substituted by ethanol, resulting in desorption of MB and regeneration of the adsorbent. As for the much better desorption performance in ethanol than in water, it can mainly be attributed to the existence of a significant amount of hydrophobic groups (such as C–H, C=C, and C–C) on surface of the adsorbent, in addition to the hydroxyl groups. These hydrophobic groups on the adsorbent's surface would resist integrating water with the adsorbent's surface [50]. Instead, the organic ethanol molecules are more favorable for approaching the adsorbent's surface owing to their surface hydrophobic groups [95]. Actually, this is in good agreement with the previous report about desorption of methyl orange in ethanol [50]. In addition, the electrostatic interactions are weak in the desorption process due to the much lower pH value (5.0) [35], leading to almost complete (~90%) desorption of MB in ethanol.

In a word, the specific surface area of the Fe_3S_4/C composites is not high but the adsorption capacity to MB (~95 mg/g) is not low, but higher than that of some activated carbons reported previously [96]. Although many activated carbons and carbon nanotubes have larger adsorption capacity to MB (>100 mg/g) [97–99], they cannot be separated from the solution easily and quickly by magnetic field after use. Comparatively, the Fe_3S_4/C composites can expediently be collected due to the magnetic property and prepared by one-step hydrothermal procedure with low cost and high productivity [85]. Therefore, the micro/nanostructured tremella-like magnetic Fe_3S_4/C composites could be a good candidate of adsorbent for water treatment, due to their easy availability, and especially, the easy separation by magnet and good recycling.

6.4.2 Micro/Nanostructured Porous Fe_3O_4 Nanofibers As an Effective and Broad-Spectral Adsorbent

In Section 2.3, we described fabrication of the magnetite (Fe_3O_4) porous NFs based on protein-assisted hydrothermal method in citric aqueous solution. Such Fe_3O_4 NFs are nearly cylindrical in shape, several tens of micrometers in length and 120 nm in mean diameter. The whole single fibers are packed by the ultrafine Fe_3O_4 nanoparticles (<5 nm in size) with similar crystal orientation (along [111] direction) and show porous structure with pore size below 5 nm and high specific surface area (123 m^2/g), exhibiting good magnetic property, as mentioned in Section 2.3 [100]. It has been found that, due to the high surface area, strong magnetic property, and easy separation, such ultrafine magnetite nanoparticle–built porous fibers can be used as an effective adsorbent for removal of some toxic chemicals, which are usually difficult to remove, not only heavy metal anions and cations with good recycling performance, but also some nonpolar contaminant molecules in solution, and have exhibited significantly structurally enhanced adsorption and very high removal performance of Cr(VI) (anions), Hg(II) (cations), and polychlorinated biphenyl (PCB-77, or $C_{12}H_6Cl_4$) (nonpolar molecules) from the contaminated solutions. This is a new and very promising environmental material and makes it possible to develop a lower magnetic separation process to efficiently enrich and remove the toxic chemicals in one step, which is particularly important in environment remediation. The details are discussed in this section.

6.4.2.1 Highly Effective Removal of PCBs

PCB-77 is a kind of persistent organic pollutant and very deleterious to human, even in a trace amount [5,101]. There exist PCBs in some soils, which can be washed out and dissolved in organic solutions such as acetone and n-hexane (C_6H_{14}). How to effectively enrich the trace of PCBs in the organic solutions is still not known and is challenging. Here, we discuss the removal behavior of PCB-77 in the n-hexane solution by the magnetite porous NFs as shown in Figures 2.20 and 2.21. A certain amount of porous NFs was initially added to a certain amount of n-hexane solution with PCB-77 in a beaker. Then, the beaker was sealed in the dark, put on a bed, and vibrated for a certain time period. A magnet of about 20 Gauss was used for collecting the magnetite NFs from the solution. Finally, the content of PCB-77 in the solution was measured by gas chromatographic method.

Figure 6.22a shows the PCB-77 content in the solution versus function of the amount of the NFs, added in 100 mL solution with initial PCB-77 concentration of 0.1 μmol/L (a very low concentration) in 5 minutes. The content of PCB-77 in the solution, after adsorption for 5 minutes, decreases nearly linearly with increasing

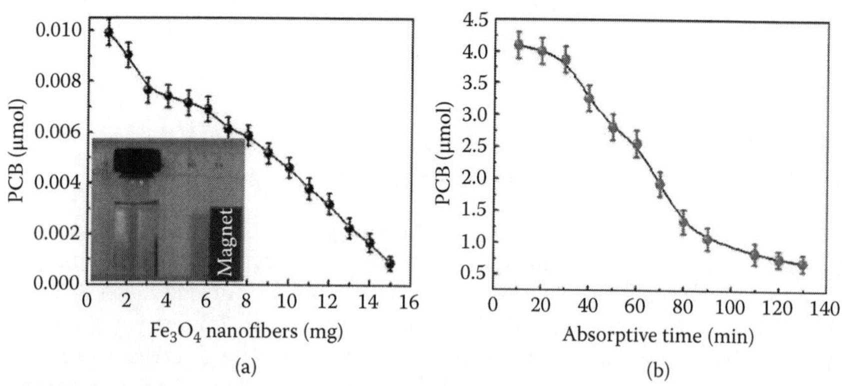

FIGURE 6.22 Adsorption of polychlorinated biphenyl (PCB-77) on the magnetite porous nanofibers, shown in Figure 2.19, in the n-hexane solution. (a) PCBs-77 content in the 100 mL n-hexane solution, with initial PCB-77 concentration 0.1 μmol/L, as a function of addition amount of the fibers (soaking in the solution for 5 minutes). The bottom-left inset is a photo showing the enrichment of the magnetite fibers on the bottle wall close to the magnet at about 15 cm distance from the solution. (b) The evolution of PCB-77 content in the solution with adsorption time, for 1.00 g magnetite fibers in 1000 mL solution with initial PCB-77 concentration 4.1 μmol/L. (Han et al., *J. Mater. Chem.*, 21, 11188–96, 2011. Reproduced by permission of The Royal Society of Chemistry.)

magnetite amount. With the addition of 15 mg porous NFs to the solution, more than 90% PCB-77 can be removed in 5 minutes. Figure 6.22a presents the time evolution of PCB-77 content in 1000 mL solution with much higher initial PCB-77 concentration (4.1 μmol/L) after addition of 1.0 g magnetite porous NFs. An 82% PCB-77 in the solution can be removed within 2 hours. The removal capacity of the porous NFs in this case could be estimated to be more than 1 mg PCB-77 per 1 g fibers. It means that the as-prepared porous magnetite NFs can effectively and expediently remove a trace of PCBs-77 in the solution.

Such high removal capacity of PCB-77 could be attributed to the porous structure and high specific surface area of the NFs. Compared with the normal adsorbents, such as activated carbon, an obvious advantage of this porous fibers as adsorbent is easy magnetic separation from the solution, as shown in the bottom-left inset of Figure 6.22a. Similarly, for the other nonpolar contaminant molecules, such as methyl orange and methyl blue, the porous NFs also exhibit high removal capacities [100].

6.4.2.2 Strong Adsorption of Heavy Metal Cations and Anions

Furthermore, such magnetite porous NFs can also be used as broad-spectral adsorbent for high effective removal of heavy metal ions, both cations, such as Pb^{2+}, Cd^{2+}, Cu^{2+}, and Hg^{2+}, and anions, such as, Cr(VI) (or $Cr_2O_7^{2-}$), from wastewater.

For cations, typically, the Hg^{2+}-removal results are shown in Figure 6.23 for the porous NFs with different amounts in 100 mL aqueous solution with initial Hg content of 100 mg/L (pH = 5.0). The removal capacity is more than 500 mg Hg per 1 g NFs.

For anions, representatively, Cr(VI) was taken as the adsorbate. Figure 6.24a gives the results of Cr(VI) removal for the porous NFs with 25, 45, and 60 mg, respectively, in 100 mL aqueous solution with 10 mg Cr(VI) (pH = 5). The removal

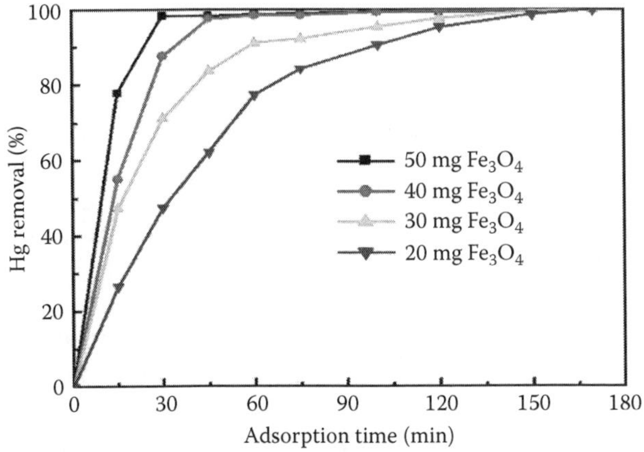

FIGURE 6.23 Hg removal as a function of adsorption time, for the magnetite nanofibers, with different addition amounts, in 100 mL aqueous solution with initial Hg(II) concentration of 100 mg/L. (Han et al., *J. Mater. Chem.*, 21, 11188–96, 2011. Reproduced by permission of The Royal Society of Chemistry.)

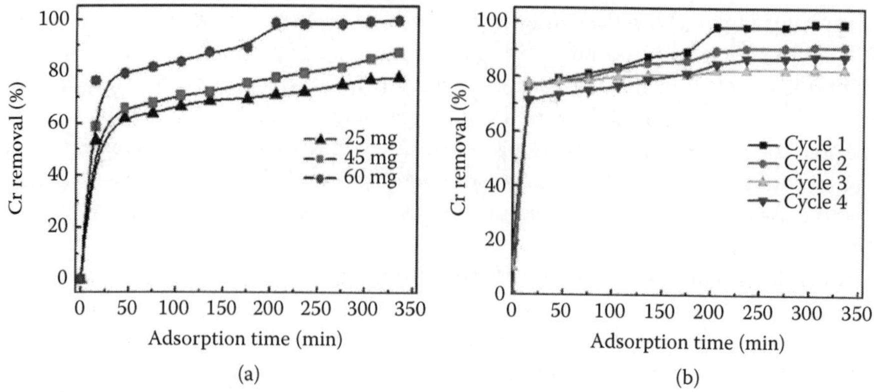

FIGURE 6.24 Cr removal as a function of adsorption time in 100 mL aqueous solution with initial Cr(VI) concentration of 100 mg/L. (a) For the magnetite nanofibers with different amounts. (b) Cr-removal performance in four cycles for the magnetite nanofibers with 60 mg. (Han et al., *J. Mater. Chem.*, 21, 11188–96, 2011. Reproduced by permission of The Royal Society of Chemistry.)

capacity depends on the amount of the fiber addition and is estimated to be more than 250 mg Cr per 1 g NFs. This is much higher than the reported values of the other nanomaterials as Cr(VI)-removal adsorbents, such as PAN/ferrous chloride composite porous fibers (11.7 mg Cr per 1 g fibers) [51], ceria hollow nanospheres (15.4 mg Cr per 1 g CeO_2) [52], magnetite Fe_3O_4 nanoparticles (2.95 mg Cr per 1 g Fe_3O_4) [53], γ-Fe_2O_3 nanoparticles or mesoporous γ-Fe_2O_3 (14.6–15.6 mg Cr per 1 g γ-Fe_2O_3) [54], and the nanosized Akaganéite (80 mg Cr/g) [45]. In addition, those nanomaterials should be centrifugalized for the separation from the contaminated solution after treatment.

6.4.2.3 Recycling Performance

Furthermore, recycling measurement was carried out for Cr(VI). The porous Fe_3O_4 NFs, which were immersed in the Cr(VI) aqueous solution for adsorption for a determined time period, were first collected by a magnet. The collected fibers were ultrasonically cleaned in deionized water for 5 minutes, washed with 0.1 mol/L HCl and then distilled water, and rinsed with 0.1 mol/L NaOH followed by distilled water. Such washing procedures were repeated until the rinsed water turned colorless. The collected fibers can also be cleaned in alcohol aqueous solution The cleaned NFs were then dried at 80°C for 24 hours before reuse as adsorbent. It has been shown that the magnetite porous fibers are of good recycling performance as an adsorbent of heavy metal ions [Cr(VI)]. The ions adsorbed on the fibers could be washed out by dilute acid solution or alcohol aqueous solution and the magnetite fibers can be reused. Figure 6.24b illustrates the recycling results of Cr removal for the porous NFs with 60 mg in 100 mL aqueous solution with initial Cr(VI) content of 100 mg/L. It can be seen that the Cr-removal rate is still up to 80% after four cycles.

To be brief, the porous magnetite NFs possess significantly structurally enhanced adsorption performance and can thus be used as a highly effective adsorbent for removal of some toxic chemicals that are usually difficult to remove, not only heavy metal anions and cations with good recycling performance, but also some nonpolar contaminant molecules in solution. This could be attributed to the porous structure and high specific surface area of the fibers. Further work about adsorption of the fibers to contaminants in solutions is needed. This material aids lower magnetic separation process and effectively removes toxic chemicals in one step, which is particularly important in environment remediation.

6.5 MICRO/NANOSTRUCTURED POROUS METAL SILICATE HOLLOW SPHERES AS EFFICIENT ADSORBENTS

As mentioned in the foregoing sections of this chapter, nanostructured material–based technologies are promising for environmental remediation. The hierarchical micro/nanostructured materials can avoid aggregation and maintain high specific surface areas, which are important in enhancing the accessibility of adsorbates to the reactive sites [9]. Micro/nanostructured hollow colloidal particles represent a promising type of nanomaterials, and the special structure contributes to the hollow spheres with low density, high specific surface areas, void properties, and so on. These spheres have proven to be excellent in widespread applications, including lithium ion batteries, catalysis, sensors, drug delivery, and controlled drug release [102–105].

Silicates are one of the most interesting and complicated class of minerals by far. Approximately 30% of all minerals are silicates and 90% of the earth's crust is made up of silicates. The basic unit of silicates is a tetrahedron-shaped anionic group with a negative four charge, which can be linked to each other in different modes and form as single units, double units, chains, sheets, rings, and framework structures. The structure offers attractive chemical and physical properties in the field of adsorption. For instance, sepiolite has always been used as efficient adsorbents due to their charged surface. In this section, we discuss the adsorption performances of the micro/nanostructured porous metal silicate hollow spheres. Such metal silicate hollow spheres can be produced by the template-etching strategies, as described in Section 3.2.

6.5.1 Magnesium Silicate Micro/Nanostructured Hollow Spheres

The chemical-template-synthesized magnesium silicate hollow spheres possess a porous surface with a morphology and structure similar to sepiolite, as shown in Figure 3.5. Such as-prepared porous magnesium silicate hollow spheres have exhibited good adsorption ability in the removal of cationic dyes and heavy metal ions from model wastewater [106]. The magnesium ions in water are not harmful to the environment. So, the prepared magnesium silicate hollow spheres could be used as an environmentally benign and efficient adsorbent.

Here, the adsorption performance of the porous magnesium silicate hollow microspheres in model wastewater is discussed. MB, a common cationic dye in the textile industry, was chosen as a model organic pollutant in the adsorption experiments. A certain amount of magnesium silicate hollow spheres (20 mg) was mixed with MB solution (40 mL) of concentration 75 mg/L at room temperature with no other additives. After several hours, the magnesium silicate hollow spheres were deposited at the bottom of the bottle under gravitation, and the solution was characterized by optical absorbance spectroscopy. Figure 6.25 shows the MB concentration evolution in the solution with adsorption time. More than 90% of MB in the solution could be removed in less than 10 minutes. No significant variation in the residual dye concentration was detected after adsorption for 30 minutes. The preceding results primarily show that the magnesium silicate hollow spheres could be used as a fast and efficient adsorbent.

The adsorption isotherm of MB on the as-prepared magnesium silicate hollow spheres is shown in curve 1 of Figure 6.26a and could well be described by Langmuir model or Equation 6.2. The maximum adsorption capacity was found to be 207 mg/g for MB, which is much higher than the adsorption capacity of the treated sepiolite (60 mg/g) under the same conditions [107] and indicates a better adsorption performance for such as-prepared magnesium hollow spheres. After adsorption of MB, this material could be regenerated by combustion at 400°C in air for 4 hours, and the regenerated magnesium silicate hollow spheres still exhibit good adsorption performance, as shown in curves 2–4 in Figure 6.26a. It should be pointed out that the removal of pollutants in water by the magnesium silicate hollow spheres is not limited to MB. The other cationic dyes, such as methyl violet, can also be removed effectively, as seen from the adsorption isotherm, in Figure 6.26b, with an adsorption capacity of 180 mg/g, which is larger than the adsorption capacity of sepiolite (68 mg/g) [107].

In addition, the as-prepared magnesium silicate hollow spheres could also be used as absorbents to remove heavy metal ions. Here, lead ions were selected as an example. Figure 6.27 shows the corresponding adsorption isotherm, which can also be

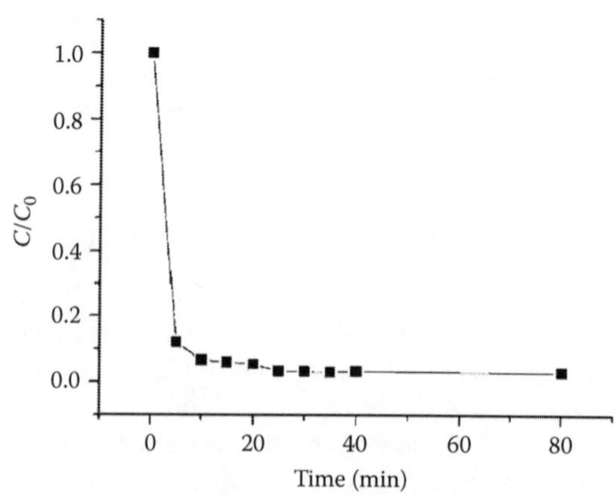

FIGURE 6.25 Evolution of the normalized methylene blue (MB) concentration in the solution with adsorption time (20 mg silicate hollow spheres in 40 mL solutions with 75 mg/L MB). (Wang et al., *Chem. Eur. J.* 2010. 16. 3497–503. Copyright Wiley-VCH Verlag GmbH & Co. KGaA. Reproduced with permission.)

well fitted by Langmuir model. Most lead ions could be removed at low concentrations, and the maximum adsorption capacity reached about 300 mg/g for lead ions, which was several times higher than that of acid-treated sepiolite (94 mg/g) [108]. The adsorption of metal ions onto magnesium hollow spheres could also be extended to other metal ions, like copper ions (see the inset of Figure 6.27).

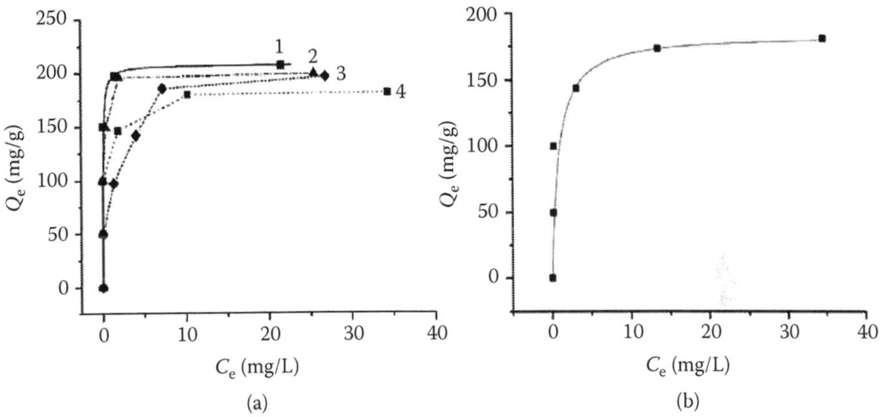

FIGURE 6.26 (a) Adsorption isotherms of methylene blue on the new as-prepared magnesium silicate hollow spheres (1) or those regenerated once (2), twice (3), or three times (4). (b) Adsorption isotherm of methyl violet on the as-prepared magnesium silicate hollow spheres at room temperature. (Wang et al., *Chem. Eur. J.* 2010. 16. 3497–503. Copyright Wiley-VCH Verlag GmbH & Co. KGaA. Reproduced with permission.)

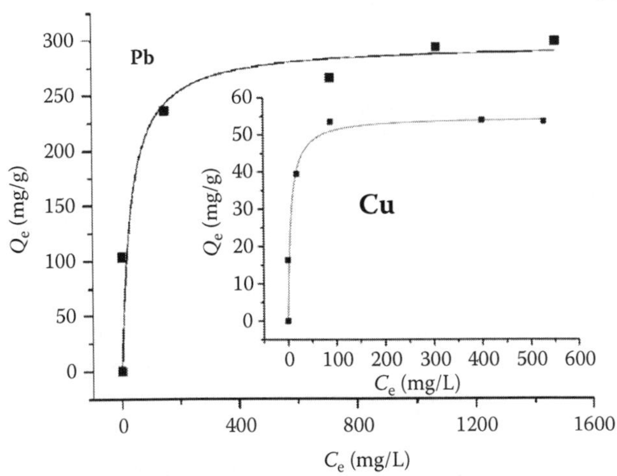

FIGURE 6.27 Adsorption isotherms of lead ions and copper ions (the inset) on the as-prepared magnesium silicate hollow spheres at room temperature. (Wang et al., *Chem. Eur. J.* 2010. 16. 3497–503. Copyright Wiley-VCH Verlag GmbH & Co. KGaA. Reproduced with permission.)

Compared with the sepiolite adsorbent, the micro/nanostructured magnesium silicate hollow spheres are promising materials for the removal of weakly biodegradable pollutants and toxic metal ions from wastewater. The high adsorption performance can be explained as follows. As we know, the sepiolite is an attracting adsorbent due to its special structure, which is formed by alternation of blocks and tunnels that grow up in the microfiber direction; the blocks have talc structure [109]. The magnesium silicate hollow spheres are composed of narrow and nanoscaled lamellae. Comparing both in structure, we found that the separate narrow and nanoscaled lamellae of the magnesium silicate hollow spheres could be seen as the blocks of sepiolite and the space between lamellae could be seen as the tunnels in sepiolite. Thus, a single magnesium silicate hollow sphere seemed to be a big "sepiolite" particle. The nitrogen adsorption has revealed that the specific surface area is much higher for the magnesium silicate hollow spheres than that of sepiolite. On the other hand, the size of the pores between the lamellae was larger than that of tunnels, which favors the fast diffusion of dye molecules. Thus, the porous structure of the artificial magnesium silicate hollow spheres could increase the amount of surface active adsorption sites and enhance the adsorption capacity and rate greatly. Taking into consideration that natural sepiolite often needs to be activated with complex physical and chemical processes for high adsorption performance, these artificial magnesium silicate hollow microspheres could be potential efficient absorbents for the removal of some toxic pollutants from water due to their advantageous structure.

6.5.2 Copper Silicate Micro/Nanostructured Hollow Spheres

The copper silicate micro/nanostructured hollow spheres are built of a lot of nanotubes and most of the nanotubes stand vertically on the surface of the spheres with open ends, as illustrated in Figures 3.2 and 3.3. Such copper silicate hollow microspheres have a specific surface area of 270 m^2/g and pore size of 3.2 nm of the nanotubes. Combination of the large specific surface area and thermal stability of silicate would enhance the potential applications of the copper silicate hollow spheres as an efficient adsorbent. It has been found that these hollow spheres exhibited excellent dye removal capability [93].

Here, the application possibility of the micro/nanostructured copper silicate hollow spheres in wastewater treatment is demonstrated. The adsorption isotherm of MB was obtained by varying its initial concentration in the solution without any additives, as illustrated in curve 1 of Figure 6.28. We can see that 1 g as-prepared copper silicate hollow micorspheres can remove 162 mg MB, which is much higher than that of the natural sepiolite with an adsorption capacity of 58 mg/g [107]. Furthermore, the copper silicate hollow spheres containing MB could be renewed by combustion at 300°C in air for 4 hours, and the renewed copper silicate hollow spheres still exhibit a large adsorption performance as shown by curve 2 in Figure 6.28. These results show that the micro/nanostructured copper silicate hollow sphere is a novel material for the removal of weakly biodegradable pollutants, which indicates a great potential application in the practical dye removal treatment.

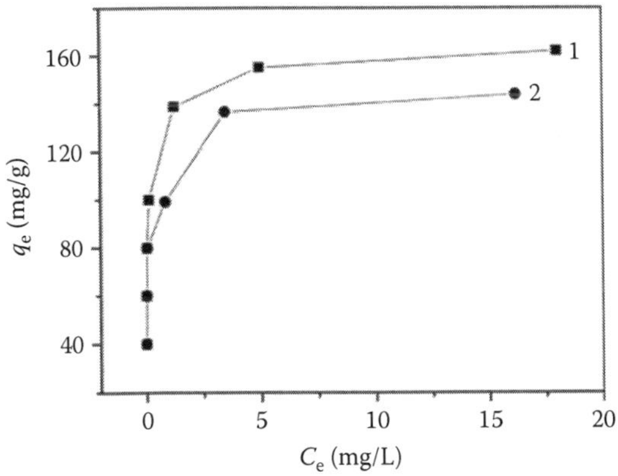

FIGURE 6.28 Adsorption isotherm of methylene blue on the copper silicate hollow microspheres. Curves 1 and 2: the as-prepared and renewed hollow microspheres, respectively. (Wang et al., *Chem. Commun.*, 6555–7, 2008. Reproduced by permission of The Royal Society of Chemistry.)

6.5.3 Nickel Silicate Micro/Nanostructured Hollow Spheres

Nickel silicate micro/nanostructured hollow spheres could be fabricated based on the template-etching strategy, as described in Section 3.2. The nickel silicate hollow spheres have a shell thickness of 100 nm and porous shell layer, which consists of nanoscaled lamella, as illustrated in Figures 3.9 and 3.10. Such hollow spheres are of high specific surface area (about 350 m^2/g). Like the metal silicate materials mentioned in Sections 6.5.1 and 6.5.2, this material also exhibits good adsorption performance.

Here, MB is used as the adsorbate to demonstrate the adsorption performance. Figure 6.29 gives the corresponding adsorption isotherm of MB on the nickel silicate hollow microspheres (curve 1), which can also be well described by Langmuir

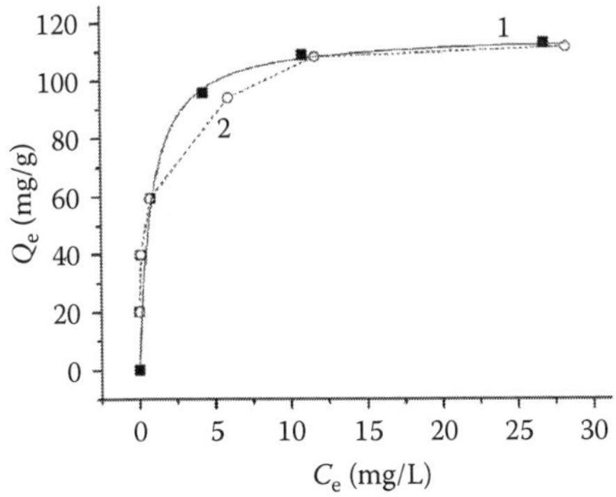

FIGURE 6.29 Adsorption isotherms of methylene blue on the nickel silicate micro/nanostructured hollow spheres. Curve 1: on the fresh microspheres. Curve 2: on the regenerated microspheres. (Reprinted with permission from Wang et al., 2010, 14830–4. Copyright 2010 American Chemical Society.)

adsorption model or Equation 6.2. The maximum adsorption is found to be 113 mg/g, which is also much higher than 58 mg/g of the sepiolite in the same conditions [107]. This indicates that the porous nickel silicate hollow microspheres have a good adsorption performance. In addition, the porous nickel silicate hollow spheres could be refreshed by combustion at 400°C in air for 4 hours, and the regenerated ones still exhibit large adsorption performance, as shown in curve 2 of Figure 6.29. These results show that the nickel silicate micro/nanostructured hollow spheres are also a dye removal agent for the removal of pollutants in weakly biodegradable materials.

6.6 BRIEF SUMMARY

In this chapter, we have demonstrated the good adsorption performances of some typical micro/nanostructured materials, including oxide semiconductors (ZnO), magnetic compounds (Fe_3O_4, FeS/C), composite NFs, and metal silicates, which could be produced by three main synthetic strategies: (1) solvothermal/hydrothermal methods, (2) template-etching strategies, and (3) electrospinning method, as described in the Chapters 2 through 4. Since the micro/nanostructured materials possess large specific surface area and high structural stability (anti-aggregation), and can maintain high active surface area during use, they exhibit much stronger adsorption capacity, and hence much higher contaminant-removal ability than the normal nanopowder materials, showing significantly structurally enhanced performances. So, these micro/nanostructured materials are really the promising materials for contaminant removal and environmental remediation.

It should be mentioned that surface functionalization of the micro/nanostructured materials is still a challenge. This problem could be partially overcome by surface modification. More importantly, their structurally enhanced performances for environmental remediation are attracting more and more attention. The advantage of micro/nanostructured materials lies in their high surface area, aggregation resistance, and easy collection. The application potentials of these aspects, although still in infancy, offer exciting opportunities and challenges for the newcomers.

REFERENCES

1. Ru, J.; Liu, H.; Qu, J.; Wang, A.; and Dai, R. 2007. Removal of dieldrin from aqueous solution by a novel triolein-embedded composite adsorbent. *J. Hazard Mater.* 141: 61–9.
2. Schwarzenbach, R. P.; Escher, B. I.; Fenner, K.; Hofstetter, T. B.; Johnson, C. A.; von Gunten, U.; and Wehrli, B. 2006. The challenge of micropollutants in aquatic systems. *Science* 313: 1072–7.
3. Shannon, M. A.; Bohn, P. W.; Elimelech, M.; Georgiadis, J. G.; Marinas, B. J.; and Mayes, A. M. 2008. Science and technology for water purification in the coming decades. *Nature* 452: 301–10.
4. Theron, J.; Walker, J. A.; and Cloete, T. E. 2008. Nanotechnology and water treatment: Applications and emerging opportunities. *Crit. Rev. Microbiol.* 34: 43–69.
5. Varanasi, P.; Fullana, A.; and Sidhu, S. 2007. Remediation of PCB contaminated soils using iron nano-particles. *Chemosphere* 66: 1031–8.
6. Peng, X.; Li, Y.; Luan, Z.; Di, Z.; Wang, H.; Tian, B.; and Jia, Z. 2003. Adsorption of 1,2-dichlorobenzene from water to carbon nanotubes. *Chem. Phys. Lett.* 376: 154–8.

7. Kostal, J.; Mulchandani, A.; Gropp, K. E.; and Chen, W. 2003. A temperature responsive biopolymer for mercury remediation. *Environ. Sci. Technol.* 37: 4457–62.
8. Hu, J. S.; Zhong, L. S.; Song, W. G.; and Wan, L. J. 2008. Synthesis of hierarchically structured metal oxides and their application in heavy metal ion removal. *Adv. Mater.* 20: 2977–82.
9. Zhong, L. S.; Hu, J. S.; Liang, H. P.; Cao, A. M.; Song, W. G.; and Wan, L. J. 2006. Self-assembled 3D flowerlike iron oxide nanostructures and their application in water treatment. *Adv. Mater.* 18: 2426–31.
10. Gao, C.; Zhang, W.; Li, H.; Lang, L.; and Xu, Z. 2008. Controllable fabrication of mesoporous MgO with various morphologies and their absorption performance for toxic pollutants in water. *Cryst. Growth Des.* 8: 3785–90.
11. Zhong, L.-S.; Hu, J.-S.; Cao, A.-M.; Liu, Q.; Song, W.-G.; and Wan, L.-J. 2007. 3D flowerlike ceria micro/nanocomposite structure and its application for water treatment and CO removal. *Chem. Mater.* 19: 1648–55.
12. Ma, T.-Y.; Zhang, X.-J.; and Yuan, Z.-Y. 2009. Hierarchical meso-/macroporous aluminum phosphonate hybrid materials as multifunctional adsorbents. *J. Phys. Chem. C* 113: 12854–62.
13. Yongqiang, W.; Guozhong, W.; Hongqiang, W.; Weiping, C.; Changhao, L.; and Lide, Z. 2009. Template-induced synthesis of hierarchical SiO_2@γ-AlOOH spheres and their application in Cr(VI) removal. *Nanotechnology* 20: 155604.
14. Lu, F.; Cai, W.; and Zhang, Y. 2008. ZnO hierarchical micro/nanoarchitectures: Solvothermal synthesis and structurally enhanced photocatalytic performance. *Adv. Funct. Mater.* 18: 1047–56.
15. Zeng, H.; Cai, W.; Liu, P.; Xu, X.; Zhou, H.; Klingshirn, C.; and Kalt, H. 2008. ZnO-based hollow nanoparticles by selective etching: Elimination and reconstruction of metal–semiconductor interface, improvement of blue emission and photocatalysis. *ACS Nano* 2: 1661–70.
16. Han, N.; Chai, L.; Wang, Q.; Tian, Y.; Deng, P.; and Chen, Y. 2010. Evaluating the doping effect of Fe, Ti and Sn on gas sensing property of ZnO. *Sens. Actuators B* 147: 525–30.
17. Zhang, Q.; Dandeneau, C. S.; Candelaria, S.; Liu, D.; Garcia, B. B.; Zhou, X.; Jeong, Y.-H.; and Cao, G. 2010. Effects of lithium ions on dye-sensitized ZnO aggregate solar cells. *Chem. Mater.* 22: 2427–33.
18. Wang, Z. L. 2009. ZnO nanowire and nanobelt platform for nanotechnology. *Mater. Sci. Eng. R* 64: 33–71.
19. Meyer, B.; Rabaa, H.; and Marx, D. 2006. Water adsorption on ZnO(10[1 with combining macron]0): From single molecules to partially dissociated monolayers. *Phys. Chem. Chem. Phys.* 8: 1513–20.
20. Gorria, P.; Sevilla, M.; Blanco, J. A.; and Fuertes, A. B. 2006. Synthesis of magnetically separable adsorbents through the incorporation of protected nickel nanoparticles in an activated carbon. *Carbon* 44: 1954–7.
21. Wang, X.; Cai, W.; Lin, Y.; Wang, G.; and Liang, C. 2010. Mass production of micro/nanostructured porous ZnO plates and their strong structurally enhanced and selective adsorption performance for environmental remediation. *J. Mater. Chem.* 20: 8582–90.
22. David, R. L. 2003–2004. *CRC Handbook of Chemistry and Physics*, 84th edition, CRC Press, Boca Raton, FL, Chapter 8, p. 112.
23. Yantasee, W.; Lin, Y.; Fryxell, G. E.; Alford, K. L.; Busche, B. J.; and Johnson, C. D. 2004. Selective removal of copper (II) from aqueous solutions using fine-grained activated carbon functionalized with amine. *Ind. Eng. Chem. Res.* 43: 2759–64.
24. Freundlich, H. and Heller, W. 1939. On adsorption in solution. *J. Am. Chem. Soc.* 61: 2228.
25. Langmuir, I. 1918. The adsorption of gases on plane surfaces of glass, mica and platinum. *J. Am. Chem. Soc.* 40: 1361–403.

26. Xu, Y.-J.; Weinberg, G.; Liu, X.; Timpe, O.; Schlögl, R.; and Su, D. S. 2008. Nanoarchitecturing of activated carbon: Facile strategy for chemical functionalization of the surface of activated carbon. *Adv. Funct. Mater.* 18: 3613–9.
27. Ballerini, G.; Ogle, K.; and Barthés-Labrousse, M. G. 2007. The acid–base properties of the surface of native zinc oxide layers: An XPS study of adsorption of 1,2-diaminoethane. *Appl. Surf. Sci.* 253: 6860–7.
28. Moretti, G.; Fierro, G.; Lo Jacono, M.; and Porta, P. 1989. Characterization of CuO–ZnO catalysts by X-ray photoelectron spectroscopy: Precursors, calcined and reduced samples. *Surf. Interface Anal.* 14: 325–6.
29. Hussain, Z.; Salim, M. A.; Khan, M. A.; and Khawaja, E. E. 1989. X-ray photoelectron and Auger spectroscopy study of copper-sodium-germanate glasses. *J. Non Cryst. Solids* 110: 44–52.
30. Wang, X.; Cai, W.; Liu, S.; Wang, G.; Wu, Z.; and Zhao, H. 2013. ZnO hollow microspheres with exposed porous nanosheets surface: Structurally enhanced adsorption towards heavy metal ions. *Colloids Surf. A* 422: 199–205.
31. Li, Y.-H.; Ding, J.; Luan, Z.; Di, Z.; Zhu, Y.; Xu, C.; Wu, D.; and Wei, B. 2003. Competitive adsorption of Pb2+, Cu2+ and Cd2+ ions from aqueous solutions by multiwalled carbon nanotubes. *Carbon* 41: 2787–92.
32. Nefedov, V. I.; Firsov, M. N.; and Shaplygin, I. S. 1982. Electronic structures of $MRhO_2$, $MRh2O_4$, $RhMO_4$ and $Rh2MO_6$ on the basis of X-ray spectroscopy and ESCA data. *J. Electron Spectrosc. Relat. Phenom.* 26: 65–78.
33. Światowska-Mrowiecka, J.; Zanna, S.; Ogle, K.; and Marcus, P. 2008. Adsorption of 1, 2-diaminoethane on ZnO thin films from *p*-xylene. *Appl. Surf. Sci.* 254: 5530–9.
34. Setty, M. S. and Sinha, A. P. B. 1986. Characterization of highly conducting PbO-doped Cd_2SnO_4 thick films. *Thin Solid Films* 144: 7–19.
35. Demir-Cakan, R.; Baccile, N.; Antonietti, M.; and Titirici, M.-M. 2009. Carboxylate-rich carbonaceous materials via one-step hydrothermal carbonization of glucose in the presence of acrylic acid. *Chem. Mater.* 21: 484–90.
36. Faur-Brasquet, C.; Reddad, Z.; Kadirvelu, K.; and Le Cloirec, P. 2002. Modeling the adsorption of metal ions (Cu2+, Ni2+, Pb2+) onto ACCs using surface complexation models. *Appl. Surf. Sci.* 196: 356–65.
37. Benhima, H.; Chiban, M.; Sinan, F.; Seta, P.; and Persin, M. 2008. Removal of lead and cadmium ions from aqueous solution by adsorption onto microparticles of dry plants. *Colloids Surf. B* 61: 10–6.
38. Morgan, W. E. and Van Wazer, J. R. 1973. Binding energy shifts in the x-ray photoelectron spectra of a series of related Group IVa compounds. *J. Phys. Chem.* 77: 964–9.
39. Kadirvelu, K.; Goel, J.; and Rajagopal, C. 2008. Sorption of lead, mercury and cadmium ions in multi-component system using carbon aerogel as adsorbent. *J. Hazard Mater.* 153: 502–7.
40. Sędłak, A. and Janusz, W. 2008. Specific adsorption of carbonate ions at the zinc oxide/electrolyte solution interface. *Physicochem. Probl. Miner. Process.* 42: 57–66.
41. Eary, L. E. and Rai, D. 1988. Chromate removal from aqueous wastes by reduction with ferrous ion. *Environ. Sci. Technol.* 22: 972–7.
42. Buerge, I. J. and Hug, S. J. 1997. Kinetics and pH dependence of chromium(VI) reduction by iron(II). *Environ. Sci. Technol.* 31: 1426–32.
43. Hug, S. J.; Laubscher, H.-U.; and James, B. R. 1996. Iron (III) catalyzed photochemical reduction of chromium (VI) by oxalate and citrate in aqueous solutions. *Environ. Sci. Technol.* 31: 160–70.
44. Fendorf, S. E. and Li, G. 1996. Kinetics of chromate reduction by ferrous iron. *Environ. Sci. Technol.* 30: 1614–7.

45. Lazaridis, N. K.; Bakoyannakis, D. N.; and Deliyanni, E. A. 2005. Chromium(VI) sorptive removal from aqueous solutions by nanocrystalline akaganèite. *Chemosphere* 58: 65–73.
46. Xu, Y. and Zhao, D. 2007. Reductive immobilization of chromate in water and soil using stabilized iron nanoparticles. *Water Res.* 41: 2101–8.
47. Petruzzelli, D.; Tiravanti, G.; and Passino, R. 1995. Ion exchange process for chromium removal and recovery from tannery wastes. *Ind. Eng. Chem. Res.* 34: 2612–7.
48. Zhang, D.; Wei, S.; Kaila, C.; Su, X.; Wu, J.; Karki, A. B.; Young, D. P.; and Guo, Z. 2010. Carbon-stabilized iron nanoparticles for environmental remediation. *Nanoscale* 2: 917–9.
49. Pillay, K.; Cukrowska, E.; and Coville, N. 2009. Multi-walled carbon nanotubes as adsorbents for the removal of parts per billion levels of hexavalent chromium from aqueous solution. *J. Hazard Mater.* 166: 1067–75.
50. Sun, Z. H.; Wang, L. F.; Liu, P. P.; Wang, S. C.; Sun, B.; Jiang, D. Z.; and Xiao, F. S. 2006. Magnetically motive porous sphere composite and its excellent properties for the removal of pollutants in water by adsorption and desorption cycles. *Adv. Mater.* 18: 1968–71.
51. Lin, Y.; Cai, W.; Tian, X.; Liu, X.; Wang, G.; and Liang, C. 2011. Polyacrylonitrile/ferrous chloride composite porous nanofibers and their strong Cr-removal performance. *J. Mater. Chem.* 21: 991–7.
52. Cao, C.-Y.; Cui, Z.-M.; Chen, C.-Q.; Song, W.-G.; and Cai, W. 2010. Ceria hollow nanospheres produced by a template-free microwave-assisted hydrothermal method for heavy metal ion removal and catalysis. *J. Phys. Chem. C* 114: 9865–70.
53. Begum, K. and Anantharaman, N. 2009. Removal of chromium(VI) ions from aqueous solutions and industrial effluents using magnetic Fe_3O_4 nanoparticles. *Adsorpt. Sci. Technol.* 27: 701–22.
54. Wang, P. and Lo, I. M. C. 2009. Synthesis of mesoporous magnetic γ-Fe_2O_3 and its application to Cr(VI) removal from contaminated water. *Water Res.* 43: 3727–34.
55. Melby, L. R.; Harder, R. J.; Hertler, W. R.; Mahler, W.; Benson, R. E.; and Mochel, W. E. 1962. Substituted quinodimethans. II. Anion-radical derivatives and complexes of 7,7,8,8-tetracyanoquinodimethan. *J. Am. Chem. Soc.* 84: 3374–87.
56. Padhye, M. R. and Karandikar, A. V. 1985. The effect of alkali salt on solvent–polyacrylonitrile interaction. *J. Appl. Polym. Sci.* 30: 667–73.
57. Abdel-Samad, H. and Watson, P. R. 1997. An XPS study of the adsorption of chromate on goethite (α-FeOOH). *Appl. Surf. Sci.* 108: 371–7.
58. Li, X.Q.; Cao, J.; and Zhang, W.-X. 2008. Stoichiometry of Cr(VI) immobilization using nanoscale zerovalent iron (nZVI): A study with high-resolution x-ray photoelectron spectroscopy (HR-XPS). *Ind. Eng. Chem. Res.* 47: 2131–9.
59. Biesinger, M. C.; Brown, C.; Mycroft, J. R.; Davidson, R. D.; and McIntyre, N. S. 2004. X-ray photoelectron spectroscopy studies of chromium compounds. *Surf. Interface Anal.* 36: 1550–63.
60. Manning, B. A.; Kiser, J. R.; Kwon, H.; and Kanel, S. R. 2006. Spectroscopic investigation of Cr(III)- and Cr(VI)-treated nanoscale zerovalent iron. *Environ. Sci. Technol.* 41: 586–92.
61. Craver, C. and Carraher, C. 2000. *Applied Polymer Science: 21st Century*, Elsevier B.V., Amsterdam.
62. Ma, H.; Yoon, K.; Rong, L.; Mao, Y.; Mo, Z.; Fang, D.; Hollander, Z.; Gaiteri, J.; Hsiao, B.; and Chu, B. 2010. High-flux thin-film nanofibrous composite ultrafiltration membranes containing cellulose barrier layer. *J. Mater. Chem.* 20: 4692–704.
63. Yoon, K.; Hsiao, B. S.; and Chu, B. 2009. High flux ultrafiltration nanofibrous membranes based on polyacrylonitrile electrospun scaffolds and crosslinked polyvinyl alcohol coating. *J. Membrane Sci.* 338: 145–52.

64. Yoon, K.; Kim, K.; Wang, X.; Fang, D.; Hsiao, B. S.; and Chu, B. 2006. High flux ultrafiltration membranes based on electrospun nanofibrous PAN scaffolds and chitosan coating. *Polymer* 47: 2434–41.
65. Lin, Y.; Cai, W.; He, H.; Wang, X.; and Wang, G. 2012. Three-dimensional hierarchically structured PAN@γ-AlOOH fiber films based on a fiber templated hydrothermal route and their recyclable strong Cr(VI)-removal performance. *RSC Adv.* 2: 1769–73.
66. Selvi, K.; Pattabhi, S.; and Kadirvelu, K. 2001. Removal of Cr(VI) from aqueous solution by adsorption onto activated carbon. *Bioresour. Technol.* 80: 87–9.
67. Mor, S.; Ravindra, K.; and Bishnoi, N. R. 2007. Adsorption of chromium from aqueous solution by activated alumina and activated charcoal. *Bioresour. Technol.* 98: 954–7.
68. Gao, M.-R.; Zhang, S.-R.; Jiang, J.; Zheng, Y.-R.; Tao, D.-Q.; and Yu, S.-H. 2011. One-pot synthesis of hierarchical magnetite nanochain assemblies with complex building units and their application for water treatment. *J. Mater. Chem.* 21: 16888–92.
69. Wang, X.; Liu, J.; Xu, W.; Cao, T.; Song, X.; and Cheng, C. 2012. Preparation of carbon microstructures by thermal treatment of thermosetting/thermoplastic polymers and their application in water purification. *Micro Nano Lett.* 7: 918–22.
70. Wang, X.; Liu, J.; and Xu, W. 2012. One-step hydrothermal preparation of amino-functionalized carbon spheres at low temperature and their enhanced adsorption performance towards Cr (VI) for water purification. *Colloids Surf. A* 415: 288–94.
71. Mendez, A.; Fernández, F.; and Gascó, G. 2007. Removal of malachite green using carbon-based adsorbents. *Desalination* 206: 147–53.
72. Vasu, A. E. 2008. Adsorption of Ni (II), Cu (II) and Fe (III) from aqueous solutions using activated carbon. *Eur. Chem. J.* 5: 1–9.
73. Pokonova, Y. V. 1996. Production of carbon adsorbents from brown coal. *Carbon* 34: 411–5.
74. Gorria, P.; Sevilla, M.; Blanco, J. A.; and Fuertes, A. B. 2006. Synthesis of magnetically separable adsorbents through the incorporation of protected nickel nanoparticles in an activated carbon. *Carbon* 44: 1954–7.
75. Rudge, S. R.; Kurtz, T. L.; Vessely, C. R.; Catterall, L. G.; and Williamson, D. L. 2000. Preparation, characterization, and performance of magnetic iron–carbon composite microparticles for chemotherapy. *Biomaterials* 21: 1411–20.
76. Oliveira, L. C. A.; Rios, R. V. R. A.; Fabris, J. D.; Garg, V.; Sapag, K.; and Lago, R. M. 2002. Activated carbon/iron oxide magnetic composites for the adsorption of contaminants in water. *Carbon* 40: 2177–83.
77. Gupta, V. K.; Agarwal, S.; and Saleh, T. A. 2011. Chromium removal by combining the magnetic properties of iron oxide with adsorption properties of carbon nanotubes. *Water Res.* 45: 2207–12.
78. Zhang, Z. and Kong, J. 2011. Novel magnetic Fe_3O_4@C nanoparticles as adsorbents for removal of organic dyes from aqueous solution. *J. Hazard Mater.* 193: 325–9.
79. Dekkers, M. J. and Schoonen, M. A. A. 1996. Magnetic properties of hydrothermally synthesized greigite (Fe_3S_4)—I. Rock magnetic parameters at room temperature. *Geophys. J. Int.* 126: 360–8.
80. Cao, F.; Hu, W.; Zhou, L.; Shi, W.; Song, S.; Lei, Y.; Wang, S.; and Zhang, H. 2009. 3D Fe_3S_4 flower-like microspheres: High-yield synthesis via a biomolecule-assisted solution approach, their electrical, magnetic and electrochemical hydrogen storage properties. *Dalton Trans.* November 14: 9246–52.
81. Watson, J.; Cressey, B.; Roberts, A.; Ellwood, D.; Charnock, J.; and Soper, A. 2000. Structural and magnetic studies on heavy-metal-adsorbing iron sulphide nanoparticles produced by sulphate-reducing bacteria. *J. Magn. Magn. Mater.* 214: 13–30.
82. He, Z.; Yu, S. H.; Zhou, X.; Li, X.; and Qu, J. 2006. Magnetic-field-induced phase-selective synthesis of ferrosulfide microrods by a hydrothermal process: Microstructure control and magnetic properties. *Adv. Funct. Mater.* 16: 1105–11.

83. Devey, A. J.; Grau-Crespo, R.; and de Leeuw, N. H. 2009. Electronic and magnetic structure of Fe_3S_4: GGA+U investigation. *Phys. Rev. B* 79: 195126.
84. Vanitha, P. V. and O'Brien, P. 2008. Phase control in the synthesis of magnetic iron sulfide nanocrystals from a cubane-type Fe–S cluster. *J. Am. Chem. Soc.* 130: 17256–7.
85. Wang, X.; Cai, W.; Wang, G.; Wu, Z.; and Zhao, H. 2013. One-step fabrication of high performance micro/nanostructured Fe_3S_4/C magnetic adsorbent with easy recovery and regeneration properties. *CrystEngComm* 15: 2956–65.
86. Kortum, G. 1936. Part B: Chemistry of elementary process, structure of matter. *J. Phy. Chem.* 33: 243–64.
87. Ho, Y. S. and McKay, G. 1999. Batch lead(II) removal from aqueous solution by peat: Equilibrium and kinetics. *Process Saf. Environ. Prot.* 77: 165–73.
88. Ho, Y. S. and McKay, G. 1999. Pseudo-second order model for sorption processes. *Process Biochem.* 34: 451–65.
89. Souag, R.; Touaibia, D.; Benayada, B.; and Boucenna, A. 2009. Adsorption of heavy metals (Cd, Zn and Pb) from water using keratin powder prepared from Algerien sheep hoofs. *Eur. J. Sci. Res.* 35: 416–25.
90. Ho, Y.-S. 2006. Review of second-order models for adsorption systems. *J. Hazard Mater.* 136: 681–9.
91. Abechi, E.; Gimba, C.; Uzairu, A.; and Kagbu, J. 2011. Kinetics of adsorption of methylene blue onto activated carbon prepared from palm kernel shell. *Appl. Sci. Res.* 3: 154–64.
92. Fei, J. B.; Cui, Y.; Yan, X. H.; Qi, W.; Yang, Y.; Wang, K. W.; He, Q.; and Li, J. B. 2008. Controlled preparation of MnO_2 hierarchical hollow nanostructures and their application in water treatment. *Adv. Mater.* 20: 452–6.
93. Wang, Y.; Wang, G.; Wang, H.; Cai, W.; and Zhang, L. 2008. One-pot synthesis of nanotube-based hierarchical copper silicate hollow spheres. *Chem. Commun.* 6555–7.
94. Yang, R. T. 2003. *Adsorbents: Fundamentals and Applications*, John Wiley & Sons Inc., Hoboken, NJ, 208–15.
95. Miyake, Y.; Yumoto, T.; Kitamura, H.; and Sugimoto, T. 2002. Solubilization of organic compounds into as-synthesized spherical mesoporous silica. *Phys. Chem. Chem. Phys.* 4: 2680–4.
96. Ho, Y.-S.; Malarvizhi, R.; and Sulochana, N. 2009. Equilibrium isotherm studies of methylene blue adsorption onto activated carbon prepared from *Delonix regia* pods. *J. Environ. Prot. Sci.* 3: 111–6.
97. Hameed, B. H.; Ahmad, A. L.; and Latiff, K. N. A. 2007. Adsorption of basic dye (methylene blue) onto activated carbon prepared from rattan sawdust. *Dyes Pigments* 75: 143–9.
98. El Qada, E. N.; Allen, S. J.; and Walker, G. M. 2006. Adsorption of methylene blue onto activated carbon produced from steam activated bituminous coal: A study of equilibrium adsorption isotherm. *Chem. Eng. J.* 124: 103–10.
99. Shahryari, Z.; Goharrizi, A. S.; and Azadi, M. 2010. Experimental study of methylene blue adsorption from aqueous solutions onto carbon nanotubes. *Int. J. Water Res. Environ. Eng.* 2: 16–28.
100. Han, C.; Cai, W.; Tang, W.; Wang, G.; and Liang, C. 2011. Protein assisted hydrothermal synthesis of ultrafine magnetite nanoparticle built-porous oriented fibers and their structurally enhanced adsorption to toxic chemicals in solution. *J. Mater. Chem.* 21: 11188–96.
101. Wang, C.-B. and Zhang, W.-X. 1997. Synthesizing nanoscale iron particles for rapid and complete dechlorination of TCE and PCBs. *Environ. Sci. Technol.* 31: 2154–6.
102. Wang, J.; Xiao, Q.; Zhou, H.; Sun, P.; Yuan, Z.; Li, B.; Ding, D.; Shi, A. C.; and Chen, T. 2006. Budded, mesoporous silica hollow spheres: Hierarchical structure controlled by kinetic self-assembly. *Adv. Mater.* 18: 3284–8.

103. Wu, C.; Xie, Y.; Lei, L.; Hu, S.; and OuYang, C. 2006. Synthesis of new-phased VOOH hollow "dandelions" and their application in lithium ion batteries. *Adv. Mater.* 18: 1727–32.
104. Choi, W. S.; Koo, H. Y.; Zhongbin, Z.; Li, Y.; and Kim, D. Y. 2007. Templated synthesis of porous capsules with a controllable surface morphology and their application as gas sensors. *Adv. Funct. Mater.* 17: 1743–9.
105. Kim, S.-W.; Kim, M.; Lee, W. Y.; and Hyeon, T. 2002. Fabrication of hollow palladium spheres and their successful application to the recyclable heterogeneous catalyst for Suzuki coupling reactions. *J. Am. Chem. Soc.* 124: 7642–3.
106. Wang, Y.; Wang, G.; Wang, H.; Liang, C.; Cai, W.; and Zhang, L. 2010. Chemical-template synthesis of micro/nanoscale magnesium silicate hollow spheres for waste-water treatment. *Chem. Eur. J.* 16: 3497–503.
107. Doğan, M.; Özdemir, Y.; and Alkan, M. 2007. Adsorption kinetics and mechanism of cationic methyl violet and methylene blue dyes onto sepiolite. *Dyes Pigments* 75: 701–13.
108. Bektaş, N.; Ağım, B. A.; and Kara, S. 2004. Kinetic and equilibrium studies in removing lead ions from aqueous solutions by natural sepiolite. *J. Hazard Mater.* 112: 115–22.
109. Shin, S. and Jang, J. 2007. Thiol containing polymer encapsulated magnetic nanoparticles as reusable and efficiently separable adsorbent for heavy metal ions. *Chem. Commun.* 4230–2.
110. Wang, Y.; Tang, C.; Deng, Q.; Liang, C.; Ng, D. H.; Kwong, F.-l.; Wang, H.; Cai, W.; Zhang, L.; and Wang, G. 2010. A versatile method for controlled synthesis of porous hollow spheres. *Langmuir* 26: 14830–4.

Section II

Hierarchical Micro/ Nanostructured Arrays

7 Micro/Nanostructured Block-Built Arrays

7.1 INTRODUCTION

As we know, hierarchical micro/nanostructured materials have many unique performances and hence the potential applications in many fields. If the hierarchical micro/nanostructured units, or the microsized objects (building blocks) with nanostructure, are periodically or regularly arranged on a substrate according to some rules, the micro/nanostructured arrays are thus formed, which will be very helpful to devise new micro/nanodevices. Therefore, the hierarchical micro/nanostructured arrays have attracted much interest due to their important applications in optoelectronic devices, microfluidic devices, nanogenerators, sensors, field emitters, and so on [1–11]. As mentioned in the Chapters 2 through 4, the hierarchical micro/nanostructured powders could be synthesized by solvothermal routes [12], template-etching strategies [13], electrospinning, and in situ conversion [14]. However, it is difficult to make them to form periodic arrays by self-assembly due to limitations of the geometric configuration once the hierarchical structured units have been prepared. The conventional methods to fabricate periodic hierarchical micro/nanostructured arrays are generally divided into two steps. Microsized structure arrays are first created by traditional lithographic techniques (e.g., photolithography, electron-beam lithography, ion beam lithography, and x-ray lithography) [15–18] or soft lithography (e.g., the techniques of microcontact printing, replica molding, and micromolding in capillaries) [19–22], the nanostructures are then modified or transferred on the microsized units in array [23], and thus hierarchical micro/nanostructured arrays can be finally obtained. However, they cannot be afforded owing to time consuming and the high costs in the most laboratories. In the last decades, template technique based on the monolayered colloidal crystals (or called colloidal monolayers) has been well developed, and so-called monolayer colloidal crystals are periodic colloidal sphere arrays with hexagonal close-packed (hcp) arrangement on substrates, which can be fabricated by self-assembly [24–29]. Figure 7.1 shows the typical monolayer polystyrene (PS) colloidal crystal with hcp arrangement on a substrate. Colloidal monolayer templates can be used to prepare periodic structure arrays, for instance, nanoparticle arrays [30–33], nanopore arrays [34–43], and hollow sphere arrays [44–47] assisted by other techniques. It has been proved that it is a flexible, facile approach to create the periodic micro/nanostructure arrays. Besides these periodic structure arrays, the colloidal monolayer template also can be used to prepare hierarchical micro/nanostructured arrays. If colloidal monolayers composed of microsized spheres are selected as templates or substrates, and then the nanosize structured units are introduced on microsized units in arrays, the hierarchical micro/nanostructured arrays

FIGURE 7.1 Morphology of a monolayer polystyrene (PS) colloidal crystal on glass substrate based on self-assembly. The inset is the magnified image in the edge region of the monolayer. PS spheres are 1000 nm in diameter.

could be easily obtained. For example, combining colloidal monolayer templates with basically chemical reaction, the hierarchical micro/nanostructured PS sphere/carbon nanotubes (CNTs) composite arrays were obtained by wet chemical self-assembling [48,49], hierarchical microsized PS sphere/silver nanoparticle composite arrays or microsized pore/silver nanoparticle arrays were made by thermal deposition of silver precursor [50,51], hierarchically $Ni(OH)_2$ or silver nanoplate–built monolayer hollow-sphere arrays or silver nanosheet were prepared by direct electrodeposition route based on colloidal templates [52,53], gold hierarchical micro/nanostructured particle arrays were created by electrochemical deposition based two-step replication of colloidal monolayer template [54], and so on. Additionally, the hierarchical micro/nanostructured arrays can also be fabricated by colloidal monolayer template combining physical deposition (pulsed laser deposition [PLD], sputtering, etc.) [55–59]. In this chapter, we mainly introduce the recent work to create micro/nanostructured arrays, including zero-dimensional (0D) object-built arrays, one-dimensional (1D) nanoobject-built arrays and two-dimensional (2D) nanoobject-built arrays, based on the PS colloidal templates with different routes, for example, chemical reaction and physical deposition.

7.2 ZERO-DIMENSIONAL OBJECT-BUILT ARRAYS

7.2.1 Nanoparticle Array and Laser Morphological Manipulation

Nanoparticles can be periodically arranged on solid supports by nanosphere lithography [60,61], which is a general, simple, and low-cost method. Here, we introduce the gold nanoparticle array based on the thermal evaporation on the PS monolayer template [30]. Briefly, the indium tin oxide (ITO) substrate coated with the 2D colloidal crystal, as shown in Figure 7.1, was mounted on a sample holder and transferred in an ultrahigh vacuum chamber. Gold was thermally evaporated under a base pressure of 10^{-6} Pa and deposited at a rate of ~0.2 nm/min. After deposition to a 70 nm thickness, the samples were immersed in methylene chloride under sonication to remove the nanosphere mask leaving a highly ordered array of gold particles on the surface.

7.2.1.1 Morphology and Evolution

Figure 7.2 shows the as-prepared Au nanoparticle array on an ITO substrate. The particles are hexagonally arranged with a P6mm symmetry and have triangular cross section with a height of about 70 nm. Such morphology is formed only due to the template geometry and slow evaporation deposition.

As we know, the intrinsic properties of such nanoparticle array are determined by factors such as the size, shape, crystallinity, and composition of the nanoparticles as well as the geometry and interparticle spacing of the array. Recent work has shown that the optical properties of the array are particularly sensitive to the nanoparticle shape [61]. Typically morphologies are controlled during the formation process through the size of the nanospheres in the mask or by angle-controlled deposition in the nanosphere lithography [62]. However, the direct morphology manipulation of the as-prepared array is expected to provide a more flexible way to control the morphologies of the nanoparticle arrays and hence their properties.

In general, thermal annealing and laser irradiation are two suitable techniques to manipulate the morphologies of nanoparticles. Thermal annealing has been used to change the shapes of particles in solid matrixes or in arrays on solid supports. Numerous groups have induced a spherical particle morphology of gold [63], silver [62], nickel [64], and germanium [65] particles by heating the corresponding triangular particle arrays at elevated temperatures. This method, however, can only

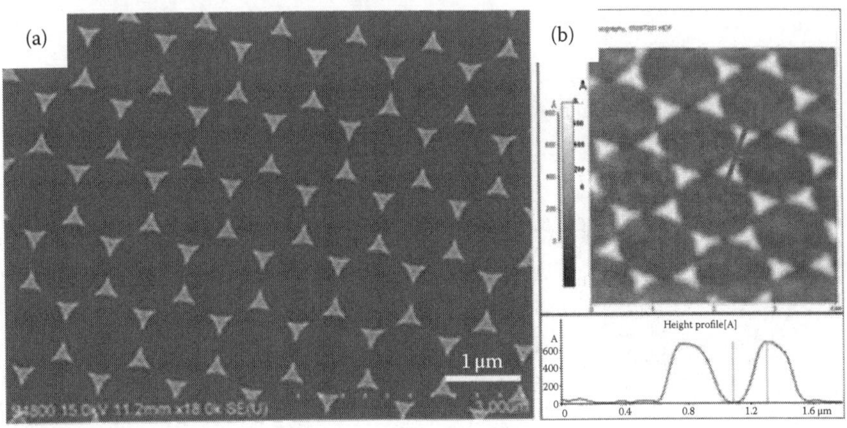

FIGURE 7.2 (a) Field emission scanning electron microscopy (FESEM) and (b) atomic force microscope images of a gold particle array fabricated with 1000-nm-diameter PS sphere colloidal monolayer before laser irradiation. (With kind permission from Springer Science+Business Media: *Appl. Phys. B*, Laser morphological manipulation of gold nanoparticles periodically arranged on solid supports, 81, 2005, 765–8, Sun, F et al.)

treat the whole sample, but cannot manipulate the particles in a selected area of the sample. In contrast, laser irradiation is able to directly manipulate the morphologies of the nanoparticles in a well-defined area of the sample. It has been shown that metal particles and their aggregates, dispersed in colloidal solution [66–72], in a glass matrix [73,74], or disorderly arranged on solid supports [75,76] can be transformed into smaller or larger, and from nonspherical into spherical shapes, by laser irradiation. Here, we introduce the direct morphology manipulation of the as-prepared array by laser irradiation.

The gold nanoparticle array shown in Figure 7.2 was then irradiated by laser pulses from a Nd:YAG laser operating at 1 Hz at the third harmonic wavelength of 355 nm with a nominal pulse width of 7 ns. The laser pulses were unfocused with an energy density of 15 mJ/cm^2 [30]. Figure 7.3 shows the morphological evolution versus the number of laser pulses. After about 40 laser pulses, the three sharp corners of each particle become separated from the main body of the particle and three nanogaps of about 30 nm are formed in each particle, as demonstrated in Figure 7.3a. This morphology is particularly intriguing because it might be possible for such an array to be used as a substrate for molecular switching devices [77]. As the number of laser pulses is increased to 100, the nanoparticles at the corners become smaller and almost disappear, while the main body of the particle evolves from a polyhedron to a rounded and finally to a nearly fully circular shape, as illustrated in Figure 7.3b through d (from top view). This demonstrates that the morphology of the nanostructured arrays can be manipulated by laser radiation through appropriate selection

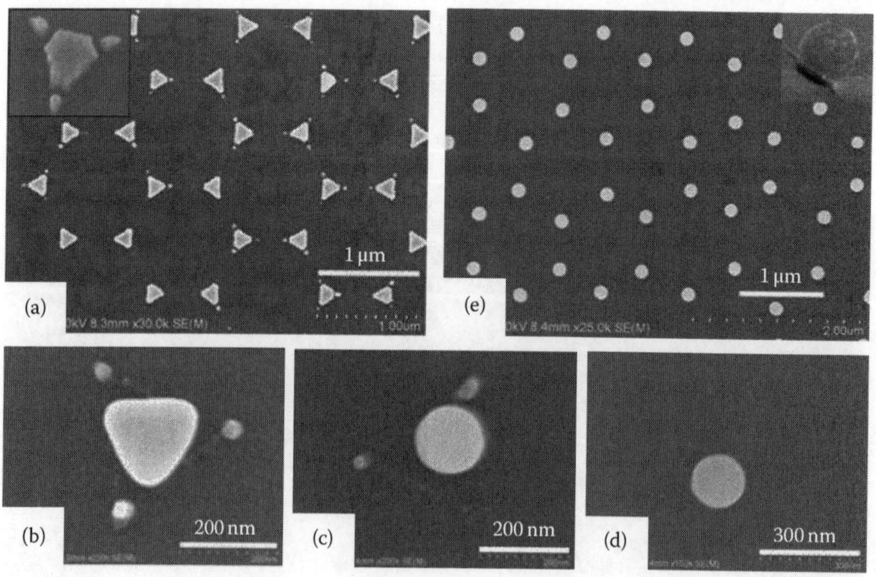

FIGURE 7.3 Morphology of the gold particle array on indium tin oxide (ITO) substrates after 355 nm laser irradiation (15 mJ/cm^2 per pulse) for different numbers of pulses: (a) 40 pulses; (b) 60 pulses; (c) 80 pulses; (d) 100 pulses, and (e) 500 pulses. (With kind permission from Springer Science+Business Media: *Appl. Phys. B*, Laser morphological manipulation of gold nanoparticles periodically arranged on solid supports, 81, 2005, 765–8, Sun, F et al.)

of the number of pulses. Applying more than 100 laser pulses did not induce any further changes but the complete disappearance of the nanoparticles at the corners and the edge sides of the original particles. Figure 7.3e gives the result irradiated by more than 500 pulses, and its morphology is similar to that irradiated for about 100 pulses indicating that the particle has reached its equilibrium shape after 100 pulses. Further, the final particles are nearly spherically shaped, as demonstrated in the inset of Figure 7.3e, which is also consistent with Kawasaki et al.'s reports [78]. Similar morphological evolutions were also observed for gold particles on quartz substrates irradiated with 355 and 532 nm laser wavelength pulses, respectively, suggesting a generality to this technique. Finally, it is noted that many of the particles seem to move slightly during irradiation, or the final spherical particles are not exactly located at the centers of the triangles (see Figure 7.3c and d).

Transmission electron microscopy (TEM) examination has shown that the individual gold particles in the array are polycrystalline prior to irradiation. Laser irradiation, however, leads to transformation from polycrystalline to a single crystal structure, as illustrated by selected area diffraction patterns of the single particles scraped from the substrates shown in Figure 7.4. It means that laser irradiation modify not only the morphology but also the structure of the particles, indicating that some new properties sensitive to the structure will appear.

7.2.1.2 Laser-Induced Spheroidization

On the basis of particle–laser interaction, the above morphology evolution could be understood. The nanoparticle array mentioned above was prepared by physical vapor deposition on the PS monolayer. It is well known that such vapor deposition usually leads to the formation of triangular noncompact particles (with a porosity) consisting of ultrafine nanoparticles or grains, which exhibit polycrystalline electronic diffraction, as shown in Figure 7.4a. Obviously, laser irradiation will heat and

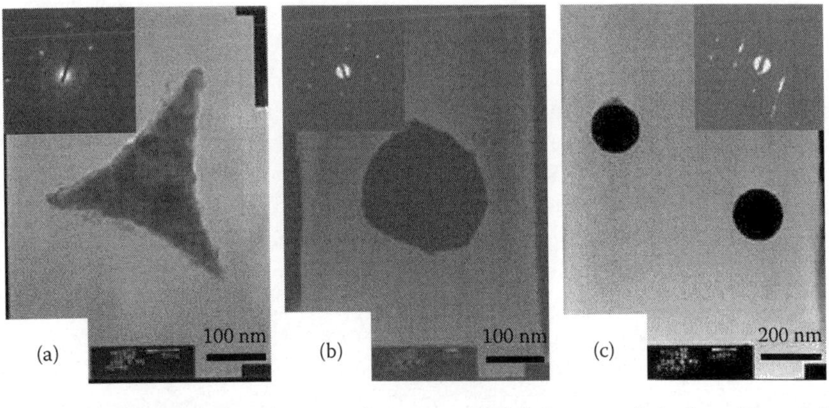

FIGURE 7.4 Transmission electron microscopy (TEM) images of single gold particles. (a) As-prepared triangular gold particle. (b) and (c) After irradiation at 100 and 500 laser pulses, respectively. The insets show the corresponding selected-area electron diffraction (SAED) patterns. (With kind permission from Springer Science+Business Media: *Appl. Phys. B*, Laser morphological manipulation of gold nanoparticles periodically arranged on solid supports, 81, 2005, 765–8, Sun et al.)

sinter the triangular particles. After laser irradiation for a short time (e.g., 40 pulses), the main body of a triangular particle will become compact due to heating-induced sintering of ultrafine nanoparticles. Such compactness will induce contraction of the triangular particles, and the contraction will form nanogaps and nanoparticles at the corners and the edge sides of the triangular particles because of the edge effect and the interaction between the particles and the substrate (see Figure 7.3). With laser irradiation going on, the compact nanoparticles will be spheroidized by surface atomic diffusion and the grains in the particles will grow due to heating [79], leading to final spherical particles with single crystal structure in TEM observation, whereas the small nanoparticles at corners and edges will get smaller and smaller by local evaporation and/or ripening process [79], and finally disappear. In addition, since the nanogaps were not always formed at the symmetrical sites of a triangle during initial irradiation, some of the final spherical particles slightly deviate from the centers of the triangles (see Figure 7.3e).

All in all, the pulsed laser irradiation could induce a morphology evolution from triangularly to nearly spherically shaped particles and a structural evolution from poly to single crystal. This method not only provides a good way to control the morphology of nanostructured materials and hence their properties but also introduces a new tool for the fabrication of specific future nanodevices by area-selective treatment.

7.2.2 Nanoparticles on Microsized PS Sphere Arrays

Recently, it has been found that uniform silver nanoparticles can be formed on the substrate by thermal decomposition of silver acetate (AgAc) at low temperatures [50]. The silver nanoparticle on the PS colloidal monolayer can thus be fabricated to get hierarchical micro/nanostructured arrays. The fabrication process is illustrated in Figure 7.5. The PS monolayer colloidal crystal with the area of about 2 cm^2 was prepared on a glass substrate by self-assembling process. Subsequently, AgAc aqueous solution was dripped onto the colloidal crystals, forming a thin AgAc coating on the PS monolayer colloidal crystal. The colloidal crystal with the AgAc coating was heated in an oven, leading to the formation of comparatively uniform decoration of silver nanoparticles on the surfaces of the PS spheres. Consequently, hierarchical structures consisting of ordered PS microspheres and silver nanoparticles were created [50].

Figure 7.6 shows field emission scanning electron microscopy (FESEM) image of the synthesized hierarchical micro/nanostructures. The monolayer colloidal crystal has the periodicity of 5 μm and the nanoparticles on the colloidal crystal have an average size of 180 nm. The hierarchical structure was fabricated with a precursor solution of 0.5 M and at a heating temperature of 200°C for 3 hours, which was the optimized experimental condition for the fabrication.

In the presented method, the distribution density of nanoparticles can be controlled by changing the concentration of AgAc precursor. The nanoparticle density decreased with decreasing the precursor concentration. This is because the lower precursor concentration induced thinner AgAc coating on the PS spheres. If the precursor concentration was low (e.g., 0.3 M), silver nanoparticles were not completely

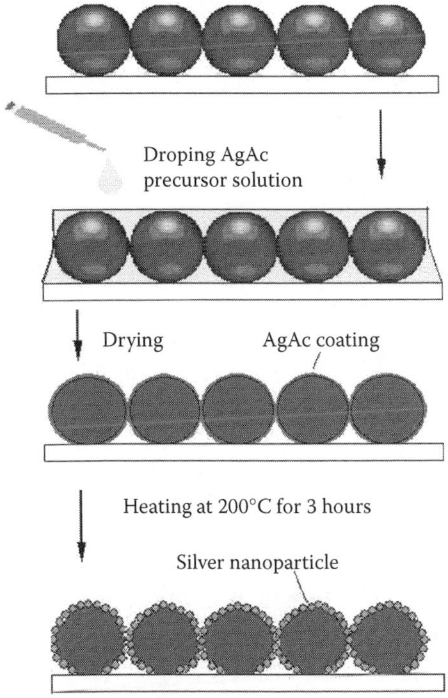

FIGURE 7.5 Scheme for the fabricating process for the hierarchical microsphere/nanoparticle composite arrays. (Data from Li, Y., et al., *J. Phys. Chem. C*, 111, 14813–7, 2007.)

FIGURE 7.6 FESEM images of hierarchical micro/nanostructured arrays fabricated using the colloidal monolayer (the diameter of PS sphere is 5 μm) with silver acetate (AgAc) coating after heating at 200°C for 3 hours. (Reprinted with permission from Li et al., 2007, 14813–7. Copyright 2007 American Chemical Society.)

coated on the PS spheres (Figure 7.7) and furthermore some PS spheres were melted in the heating process because of the incomplete coating of AgAc on the PS spheres (Figure 7.7b). Moreover, a well-controlled heating temperature was also crucial for the successful synthesis of the micro/nanostructured array surfaces. Heating temperature below 180°C was insufficient to convert AgAc precursor on the PS spheres to silver nanocrystals. At a heating temperature above 220°C, PS spheres were gradually decomposed and the microsized structures collapsed (glass transition temperature T_g of PS is nearly 100°C). The experimental results showed that the temperature in the range of 195°C–215°C was optimum value for the formation of the hierarchical micro/nanostructured arrays. Heating treatment has two important roles: first, decomposition of AgAc coating into silver nanoparticles and second, enhancing the mechanical stability of surfaces. A bare monolayer colloidal crystal can be easily peeled off from the substrate. However, the PS spheres can be slightly melted during the heating process (the temperature is higher than T_g of PS), resulting that the PS spheres adhere tightly to the substrate and that the silver nanoparticles produced in the heating process are embedded into PS spheres. The embedment of silver nanoparticles on PS microspheres is clearly seen in Figure 7.7b. Consequently, the heating process leads to the formation of highly durable hierarchical micro/nanostructured films: the organized films were not detached from a substrate even when the hierarchical surface was ultrasonically washed in water for 1 hour.

The periodicity of the microstructures can be tuned by changing the PS sphere size in colloidal monolayer. For example, Figure 7.8 shows a typical morphology of the hierarchical micro/nanostructured arrays synthesized from the PS colloidal monolayer of 1.3 μm sphere size.

Additionally, the silver micro/nanobowl arrays can also be obtained by colloidal monolayer combined with thermal decomposition of precursor [51]. For instance, using a colloidal monolayer with PS sphere of 5 μm as the template and AgAc as precursor after heating at 360°C for 3 hours, silver-ordered micro/nanobowl array was fabricated on the substrate, as shown in Figure 7.9. This ordered array with the periodicity of 5 μm exhibits a hexagonal alignment. Each unit takes on a bowl-like structure (Figure 7.9a and b) in the array and the whole bowl-like array has

FIGURE 7.7 (a) FESEM images of a colloid monolayer covered with a thin silver coating. Precursor concentration: 0.3 M AgAc. (b) A high magnification of (a). (Reprinted with permission from Li et al., 2007, 14813–7. Copyright 2007 American Chemical Society.)

FIGURE 7.8 FESEM image of a hierarchical surface structure fabricated using the colloidal monolayer with the PS diameter of 1.3 μm and 0.5 M AgAc at the heating temperature of 200°C for 3 hours. (Reprinted with permission from Li et al., 2007, 14813–7. Copyright 2007 American Chemical Society.)

rough inner walls composed of nanoparticles with an average size of ca. 135 nm, as shown in the inset of Figure 7.9b, which clearly shows that most silver nanoparticles are welded with neighbors because of the surface melt during the heating process. The heating process leads to the formation of highly durable micro/nanostructured arrays: these structures were not destroyed and the whole hierarchical arrays were not detached from a substrate even when the substrate was ultrasonically washed in water for 30 minutes.

The formation of silver nanoparticles is discussed and it is closely related to the heating temperature. The AgAc can be decomposed into silver at a low temperature of 360°C. In this decomposition process, silver nucleation is formed and subsequently silver particles grow up. If the temperature increases much (e.g., 500°C for 3 hours), the as-produced silver nanoparticles will be melted again and larger particles or film will incline to be generated. If the temperature is lower (e.g., 260°C for 3 hours), although the silver nanoparticles are formed, the hierarchical bowl-like structures cannot be obtained because of incomplete decomposition of PS sphere. These results indicate that the suitable temperature is very important for the formation of the silver nanoparticle. In this route, heating at 360°C for 3 hours has two roles: one is to make the AgAc decompose into silver nanoparticles and another is to remove the PS spheres of colloidal monolayer template by burning it out. These results displayed that the synthesized ordered array possessed hierarchical micro/nanostructure, microstructure was supplied by ordered bowl-like array caused by colloidal monolayer template, and nanostructure was resulted from the decomposition of AgAc.

Interestingly, such hierarchical micro/nanostructured arrays also can be fabricated on the curved surface based on the colloidal monolayer transferring strategy [50].

FIGURE 7.9 Morphology of as-prepared silver hierarchical bowl-like array film. (a) FESEM image with low magnification. (b) A magnified image of a feature bowl-like unit. The inset: An image of bowl bottom with larger magnification. (Reprinted with permission from Li et al., 2007, 9802–7. Copyright 2007 American Chemical Society.)

It has been found that polymer colloidal monolayer on a substrate can be transferred onto another substrate while retaining its integrality. On this basis, the hierarchical micro/nanostructured arrays were created on the curved glass tubes. The fabrication process of the hierarchical micro/nanostructured coating on a curved substrate is illustrated in Figure 7.10. In previous studies, it was found that polymer colloidal monolayer on a substrate can be transferred onto another substrate. This technique

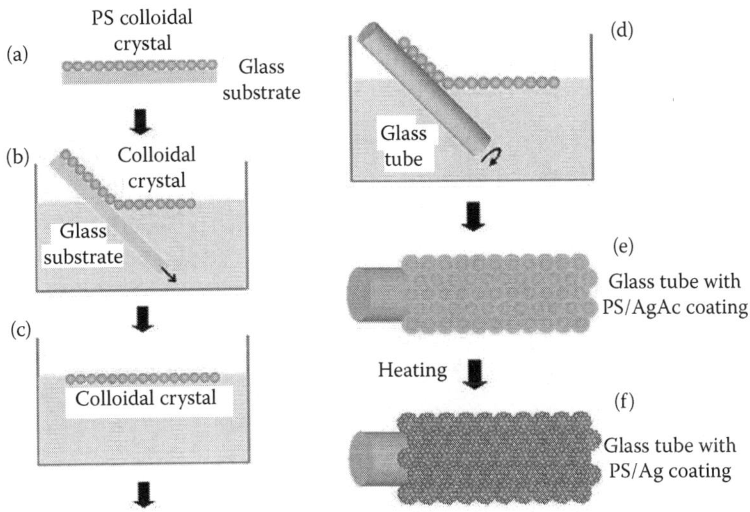

FIGURE 7.10 Scheme for fabricating a superhydrophobic coating on a curved surface. (a) A PS colloidal monolayer is fabricated on a flat substrate by self-assembling. (b) The colloidal monolayer on the substrate is gradually dipped into the AgAc solution and then the monolayer is peeled off from the substrate. (c) The colloid monolayer floats on the solution surface. (d) The monolayer is picked by a glass tube with curved surface. (e) The PS colloidal monolayer with AgAc coating is formed on the curved surface. (f) After heating, the hierarchical structure consisting of micrometer-sized PS spheres and silver nanoparticles is prepared on the curved surface. (Reprinted with permission from Li et al., 2007, 14813–7. Copyright 2007 American Chemical Society.)

was used for the fabrication of the hierarchical micro/nanostructured arrays on a curved surface using a precursor solution as a medium. At first, a monolayer PS colloidal crystal is prepared on a glass substrate by self-assembly. Subsequently, the colloidal monolayer on the substrate is gradually dipped into the AgAc solution with an inclination angle of about 30°. As a result, the colloidal monolayer is peeled off from the glass substrate and floated on the AgAc solution surface while retaining its integrality. Then, the colloidal monolayer is slowly picked up with a glass tube and accordingly the colloidal monolayer coated with AgAc covers the outer surface of the tube, followed by drying the sample at room temperature. Finally, the samples were heated at 200°C for 3 hours in an oven. Glass tubes with two different outer diameters of 1.4 and 4.87 mm were used in the experiments. Due to the heating process, AgAc is transformed into silver nanoparticles and consequently hierarchical structures that consist of PS microsphere arrays coated with silver nanoparticles are created. The photographs to show each process described above are displayed in Figure 7.11.

Figure 7.12a presents a typical FESEM image of the fabricated hierarchical micro/nano structure on a convex glass substrate. Close-packed arrays of monolayer PS spheres with 5 μm diameter completely covered the glass substrates and nanoparticles with an average size of 180 nm uniformly decorated the PS spheres. Using the same strategy, such a hierarchical micro/nanostructure was also fabricated on a concave surface of the inner wall of a glass tube (Figure 7.12b).

FIGURE 7.11 Photographs corresponding to the manipulation process described in Figure 7.6. (Reprinted with permission from Li et al., 2007, 14813–7. Copyright 2007 American Chemical Society.)

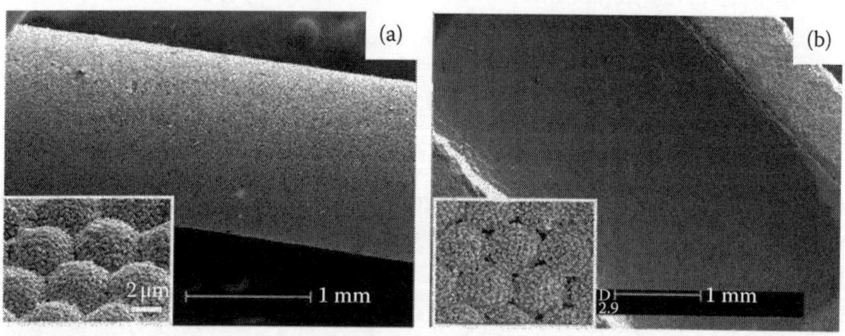

FIGURE 7.12 FESEM images of the hierarchical structure on (a) an outer surface of a glass tube (outer diameter: 1.4 mm) and (b) an inner surface of glass tube (inner diameter: 3.0 mm). The insets are magnified images of the PS spheres of 5 μm. (Reprinted with permission from Li et al., 2007, 14813–7. Copyright 2007 American Chemical Society.)

7.2.3 Two-Step Replication to Prepare Zero-Dimensional Nanostructured Materials

The colloidal monolayer was used as the first template to prepare alumina-ordered pore arrays by solution-dipping strategy and then such pore arrays were used as the second templates to prepare the hierarchical micro/nanostructured particle arrays by further electrodeposition after the removal of the second templates [4].

Figure 7.13 describes the fabrication process of the hierarchical micro/nanostructured Au particle arrays. First, PS colloidal monolayer was bonded onto a conductive ITO substrate by heating at the glass transition temperature of PS. A droplet of $Al(NO_3)_3$ aqueous solution was put on the monolayer, and amorphous Al_2O_3 ordered through-pore arrays can thus be obtained after drying, removal of PSs, and heat treatment. Finally, the Au particle array with the hierarchical structure can be obtained by electrodeposition on the substrate with the ordered through-pore Al_2O_3 template and subsequent removal of it. A solution composed of $HauCl_4$ (12 g/L), ethylenediaminetetraacetic acid (5 g/L), Na_2SO_3 (160 g/L), and K_2 HPO_4 (30 g/L) was used as the electrolyte; the pH value was 5. A graphite plate and a saturated calomel electrode were used as the auxiliary and reference electrode, respectively. The electrodeposition was carried out at 25°C and cathodic current density $J = 1$ mA/cm^2.

Figure 7.14a shows the Au hierarchical micro/nanostructured particle arrays obtained by presented method. The microsized Au particles are isolated from each other and arranged hexagonally with the periodicity of 1000 nm, corresponding to pores' arrangement in the porous alumina template. High magnification reveals that

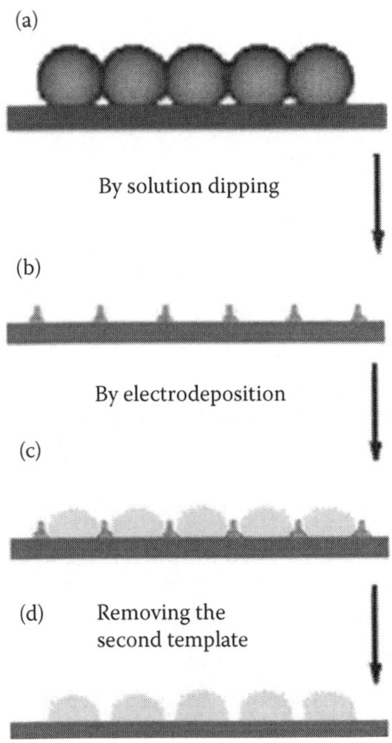

FIGURE 7.13 Fabrication procedures for hierarchical roughness gold particle arrays. (a) PS colloidal monolayer on ITO substrate; (b) ordered alumina through-pore array by solution-dipping strategy; (c) electrodeposition using the ordered alumina pore array as second template; (d) ordered gold particle array after removing the alumina template. (Duan et al., *Appl. Phys. Lett.* © 2006 IEEE.)

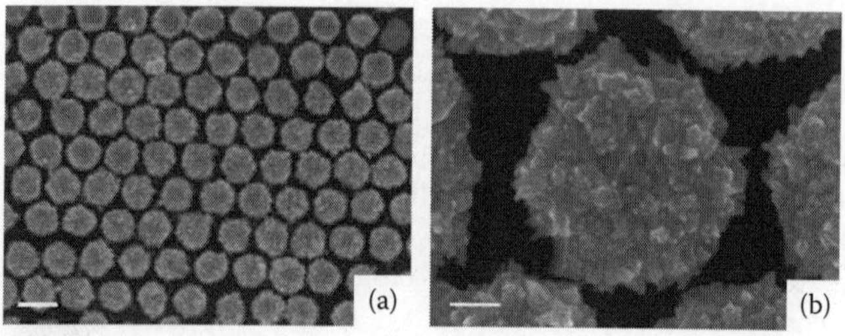

FIGURE 7.14 FESEM images of (a) Au particle array deposited for 20 minutes at 1 mA/cm^2 based on the Al_2O_3 ordered pore array film (b) a local magnification of (a). Scale bars are 1000 nm for (a) and 200 nm for (b). (Duan et al., *Appl. Phys. Lett.*© 2006 IEEE.)

the individual building blocks in the array are composed of many smaller particles with the size of about 60 nm (see Figure 7.14b), showing a nanoscaled surface roughness. Therefore, such an array exhibits the hierarchical surface roughness: the nonclose-packed microparticle array forming a microscaled roughness on the substrate and the nanoscaled surface roughness on the microparticles. Such hierarchically structured array can be used as an active substrate for surface-enhanced Raman scattering to detect organic molecule, which could be useful in molecular level detection, sensors, nanoscience, and nanotechnology.

The formation of such hierarchical micro/nanostructure is mainly attributed to the insulating porous Al_2O_3 template, which restricts Au deposition within the through-pores and leads to the formation of the isolated periodically arranged Au particles. It should be pointed that Au atoms did not electrodeposit along Al_2O_3 pore walls. Figure 7.15 presents the morphology of the as-prepared sample before removal of the Al_2O_3 second template. One can see clearly the wall of Al_2O_3 template between the neighboring Au particles. Au particles are obviously isolated from the pore walls and located within Al_2O_3 pores, showing preferential deposition on the substrate around the middle area of each pore. No growth along the pore walls was found. Because alumina is insulating, Au nucleation can only occur on the conductive ITO substrate. Such nucleation should be unselective on the substrate within Al_2O_3 pores because of ITO (or more than one nucleus on the substrate within a single pore). Also, since gold is insoluble in Al_2O_3, it is unfavorable in energy for Au nuclei on the substrate to grow along Al_2O_3 pore walls. Finally, the nuclei grow within pores and away from the walls, constituting a microparticle within each pore and leading to nanoscaled surface roughness.

Further investigation indicates that the cathodic current density J is of importance in the formation of such hierarchically rough particle arrays. With the decrease of J down to 0.3 mA/cm^2, only a low percentage (<20%) of Al_2O_3 pores is deposited with Au microparticles. If further decreasing J to 0.1 mA/cm^2, no Au deposition within alumina pores was found, indicating the Au^o atoms or clusters from the reduction of $[AuCl_4]^{-1}$ cannot nucleate on the substrate within the Al_2O_3 pores due to the nucleation barrier. Also, too high J is not appropriate because the rapid deposition rate leads to Au deposition everywhere including onto Al_2O_3 pore walls, and hence the failure to the formation of the isolated microparticles.

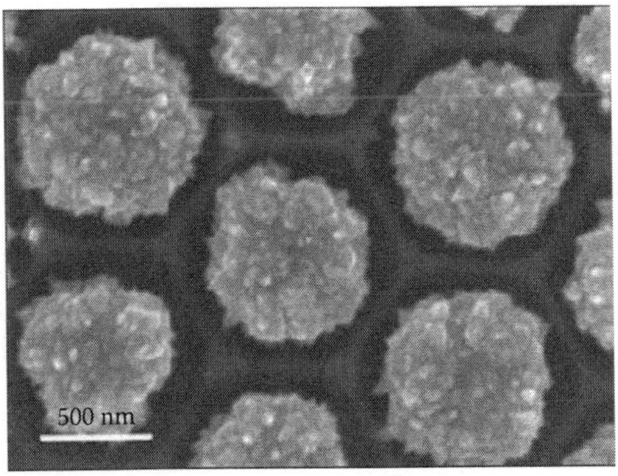

FIGURE 7.15 FESEM image of Au particle array before removal of the alumina template. (Duan et al., *Appl. Phys. Lett.* © 2006 IEEE.)

7.3 ONE-DIMENSIONAL NANOOBJECT-BUILT ARRAYS

7.3.1 Self-Assembling 1D Nanostructures

Besides creating the hierarchical micro/nanostructured arrays by decorating 0D nanostructures on the microsized PS spheres, we also presented a facile and alternative method to create hierarchically micro/nanostructured arrays by loading 1D nanostructures (CNTs) on the microsized PS spheres. The microstructure was induced by PS colloidal monolayer on a glass or silicon substrate, and the nanostructure was supplied by single-walled carbon nanotubes (SWCNTs) decorated on the microstructures by wet chemical self-assembly. The morphology and the distribution density of the nanostructure can be easily controlled by the concentration of SWCNT solution [48].

The fabrication process is illustrated in Figure 7.16. The monolayer PS colloidal crystals with square centimeter size were prepared on glass substrates by spin coating using PS colloidal microsphere suspension. The colloidal crystals were then heated at a temperature of 130°C (above than the glass transition temperature 105°C of the microspheres) for 40 minutes, which strongly increases the adherence of the PS colloidal monolayers to the substrate. Subsequently, a gold layer with thickness of 30 nm was coated on the microsphere surface by sputtering. The sample was then dipped into 0.1 mol·L^{-1} aqueous solution of mercaptoethylamine for the funtionalization of amino group on the gold surface. SWCNTs with carboxylic acid functionality (–COOH) solution was prepared by ultrasonating the raw CNTs (Iljin Nanotech Co., Ltd., Korea) in the mixture of concentrated sulfuric and nitric acids with a volume ratio of 3:1 for 6 hours, followed by redispersing the SWCNTs into acetone after filtration by sonication. When the prepared SWCNTs solution was dropped on the microsphere surface, the SWCNTs can be self-assembled on the PS spheres due to the condensation reaction between the –COOH and –NH_2 as well as the electrostatic attraction and van der Waals interactions between the CNTs. The SWCNTs were then decorated randomly on the surfaces of PS microspheres with almost uniform

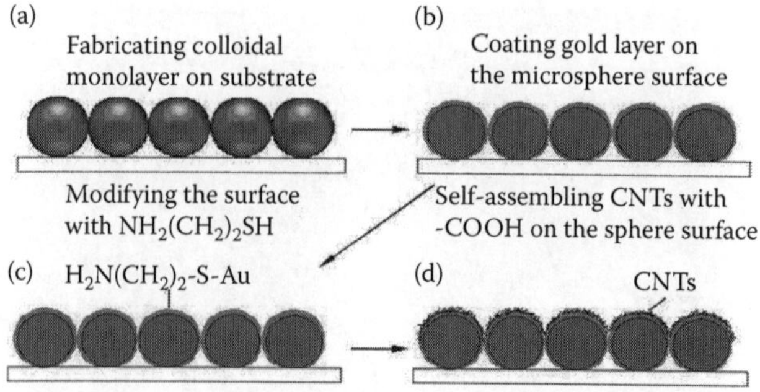

FIGURE 7.16 Schematic fabrication process of hierarchical surface with micro/nanostructures. (Reprinted with permission from Li et al., 2007, 2169–74. Copyright 2007 American Chemical Society.)

distribution density. As a result, the hierarchical micro/nanostructure arrays comprising PS microspheres and SWCNT was created.

Figure 7.17a through c show the FESEM images of the hierarchical structure fabricated by 2.0 mg·L^{-1} SWCNTs solution and the colloidal monolayer with periodicity of 5.0 μm. The images clearly show that dense SWCNTs were assembled on the surface of microspheres by the wet chemical self-assembly method described above: SWCNTs on the microspheres take on interlaced "net" structures and they tightly adhere to the microsphere surfaces. The whole hierarchical microsphere/SWCNT composite arrays take on hcp arrangement. The FESEM image of a natural lotus leaf surface was also shown for the comparison of the morphology (Figure 7.17d). It can be seen that the synthesized hierarchical structure well mimicked the surface of a lotus leaf.

7.3.2 1D Nanostructure by Pulsed Laser Deposition

A PS colloidal monolayer was first fabricated on a substrate. The desired material was then deposited on this colloidal monolayer substrate by PLD at room temperature and oxygen was introduced into the PLD chamber as the background gas. This periodic array has a special hierarchical micro/nanostructure array with an hcp arrangement, which originate from the pattern of colloidal monolayer. In this micro/nanostructure unit in array, the nanorod stands vertically on the microsized PS sphere tops, and 1D nanobranches in each nanorod grow in a radiation-like manner, perpendicular to the PS sphere surface [55].

The PS colloidal monolayers were first fabricated on cleaned Si substrates by self-assembly. The colloidal monolayer with its supporting substrate was placed in a deposition chamber of PLD, close to the target and at an off-axial position with respect to the target. A laser beam with a 355 nm wavelength from a Q-switched Nd:YAG laser, operated at 10 Hz with 100 mJ/pulse and a pulse width of 7 ns was applied and focused on the target surface with a diameter of about 2 mm. The desired target, for example, rutile-typed titanium dioxide was used for deposition. The substrate and target were rotated at 40 and 30 rpm, respectively. PLD was carried out at a base pressure of 2.66×10^{-4} Pa and a background O_2 pressure of 6.7 Pa.

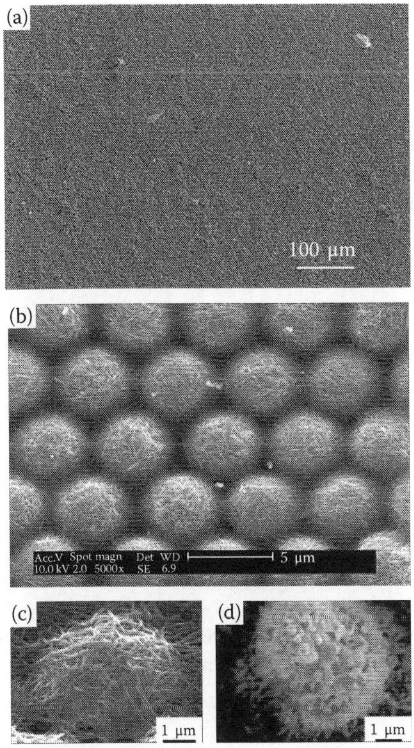

FIGURE 7.17 (a) FESEM image with large area of the hierarchical microsphere/single-walled carbon nanotubes (SWCNTs) composite array. (b) FESEM image with higher magnification. The concentration of SWCNTs was 2.0 mg·L^{-1}, (c) the feature picture of the hierarchical surface obtained with a tilting angle of 40°, and (d) is the surface microstructure of natural lotus leaf. (From Ye, C.H. et al., *Adv. Mater.*, 16, 1019–23, 2004; Reprinted with permission from Li et al., 2007, 2169–74. Copyright 2007 American Chemical Society.)

After deposition, the sample demonstrated a periodic hierarchical micro/nanorod array with an hcp arrangement, as reflected from Figure 7.18a. Each nanorod consists of two parts: a PS sphere at the bottom and a vertical nanorod on the top of the PS sphere (Figure 7.18b). The diameter of the nanorod was almost the same as that of the PS sphere, 350 nm, and its height was about 870 nm. The nanorod had a very rough structure on the surface and was composed of many 1D nanobranches, according to the high-resolution images of the side view (Figure 7.18c and d). TEM observation from the top of the nanorod arrays reflects that each nanorod consists of radiation-shaped 1D nanobranches emanating from the center (Figure 7.19a). The TEM image of a single nanorod also clearly displays that the nanorod consists of a PS sphere at the bottom and a nanorod on the sphere surface. The nanorod possesses 1D nanobranched structures, which grow almost vertically on the PS sphere surface (Figure 7.19b). The nanobranched structures indicate that the nanorod has a hierarchical, porous structure and hence has a high surface area. The selected area electron diffraction (SAED) pattern shows that the deposited materials on PS sphere surfaces by PLD are amorphous. Besides TiO_2 amorphous hcp nanocolumn arrays,

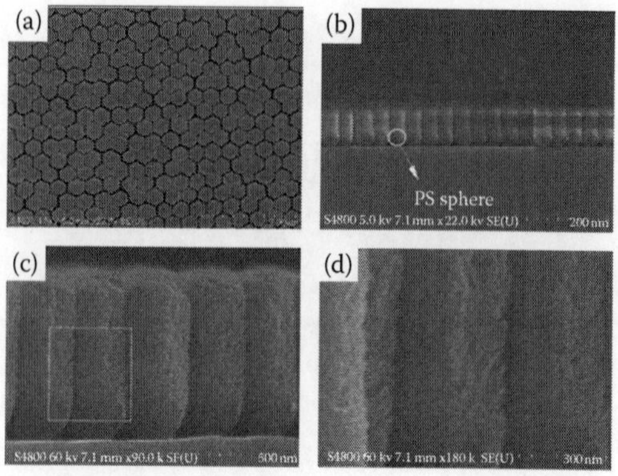

FIGURE 7.18 FESEM images of a sample obtained by pulsed laser deposition (PLD) using an Si substrate with a PS colloidal monolayer coating (PS sphere size: 350 nm; deposition time: 70 minutes). (a) Top view and (b) cross section. (c) and (d) are high-resolution images observed from the side. (d) Much higher magnification image of (c). (Reprinted with permission from Li et al., 2008, 14755–22. Copyright 2008 American Chemical Society.)

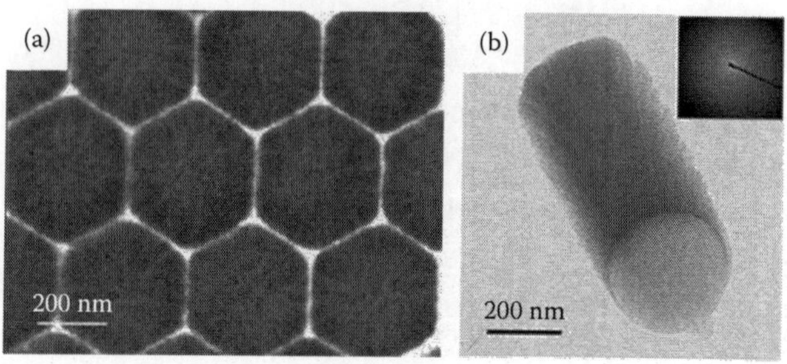

FIGURE 7.19 Corresponding TEM images of the sample in Figure 7.18. (a) Periodic nanorod array observed from the top. (b) Single nanorod observed from the side. The inset in (b) is the corresponding electron diffraction pattern. (Reprinted with permission from Li et al., 2008, 14755–22. Copyright 2008 American Chemical Society.)

the presented strategy can be extended to the fabrication of similar amorphous structures of SnO_2, WO_3, C, and so forth, just by changing the corresponding target in the PLD process.

Additionally, some materials, for example, CuO, Fe_2O_3, ZnO are easily crystalline by PLD at room temperature. If the colloidal monolayer is applied as a template, then the crystalline CuO, Fe_2O_3, ZnO, and so on, hierarchical micro/nanostructured arrays can be also obtained [56]. Figure 7.20 shows the scanning electron microscope (SEM) and TEM images of CuO crystalline hierarchical micro/nanostructured arrays using colloidal monolayers as templates by PLD. Each arrayed unit is composed of PS sphere at bottom and deposited materials on top. Deposited materials are well crystalline, they do not exhibit round shapes but radially aligned nanocolumns having tips with trigonal pyramidal shapes on the PS sphere.

The deposited CuO nanostructures can be tuned by varying ambient gas pressures during the PLD process. Figure 7.21 shows the FESEM and TEM images of samples

FIGURE 7.20 (a, b): FESEM images of a CuO hierarchical micro/nanostructured array obtained by combining the PS colloidal monolayer and PLD process. (PS sphere size: 350 nm, deposition time: 2 hours, ambient oxygen pressure during deposition: 6.7 Pa). (a) Top-view image; (b) section view. Scale bars in parts (a) and (b) indicate 500 nm. (c) and (d) TEM images of a CuO hierarchical micro/nanostructured array: (c) TEM image from the top; (d) TEM image of several separated units from the periodic array and the corresponding SAED pattern. (Reprinted with permission from Li et al., 2009, 2580–8. Copyright 2009 American Chemical Society.)

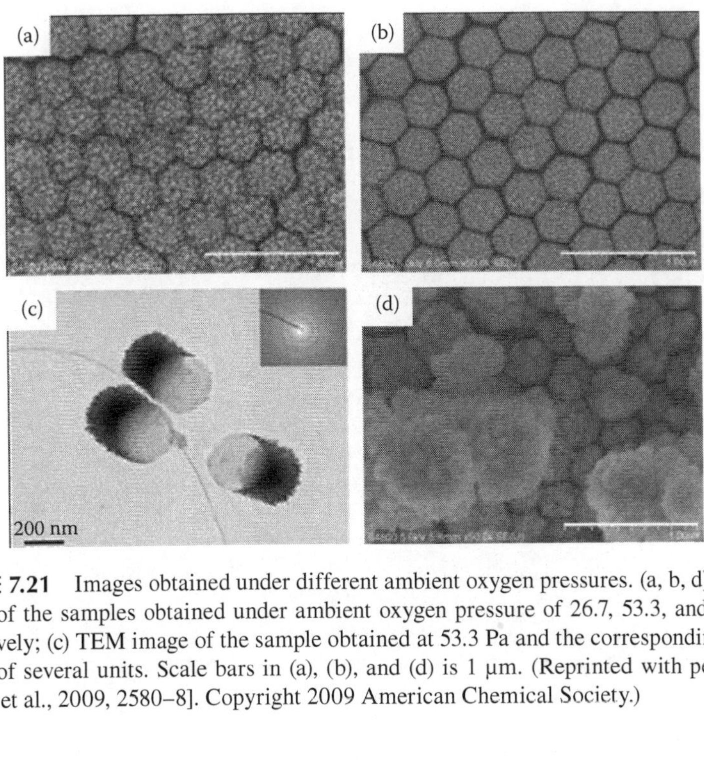

FIGURE 7.21 Images obtained under different ambient oxygen pressures. (a, b, d) FESEM images of the samples obtained under ambient oxygen pressure of 26.7, 53.3, and 79.8 Pa, respectively; (c) TEM image of the sample obtained at 53.3 Pa and the corresponding SAED pattern of several units. Scale bars in (a), (b), and (d) is 1 μm. (Reprinted with permission from Li et al., 2009, 2580–8]. Copyright 2009 American Chemical Society.)

achieved by PLD under higher ambient gas pressures during the PLD process using the colloidal monolayers as substrates. When oxygen pressure increased from 6.7 to 26.7 Pa, the morphology did not appreciably change and exhibited similar hierarchical structures as before (Figure 7.21a). However, when the oxygen pressure increased to 53.3 Pa, the morphology completely changed and was very different from those at lower pressures. The nanocolumn tips on the PS sphere demonstrated imperfect trigonal pyramid shapes, and the tip sizes became much smaller (Figure 7.21b). According to the corresponding TEM image and SAED pattern (Figure 7.21c), we find that hierarchical micro/nanostructures were still observed at such high oxygen pressure, but the crystallization of deposited aligned nanocolumns on the PS sphere becomes worse than those obtained at lower oxygen pressure. When the gas pressure increased to as high as 79.8 Pa, similar hierarchical micro/nanostructured array was not obtained, and many aggregates of small particles were produced on the colloidal monolayer template (Figure 7.21d).

The x-ray diffraction (XRD) spectra of the samples obtained under different oxygen pressures are shown in Figure 7.22. Strong preferential orientation growth along (002) was observed at the gas pressure of 26.7 Pa. Increasing oxygen pressure led to weakening of this preferential orientation and broadening of XRD peaks. This result reflects that deposited materials gradually changed to small nanoparticles from aligned nanocolumn arrays and the particles became much smaller with increasing oxygen pressure during PLD, agreeing with FESEM images. When the oxygen pressure increased to very high value, 79.8 Pa, the deposited material completely

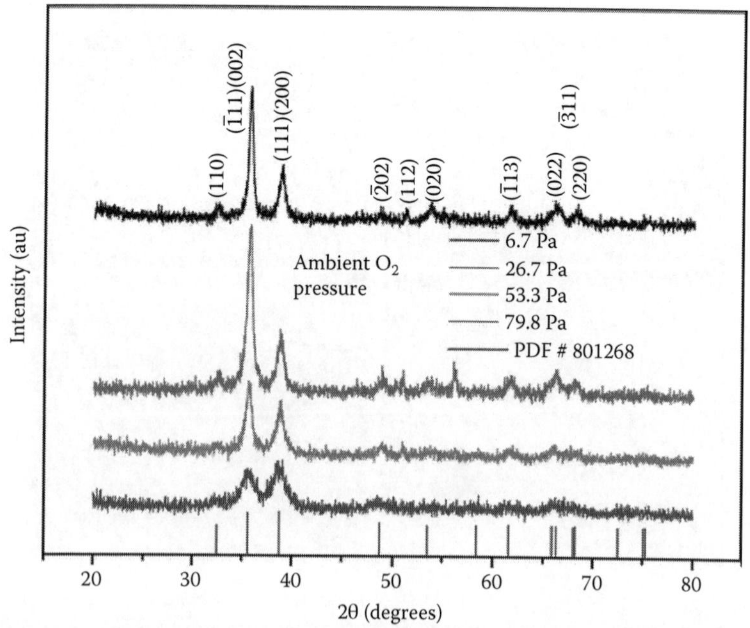

FIGURE 7.22 X-ray diffraction (XRD) patterns of the samples obtained under different oxygen pressures. (Reprinted with permission from Li et al., 2009, 2580–8. Copyright 2009 American Chemical Society.)

consisted of small nanoparticles or the aggregates of small nanoparticles, and there was no preferential orientation growth. Because when the gas pressure increases to a high value, the plume is compressed into a smaller space in PLD process, and the possibility of collision among ions or atoms in plasma is greatly enhanced, further resulting in a kinetic energy decrease of ions or atoms, which leads to less crystallization and smaller nanoparticle formation.

Similar crystalline hierarchical micro/nanostructured arrays of Fe_2O_3 and ZnO can be also created by the same route, as shown in Figure 7.23. Fe_2O_3 nanobelts or ZnO nanocolumns were well aligned on the PS sphere tops, like those of CuO. However, the Fe_2O_3 nanobelt or ZnO nanocolumn tops were not like those of CuO. The slight differences among CuO, Fe_2O_3, and ZnO fine nanostructures are determined mainly by their various chemical and physical properties: a crystal facet of the interface with different energies, and so on.

In this strategy, the height of the micro/nanostructured unit can be obviously controlled by varying the deposition time during the PLD process; the height will increase with the increase of PLD time [55]. It has been found that the unit height increases by increasing the deposition time from 30 to 60 minutes, as demonstrated in Figure 7.24. However, if the deposition time is too long, for example, 180 minutes, the tops of micro/nanostructured units will aggregate with each other due to strong van der Waals attraction among units in the deposition process, as shown in Figure 7.24c, c', and c''. Additionally, the top of the micro/nanostructured

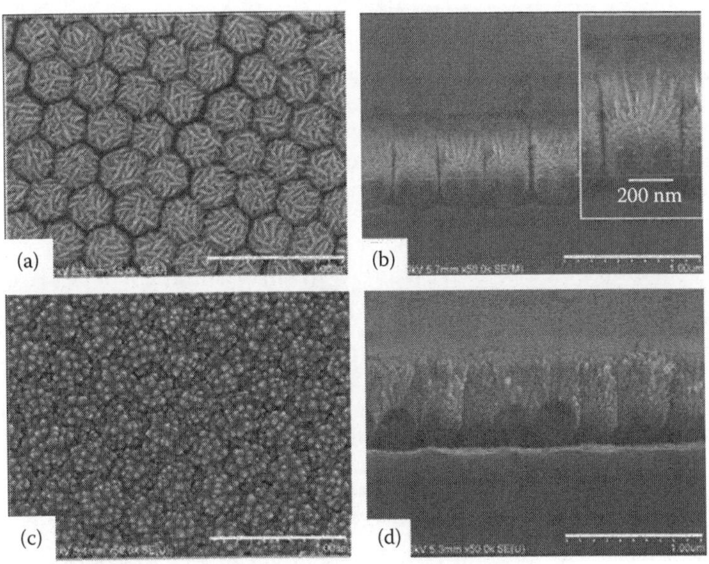

FIGURE 7.23 FESEM images of hierarchical micro/nanostructured arrays of Fe_2O_3 and ZnO. (a, b) Fe_2O_3, oxygen pressure 6.7 Pa, deposition time 1.5 hours; (c, d) ZnO, oxygen pressure 6.7 Pa, deposition time 40 minutes. (a) and (c) are top views and (b) and (d) are side views. The inset in (b) is the high magnification image of a single Fe_2O_3 hierarchical micro/nanostructure. (Reprinted with permission from Li et al., 2009, 2580–8. Copyright 2009 American Chemical Society.)

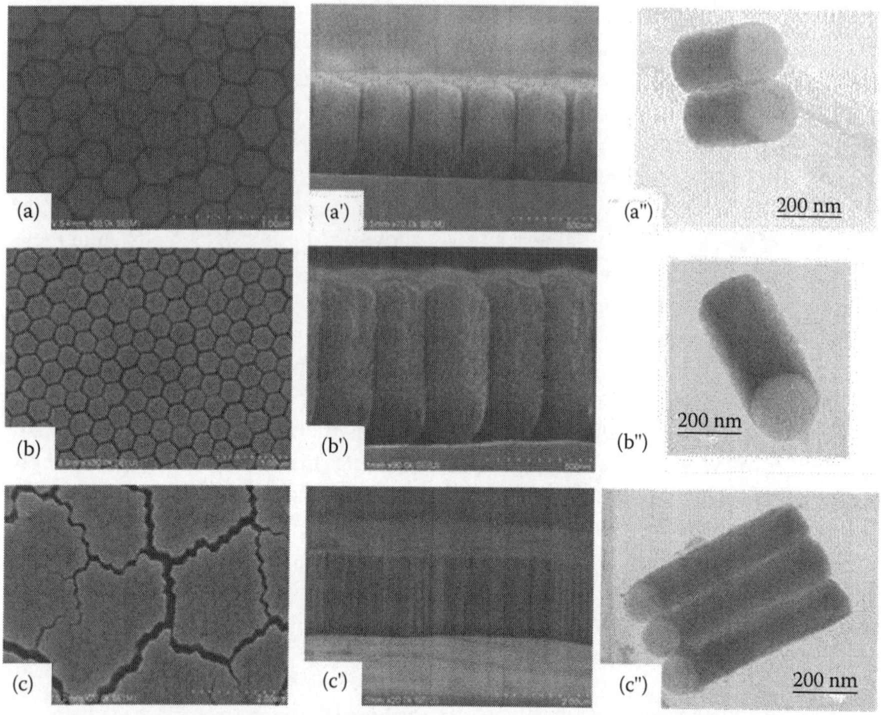

FIGURE 7.24 The height changes of micro/nanostructured unit with increase deposition time. Deposition time: 30 minutes in (a), (a'), and (a"); 60 minutes in (b), (b'), and (b"); 180 minutes in (c), (c'), and (c"). (a), (b), and (c) are scanning electron microscope (SEM) images of top view; (a'), (b') and (c') are SEM images of cross section; (a"), (b"), and (c") are TEM images of micro/nanostructure units. (Reprinted with permission from Li et al., 2008, 14755–22. Copyright 2008 American Chemical Society.)

unit gradually flattens from convex shape with increasing deposition time, resulting in a weakening shadow effect. Therefore, a continuous film might be formed at the top of hierarchical micro/nanostructured array if further increasing deposition time after 180 minutes.

The as-prepared hierarchical micro/nanostructured units in periodic arrays are composed of a PS sphere at the bottom and a micro/nanoparticle or rod on the top of the PS sphere. If the PS colloidal template is dissolved by an organic solution (CH_2Cl_2), this periodic array could retain its integrity while being peeled from the substrate due to the van der Waals force between the neighboring micro/nanostructured units suspended in the solution. It could then be transferred to any desired substrate (e.g., TEM copper grid) by picking it up using another substrate, as illustrated in Figures 7.25 and 7.26. The transferability avoids restrictions on substrates in the fabrication process of hierarchical micro/nanostructured arrays, which is helpful in the design and fabrication of new micro/nanodevices on any desired substrates.

The formation process was traced by PLD using colloidal clusters with different PS spheres as templates. Herein, the TiO_2 was selected as desired material and colloidal monolayer with PS sphere size of 350 nm as template to demonstrate the

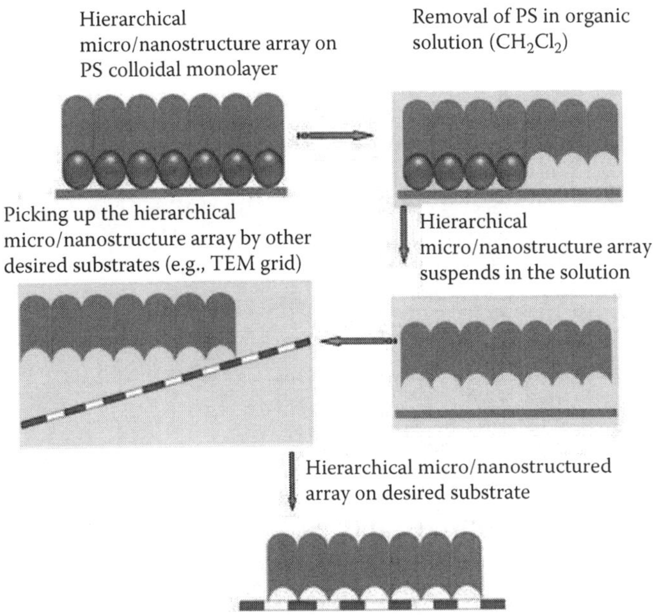

FIGURE 7.25 Schematic illustration of transferability of hexagonal close-packed (hcp) hierarchical micro/nanostructured arrays. (Reprinted with permission from Li et al., 2008, 14755–22. Copyright 2008 American Chemical Society.)

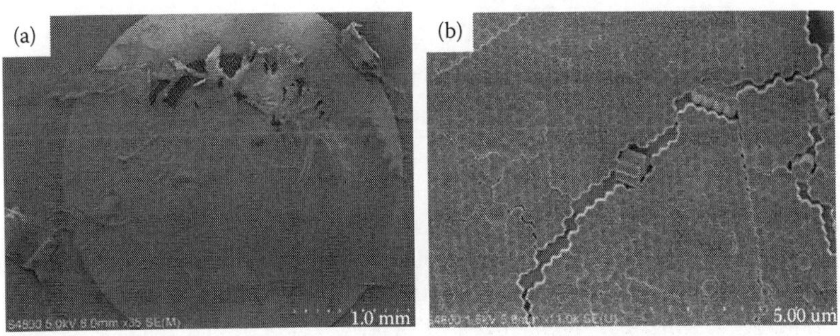

FIGURE 7.26 FESEM images of transferred TiO_2 micro/nanostructured arrays from a silicon substrate on a TEM grid. (a) Low- and (b) high-magnification images of array film on a TEM grid. (Reprinted with permission from Li et al., 2008, 14755–22. Copyright 2008 American Chemical Society.)

formation process of hierarchical micro/nanostructured arrays. The colloidal clusters with different PS sphere were fabricated by spin coating with a higher rotation speed (2000 rpm) and lower concentration (1.0 wt%) of PS colloidal microsphere suspension. For example, a single PS sphere or PS sphere clusters with different sphere numbers (2, 3, 4, …) can be easily created on the substrate by above route, as indicated in column (A) of Figure 7.27. After PLD, morphologies observed from the top were compared to those before PLD, as demonstrated in column (B) of Figure 7.27.

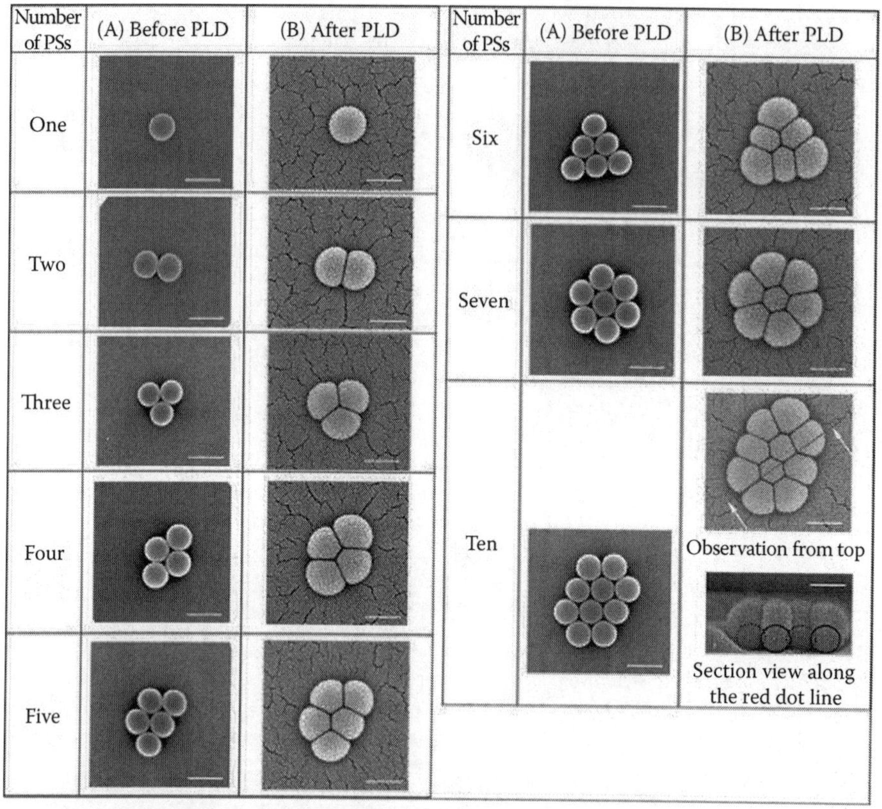

FIGURE 7.27 Morphologies of before and after PLD on the PS sphere surface. (Scale bars are 500 nm.) (Reprinted with permission from Li et al., 2008, 14755–22. Copyright 2008 American Chemical Society.)

For a single PS microsphere, the shape kept spherical but that the size increased from 350 nm (PS sphere size) to 500 nm after PLD. For the PS sphere clusters with sphere number from two to six, each unit size in the sphere cluster still increased, but could not maintain the spherical shape after PLD. Growth of deposited TiO_2 was restricted at the contact point of two neighboring PS spheres, the contact between the neighboring units changed from a quasidot contact to a facet contact before (PS sphere cluster) and after (PS sphere cluster with deposited materials on the surface) PLD. If a PS sphere in sphere cluster was completely surrounded by others, for example, the central sphere in an hcp sphere cluster of seven, its size after the deposition was almost the same as before PLD and the morphology was slightly changed from spherical shape to hexagonal one. A section of a PS sphere cluster of 10 spheres with the hcp arrangement after PLD displays that the hierarchical micro/nanorods have formed on the two spheres completely surrounded by the others and that hierarchical rod cannot be formed on the spheres at the edge of the sphere cluster. This implies that a hierarchical micro/nanostructured array will be easily produced after PLD if a colloidal monolayer with a large scale is applied in the PLD process. Additionally, if the desired materials are deposited on a bare silicon substrate without any PS spheres by PLD, nanocolumns grow vertically on the Si substrate.

Generally, nanocolumns prefer to grow in the normal direction on the substrate during the PLD process. In the PLD process, the desired target (TiO_2) is irradiated by a laser beam using an energy level exceeding its threshold in vacuum environment; plasma including ions (Ti^{4+}, O^{2-}, etc.), molecules, electrons, and clusters are released into the PLD chamber from the target. However, if a background gas with high pressure is introduced into the chamber, the movement direction of ions or electrons will be changed from an almost uniform direction to multidirection due to collisions between the ions, electrons, molecules, and clusters of the ejected species and the background gas. According to the above facts, the formation mechanism of hierarchical hcp nanocolumn arrays can be easily understood, as displayed in Figure 7.28. If a substrate without PS spheres is used in the PLD process, a film consisting of vertical nanocolumns of small diameter will be formed. If a single PS sphere exists on the

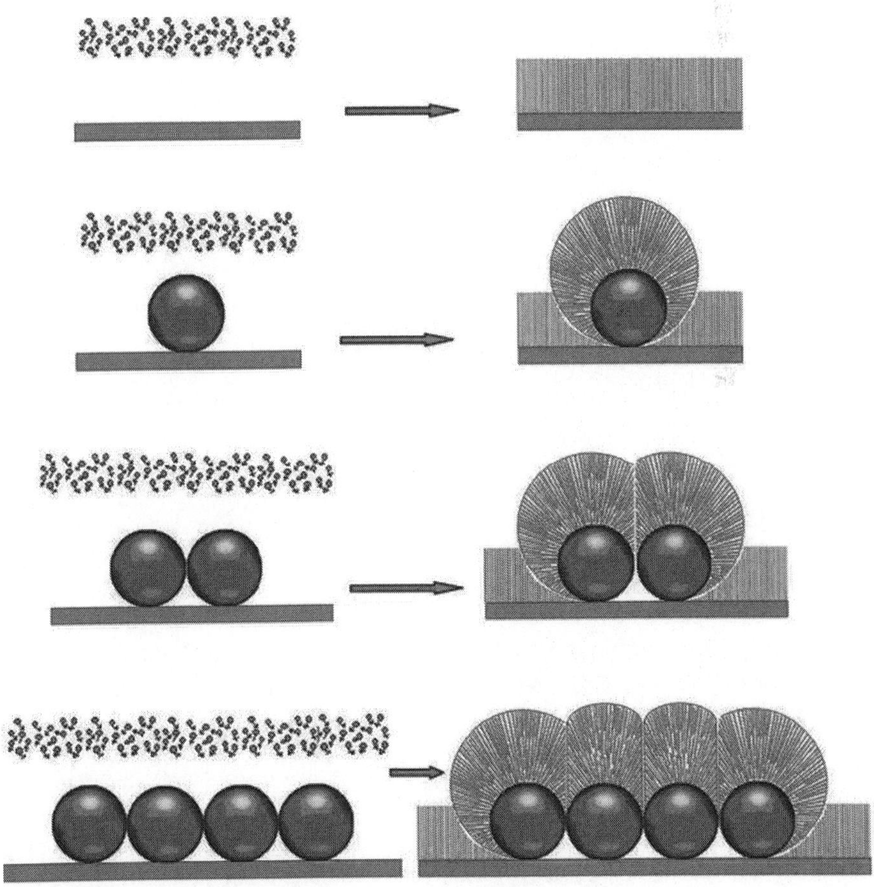

FIGURE 7.28 Schematic illustration of the formation mechanism of hcp hierarchical micro/nanostructured arrays. (Reprinted with permission from Li et al., 2008, 14755–22. Copyright 2008 American Chemical Society.)

substrate, a composite of a PS sphere at the bottom and a shell composed of TiO_2 radiation-shaped nanobranches on sphere top will turn up due to preferential vertical growth along the normal direction of the supporting surface and multidirectional deposition. For a PS sphere cluster (more than one sphere) on the substrate, a shadow effect will be produced in the deposition between any two neighboring spheres. If one sphere in the sphere cluster is completely surrounded by six other spheres as in the case of hcp arrangement, one rod with hierarchical micro/nanostructure will grow on this sphere top. If a colloidal monolayer with a large scale is adopted, this route can easily fabricate an hcp hierarchical micro/nanostructured array. In this strategy, an off-axis configuration is adopted where the target and substrate are perpendicularly placed. It is similar to the glancing angle deposition (GLAD) or oblique angle deposition in which there is a large angle between the deposition direction and the normal direction of the substrate. In the traditional GLAD method, atoms from the target obliquely arrive and condense on the substrate, and the tilted and separated nanowire or nanopillar array with a porous structure are gradually produced due to the shadow effect of the initial deposited nanoparticles under high-vacuum conditions. The critical difference between this route and GLAD is the background gas pressure during deposition, which converts the directional flow of ejected species in a vacuum into a multidirectional one at higher pressure. Therefore, this multidirection deposition and shadow effect are the principal reasons why a vertical hierarchical micro/nanostructured array with hcp alignment is formed on the colloidal monolayer. This can be further verified by varying the angles between substrates and target in PLD process. If these experiments were carried out in a vacuum, tilted rod-like structured arrays with different angles would be obtained on different substrates. However, from these results, the rod-like morphologies are independent of the angle between the substrate and target but the growth rates are different for different angles because of the plume shape in PLD, and they always grow vertically on the substrate due the multiple direction deposition combined with shadow effect of neighboring colloidal sphere.

7.3.3 1D Nanostructure by Magnetron Sputtering

A PS colloidal monolayer on a substrate is placed into the chamber of radio frequency magnetron sputtering for material deposition at room temperature and Ar is introduced as the background gas. A unique hierarchical micro/nanostructured array is formed due to PS-templated plasma etching/deposition in a relatively high vacuum (0.06 Pa) [80]. The features of this technique are (1) low-pressure sputtering; (2) PS-templated sputtering, which guarantees a periodic arrangement, and (3) plasma etching/deposition, which eventually produces the unique hierarchical micro/nanostructure.

Figure 7.29 presents FESEM and TEM images of hierarchical alumina micro/nanostructured arrays by sputtering target alumina using a PS colloidal template with a sphere size of 750 nm in a relatively high vacuum (0.06 Pa). They show the following three unique features. First, it is a periodic nanocolumn array cushioned by a semishell in an hexagonal nonclose packed arrangement. Each nanocolumn is composed of two parts: a semishell-shaped cushion at the bottom and a vertical nanocolumn on the top of the cushion. Such nanocolumn possesses a very rough structure on the surface and seems to be composed of many minicolumns, indicating that the

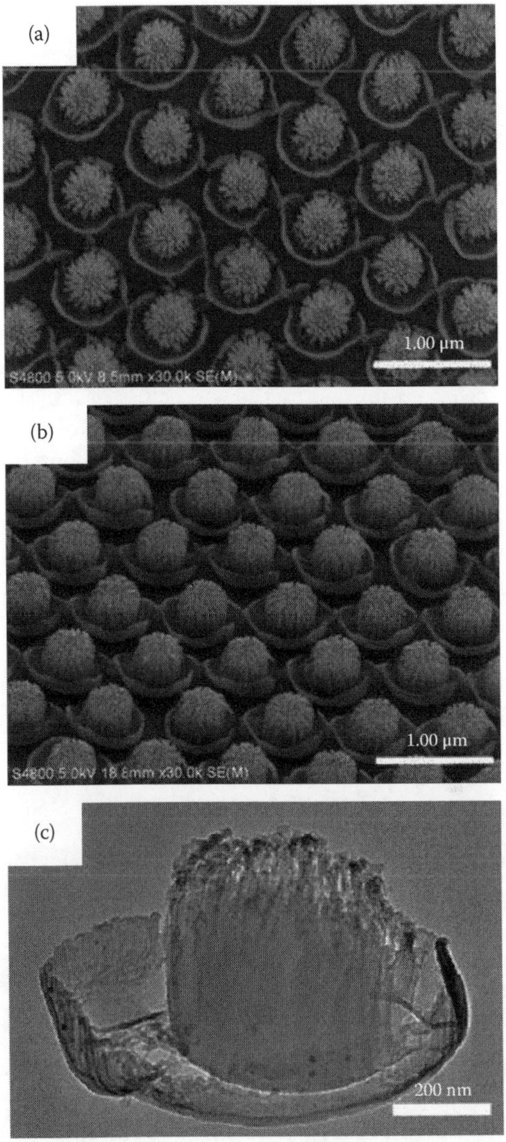

FIGURE 7.29 (a, b) FESEM and (c) TEM images of a sample obtained by sputtering using a PS colloidal monolayer as the substrate (PS sphere size: 750 nm; deposition time: 2 hours). (a) Top view. (b) Titled view with 45° angle. The scale bar in (a) and (b) is 1 μm. (c) TEM image of one unit. (Gao et al., *J. Mater. Chem.*, 21, 2087–90, 2011. Reproduced by permission of The Royal Society of Chemistry.)

sample possesses a hierarchical, porous structure and hence has a high surface area. Second, the periodicity was 750 nm, matching well with the initial size of the PS spheres. It is very evident that the sizes of the cushion and the central columns were reduced by about 15% and 45% compared to the original size of the PS template.

The formation of such hierarchical micro/nanostructured arrays is traced by the different sputtering deposition time, as demonstrated in Figure 7.30. With the increase of deposition time, the PS sphere size gradually decreases and the alumina columns grow vertically in the center, and finally the columnar structures and a salver-shaped semishell are formed. Generally, thin alumina continuous film is formed on bare substrate without PS spheres due to the strong ion energy and subsequent rapid surface migration under such a low sputtering pressure. In the case with PS sphere array on the substrate, alumina components sputtered from the target are impinged and implanted into the PS due to the strong ion energy and soft nature of PS. The part of PS sphere is also continuously etched away by argon ions and part of deposited alumina is also etched away, but remaining part will gradually form a structure. Additionally, the PS colloidal monolayer supplies the periodic array template. Merging these two aspects into one forms a unique hierarchical micro/nanostructured arrays. The PS spheres become smaller with plasma etching and a salver-shaped semishell gradually appear. Further sputtering causes the PS spheres to be etched more significantly, and the species generated from the target deposit perpendicularly onto the template (both the center and the semishell part), thus forming a column structure and salver-shaped semishell. Implanted components of aluminum and oxygen into PS sphere will be linked together by continuous etching of PS. But at the side edge of the spheres, the amount of PS is not much and easily etched away to form aluminum oxide film,

FIGURE 7.30 FESEM images of samples obtained at different deposition times. (a) 10 minutes, (b) 25 minutes, (c) 30 minutes, and (d) 60 minutes. The insets are the high-magnification images of one unit in each figure. Scale bars: 1 μm. (Gao et al., *J. Mater. Chem.*, 21, 2087–90, 2011. Reproduced by permission of The Royal Society of Chemistry.)

resulting in the cushion shell. The amount of PS at the center part is much more and even by continuous etching a film cannot be formed and rod-like structures are generated. Further sputtering continues etching the PS spheres until the final unique hexagonal nonclose packed hierarchical structure forms. The formation process of this unique hierarchical micro/nanostructure is schematically illustrated in Figure 7.31.

To further confirm the above processes, the pressures of Ar were adjusted from low level (0.06 Pa) to 0.13 and final 6.7 Pa. Figure 7.32 shows the corresponding results. With increase of the background gas pressure from 0.06 Pa to 0.13 and 6.7 Pa, the collision probability between the ejected species and Ar molecules increases; thus, the PS spheres are more significantly etched and no semishell can be formed. Therefore, only columnar

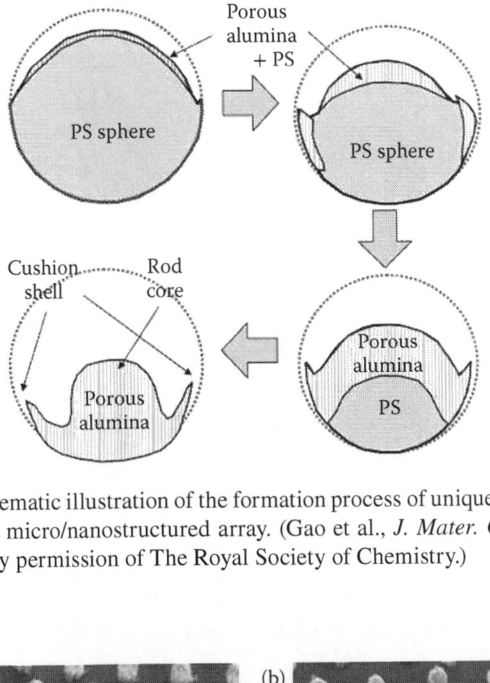

FIGURE 7.31 Schematic illustration of the formation process of unique hexagonal nonclose packed hierarchical micro/nanostructured array. (Gao et al., *J. Mater. Chem.*, 21, 2087–90, 2011. Reproduced by permission of The Royal Society of Chemistry.)

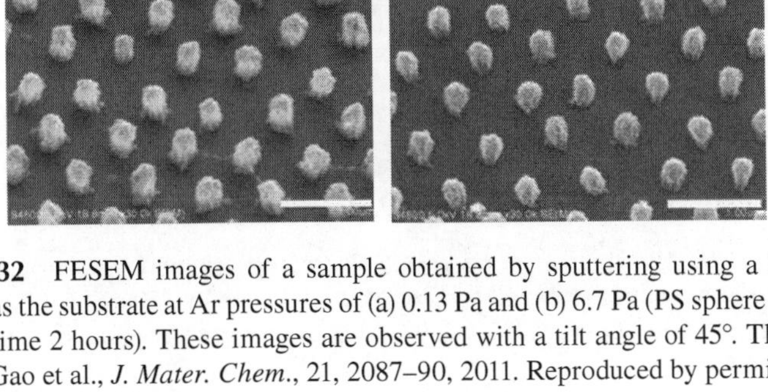

FIGURE 7.32 FESEM images of a sample obtained by sputtering using a PS colloidal monolayer as the substrate at Ar pressures of (a) 0.13 Pa and (b) 6.7 Pa (PS sphere size 750 nm; deposition time 2 hours). These images are observed with a tilt angle of 45°. The scale bars are 1 μm. (Gao et al., *J. Mater. Chem.*, 21, 2087–90, 2011. Reproduced by permission of The Royal Society of Chemistry.)

structures are obtained. The amounts of deposited materials in the intercolumnar structures are negligible probably due to the blocking effects of gaseous species emitted from decomposed PS spheres during sputtering. These results firmly prove that a relatively high vacuum condition subsequently induces mild plasma etching/deposition.

Besides hexagonal nonclose packed alumina micro/nanostructured arrays with a periodicity of 750 nm, novel hierarchical arrays with periodicities of 350 nm, 1 μm, and 2 μm could be also created by colloidal monolayers with different PS sphere sizes during sputtering at the same pressure of Ar, as typically shown in Figure 7.33. Besides alumina, the hierarchical arrays of the other materials including Au/Al_2O_3 composite, CuO, and NiO can also be fabricated by the presented one-step plasma etching. Some of the results are presented in Figure 7.34. In this method, only the inorganic materials, which are much stronger to plasma than PS

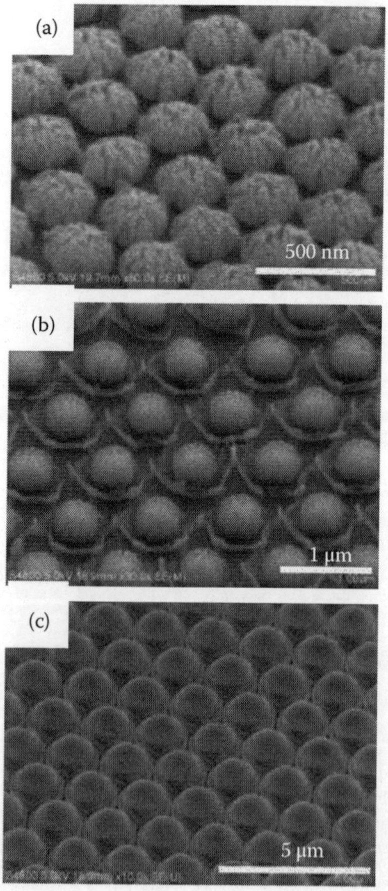

FIGURE 7.33 FESEM images of the arrays obtained by sputtering using a PS colloidal monolayer template with different PS sphere sizes at 0.06 Pa: (a) 350 nm, (b) 1 μm, and (c) 2 μm. (Gao et al., *J. Mater. Chem.*, 21, 2087–90, 2011. Reproduced by permission of The Royal Society of Chemistry.)

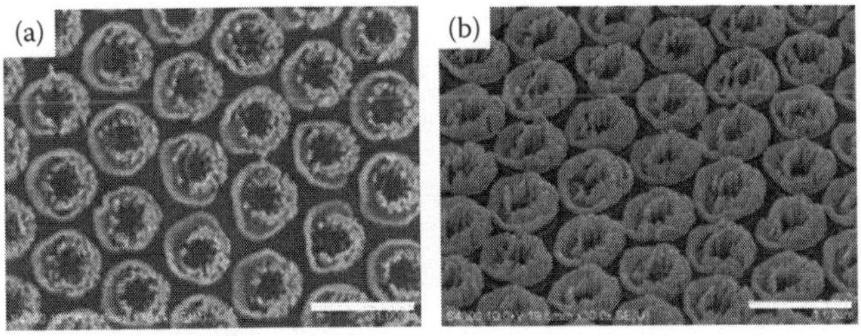

FIGURE 7.34 FESEM images of periodic Au/Al_2O_3 nanocomposite arrays obtained by cosputtering multiple targets consisting of an Al_2O_3 target and Au sheets and using a PS colloidal monolayer as the substrate (PS sphere size 750 nm; deposition time 2 hours). (a) Image observed from the top. (b) Image with a tilt angle of 45°. The scale bars are 1 μm. (Gao et al., *J. Mater. Chem.*, 21, 2087–90, 2011. Reproduced by permission of The Royal Society of Chemistry.)

can be used as the deposited materials. Otherwise, the deposition cannot be guaranteed because of the subsequent etching.

7.4 BRIEF SUMMARY

In summary, the recent work to create micro/nanostructured arrays have been introduced, including nanoparticle array and laser morphological manipulation, 0D object-built arrays, 1D nanoobject-built arrays, and 2D nanoobject-built arrays, based on the PS colloidal templates with different routes, for example, chemical reaction, physical deposition (PLD and magnetron sputtering). Such periodically arranged arrays on solid supports can be fabricated by nanosphere lithography, which is a general, simple, and low-cost method. The morphology of these arrays can be well controlled by the experiment conditions, for instance, diameter of colloidal spheres, the deposition conditions, and/or laser manipulation. These nanostructured arrays would exhibit some new properties in optics, wettability, magnetics, surface-enhanced Raman scattering, gas sensitivity, and so on, and hence have the important potential applications in energy storage or conversion, data storage, next-generation integrated nanophotonics devices, biomolecular labeling and identification, microfluidic devices, catalysts, microreactor devices, miniaturized optical components, self-cleaning surface, and so on.

REFERENCES

1. Morariu, M. D.; Voicu, M. N.; Schäffer, E.; Lin, Z.; Russell, T. P.; and Steiner, U. 2003. Hierarchical structure formation and pattern replication induced by an electric field. *Nat. Mater.* 2: 48–52.
2. Ye, C. H.; Zhang, L. D.; Fang, X. S.; Wang, Y. H.; Yan, P.; and Zhao, J. W. 2004. Hierarchical structure: Silicon nanowires standing on silica microwires. *Adv. Mater.* 16: 1019–23.

3. Dorozhkin, S. V. 2007. A hierarchical structure for apatite crystals. *J. Mater. Sci. Mater. Med.* 18: 363–6.
4. Duan, G.; Cai, W.; Luo, Y.; Li, Y.; and Lei, Y. 2006. Hierarchical surface rough ordered Au particle arrays and their surface enhanced Raman scattering. *Appl. Phys. Lett.* 89: 181918.
5. Lao, J. Y.; Wen, J. G.; Wang, D. Z.; and Ren, Z. F. 2002. Hierarchical ZnO nanostructures. *Nano Lett.* 2: 1287–91.
6. Gao, P. X.; Ding, Y.; and Wang, Z. L. 2003. Crystallographic orientation-aligned ZnO nanorods grown by a tin catalyst. *Nano Lett.* 3: 1315–20.
7. Yan, R.; Gargas, D.; and Yang, P. 2009. Nanowire photonics. *Nat. Photonics* 3: 569–76.
8. Roduner, E. 2006. Size matters: Why nanomaterials are different. *Chem. Soc. Rev.* 35: 583–92.
9. Cobley, C. M.; Chen, J.; Cho, E. C.; Wang, L. V.; and Xia, Y. 2011. Gold nanostructures: A class of multifunctional materials for biomedical applications. *Chem. Soc. Rev.* 40: 44–6.
10. Wang, Z. L. 2008. Towards self-powered nanosystems: From nanogenerators to nanopiezotronics. *Adv. Funct. Mater.* 18: 3553–67.
11. Varin, R. A.; Zbroniec, L.; Polanski, M.; and Bystrzycki, J. 2011. A review of recent advances on the effects of microstructural refinement and nano-catalytic additives on the hydrogen storage properties of metal and complex hydrides. *Energies* 4 (1): 1–25.
12. Lu, F.; Cai, W. P.; and Zhang, Y. G. 2008. ZnO hierarchical micro/nanoarchitectures: Solvothermal synthesis and structurally enhanced photocatalytic performance. *Adv. Funct. Mater.* 18: 1047–56.
13. Zeng, H. B.; Cai, W. P.; Liu, P. S.; Xu, X. X.; Zhou, H. J.; Klingshirn, C.; and Kalt, H. 2008. ZnO-based hollow nanoparticles by selective etching: Elimination and reconstruction of metal–semiconductor interface, improvement of blue emission and photocatalysis. *ACS Nano* 2 (8): 1661–70.
14. He, H.; Cai, W. P.; Lin, Y. X.; and Dai, Z. F. 2011. Silver porous nanotube built three-dimensional films with structural tunability based on the nanofiber template-plasma etching strategy. *Langmuir* 27 (5): 1551–5.
15. Smith, H. I. and Schattenburg, M. L. 1993. X-ray lithography from 500 to 30 nm: X-ray nanolithography. *IBM J. Res. Dev.* 37: 319–29.
16. Stroscio, J. A. and Eigler, D. M. 1991. Molecular self-assembly and nanochemistry: A chemical strategy for the synthesis of nanostructures. *Science* 254: 1312–9.
17. Liu, G.-Y.; Xu, S.; and Qian, Y. 2000. Nanofabrication of self-assembled monolayers using scanning probe lithography. *Acc. Chem. Res.* 33: 457–66.
18. Piner, R. D.; Zhu, J.; Xu, F.; Hong, S.; and Mirkin, C. A. 1999. "Dip-Pen" nanolithography. *Science* 283: 661–3.
19. Xia, Y. N. and Whitesides, G. M. 1997. Extending microcontact printing as a microlithographic technique. *Langmuir* 13: 2059–67.
20. Kumar, A. and Whitesides, G. M. 1993. Features of gold having micrometer to centimeter dimensions can be formed through a combination of stamping with an elastomeric stamp and an alkanethiol "ink" followed by chemical etching. *Appl. Phys. Lett.* 63: 2002.
21. Quist, A. P.; Pavlovic, E.; and Oscarsson, S. 2005. Recent advances in microcontact printing. *Anal. Bioanal. Chem.* 381: 591–600.
22. Kim, E.; Xia, Y.; and Whitesides, G. M. 1995. Polymer microstructures formed by moulding in capillaries. *Nature* 376: 581–4.
23. Gao, L. and McCarthy, T. J. 2006. The "lotus effect" explained: Two reasons why two length scales of topography are important. *Langmuir* 22: 2966–7.
24. Denkov, N. D.; Velev, O. D.; Kralchevsky, P. A.; Ivanov, I. B.; Yoshimura, H.; and Nagayama, K. 1993. 2-Dimensional crystallization. *Nature* 361: 26.

25. Ozin, G. A. and Yang, S. M. 2001. The race for the photonic chip: Colloidal crystal assembly in silicon wafers. *Adv. Funct. Mater.* 11: 95–104.
26. Jiang, P. and McFarland, M. J. 2004. Large-scale fabrication of wafer-size colloidal crystals, macroporous polymers and nanocomposites by spin-coating. *J. Am. Chem. Soc.* 126: 13778–86.
27. Wang, D. and Möhwald, H. 2004. Rapid fabrication of binary colloidal crystals by stepwise spin-coating. *Adv. Mater.* 16: 244–7.
28. Li, Y.; Cai, W. P.; Duan, G. T.; Sun, F. Q.; Cao, B. Q.; and Lu, F. 2005. Synthesis and optical absorption property of ordered macroporous titania film doped with Ag nanoparticles. *Mater. Lett.* 59: 276–9.
29. Im, S. H.; Kim, M. H.; and Park, O. O. 2003. Thickness control of colloidal crystals with a substrate dipped at a tilted angle into a colloidal suspension. *Chem. Mater.* 15: 1797–802.
30. Sun, F.; Cai, W.; Li, Y.; Duan, G.; Nichols, W. T.; Liang, C.; Koshizaki, N.; Fang, Q.; and Boyd, I. W. 2005. Laser morphological manipulation of gold nanoparticles periodically arranged on solid supports. *Appl. Phys. B* 81: 765–8.
31. Jensen, T. J.; Duval, M. L.; Kelly, K. L.; Lazarides, A. A.; Schatz, G. C.; and Van Duyne, R. P. 1999. Nanosphere lithography: Effect of the external dielectric medium on the surface plasmon resonance spectrum of a periodic array of silver nanoparticles. *J. Phys. Chem. B.* 103: 9846–53.
32. Tan, B. J. Y.; Sow, C. H.; Koh, T. S.; Chin, K. C.; Wee, A. T. S.; and Ong, C. K. 2005. Fabrication of size-tunable gold nanoparticles array with nanosphere lithography, reactive ion etching, and thermal annealing. *J. Phys. Chem. B.* 109: 11100–9.
33. Sort, J.; Glaczynska, H.; Ebels, U.; Dieny, B.; Miersig, M.; and Rybczynski, J. 2004. Exchange bias effects in submicron antiferromagnetic-ferromagnetic dots prepared by nanosphere lithography. *J. Appl. Phys.* 95: 7516–8.
34. Sun, F. Q.; Cai, W. P.; Li, Y.; Cao, B. Q.; Lei, Y.; and Zhang, L. D. 2004. Morphology-controlled growth of large-area two-dimensional ordered pore arrays. *Adv. Funct. Mater.* 14: 283–8.
35. Sun, F. Q.; Yu, J. C.; and Wang, X. C. 2006. Construction of size-controllable hierarchical nanoporous TiO_2 ring arrays and their modifications. *Chem. Mater.* 18: 3774–9.
36. Li, Y.; Cai, W. P.; Duan, G. T.; Sun, F. Q.; and Lu, F. 2005. Large area In_2O_3 ordered pore arrays and their photoluminescence properties. *Appl. Phys. A* 81: 269–273.
37. Li, Y.; Cai, W. P.; Cao, B. Q.; Duan, G. T.; Li, C. C.; Sun, F. Q.; and Zeng, H. B. 2006. Morphology-controlled 2D ordered arrays by heating-induced deformation of 2D colloidal monolayer. *J. Mater. Chem.* 16: 609–12.
38. Sun, F. Q.; Cai, W. P.; Li, Y.; Cao, B. Q.; Lei, Y.; and Zhang, L. D. 2005. Morphology-controlled growth of large area ordered porous film. *Mater. Sci. Technol.* 21: 500–4.
39. Sun, F. Q.; Cai, W. P.; Li, Y.; Jia, L. C.; and Lu, F. 2005. Direct growth of mono- and multilayer nanostructured porous films on curved surfaces and their application as gas sensors. *Adv. Mater.* 17: 2872–7.
40. Sun, F. Q.; Cai, W. P.; Li, Y.; Cao, B. Q.; Lu, F.; Duan, G. T.; and Zhang, L. D. 2004. Morphology control and transferability of ordered through-pore arrays based on electrodeposition and colloidal monolayers. *Adv. Mater.* 16: 1116–21.
41. Cao, B. Q.; Cai, W. P.; Sun, F. Q.; Li, Y.; Lei, Y.; and Zhang, L. D. 2004. Fabrication of large-scale zinc oxide ordered pore arrays with controllable morphology. *Chem. Commun.* (14): 1604–5.
42. Cao, B. Q.; Sun, F. Q.; and Cai, W. P. 2005. Electrodeposition-induced highly oriented zinc oxide ordered pore arrays and its ultraviolet emissions. *Electrochem. Solid-State Lett.* 8: G237–40.
43. Duan, G. T.; Cai, W. P.; Li, Y.; Li, Z. G.; Cao, B. Q.; and Luo, Y. Y. 2006. Transferable ordered Ni hollow sphere arrays induced by electrodeposition on colloidal monolayer. *J. Phys. Chem. B* 110: 7184–8.

44. Wang, X.; Lao, C.; Graugnard, E.; Summers, C. J.; and Wang, Z. L. 2005. Large-size liftable inverted-nanobowl sheets as reusable masks for nanolithiography. *Nano Lett.* 5: 1784–8.
45. Yan, F. and Goedel, W. A. 2004. Preparation of mesoscopic gold rings using particle imprinted templates. *Nano Lett.* 4: 1193–6.
46. Duan, G. T.; Cai, W. P.; Luo, Y. Y.; Li, Z. G.; and Lei, Y. 2006. Hierarchical structured Ni nanoring and hollow sphere arrays by morphology inheritance based on ordered through-pore template and electrodeposition. *J. Phys. Chem. B* 110: 15729–33.
47. Li, Y.; Cai, W. P.; Duan, G. T.; Cao, B. Q.; and Sun, F. Q. 2005. 2D ordered polymer hollow sphere and convex structure arrays based on monolayer pore films. *J. Mater. Res.* 20: 338–43.
48. Li, Y.; Huang, X. J.; Heo, S. H.; Li, C. C.; Choi, Y. K.; Cai, W. P.; and Cho, S. O. 2007. Superhydrophobic bionic surfaces with hierarchical microsphere/SWCNT composite arrays. *Langmuir* 23: 2169–74.
49. Huang, X. J.; Li, Y.; Im, H. S.; Yarimaga, O.; Kim, H. J.; Jang, D. Y.; Cho, S. O.; Cai, W. P.; and Choi, K. Y. 2006. Morphology-controlled SWCNT/polymeric microsphere arrays by a wet chemical self-assembly technique and their application for sensors. *Nanotechnology* 17: 2988.
50. Li, Y.; Lee, E. J.; and Cho, S. O. 2007. Superhydrophobic coatings on curved surfaces featuring remarkable supporting force. *J. Phys. Chem. C* 111: 14813–7.
51. Li, Y.; Li, C. C.; Cho, S. O.; Duan, G. T.; and Cai, W. P. 2007. Silver hierarchical bowl-like array: Synthesis, superhydrophobicity, and optical properties. *Langmuir* 23: 9802–7.
52. Duan, G. T.; Cai, W. P.; Luo, Y. Y.; and Sun, F. Q. 2007. A hierarchically structured $Ni(OH)_2$ monolayer hollow-sphere array and its tunable optical properties over a large region. *Adv. Funct. Mater.* 17: 644–50.
53. Liu, G. Q.; Cai, W. P.; Kong, L. C.; Duan, G. T.; Li, Y.; Wang, J. J.; Zuo, G. M.; and Cheng, Z. X. 2012. Standing Ag nanoplate-built hollow microphere arrays controllable structural parameters and strong SERS performances. *J. Mater. Chem.* 22: 3177–84.
54. Duan, G. T.; Cai, W. P.; Luo, Y. Y.; Li, Y.; and Lei, Y. 2006. Hierarchical surface rough ordered Au particle arrays and their surface enhanced Raman scattering. *Appl. Phys. Lett.* 89: 181918.
55. Li, Y.; Sasaki, T.; Shimizu, Y.; and Koshizaki, N. 2008. Hexagonal-close-packed, hierarchical amorphous TiO_2 nanocolumn arrays: Transferability, enhanced photocatalytic activity, and superamphiphilicity without UV irradiation. *J. Am. Chem. Soc.* 130: 14755–22.
56. Li, Y.; Koshizaki, N.; Shimizu, Y.; Li, L.; Gao, S.; and Sasaki, T. 2009. Unconventional lithography for hierarchical micro-/nanostructure arrays with well-aligned 1D crystalline nanostructures: Design and creation based on the colloidal monolayer. *ACS Appl. Mater. Interfaces* 1: 2580–8.
57. Li, Y.; Sasaki, T.; Shimizu, Y.; and Koshizaki, N. 2008. A hierarchically ordered TiO_2 hemispherical particle array with hexagonal-non-close-packed tops: Synthesis and stable superhydrophilicity without UV irradiation. *Small* 4: 2286–91.
58. Li, Y.; Fang, X.; Koshizaki, N.; Sasaki, T.; Li, L.; Gao, S.; Shimizu, Y.; Sasaki, T.; Bando, Y.; and Golberg, M. 2008. Periodic TiO_2 nanorod arrays with hexagonal nonclose-packed arrangements: Excellent field emitters by parameter optimization. *Adv. Funct. Mater.* 19: 2467–73.
59. Li, L.; Li, Y.; Gao, S.; and Koshizaki, N. 2009. Ordered Co_3O_4 hierarchical nanorod arrays: Tunable superhydrophilicity without UV irradiation and transition to superhydrophobicity. *J. Mater. Chem.* 19: 8366–71.
60. Hulteen, J. C. and Van Duyne, R. P. V. 1995. Nanosphere lithography: A materials general fabrication process for periodic particle array surfaces. *J. Vac. Sci. Technol. A* 13: 1553.

61. Haynes, C. L. and Van Duyne, R. P. 2001. Nanosphere lithography: A versatile nanofabrication tool for studies of size-dependent nanoparticle optics. *J. Phys. Chem. B* 105: 5599–611.
62. Haynes, C. L.; McFarland, A. D.; Smith, M. T.; Hulteen, J. C.; and Van Duyne, R. P. 2002. Angle-resolved nanosphere lithography: Manipulation of nanoparticle size, shape, and interparticle spacing. *J. Phys. Chem. B* 106: 1898–902.
63. Burmeister, F.; Schafle, C.; and Leiderer, P. 1997. Colloid monolayers as versatile lithographic masks. *Langmuir* 13: 2983–7.
64. Kempa, K.; Kimball, B.; Rybczynski, J.; Huang, Z. P.; Wu, P. F.; Steeves, D.; Sennett, M. et al. 2003. Photonic crystals based on periodic arrays of aligned carbon nanotubes. *Nano Lett.* 3: 13–8.
65. Li, N. and Zinke-Allmang, M. 2002. Size-tunable Ge nano-particle arrays patterned on Si substrates with nanosphere lithography and thermal annealing. *Jpn. J. Appl. Phys.* 41: 4626–9.
66. Kurita, H.; Takami, A.; and Koda, S. 1998. Size reduction of gold particles in aqueous solution by pulsed laser irradiation. *Appl. Phys. Lett.* 72: 789.
67. Hodak, J.; H. Henglein, A.; Giersig, M.; and Hartland, G. V. 2000. Laser-induced interdiffusion in AuAg core–shell nanoparticles. *J. Phys. Chem. B* 104: 11708–18.
68. Link, S.; Burda, C.; Nikoobakht, B.; and El-Sayed, M. A. 2000. Laser-induced shape changes of colloidal gold nanorods using femtosecond and nanosecond laser pulses. *J. Phys. Chem. B* 104: 6152–63.
69. Link, S.; Wang, Z. L.; and El-Sayed, M. A. 2000. How does a gold nanorod melt? *J. Phys. Chem. B* 104: 7867–70.
70. Fujiwara, H.; Yanagida, S.; and Kamat, P. V. 1999. Visible laser induced fusion and fragmentation of thionicotinamide-capped gold nanoparticles. *J. Phys. Chem. B* 103: 2589–91.
71. Takeuchi, Y.; Ida, T.; and Kimura, K. 1997. Colloidal stability of gold nanoparticles in 2-propanol under laser irradiation. *J. Phys. Chem. B* 101: 1322–7.
72. Mafune, F.; Kohno, J. Y.; Takeda, Y.; and Kondow, T. 2002. Dissociation and aggregation of gold nanoparticles under laser irradiation. *J. Phys. Chem. B* 106: 8555–61.
73. Stepanov, A. L.; Hole, D. E.; Bukharaev, A. A.; Townsend, P. D.; and Nurgazizov, N. I. 1998. Reduction of the size of the implanted silver nanoparticles in float glass during excimer laser annealing. *Appl. Surf. Sci.* 136: 298–305.
74. Kaempfe, M.; Rainer, T.; Berg, K.-J.; Seifert, G.; and Graener, H. 1999. Ultrashort laser pulse induced deformation of silver nanoparticles in glass. *Appl. Phys. Lett.* 74: 1200.
75. Bosbach, J.; Martin, D.; Stietz, F.; Wenzel, T.; and Träger, F. 1999. Laser-induced manipulation of the size and shape of small metal particles: Towards mono disperse clusters on surfaces. *Euro. Phys. J. D* 9: 613–7.
76. Wenzel, T.; Bosbach, J.; Goldmann, A.; Stietz, F.; and Träger, F. 1999. Shaping nanoparticles and their optical spectra with photons. *Appl. Phys. B* 69: 513–7.
77. Lahann, J.; Mitragotri, S.; Tran, T. N.; Kaido, H.; Sundaram, J.; Choi, I. S.; Hoffer, S.; Somorjai, G. A.; and Langer, R. 2003. A reversibly switching surface. *Science* 299: 371–4.
78. Kawasaki, M. and Hori, M. 2003. Laser-induced conversion of noble metal-island films to dense monolayers of spherical nanoparticles. *J. Phys. Chem. B* 107: 6760–5.
79. Verhoeven, J. D. 1975. *Fundamentals of Physical Metallurgy*, Wiley, New York.
80. Gao, S.; Koshizaki, N.; Li, Y.; and Li, L. 2011. Unique hexagonal non-close-packed arrays of alumina obtained by plasma etching/deposition with catalytic performance. *J. Mater. Chem.* 21: 2087–90.

8 Micro/Nanostructured Ordered Porous Arrays

8.1 INTRODUCTION

Micro/nanostructured ordered porous arrays are the films consisting of periodically arranged microscaled pores or network-structured porous film with hexagonally packed pores and skeleton with nanoscaled thickness. Usually, the pore size ranges from microscale to submicroscale and the thickness of skeleton (pore wall) is about several or tens of nanometers. In the past decades, these films have attracted great interest due to their unique properties, such as uniform submicrometer-scale pore sizes, highly ordered structure, and large accessible surface, and their widely potential applications in many fields, including catalysis [1], sensors [2–4], cell [5], surface-enhanced Raman scattering (SERS) [6,7], superhydrophobic or superhydrophilic surface [8,9], photonic crystal [10], optoelectronics, and optical devices [11–13]. Furthermore, many experimental and theoretical results indicate that their properties are greatly related to morphology, structure, and arrangement parameters of the porous array, hence the controlled fabrication of various ordered porous arrays with the desired structural parameters is of great importance.

In conventional technologies, the most widely used fabrication method is photolithography, with the advantages of high resolution and high throughput [14–16]. However, due to optical resolution limitations such as diffraction of light and backscattering from the substrate, it is very difficult to generate features less than 100 nm and develop nanoscale patterns. The other top-down techniques, such as electron beam lithography [17], ion beam lithography [18], nanolithography [19], X-ray lithography [20], and atomic microscopy lithography [21], are generally constrained by lower yield, high cost, and longer time fabrication. Contrarily, bottom-up approach, which is usually based on spontaneous self-assembly, has the potential to develop devices such as electronics, actuators, and sensors more efficiently and economically, becoming a natural alternative to the traditional top-down construction. Recently, numerous template-assisted bottom-up processes have focused on synthesis of nanoporous materials. Among these methods, two-dimensional (2D) colloidal crystal template method (also named "colloidal lithography") has been proven to be an effective strategy for its flexible, simple, low cost, high throughput, and universal features [22–24]. Up to now, a great deal of research has been devoted to the fabrication of 2D ordered micro/nanostructured arrays (films) based on different methods using monolayer colloidal crystal as the template, such as physical vapor deposition, sol-gel, solution-dipping, electrophoretic deposition, and electrodeposition [25–29]. After removal of the colloidal spheres, one can obtain a series of micro/nanostructured films of size in centimeters, including nanoparticle arrays, pore arrays, nanoring arrays, nanobowl arrays, ordered nanorod/nanopillar arrays, hollow sphere arrays, and so on. It is noteworthy that all the structures and morphologies of the porous array can be tuned by changing the diameter of the colloidal spheres in the template and by controlling the experimental parameters. The colloidal crystal template method opens up a flexible and promising way to produce the patterned ordered micro/nanostructured films. In this chapter, we introduce the typical micro/nanostructured porous arrays (films) based on the colloidal template technique, including fabrication of monolayer and multilayered porous arrays with homopore size or heteropore size and monolayer hollow sphere arrays.

8.2 MICRO/NANOSTRUCTURED ORDERED PORE ARRAYS BASED ON SOLUTION-DIPPING STRATEGY

As we know, monolayer polystyrene (PS) colloidal crystal can be produced by many methods based on self-assembly, such as spin coating, dipping, and gas/liquid interface assembly [27,30,31], as illustrated in Figure 8.1, corresponding to a large sized monolayer colloidal crystal by spin coating. Importantly, such monolayer can be transferred, by water or solution medium, from one substrate to another according to the requirement [27], as shown in Figure 8.2, and hence can be used as the template for fabrication of the micro/nanostructured ordered pore arrays (films).

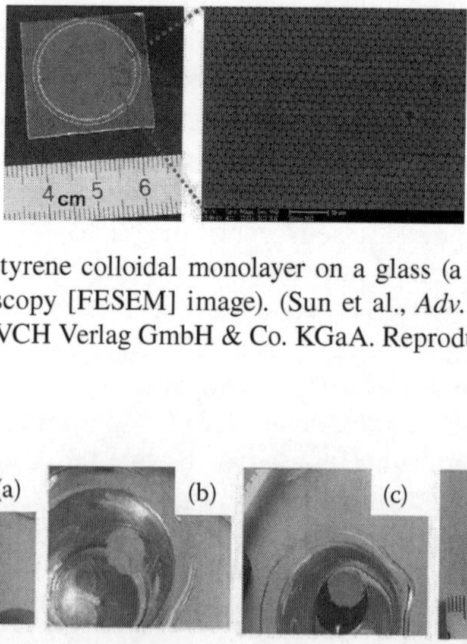

FIGURE 8.1 The polystyrene colloidal monolayer on a glass (a photo and field emission scanning electron microscopy [FESEM] image). (Sun et al., *Adv. Funct. Mater.* 2004. 14. 283–8. Copyright Wiley-VCH Verlag GmbH & Co. KGaA. Reproduced with permission.)

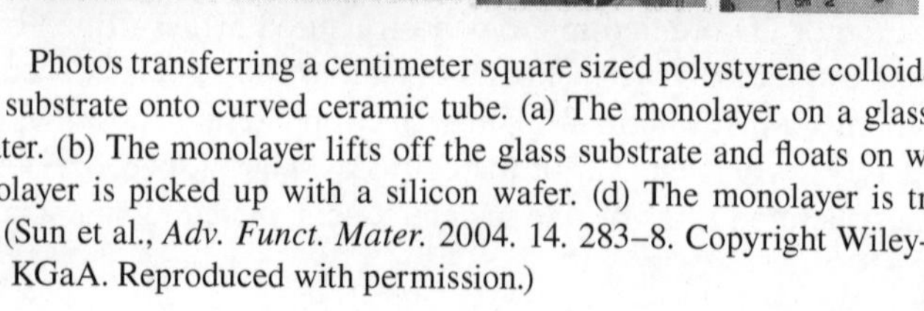

FIGURE 8.2 Photos transferring a centimeter square sized polystyrene colloidal monolayer from a glass substrate onto curved ceramic tube. (a) The monolayer on a glass substrate is dipped in water. (b) The monolayer lifts off the glass substrate and floats on water surface. (c) The monolayer is picked up with a silicon wafer. (d) The monolayer is transferred to the Si wafer. (Sun et al., *Adv. Funct. Mater.* 2004. 14. 283–8. Copyright Wiley-VCH Verlag GmbH & Co. KGaA. Reproduced with permission.)

Based on such 2D PS colloidal crystal, the morphology- and structure-controlled ordered porous array films of various metals, semiconductors, and compounds can be fabricated, by using the solution-dipping strategy. The morphology of the films can be controlled by concentration of precursor solution, heating time of the template, PS sphere (PSs) size, and so on [27,32–35]. In addition, some extension techniques have been developed based on this strategy. For example, micro/nanostructured porous films can be directly synthesized on any substrate with flat or even curved and rough surface based on transferability of the colloidal monolayer (Figure 8.2); the heterogeneous structured and heteropore-sized ordered pore arrays can be obtained by using either different precursor solutions or PS size.

8.2.1 Monolayer Ordered Pore Arrays

8.2.1.1 In Situ Solution Dipping

The corresponding synthesis route is illustrated in Figure 8.3. A monolayer colloidal crystal template with a large area (>1 cm^2) is initially fabricated on a substrate. Then, a droplet of desired precursor solution is dropped into the colloidal monolayer at the edge region, and subsequently the monolayer floats on the surface of the solution due to the surface tension of the solution. Next, on drying the template at a low temperature (less than the glass transition temperature of the PS spheres), the PS spheres would deform and the solution would precipitate on the surface of these deformed spheres and the substrate. Finally, removing it by calcination or ultrasonic cleanout, the ordered micro/nanostructured monolayer porous arrays (films) could thus be fabricated. The morphologies of the ordered porous films can be controlled

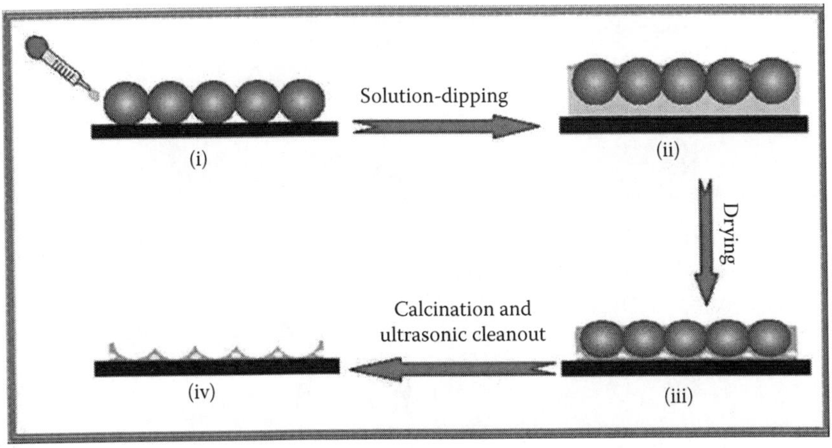

FIGURE 8.3 Schematic illustrations of the in situ template solution-dipping strategy. Step (i): Colloidal monolayer template with a large area on a substrate. Step (ii): The colloidal monolayer floats on the surface of the precursor solution. Step (iii): Integrity of the solute and colloidal spheres. Step (iv): Ordered porous film. (Sun et al., *Adv. Funct. Mater.* 2004. 14. 283–8. Copyright Wiley-VCH Verlag GmbH & Co. KGaA. Reproduced with permission.)

by the solution concentration. The ordered porous films of different materials, such as metals, semiconductor, and compounds, can be fabricated based on this strategy.

8.2.1.2 Morphology and Structure

Here, we take Fe_2O_3 as an example to demonstrate fabrication of the monolayer ordered pore arrays with controllable morphology on any substrate based on the strategy shown in Figure 8.3. $Fe(NO_3)_3$ solution was used as precursor. After drying at 80°C for 2 hours and subsequent heating at 400°C for 8 hours (the PS spheres were removed at this temperature), the α-Fe_2O_3 ordered porous film are obtained. Figure 8.4 shows the morphology of the final products, corresponding to the results from the precursor solutions with different concentrations. Clearly, the morphology of the pore array depends on concentration of the precursor. When the concentration is 0.02 M or higher, the honeycomb structures can be formed, but the shapes and diameters of pores depend on the solution concentrations. A high concentration (0.8 M) gives rise to pores with nearly circular upper-end opening. The diameter of the openings is smaller than that of PS spheres (Figure 8.4a). By decreasing the concentration, the truncated shape of the pore at the film surface gradually becomes a regular hexagon (Figure 8.4b and c). When the concentration is reduced to 0.02 M,

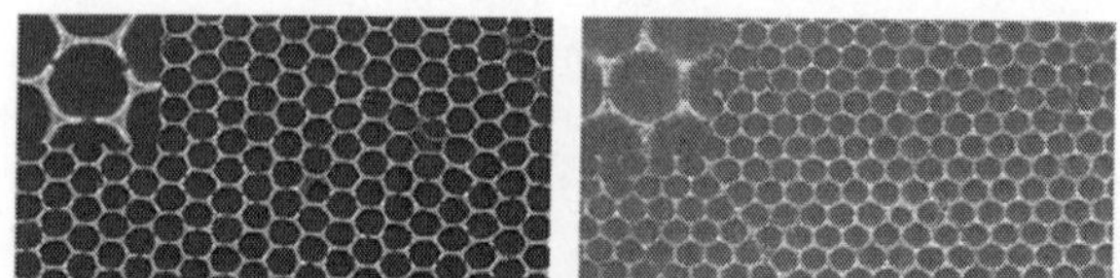

FIGURE 8.4 Scanning electron microscopy (SEM) images of α-Fe_2O_3 ordered pore arrays from the precursor solutions with different $Fe(NO_3)_3$ concentrations. (a) 0.8 M. (b) 0.08 M. (c) 0.06 M. (d) 0.02 M. (e) 0.002 M. (f) 0.8 M. (a) through (e) correspond to the template with 1000-nm polystyrene (PS) spheres, and (f) corresponds to the template with 200-nm PS spheres. (Sun et al., *Adv. Funct. Mater.* 2004. 14. 283–8. Copyright Wiley-VCH Verlag GmbH & Co. KGaA. Reproduced with permission.)

a through-pore structured film (with two open-ended pores) is formed (Figure 8.3d). However, when the concentration is decreased to a very low level (such as 0.002 M), a ring array, rather than the closely arranged pore array, is obtained (Figure 8.4e). All the rings surround the positions where the PS spheres were originally located on the substrate and are formed in hexagonal arrangement. Furthermore, for other colloidal crystal templates with different PS sphere sizes, the ordered pore array can be also formed. For instance, when a precursor solution with concentration of 0.8 M is applied onto a template with much smaller PS sphere size (e.g., 200 nm), the obtained morphology of the porous film is similar to that shown in Figure 8.4a, except for the pore size, as illustrated in Figure 8.4f.

8.2.1.3 A Solvent Evaporation–Colloidal Sphere Deformation Model

The formation of the porous films could be described by a solvent evaporation–colloidal sphere deformation model, or the deformation of PS spheres during solution evaporation, as schematically illustrated in Figure 8.5. When the precursor solution is dropped into the colloidal template, the templates float on top of the solution. At the same time, the interstitial spaces among the closely packed PS spheres are filled with the solution due to the capillary force, and a meniscus is formed on the solution surface, as shown in Figure 8.5a. During subsequent drying at 80°C, due to evaporation of the solvent, the liquid surface and the colloidal monolayer gradually descend and the PS spheres deform. When the concentration of the solution reaches to the saturation due to evaporation of the solvent, further drying will lead to solute precipitation on the sphere's surface and the substrate. Finally, a nearly triangular hole is left between two adjacent spheres and the substrate due to lack of sufficient solution compensation (Figure 8.5b–e). Apparently, the lower the solution concentration, the longer the time to reach saturation induced by evaporation at the same drying temperature, which leads to a more obvious change in the sphere's shape.

For a high precursor concentration (0.8 M), the preceding process is completed in a very short time. Thus, the deformation of PS spheres is very small before complete evaporation and there is only a small contact area between two adjacent PSs, leading to the formation of small circular holes in the pore walls and circular openings from top view (Figure 8.5b). When the solution concentration is reduced to 0.08 or 0.06 M, it takes a longer time for the solution to become saturated, leading to a larger contact area between two adjacent spheres (Figure 8.5c) or larger holes in the pore wall due to the larger deformation of PS spheres. The large contact area between two adjacent spheres induces the formation of hexagonal opening at the film surface. When the concentration is further decreased (0.02 M), before solution saturation and solute precipitation, the PS spheres have already contacted the substrate, and this contact area has become large enough due to their deformation during drying (shown in Figure 8.5d), and pores with two-ended opening are thus formed. When the concentration of the precursor becomes very low (e.g., 0.002 M), only a thin solute deposition (shell) is preferentially formed on the free surface of the PS spheres above and below the tangent-point plane (Figure 8.5e). Subsequent calcinations and ultrasonic vibration for a short time period (few minutes) can remove the colloidal template but keep the deposited shell, as illustrated in Figure 8.6a. It is worth noting that this film has two layers, and the top layer is slightly displaced with respect to

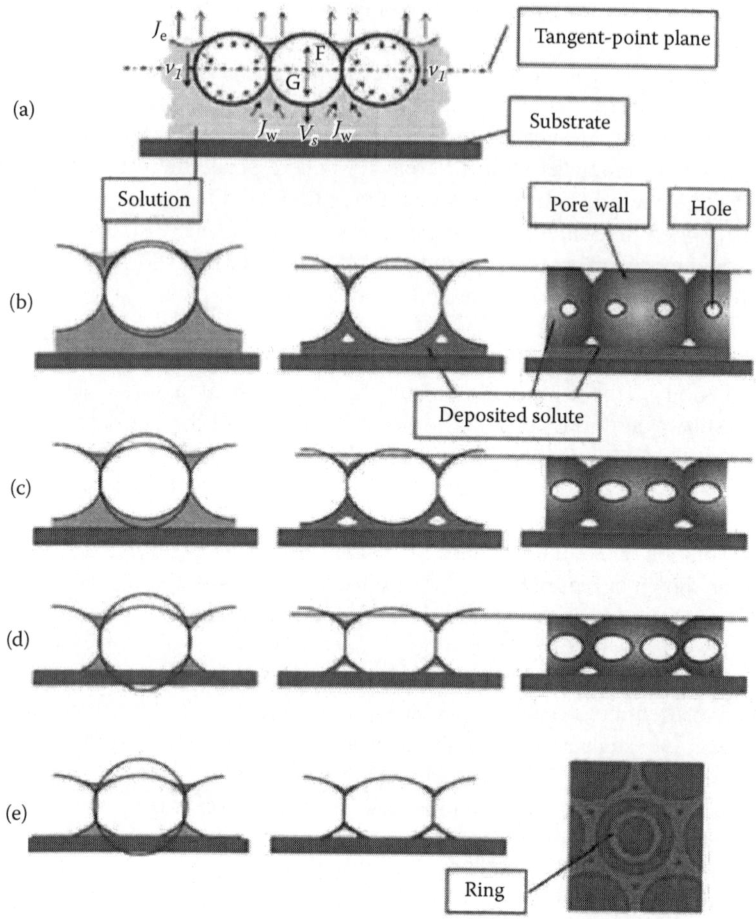

FIGURE 8.5 Schematic illustrations of the formation process of ordered pore array film for different concentrations of precursor solutions. (a) Latex spheres are floating on the solution. Letters "F" and "G" denote the flotage and gravitation of the colloidal spheres, respectively. J_e is the water evaporation flux, J_w is the water influx, and v_1 and V_s are the descending rate of the solution surface and the colloid monolayer, respectively. (b) High concentration (0.8 M). (c) 0.08 and 0.06 M. (d) Low concentration (0.02 M). (e) Very low concentration (0.002 M). Left column: Solutions reach saturation. Middle column: Complete evaporation of solvent. Right column: After removal of the template by dissolution. The illustration for (e) in this column is a top view. (Sun et al., *Adv. Funct. Mater.* 2004. 14. 283–8. Copyright Wiley-VCH Verlag GmbH & Co. KGaA. Reproduced with permission.)

the bottom layer. After ultrasonically washing for some more time (0.5–1 hour), the top layer can be cleared and the bottom layer, a ring array, can be seen (Figure 8.4e). Additionally, further experiments indicate that the noncircular shape of the rings results from calcinations at 400°C and ultrasonic cleanout. Ultrasonic washing after drying at 80°C, without calcinations, shows an ordered array of circular rings, as shown in Figure 8.6b.

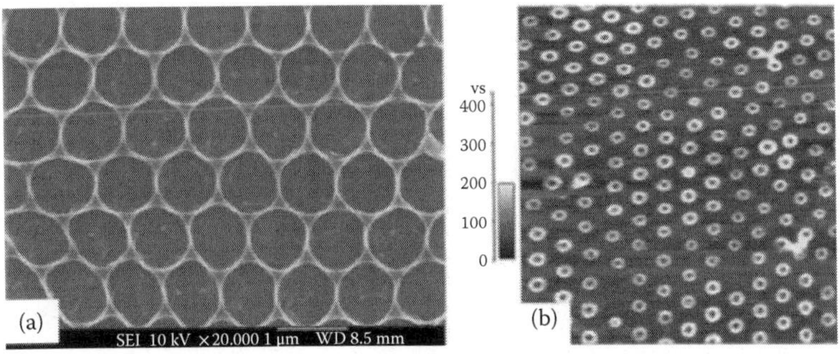

FIGURE 8.6 The products from the precursor solution. (a) After final heating at 400°C and ultrasonic vibration for a few minutes (instead of 0.5 hours). (b) After subsequent ultrasonic washing in dichloromethane solution for 0.5 hours. (Sun et al., *Adv. Funct. Mater.* 2004. 14. 283–8. Copyright Wiley-VCH Verlag GmbH & Co. KGaA. Reproduced with permission.)

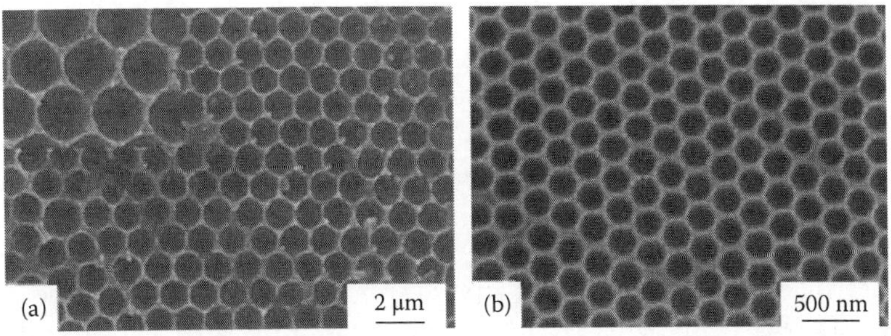

FIGURE 8.7 The ordered pore array films of CeO_2 on Si wafer (a) and Al_2O_3 on glass substrate (b) using the strategy shown in Figure 8.3. Treatment conditions, (a): $Ce(NO_3)_3$ (0.05 M), 80°C for 2 hours, and 200°C for 8 hours followed by ultrasonic cleanout for 0.5 hours. (b) $Al(NO_3)_3$ (0.05 M), 90°C for 12 hours and 300°C for 2 hours in air. (Sun et al., *Adv. Funct. Mater.* 2004. 14. 283–8. Copyright Wiley-VCH Verlag GmbH & Co. KGaA. Reproduced with permission.)

The in situ solution-dipping strategy shown here is universal and can be used for other metal or oxide ordered pore array/ring array films on any desired substrate by means of transferring the PS colloidal monolayer from one substrate to the other. A series of other morphology-controlled ordered pore arrays, such as zinc, ZnO, Co_2O_3, CuO, CeO_2, Eu_2O_3, and Dy_2O_3, could be fabricated [36–40]. Figure 8.7 shows the typical examples for CeO_2 and Al_2O_3 pore array films on silicon substrate fabricated by this strategy. Such structures may be useful in applications of energy storage or conversion, especially in next-generation integrated nanophotonics devices and biomolecular labeling and identification.

8.2.1.4 Extensions of Solution-Dipping Strategy

Based on the solution-dipping strategy, the sol/solution-dipping/soaking techniques could be developed using PS colloidal monolayer as the template [41–45]. Various ordered porous films with different morphologies, structure, and chemical composition can be fabricated. For instance, using the sols to replace solutions, 2D ordered pore films of In_2O_3, SiO_2, TiO_2, and so on, can be fabricated. Using the sol-soaking method, controllable ring arrays can be constructed.

8.2.1.4.1 Hierarchical Porous Films Based on Sol Solution-Dipping

Based on the monolayer colloidal crystal, the hierarchical porous films (orderly arranged microsized pores and disordered nanopores in its skeleton) with a high specific surface area can also be fabricated by the sol-dipping technique [46].

For example, fabrication process of the SiO_2 ordered porous film can briefly be described as follows: a mixture of tetraethyl orthosilicate, alcohol, and distilled water (molar ratio 1:4:20) was used as precursor solution. The pH value of the solution was controlled at about 1 by adding nitric acid. A droplet (about 10 μL) of the precursor solution was added onto the monolayer colloidal crystal, which could infiltrate into the interstices between the substrate and the PS colloidal monolayer. The sample was then placed into a beaker sealed with a cover (solvent can evaporate very slowly) for 1 week at room temperature to form gel and dry. Subsequently, it was ultrasonically washed in CH_2Cl_2 for 2 minutes to dissolve the template and then annealed in air at 500°C for 1 hour. The SiO_2 ordered pore array film was thus obtained, as shown in Figure 8.8a.

For further studying the skeleton structure of the porous film, the corresponding isothermal nitrogen sorption was measured, as shown in Figure 8.8b, which shows ink bottle–shaped pores in the skeleton. The specific surface area of the skeletons was thus obtained to be about 470 m^2/g. Analysis of the pore diameter has shown that the pore diameter of the skeleton is sharply distributed around 4.2 nm (see the inset in Figure 8.8b). The pores in the skeleton are disorderly arranged, interconnected, and open to ambient atmosphere. The porosity of the skeleton is estimated to be about 56%, determined by the volume of saturated sorption [47]. The isothermal nitrogen sorption, combined with corresponding field emission scanning electron microscopy (FESEM) image, indicates that the SiO_2 film is of hierarchical structure consisting

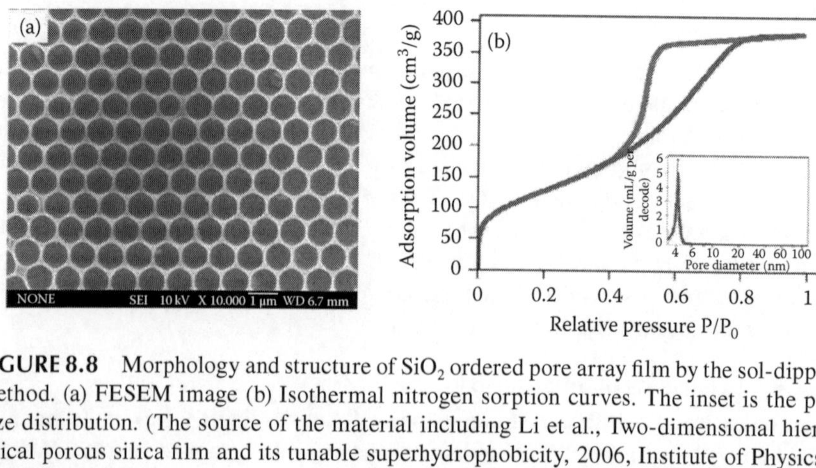

FIGURE 8.8 Morphology and structure of SiO_2 ordered pore array film by the sol-dipping method. (a) FESEM image (b) Isothermal nitrogen sorption curves. The inset is the pore size distribution. (The source of the material including Li et al., Two-dimensional hierarchical porous silica film and its tunable superhydrophobicity, 2006, Institute of Physics is acknowledged.)

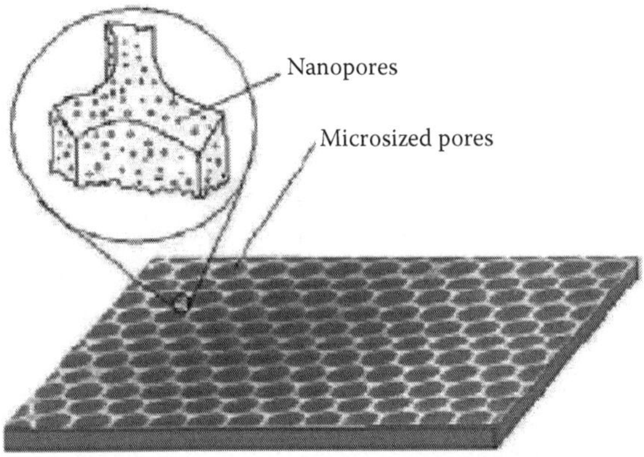

FIGURE 8.9 Schematic illustration of hierarchical structured two-dimensional SiO_2 pore array film. (The source of the material including Li et al., Two-dimensional hierarchical porous silica film and its tunable superhydrophobicity, 2006, Institute of Physics is acknowledged.)

of orderly arranged microsized pores and disordered nanopores in its skeletons, as illustrated in Figure 8.9. Such hierarchical silica film has many potential applications as adsorbents, catalysts, chromatographic supports, microseparators, and so on.

Other porous films can also be obtained by this extended strategy. For example, TiO_2 ordered porous film can be prepared by acid hydrolysis of titanium isopropoxide sol [48]; In_2O_3 ordered porous film can be created by adding $In(OH)_3$ precursor sol to the colloidal template [35]. The morphologies of these films are all similar to that of SiO_2 shown in Figure 8.8. As we know, all of these semiconductor materials possess good photocatalytic, optical, gas-sensing, and electronic properties. Such ordered porous structures would be able to increase the surface area, porosity, and so on, which will greatly improve the properties of these materials.

8.2.1.4.2 Ordered Nanopillar Array Films Based on Heating-Induced Template Deformation

Using the monolayer PS colloidal crystal as template, in addition to the ordered pore arrays and ring arrays, the patterned one-dimensional (1D) nanostructures (nanorod or nanopillar) with small aspect ratio can also be fabricated, which have potential applications, such as sensor arrays, piezoelectric antenna arrays, optoelectronic devices, interconnects, and superhydrophobic and self-cleaning surfaces [49]. Many researchers have fabricated large-area nanorod/nanopillar arrays by epitaxial approach using metal periodic nanoparticle arrays as catalysts, in which the periodic nanoparticle arrays were obtained by evaporation deposition strategy using colloidal crystal as mask.

Here, we introduce the large-scale hexagonal nanopillar arrays based on the heating-deformed colloidal monolayer template and solution-dipping method [50,51]. The fabrication route is illustrated in Figure 8.10. Briefly, the monolayer PS colloidal crystal is first heated at 120°C for a certain time period. The colloidal monolayer is deformed due to heating and the triangular prism channels are thus formed at the

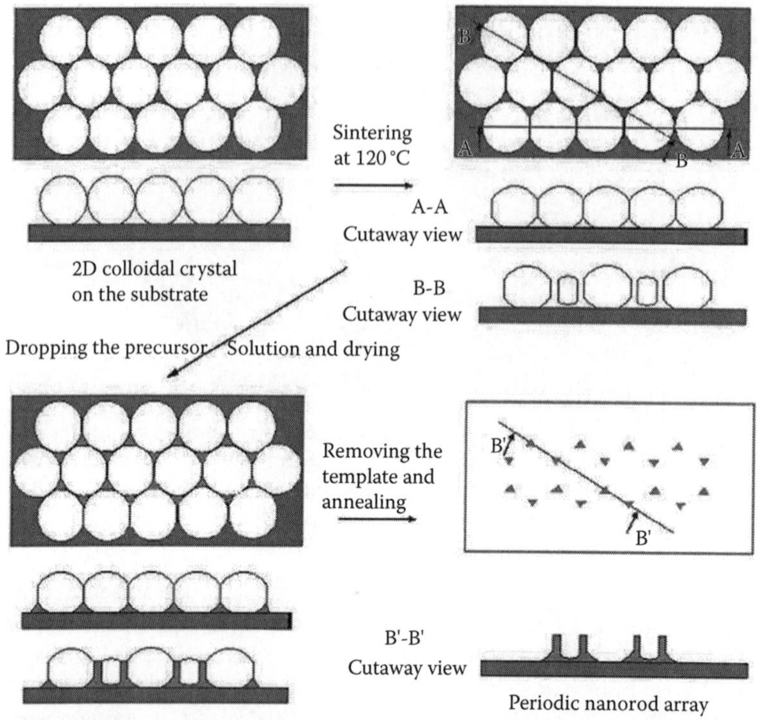

FIGURE 8.10 Schematic illustrations for fabrication of the periodic hexagonal nanopillar array based on heating-induced template deformation. (Reprinted from *Polymer*, 46, Li et al., Fabrication of the periodic nanopillar arrays by heat-induced deformation of 2D polymer colloidal monolayer, 12033–6, Copyright 2005, with permission from Elsevier.)

interstices among them, as shown in Figure 8.11. Subsequently, the desired precursor solution is added to the heat-treated monolayer colloidal crystals. Finally, the samples are dried and annealed to make the precursor decompose completely and the PSs template be burnt out.

Based on these templates with different deformation degrees, Fe_2O_3 ordered patterns with different morphologies were fabricated, as illustrated in Figure 8.12. Obviously, using the original template without heating (Figure 8.11a), one can only get the ordered pore array with circular pore morphology (Figure 8.12a). When such PS colloidal monolayer is sintered at 120°C for 10 minutes, its morphology exhibits a large change as shown in Figure 8.11b, and the contact between adjacent PS spheres changes from point to facet and the triangular prism channels are formed at the interstices among the adjacent colloidal spheres. Based on such deformed template, a small pillar with regular triangle prism shape appears at the node of the network (see Figure 8.12b). Clearly, its morphology is greatly different from those obtained by templates without heating, but the whole arrays also exhibit hexagonal alignment. The aspect ratio of these nanopillars is about 1:1. Here, it should be mentioned that the nanopillars do not grow directly on the substrate, but on the nodes of the skeleton network surrounding the previous bottoms of the sintered PS spheres. With increase of heating time, the interstices among the PS spheres in the colloidal monolayer become smaller and smaller from top view, as shown in Figure 8.11c. The corresponding aspect ratio of pillars becomes larger. When the heating time is increased to 20 minutes, the aspect ratio is increased to about 2:1 (see Figure 8.12c). However,

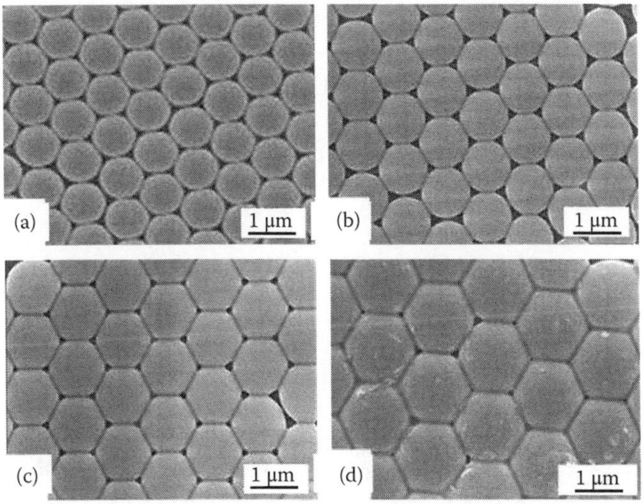

FIGURE 8.11 FESEM images of the two-dimensional colloidal monolayer after sintering at 120°C for different time periods. (a through d): 0, 10, 20, 25 minutes, respectively. (Li et al., *J. Mater. Chem.*, 16, 609–12, 2006. Reproduced by permission of The Royal Society of Chemistry.)

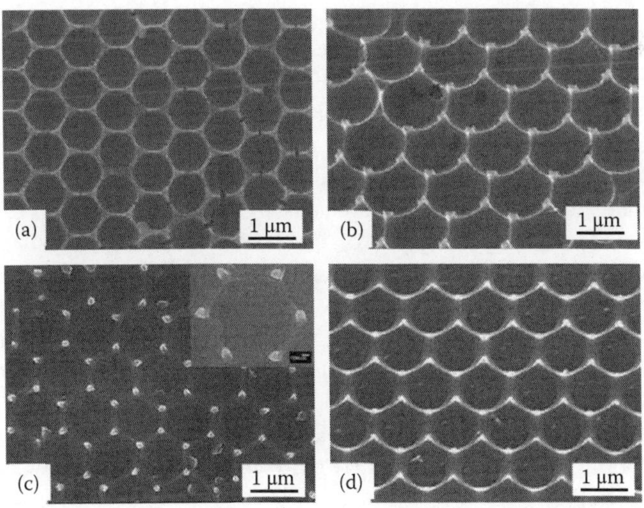

FIGURE 8.12 FESEM images of Fe_2O_3 ordered patterns fabricated by different deformed colloidal monolayers. Images (a) through (d) were obtained by the templates shown in Figure 8.11a through d, respectively. The inset in (c) is the local magnification. (Li et al., *J. Mater. Chem.*, 16, 609–12, 2006. Reproduced by permission of The Royal Society of Chemistry.)

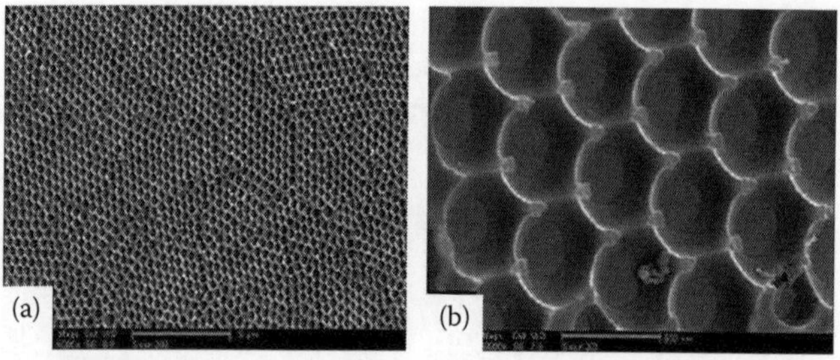

FIGURE 8.13 The morphology of Fe_2O_3 nanopillar array fabricated from the colloid monolayer with 500-nm polystyrene spheres sintered at 120°C for 10 minutes. (b) The local magnification of (a). (Li et al., *J. Mater. Chem.*, 16, 609–12, 2006. Reproduced by permission of The Royal Society of Chemistry.)

if the colloidal monolayer (e.g., 25 minutes) is overheated, the channels will almost disappear in the templates (Figure 8.11c). Correspondingly, no pillars will be formed due to nonchannels in the overheated template and the morphology takes on regular network, as shown in Figure 8.12d.

Apparently, the appropriate heating time of the PS colloidal templates is a key factor to fabricate the periodic nanopillar arrays by this strategy. In addition, for smaller PS spheres (500 in diameter), higher density of Fe_2O_3 nanopillar arrays can be prepared, as displayed in Figure 8.13. Besides Fe_2O_3, other material (such as SiO_2) ordered nanopillar arrays could also be fabricated by the strategy presented here. Such nanopillar arrays with small aspect ratio have potential applications in sensor arrays, piezoelectric antenna arrays, and optoelectronic devices [52].

8.2.1.4.3 Ordered Nanoring Arrays Based on Sol-Soaking Strategy

Recently, ringlike structures have received much attention because of their unique properties and applications in magnetic, optical, and optoelectronic devices [53,54]. Usually, these properties are size dependent, thus the development of methods that allow easy control of the ring size is highly desirable. To obtain the ring with controlled discrete size and ordered structure, Sun et al. [41] developed an annealed template-induced sol-soaking strategy, in which TiO_2 was used as a model material. The TiO_2 sol was prepared by acid hydrolysis of titanium isopropoxide. The glass substrate covered with PS spheres was put in an oven and heated at 120°C, which resulted in the formation of plane contacts between the spheres and the substrate. Then, the heated substrate was soaked in TiO_2 sol for at least 30 minutes. Next, the monolayer was removed from the sol, and the excess sol was removed with a ChemWipe. After heating at 120°C for 1 hour in an oven, the sample was subsequently sonicated in dichloromethane for about 1 hour. Finally, calcining at 400°C for 2 hours, uniform TiO_2 ring arrays were obtained, as illustrated in Figure 8.14.

Importantly, the ring size can easily be modulated by the concentration of the precursor and/or the heating time of colloidal monolayer. The images in Figure 8.14a and b correspond to the arrays heated at 120°C for 5 minutes but with different precursor concentrations (0.1 and 0.05 M, respectively). Apparently, the inner

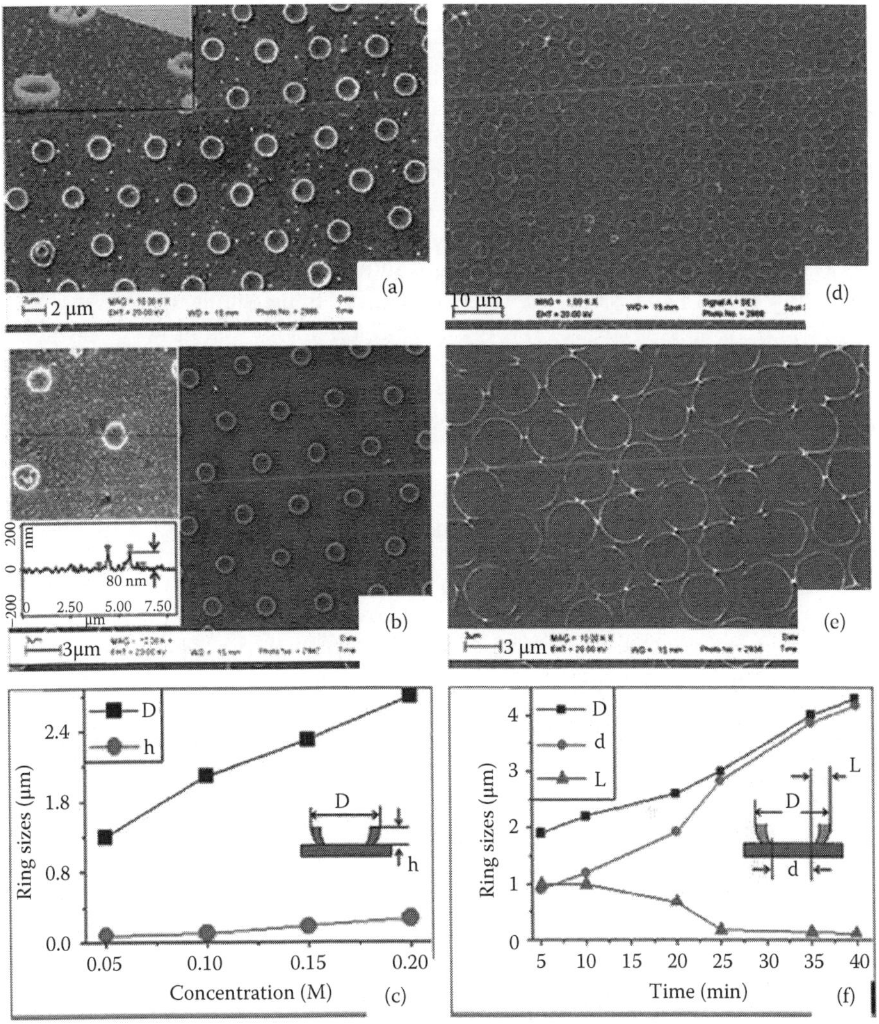

FIGURE 8.14 SEM images of varied sized TiO_2 ring arrays and size statistics. (a) and (b) are the products from the sol solution with 0.1 and 0.05 M of TiO_2 (after annealing at 120°C for 5 minutes), respectively. The concentration of the TiO_2 sol in frames (a), (d), and (e) is 0.1 M, but the annealing time periods are 5, 25, and 40 minutes, respectively. (c) and (f) are the ring size statistics. (Reprinted with permission from Sun et al., 2006, 3774–9. Copyright 2006 American Chemical Society.)

diameters of all the rings are around 0.8 μm, while the outer diameter and the height depend on concentration of the precursor. Furthermore, the concentration dependence can be seen in Figure 8.14c. When the concentration is increased from 0.05 to 0.2 M, the outer diameter increases from 1.2 to 2.8 μm, and the corresponding height varies from 80 to about 270 nm. This phenomenon is because of the increase of the solute due to higher precursor concentration, which would lead to larger outer diameter, greater height, and thicker walls of the ring. On the contrary, the heating time has a different influence on the ring size, as shown in Figure 8.14a, d, and e. When fixing the concentration at 0.1M, the inner diameter increases from 0.8 to 4.2 μm with prolonged heating time. Correspondingly, the outer diameter increases from 2.0 to 4.3 μm, while the thickness of the ring walls reduces from 600 to about 60 nm. This could be attributed to the ever-increasing contact areas and the rings' inner diameter with increasing annealing time. At the same time, the interstices in the colloidal monolayer shrink during heating, and hence less precursor molecules are available and the thickness of the ring is thinner. The formation mechanism of these ring arrays is the same as that described in the preceding subsections. The rings can overcome the restrictions of surface morphology and grow on any solid support. The other mesoscopic ring arrays including ZrO_2, SnO_2, Fe_2O_3, and ZnO can also be synthesized from appropriate precursors. These ring arrays could find applications in photoelectric devices, photocatalytic surfaces, magnetic elements, and gas sensors.

8.2.2 Multilayer Ordered Pore Arrays

As described in Section 8.2.1, various monolayer ordered porous films with controllable morphology, structure, and chemical composition can be fabricated by the solution-dipping, sol-dipping/sol-soaking, and so on based on the PS colloidal monolayer [35,41]. All of these porous films are generally fabricated directly on flat substrates. However, in most situations, for practical application and scientific research, the monolayer or multilayer porous films must be constructed on a substrate with curved surface (such as cylindrical, spherical, or concave surface). The fabrication of micro/nanostructured porous films on curved surfaces can greatly expand the applicability of ordered porous films, especially for construction of the micro/nanostructured devices, such as gas sensors on ceramic tubes.

8.2.2.1 Direct Synthesis of Homopore Sized Porous Films on Any Surface

8.2.2.1.1 Solution-Dipping and Template-Transfer Strategy

As mentioned in the beginning of Section 8.2, the PS colloidal crystal template can be peeled off and transferred to another substrate by means of distilled water as medium. Actually, such colloidal crystal template on glass substrate can also be stripped off in the precursor solutions while still retaining its integrity [4]. On this basis, a simple and effective strategy, based on the route of solution-dipping and template-transfer, was presented to directly fabricate the monolayer or multilayer porous films on any needed substrate with flat, rough, or even curved surfaces, as shown in Figure 8.15 [4].

8.2.2.1.2 Porous Films on Different Substrates

The PS colloidal crystal template on a glass slide is first slowly immersed in a desired precursor solution. Then, it can be stripped off from the substrate and made to float on the precursor solution [55]. The interstitial space between the close-packed spheres is also filled with the solution due to the capillary force. The floating colloidal template is then picked up by using a desired substrate with flat or

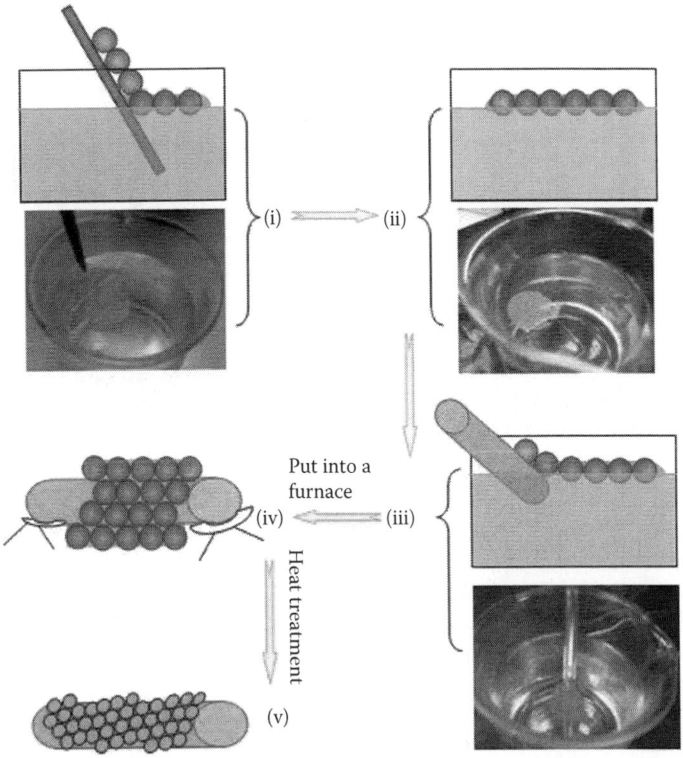

FIGURE 8.15 Schematic illustration of solution-dipping and template-transfer strategy. Step (i): A flat glass substrate covered with a polystyrene colloidal monolayer is dipped in the solution. Step (ii) The colloidal monolayer floats on the precursor solution surface. (iii) The monolayer is picked up using a glass rod with a curved surface. Step (iv) The rod with the monolayer and the solution is heated in a furnace. Step (v) An ordered pore array film is formed on the curved surface after heat treatment and removal of the colloidal monolayer. The lower frames in steps (i)–(iii) are photos corresponding to the manipulations described. (Sun et al., *Adv. Mater.* 2005. 17. 2872–7. Copyright Wiley-VCH Verlag GmbH & Co. KGaA. Reproduced with permission.)

curved surface. The colloidal template covers the new substrate. At the same time, the interstitial space is still filled with the solution. Subsequently, the substrate covered with PS monolayer is dried at a temperature slightly above the glass transition of PS spheres. With the solvent evaporating, the solute or hydrolyzate gradually deposits on the PS spheres' surfaces and the substrate. Finally, the sample is calcined to remove the colloidal template, and an ordered porous film is thus formed on the surface of the new substrate. It should be noted that the porous film grows directly on the substrate surface during the heat treatments and the adherence between the film and the substrate is very strong. By this method, the ordered porous film can be fabricated on any substrate with flat, rough, and even curved surfaces.

FIGURE 8.16 FESEM images of SnO_2 ordered pore array films on the substrates with curved surfaces. (a) On the outer surface of a glass tube (diameter: 1.2 mm), (b) on the inner surface of a glass tube (diameter: 1.0 mm), (c) on the surface of a steel sphere (diameter: 2.5 mm), and (d) on a flat surface. The insets show the corresponding low-magnification images. (Sun et al., *Adv. Mater.* 2005. 17. 2872–7. Copyright Wiley-VCH Verlag GmbH & Co. KGaA. Reproduced with permission.)

Here, tin dioxide (SnO_2) was chosen as a model material to demonstrate the synthesis strategy. Using 0.1 M $SnCl_4$ precursor solution and colloidal crystal with PS diameter of 1000 nm as template, the SnO_2 ordered porous films were fabricated on different substrates, including the outer and inner surfaces of a glass tube, spherical steel surface, as well as a flat surface, as shown in Figure 8.16. Apparently, the morphology of all the ordered porous films on the curved surfaces is similar. The pore openings at the film surfaces are nearly circular. For such films, after heating at 500°C for 1 hour, they adhere very strongly to the curved surfaces, which cannot be removed or destroyed, even by ultrasonic washing [4].

Similarly, this technique is also universal and very flexible and can be used for fabrication of many other oxide ordered porous films on any desired substrate with flat or curved surface, for example, Fe_2O_3, TiO_2, ZnO, WO_3, and In_2O_3. Furthermore, doped porous films are also easily obtained by adding corresponding dopant ions into the precursor solution to improve the sensing properties.

8.2.2.1.3 Multilayer Ordered Porous Films

Furthermore, if we repeat the procedures shown in Figure 8.15 and use the freshly formed ordered porous film as the new substrate, the multilayer porous film can be obtained. Figure 8.17 shows some typical multilayer SnO_2 ordered porous films with different pore sizes from 200 to 1000 nm. These films were fabricated by repeating the procedures shown in Figure 8.15, including double layer (Figure 8.17a) and four layer (Figure 8.17b–d).

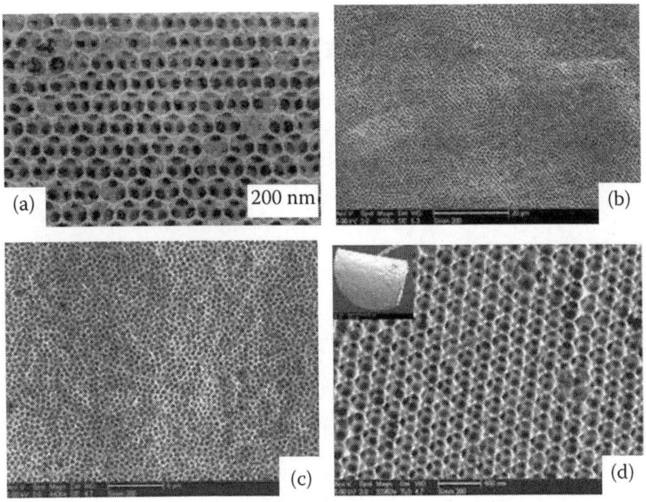

FIGURE 8.17 FESEM images of SnO_2 ordered pore array films on the ceramic tube. (a) Double-layer porous film with pore size of 200 nm and (b) through (d) four-layer porous films with pore sizes of 1000, 350, and 200 nm, respectively. (Sun et al., *Adv. Mater.* 2005. 17. 2872–7. Copyright Wiley-VCH Verlag GmbH & Co. KGaA. Reproduced with permission.)

8.2.2.2 Multilayer Heteropore Sized Porous Films

8.2.2.2.1 Layer-by-Layer Template-Transfer Strategy

In the fabrication of the porous films based on colloidal lithography strategy, emphases are mostly focused on the single material, unitary pore sized and shaped periodically structured arrays [56], which greatly restrict the applications of porous films in many fields, such as catalysis, sensing, and SERS. For instance, the porous films with small pores have many active sites on their surface, as a result of their high specific surface area, and thus enhance the detection signal for device applications, but small pore size is unfavorable for the transportation and diffusion of the detected molecules, leading to a low response rate, or vice versa for the large pore films [4,57]. Obviously, if we superpose the porous films with big and small pore sizes together to form a heteropore sized porous film, such film would combine the advantages of both small and large pores and enhance the performance of devices. Thus, the fabrication of the micro/nanostructured porous films with different pore sizes is of significance for the realization of next generation of nanostructured devices.

Here, we discuss a layer-by-layer template-transfer strategy for fabrication of heteropore sized porous films with hierarchical micro/nanoarchitectures on a desired substrate by alternately using the monolayer colloidal crystal with different sizes of PS spheres as templates [58,59]. Figure 8.18 schematically shows the strategy. The colloidal monolayer with larger PS spheres is dipped in the solution and made to float on the surface of the solution (as shown in Figure 8.18a and b). Then, the monolayer is picked up by using a desired substrate and dried (Figure 8.18c and d).

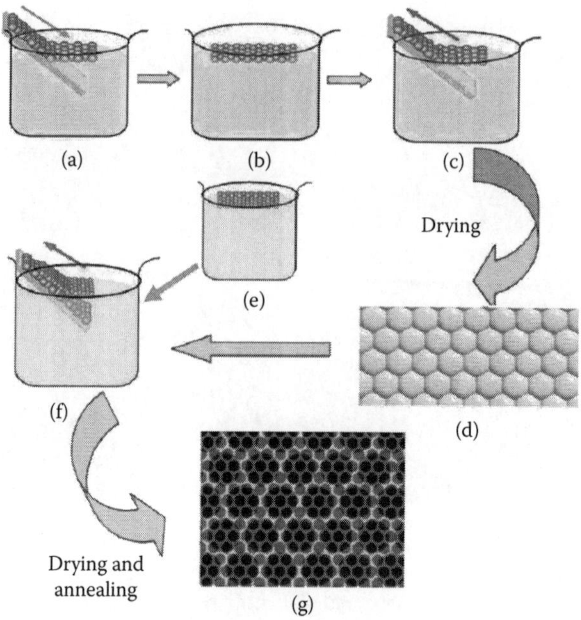

FIGURE 8.18 Outline of the layer-by-layer synthesis strategy for hierarchically micro/nanostructured ordered porous films. (a) A flat glass substrate covered with colloidal monolayer of 1000-nm polystyrene (PS) spheres is dipped in the precursor solution; (b) the colloidal monolayer floats on the surface of the solution; (c) the monolayer is picked up using a new substrate; (d) after drying the substrate covered with the monolayer, in a furnace (120°C); (e) the colloidal monolayer of smaller PS spheres (<1000 nm) floats on the surface of the solution [similar to (b)]; (f) using the dried substrate covered with the monolayer (1000-nm PSs) to pick up the colloidal monolayer of smaller PS spheres floating on the solution; (g) the hierarchical micro/nanostructured ordered porous film is obtained on the substrate after heat treatment and removal of PS spheres. (Reprinted with permission from Jia et al., 2009, 2697–705. Copyright 2009 American Chemical Society.)

Subsequently, using the dried substrate as a new substrate pick up another colloidal monolayer with smaller PS spheres from the solution (Figure 8.18e and f). Finally, dry and anneal the sample to remove the PS spheres, and the heteropore sized porous film with double layer are obtained (Figure 8.18g).

8.2.2.2.2 Double-Layer Heteropore Sized Porous Films

Typically, In_2O_3 was chosen as an example. Figure 8.19a shows the morphology of In_2O_3 double-layer heteropore sized porous films on an ordinary glass slide using 0.1 M $In(NO_3)_3$ as precursor solution. The PS spheres for the bottom (or first) and top layers are 1000 and 200 nm in diameter, respectively (or 1000/200 nm colloidal monolayers). Clearly, the pores in the film are hexagonally arranged, exhibiting bi-periodic ordered structure with periodicities of 1000 and 200 nm, respectively. The skeleton network of the first layer uploads the top layer and the skeleton of each pore of the first layer, together with its inner through-holes in the top layer, can form a window-shaped cell, as clearly shown in the inset of Figure 8.19a. Importantly,

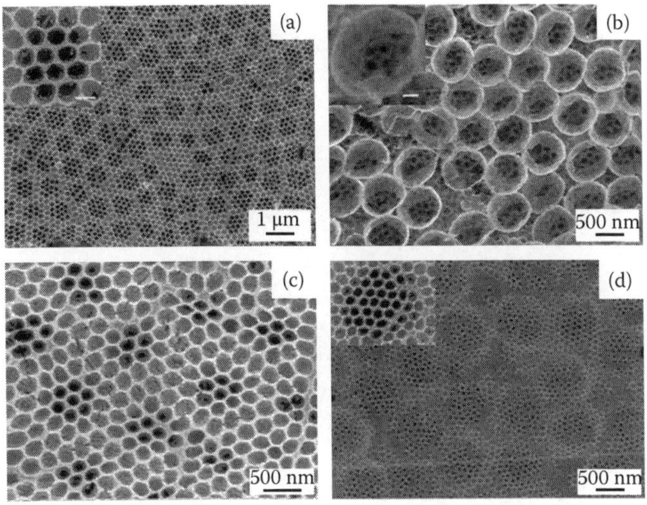

FIGURE 8.19 FESEM images of In_2O_3 hierarchical porous films with different structures on glass substrate. (a) 1000/200 nm, (b) the backside of the film shown in (a), (c) 1000/350 nm, and (d) 1000/100 nm. The insets: local magnification. (Reprinted with permission from Jia et al., 2009, 2697–705. Copyright 2009 American Chemical Society.)

this hierarchically structured porous film can be lifted off by blade with integrity, showing a good freestanding property. Figure 8.19b exhibits the morphology of the backside of the film shown in Figure 8.19a. We can see that the pores in the first layer are of hollow-spherical shape with truncated tops. The spherical shells or pore walls are very thin (about 30 nm in thickness), and the small holes in the top layer can also be clearly seen from the truncated pores. Furthermore, the through-holes embedded in the "window" can be modulated by changing the PS spheres' diameter of the top layer in the colloidal crystal. When the pore size of the top layer changes from 350 to 100 nm, the number of holes in the "window" increases with the decrease of the pore size in the top layer, as illustrated in Figure 8.19c and d. This demonstrates a modulation effect in the gas-sensing films or molecular separation films.

8.2.2.2.3 Double-Layer Heterostructured Porous Films

The strategy shown in Figure 8.18 is universal. If using different precursor solutions alternately according to such strategy, multilayer heterostructured porous films with alternate chemical composition in each layer can be easily constructed. Choosing 0.1 M $SnCl_4$ and 0.05 M $Fe(NO_3)_3$ as precursor solutions, SnO_2-Fe_2O_3 heterostructured porous film can be obtained, as demonstrated in Figure 8.20, in which the first layer is SnO_2 and the top layer is Fe_2O_3. Similarly, this bilayer heterostructured porous film is also of honeycomb structure.

To further confirm the heterostructure, transmission electron microscopy (TEM) examination and energy-dispersive X-ray (EDX) analysis were conducted, as shown in Figure 8.21. The bilayer structure of the porous film can be clearly observed (Figure 8.21a). The selected-area electron diffraction (SAED) patterns of each layer

FIGURE 8.20 FESEM image of SnO_2-Fe_2O_3 heterostructured porous film on glass substrate. (Reprinted with permission from Jia et al., 2009, 2697–705. Copyright 2009 American Chemical Society.)

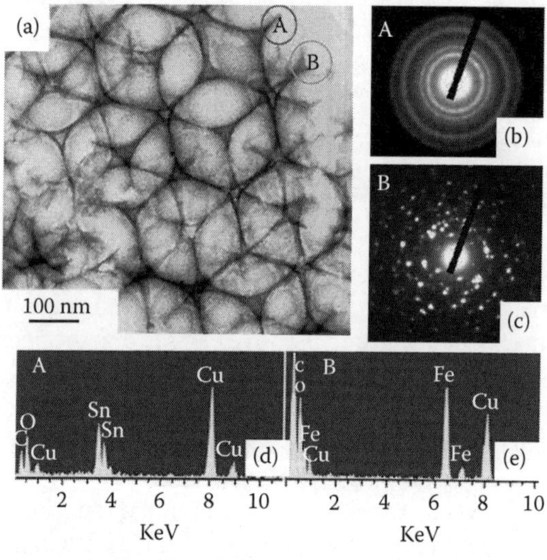

FIGURE 8.21 Structural and composition analysis for the SnO_2-Fe_2O_3 heterostructured porous film. (a) TEM image of the film, (b) and (d) the selected-area electron diffraction (SAED) and energy-dispersive x-ray (EDX) patterns for SnO_2, (c) and (e) the SAED and EDX patterns for Fe_2O_3. (Jia et al., *J. Mater. Chem.*, 19, 7301–7, 2009. Reproduced by permission of The Royal Society of Chemistry.)

indicate that the under layer (marked with A) is polycrystalline, while obvious different diffraction patterns of the top layer (marked with B) can be clearly seen (see Figure 8.21b and c). Moreover, the EDX spectrum recorded from the under layer of the porous film indicates the presence of Sn and O, while the top layer shows the existence of Fe and O (Figure 8.21d and e).

As demonstrated in this section, the ordered porous films with monolayer or multilayer can be fabricated by the solution routes based on the PS colloidal templates. In addition, owing to the stepwise feature of the fabrication routes, we can also design films with complex compositions or structures. These free-standable and hierarchical porous films have high surface-volume ratio, controlled pore size and chemical composition, and homogeneous thickness and will be advantageous for fundamental researches related to micro/nanosystem and their device applications.

8.3 MICRO/NANOSTRUCTURED ORDERED PORE ARRAYS BASED ON ELECTRODEPOSITION STRATEGY

In addition to the solution routes mentioned in the preceding discussion, electrodeposition strategy is also an important way to fabricate the micro/nanostructured ordered pore arrays (films). It is conducted by passing an electric current between two or more electrodes separated by an electrolyte [60]. For this technique, the synthetic processes mainly take place at the electrode–electrolyte interface. Comparing with the other synthetic techniques, electrodeposition has the advantages of low equipment cost, low experimental temperature, and easy manipulation and controllability of surface morphology of films. Taking colloidal template as mask, the orderly metal or semiconductor or oxide micro/nanostructured porous films can easily be fabricated [6,7,11,29,57,61].

8.3.1 Metal Micro/Nanostructured Ordered Through-Pore Arrays

The noble metal (Au, Ag, and Pt) ordered porous films were reported using the colloidal monolayer as template [62]. The pores possess bowl-like shape and diameters of the openings at the film surface can be tuned by the electrodeposition parameters. Here, we discuss a strategy, the heated template–directed electrodeposition, for morphology-controlled ordered through-pore arrays with one pore thickness. The film morphology can be controlled by heating colloidal monolayer-coated substrate and electrodeposition. Using this strategy, a series of metal, oxide, and semiconductor through-pore arrays with controlled morphologies could be obtained. More interestingly, the deposited film can be transferred onto any desired substrate, especially to an insulating substrate or curved surface, which cannot be used for electrodeposition. Here, gold is chosen as an example to demonstrate the morphology-controlled fabrication of ordered pore arrays based on the electrodeposition on the PS colloidal monolayer.

8.3.1.1 Electrodeposition Strategy Based on PS Monolayer

Figure 8.22 schematically shows the strategy [63]. A large area PS colloidal monolayer on a glass substrate is first transferred onto a conducting substrate (indium tin oxide [ITO]-coated glass or Si wafer) by a floating-transferring method as mentioned in Section 8.2 (Figure 8.22a and b). After being dried in air at room temperature, the substrate covered with the monolayer was heated at 110°C in an oven for a given time to bond the monolayer to the substrate (Figure 8.22c). Such heating will induce an area contact, instead of point contact, between PS spheres and the substrate, depending on the time period. And the enlarged area contact will increase the binding force (van der Waals force) between PS spheres and the substrate. Finally, such substrate was used as a working electrode and set in a three-electrode electrolytic cell.

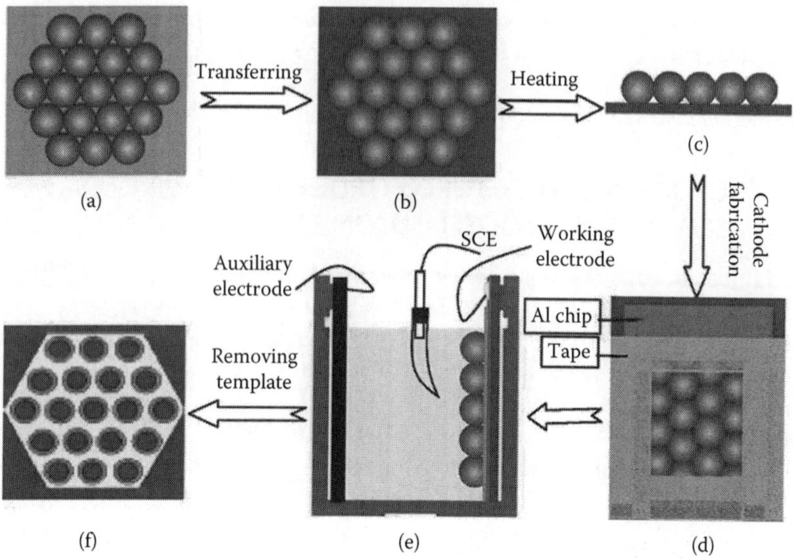

FIGURE 8.22 Illustration of the electrodeposition strategy based on polystyrene (PS) colloidal monolayer crystal. (a) PS colloidal monolayer on glass substrate. (b) The monolayer on the conductive substrate (here it is indium tin oxide [ITO] glass). (c) The monolayer bonded to the ITO glass by heating. (d) Design of the cathode. (e) Electrodeposition in a cell. (f) Ordered through-pore array after removal of the monolayer. (Sun et al., *Adv. Mater.* 2004. 16. 1116–21. Copyright Wiley-VCH Verlag GmbH & Co. KGaA. Reproduced with permission.)

A graphite plate is used for the auxiliary electrode, a saturated calomel electrode (SCE) for the reference electrode, and $HAuCl_4$ aqueous solution for the electrolyte (Figure 8.22d). After electrodeposition at the invariable electric potential for a certain time period, an ordered through-pore array is thus acquired by removing the monolayer in methylene chloride (CH_2Cl_2) solution (PS spheres can be dissolved in CH_2Cl_2 solution quickly) (Figure 8.22e and f). Obviously, morphology of the film can be controlled by heating the monolayer-coated substrate followed by electrodeposition. The bottom and top diameters of the through-pores and pore shapes depend on the heating time, and the film thickness is controlled by the deposition time. Based on the pore sizes at bottom and surface of the film, it is easy to roughly obtain the film thickness without consideration of curvature change of sphere during heating before deposition.

8.3.1.2 Ordered Through-Pore Arrays

Figure 8.23 shows the morphologies of the products or the through-pore arrays after substrate heating followed by electrodeposition for different time periods. All samples were electrodeposited under a potential of 0.7 V versus SCE. By increasing the heating time of the monolayer-covered substrate before deposition (PS spheres were 1000 nm in diameter), the openings at the film bottom evolve from irregular shape (Figure 8.23a and b) to circles (Figure 8.23c and d), and their sizes

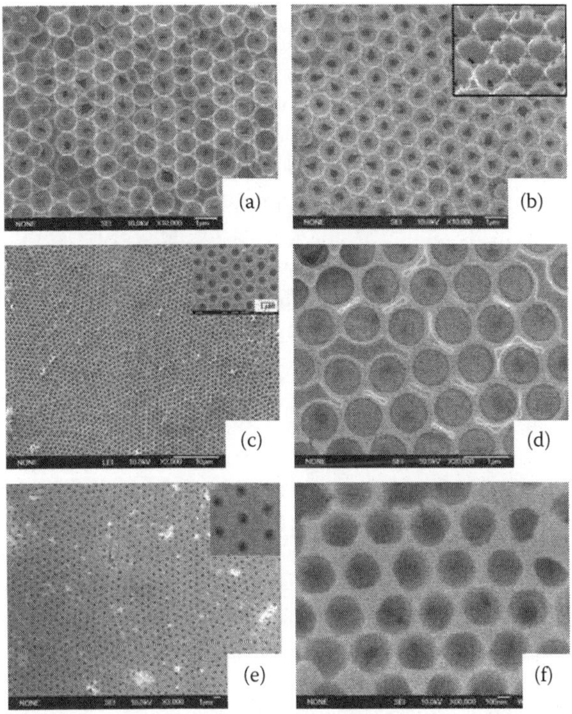

FIGURE 8.23 Gold ordered through-pore arrays induced by the strategy shown in Figure 8.22. The polystyrene spheres' sizes in monolayer are 1000 nm for (a) through (e) and 350 nm for (f). The electrodeposition potential is 0.7 V versus saturated calomel electrode. The heating time and the deposition time are (a) 2 and 15 minutes, (b) 5 and 15 minutes, (c) 16 and 8 minutes, (d) 40 and 8 minutes, (e) 16 and 30 minutes, and (f) 3 and 5 minutes, respectively. The insets in (c) and (e) are the corresponding magnified images. The inset in (b) is a corresponding tilt view. (Sun et al., *Adv. Mater.* 2004. 16. 1116–21. Copyright Wiley-VCH Verlag GmbH & Co. KGaA. Reproduced with permission.)

also increase due to heating-induced rise in contact area between PS spheres and substrate. When the heating time was up to 40 minutes, the diameter of the pores at the bottom were about 800 nm, near that of PS spheres, and the pores were nearly cylindrical-like, as shown in Figure 8.23d. The film thickness and openings at the film surface can be controlled by the electrodeposition time. When electrodeposition lasted for 15 minutes, the pore diameters at the film surface are nearly equal to those of PS spheres, and all the circles at the film surface nearly contact with each other, as illustrated in Figure 8.23a and b. The film thickness is estimated to be about 310 and 320 nm, respectively, based on the simple spherical geometry. When the electrodeposition time was shorter (e.g., 8 minutes), the film was thinner in thickness, and pore sizes at the film surface got smaller than those of PS spheres and so they were separated from each other from top view due to limitation of the template geometry, as shown in Figure 8.23c (about 160 nm in thickness). However, if the electrodeposition time is long enough, the film thickness will be thicker than the distance between the contact point plane of adjacent PS spheres and the substrate, and pore diameters at the film surface will be also smaller than those of PS spheres, as illustrated in Figure 8.23e (about 940 nm in thickness). The samples in Figure 8.23c and e were heated for the same time period (16 minutes) before deposition, but the sample in Figure 8.23e was deposited for longer (30 minutes). Both show similar morphology from top view. However, there should exist a tunnel between two adjacent pores for the latter, which is induced by the contact area between two adjacent PS spheres due to heating before deposition [64]. Similar results were obtained for the monolayer templates composed of much smaller PS spheres. Figure 8.23f shows the ordered through-pore array fabricated from the 350-nm PS monolayer. The corresponding heating and electrodeposition time are 3 and 5 minutes, respectively. It exhibits the same regularity as those acquired from the 1000-nm PS monolayer except for the pore sizes.

Interestingly, such ordered porous film can be transferred integrally from one substrate to others by picking it up from the water surface with another substrate, as shown in Figure 8.24. Substrates can be flat plane or curved surface, and the front and back surfaces of films can arbitrarily be chosen according to our requirements. Figure 8.25a and b illustrate the ordered pore arrays on mica substrate. They correspond to the back surfaces of the deposited films shown in Figure 8.23b and c. We can thus see more clearly the morphologies of the openings at pore bottom for the original through-pore films in Figure 8.23b and c, and the sizes of the bottom openings are about 300 and 400 nm, respectively. Figure 8.25c and d show the ordered porous film on a ceramic tube used as gas sensor. The film corresponds to that shown in Figure 8.23b. The inset in Figure 8.25c is a photo of the tube covered with gold ordered through-pore array film and Figure 8.25d is an enlarged image of the chosen zone in Figure 8.25c. The film had covered the curved surface and the microstructure was not changed except that it has a tilted view as shown in the inset of Figure 8.23b due to the curved surface. This is undoubtedly of importance. It means that the film could be transferred to any substrate, especially, to a desired but insulating substrate like mica or a curved surface, on which the film cannot be electrodeposited, avoiding the restriction of electrodeposition on the conducting substrate.

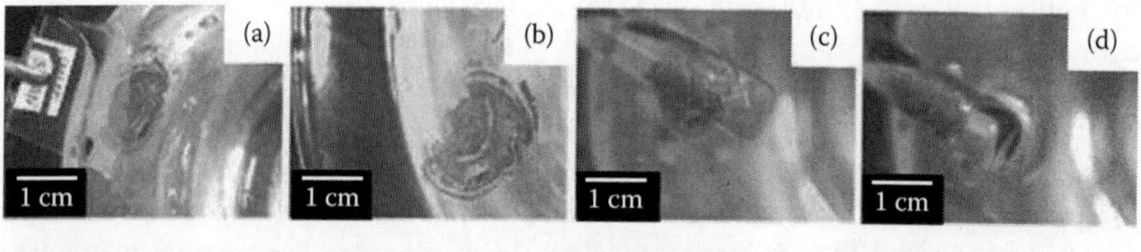

FIGURE 8.24 Photos transferring the gold ordered porous film on indium tin oxide (ITO) glass to a curved surface (glass rod). (a) ITO glass covered with gold ordered through-pore array film is dipped in water. (b) The gold film is floating on water. (c) The gold film is being picked up with a glass rod. (d) The gold film is transferred on the curved surface of the glass rod. (Sun et al., *Adv. Mater.* 2004. 16. 1116–21. Copyright Wiley-VCH Verlag GmbH & Co. KGaA. Reproduced with permission.)

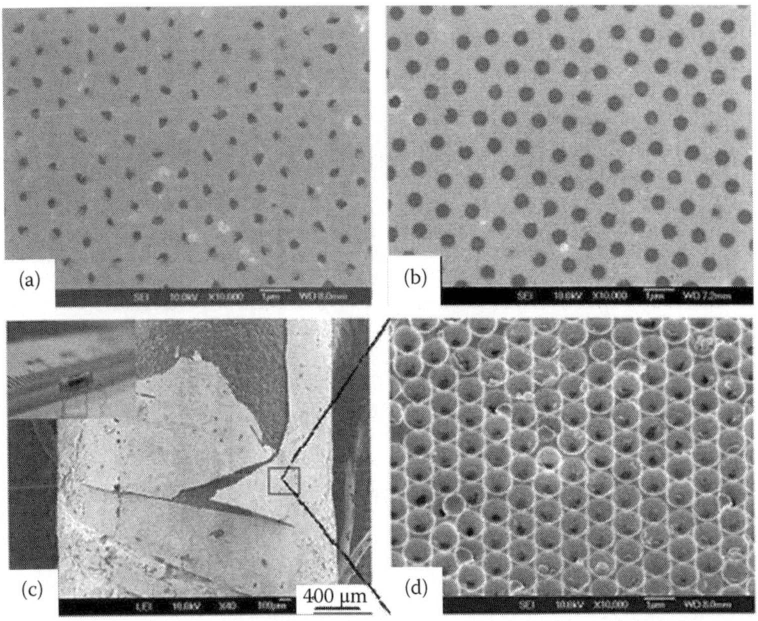

FIGURE 8.25 Ordered gold porous films on a mica substrate and on the curved surface of a ceramic tube by lifting off the films from indium tin oxide glass. (a) The back surface of the film, shown in Figure 8.23b, on mica. (b) The back surface of the film, shown in Figure 23c, on mica. (c) A low-magnified FESEM image of the gold film, shown in Figure 8.23b, on the curved surface of a ceramic tube; the inset is a photo of the tube covered with the gold film. (d) A magnified image of the zone shown in (c). (Sun et al., *Adv. Mater.* 2004. 16. 1116–21. Copyright Wiley-VCH Verlag GmbH & Co. KGaA. Reproduced with permission.)

8.3.1.3 Molecule Adsorption and Interface-Weakening Model

Transferability is related to the electrodeposition rate (or deposition potential) and additive agents in the electrolyte, or mainly determined by film-growth mechanism depending on the property of electrolyte. Here, the electrolyte includes $HAuCl_4$, Na_2SO_3, $C_{10}H_{14}N_2O_8Na_2$ $2H_2O$ (Ethylenediaminetetraacetic acid [EDTA] disodium salt), and K_2HPO_4. $HAuCl_4$ is the source of the initial Au(III), some of which immediately form other possible valence states: Au(I) [65,66]. Na_2SO_3 acts as a kind of complexing agent. EDTA can improve the mechanical property of final films. And K_2HPO_4 mainly acts as an auxiliary complexing agent and buffering agent. When the ITO glass covered with PS monolayer is dipped in the electrolyte, HPO_4^{2-} would be adsorbed onto the substrate before deposition. However, the substrate is structured with PS spheres. The PS spheres possess negative charges (surfaces of the PS spheres are dressed with SO_4^{2-}) and influences the property of ITO around the spheres. So, the HPO_4^{2-} will adsorb on the substrate but away from the area where PS spheres are situated, as illustrated in Figure 8.26a. Since the adsorbates on the substrate impede the growth of film, Au crystal nuclei will first be formed in the wedge-shaped regions (or corners) between PS spheres and substrate due to the lowest energy barrier of

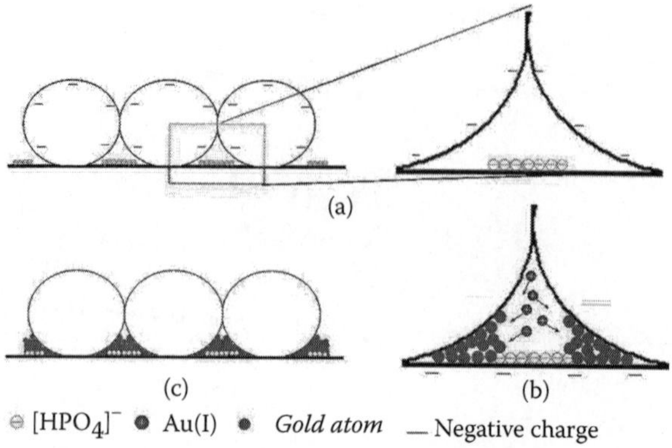

FIGURE 8.26 Schematic illustrations of molecule adsorption and interface-weakening model. (a): Monolayer polystyrene-coated indium tin oxide glass in the electrolyte before electrodeposition. (b) During initial electrodeposition. (c) After electrodeposition. (Sun et al., *Adv. Mater.* 2004. 16. 1116–21. Copyright Wiley-VCH Verlag GmbH & Co. KGaA. Reproduced with permission.)

nucleation in this area (see Figure 8.26b). In addition, the applied voltage (0.7 V vs. SCE) during electrodeposition was not high enough to make the adsorbates desorb from the substrate. So, the gold grows along the substrate and the PS spheres from the wedge-shaped region (or corner) (also covering the adsorbed area on substrate). So, film at the edge of pores (at film surface) will be always thicker than that in the interstitial among three closely packed PS spheres, as clearly illustrated by the tilted view in the inset of Figure 8.23b.

To further understand the growth process, an edge region of the sample shown in Figure 8.23d was observed, as illustrated in Figure 8.27. The arrangements of pores are not ordered. However, few details of the film growth were observed. There is no growth of film in PS spheres' area, but growth was observed in the area surrounding the PS spheres. When a single PSs was situated on the substrate and isolated with the other PSs (zone A in Figure 8.27), the gold grows only around the sphere and forms a hemisphere like shell with certain thickness. However, if two PS spheres are arranged closely (zone B), the property of the region between the two spheres and the substrate is obviously different from the others due to the surface charge of the PS spheres. Naturally, the number of adsorbates is less than those on the others, and therefore, growth in such region will be quicker. Similarly, when three spheres are arranged closely (zone C), growth in the region between the three PS spheres and the substrate is also quicker than other regions. A more obvious example is that when PS spheres are not arranged very closely but not far from each other (zone D), pores (from top view) are connected by the film with a certain thickness, and the thickness of the film between pores is smaller than that surrounding the pores. On this basis, it is easy to understand formation of the closely packed pore array shown in Figure 8.23 if all PS spheres are closely packed in monolayer.

The film could be stripped off by the surface tension of water when it was dipped in water, due to existence of adsorption layer in some areas between Au film and

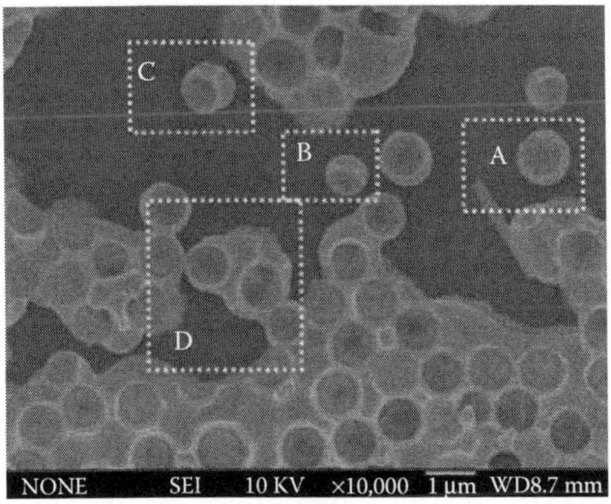

FIGURE 8.27 Morphology in the edge region of the sample shown in Figure 8.23d. (A) The pore from one polystyrene (PS) sphere. (B) Pores from two PS spheres in close contact. (C) Pores from three PS spheres in close contact. (D) Pores from PS spheres not in very close contact. (Sun et al., *Adv. Mater.* 2004. 16. 1116–21. Copyright Wiley-VCH Verlag GmbH & Co. KGaA. Reproduced with permission.)

substrate (see Figure 8.26a), which weakens the adhesive force between the film and the substrate. This was more clearly demonstrated by increasing the electrodeposition rate (or potential). When the potential was increased up to 1.7 V versus SCE, the film could not be lifted off the substrate when dipped in water. Such potential should be high enough to remove the adsorbates quickly during initial deposition, and so the gold film had much stronger adhesive force to the substrate [67] and could not be stripped off. For the same reason, without addition of K_2HPO_4 to the electrolyte, the film could not be lifted off at any potential due to nonadsorption on the substrate. The same was applicable for electrodeposition of copper.

8.3.1.4 Universality of the Strategy

It should be pointed out that the strategy shown in Figure 8.22 is of good universality and can be used for synthesis of the other ordered through-pore arrays, including metals, semiconductors, and compounds on any conducting substrate (see Section 8.3.2) or the insulating substrate with flat or curved surface by lifting off and transferring. Furthermore, if the PS monolayer-coated substrate was electrodeposited in two different electrolyte solutions separately, bimetal (or bilayer) ordered porous films can be obtained. Typically, Figure 8.28 shows the Au/Cu bilayer ordered pore array induced by electrodeposition of Cu on the Au ordered pore array film.

8.3.2 Semiconductor Micro/Nanostructured Ordered Pore Arrays

As mentioned in the preceding discussion, the strategy shown in Figure 8.22 is universal and can be used for fabrication of the other micro/nanostructured ordered pore arrays, including semiconductors and compounds on any conducting substrate, in addition to metals. Here, ZnO is chosen as a model material to demonstrate the

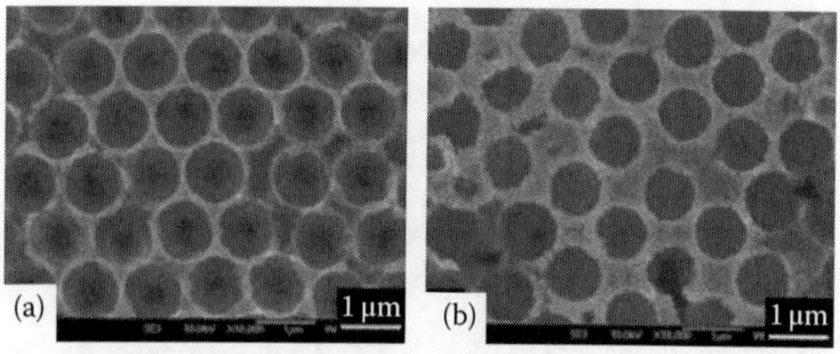

FIGURE 8.28 FESEM images of Au/Cu bilayer ordered pore array film. (a) and (b) After electrodeposition of Cu on the Au ordered pore array film for 2 and 5 minutes, respectively. (Cao et al., *Chem. Commun.*, 1604–5, 2004. Reproduced by permission of The Royal Society of Chemistry.)

ordered porous films. A quantity of 0.05 M zinc nitrate aqueous solution was used as electrolyte and a zinc sheet (99.99% purity) was used as the counter electrode. The deposition was performed at 355 K in a water bath. Galvanostatic and potentiostatic cathodic depositions were employed on PS-covered Au/Si substrate and ITO glass substrates, respectively, before removal of the PS spheres [68,69].

8.3.2.1 Morphology and Structure

Figure 8.29a shows the morphology for the sample deposited on ITO glass substrate under a deposition potential of 1.0 V. The pores are hemispherical in shape and orderly arranged. Further experiments show that the array morphology depends on the deposition potential. The higher potential induces thinner skeleton among the pores. When the potential was increased to 1.4 V, the morphology of the porous film took a well-like structure, as illustrated in Figure 8.29b. X-ray diffraction (XRD) measurements indicate that all diffraction peaks can be identified as hexagonal wurtzite ZnO (JCPDS 36-1451) and slightly preferred orientation of (0002) [68].

Contrarily, when the PS-covered Au/Si wafer was used as substrate, the deposition currents have only little influence on the morphology of the film, showing similar morphology in a current range of 0.7–1.2 mA. The corresponding XRD shows that the film is only of two peaks at 34.4° and 38.2°, indexed by ZnO(002) and face-centered cubic (fcc) gold (111) peaks, respectively, as illustrated in Figure 8.30. This indicates that the film is highly oriented in crystal structure and most of ZnO(001) and Au(111) planes are parallel to the substrate surface (the peak of Si(400) is at 69°). Furthermore, Figure 8.31 shows the FESEM images of the film. The pores are highly orderly arranged in the film, which are reverse replica of the PS colloidal monolayer template. The skeleton seems to be packed by the block units ZnO nanosheets parallel to the substrate surface. In other words, the morphology shows step structure from top view. However, the cross-sectional observation (Figure 8.31b) has revealed that the skeleton is composed of densely aligned ZnO hexagonal nanocolumns with a small difference in heights, which is consistent with the ZnO wurtzite crystal structure and indicates that such films are of good crystal quality.

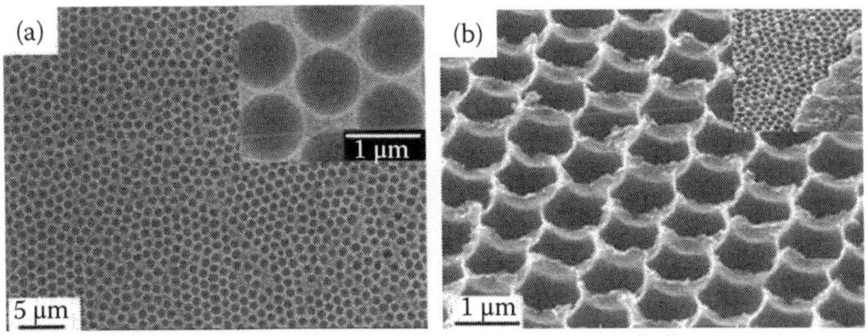

FIGURE 8.29 FESEM images of ZnO porous arrays on indium tin oxide substrate under the potential of 1.0 V (a) and 1.4 V (b). Inset in (a) is the local magnified image and inset in (b) is ultrasonic washing for a shorter time after deposition. (Cao et al., *Chem. Commun.*, 1604–5, 2004. Reproduced by permission of The Royal Society of Chemistry.)

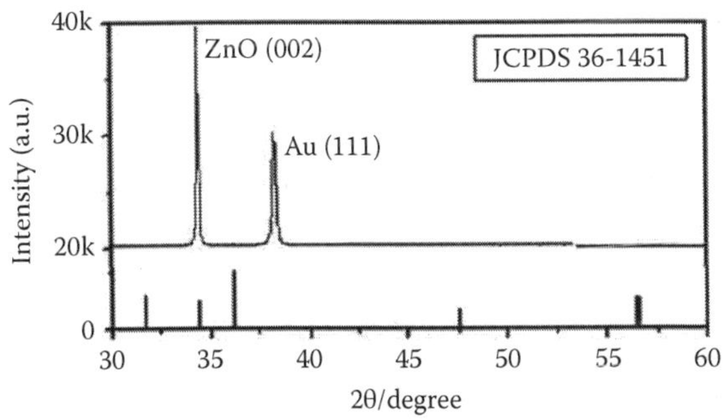

FIGURE 8.30 X-ray diffraction pattern of as-synthesized film electrodeposited on polystyrene-covered Au/Si substrate (0.9 mA, 120 minutes). The line spectrum is the standard diffraction of ZnO powders. (From Cao et al., *Electrochem. Solid-State Lett.*, 8, 237–40.)

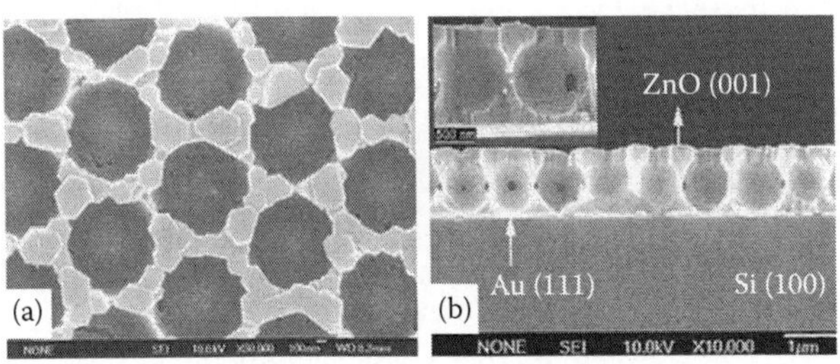

FIGURE 8.31 FESEM images of ZnO ordered pore arrays on Au/Si substrate under the deposition current of 0.9 mA for 120 minutes. (a) Top view. (b) Cross-section image of (a). The inset: Local magnified image. (From Cao et al., *Electrochem. Solid-State Lett.*, 8, 237–40.)

8.3.2.2 Oriented Substrate-Induced Oriented Growth

The oriented porous ZnO films shown in Figure 8.31 can be mainly attributed to the oriented substrate, which leads to formation of oriented nuclei due to the lattice match between the substrate and the ZnO. The electrodeposition conditions on the whole working electrode (substrate) in the electrolyte are homogeneous; therefore, nucleation can occur at any sites on the substrate that is not covered with PS spheres. Because the ITO glass substrate is amorphous in structure and there is no epitaxial influence during the initial nucleation period, the crystal nuclei of ZnO will be randomly oriented on the substrate, leading to the formation of ZnO skeleton without obvious preferred orientation. The slightly preferred orientation of (0002) can be attributed to the ZnO polar (0001) crystal plane, which has higher surface free energy compared with the other basal planes of $(01\bar{1}0)$ and $(2\bar{1}\bar{1}0)$ [70]. For the (111)-oriented Au/Si substrate (see Figure 8.30), however, preferentially oriented ZnO nuclei will be formed on the substrate to reach the lowest interface energy between the ZnO and the gold film. It is well known that the interface energy is directly related to the lattice mismatch of interfaces. To lower the interface energy, the orientation relationship between the ZnO nuclei and the Au(111)/Si substrate should be ZnO(0001) $[11\bar{2}0]$//Au(111) $[\bar{1}10]$. Figure 8.32 shows this relationship, which results in the smallest lattice mismatch between the ZnO film and the substrate. The lattice mismatch along ZnO $<11\bar{2}0>$ and Au $<\bar{1}10>$ is about 12.7%. In addition, fast growth along the *c*-axis direction is energetically favorable due to the higher surface free energy of polar (0001) crystal planes. Thus, the (001)-oriented ZnO ordered pore arrays with flat top surfaces and hexagonal nanocolumns were formed on the (111)-oriented Au/Si substrates by preferentially oriented nucleation and subsequent thermodynamically favored growth.

8.3.2.3 Other Materials

In addition to zinc oxide, other semiconductors or ion-doped semiconductors or heterostructured ordered pore array films can also be prepared by the electrodeposition strategy shown in Figure 8.22 [71–73]. For example, Yeo et al. [73] fabricated CdSe macroporous nanostructure based on galvanostatic electrodeposition, using ITO glass substrate covered by PS colloidal template as cathode, platinum sheet as anode, and the solution consisting of 3×10^{-3} M of SeO_2 and 3×10^{-1} M of $CdSO_4$ with a pH of 3 as electrolyte. Figure 8.33 illustrates the infilling behavior of CdSe material under different conditions. They found that 2D and three-dimensional (3D) porous structures consist of interconnected close-packed arrays of pores. When the film is less than one-third of a PS sphere's diameter in thickness, it presents a monolayer of circular pores (Figure 8.33a). As for the film thickness close to the diameter of the PS sphere, the pores are irregular (Figure 8.33b). When the film thickness is more than one layer of the PS colloidal template, the pores are spherical (Figure 8.33c).

Similarly, other pore arrays of semiconductors, such as Eu_2O_3, CdS, and Fe_2O_3, could also be fabricated using the strategy shown in Figure 8.22, as shown in Figure 8.34, and the corresponding experimental parameters are described as follows:

1. Eu_2O_3, electrolytes: mixture of $EuCl_3$ (0.1 M), sodium citrate (0.1 M), pH = 4, heating time of colloidal monolayer = 40 minutes, voltage = 1.0 V (vs. SCE), deposition time = 20 minutes, finally, sample is kept in furnace with hydrogen ambience at 400°C for 2 hours.
2. CdS, electrolytes: mixture of $CdSO_4$ (0.05 M), $Na_2S_2O_3$ (0.1 M), and CH_4N_2S (0.05 M), pH = 5, heating time of colloidal monolayer = 10 minutes, voltage = 1.0 V (vs. SCE), deposition time = 10 minutes.

3. Fe_2O_3, electrolyte: $Fe(NO_3)_3$ (0.1 M), pH = 5, heating time of colloidal monolayer = 5 minutes, voltage = 1.0 V (vs. SCE), deposition time = 30 minutes, finally, sample is kept in furnace at 500°C for 2 hours.

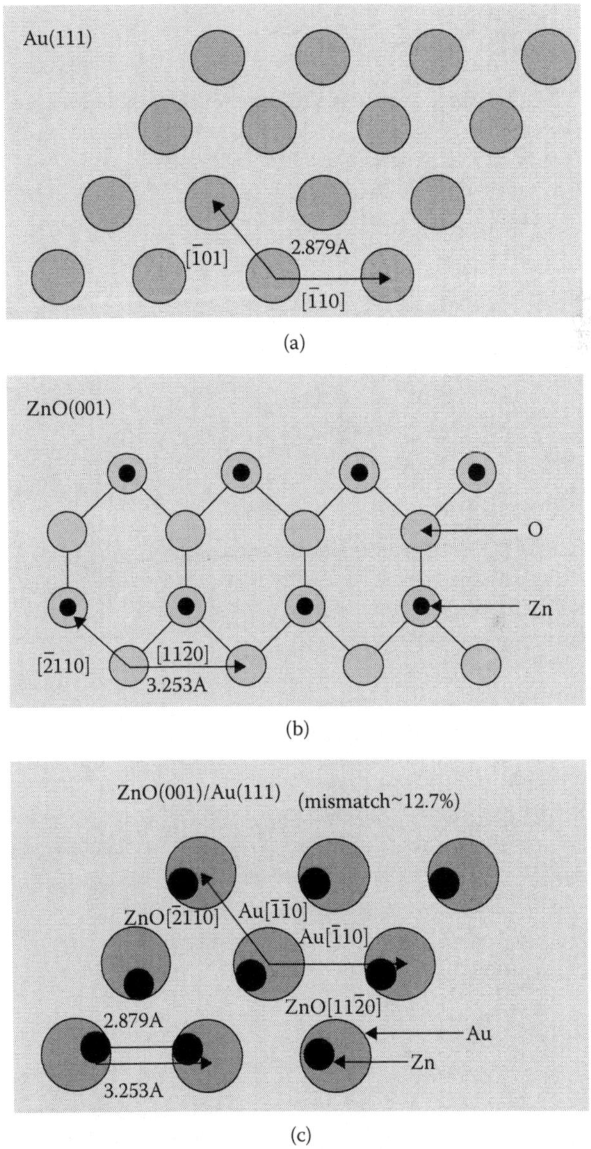

FIGURE 8.32 Schematic drawings of crystal planes (a) Au(111), (b) ZnO(001), and (c) the epitaxial relationship of ZnO(001)/Au(111). (From Cao et al., *Electrochem. Solid-State Lett.*, 8, 237–40.)

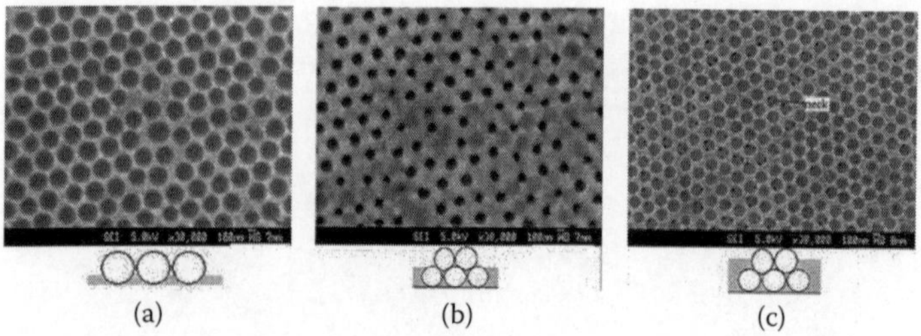

FIGURE 8.33 FESEM images of porous CdSe thin films grown with different thicknesses. (a) 2 mA/cm^2, 5 minutes, (b) 2 mA/cm^2, 20 minutes, and (c) 3 mA/cm^2, 20 minutes. (With kind permission from Springer Science+Business Media: *J. Porous Mater.*, Fabrication and characterization of macroporous CdSe nanostructure via colloidal crystal templating with electrodeposition method, 13, 2006, 281–5, Yeo et al.)

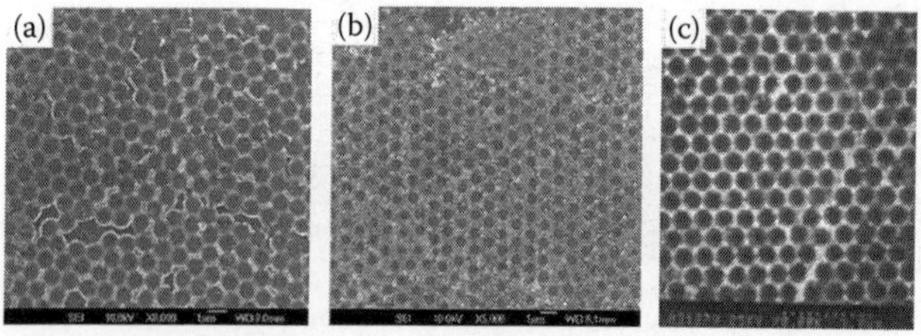

FIGURE 8.34 Ordered pore array films by the strategy shown in Figure 8.22. Polystyrene spheres are all 1000 nm in diameter. The conditions are given in the text. (a) Eu_2O_3. (b) CdS. (c) Fe_2O_3. (Sun et al., *Adv. Mater.* 2004. 16. 1116–21. Copyright Wiley-VCH Verlag GmbH & Co. KGaA. Reproduced with permission.)

8.3.3 Hollow Sphere Array Films Based on Electrodeposition

In addition to ordered pore array films, monolayer hollow sphere array films can also be prepared by the strategy shown in Figure 8.22, which should be of great importance to the next generation of nanodevice [61]. Importantly, the morphology and size of the hollow sphere can be easily controlled by the colloidal monolayer and deposition parameters. Here, we consider some examples to demonstrate the validity of the PS template-based electrodeposition strategy for fabricating the monolayer hollow sphere array films.

8.3.3.1 Ni Hollow Sphere Arrays

8.3.3.1.1 Morphology and Structure

Based on the strategy shown in Figure 8.22, Ni hollow sphere arrays could also be fabricated [61]. Figure 8.35 shows the morphology of the sample after electrodeposition for 90 minutes at a low current density (0.25 mA/cm^2) and removal of the PS template. The Ni hollow sphere arrays are formed. The thickness of the sphere shell is about 60 nm, determined from the broken hollow spheres. Additionally, it has been revealed that the deposition time is important for formation of the hollow sphere

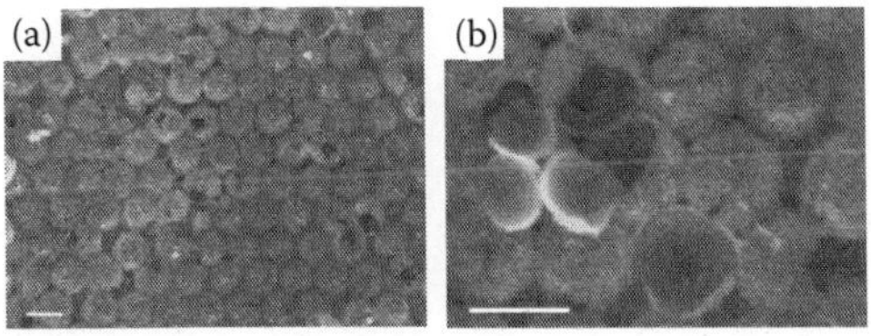

FIGURE 8.35 (a) FESEM image of the monolayer Ni hollow sphere array based on the strategy shown in Figure 8.22 after electrodeposition for 90 minutes at cathodic deposition current density of 0.25 mA/cm^2. (b) Magnification of (a). The scale bars are 1 μm. (Reprinted with permission from Duan et al., 2006, 7184–8. Copyright 2006 American Chemical Society.)

arrays under constant current. If the deposition time is decreased to 45 minutes, morphology of the film evolves from a hollow sphere to a bowl-like porous structure. When deposition time is further decreased to 15 minutes, only nearly spherical nanoparticles were obtained on the substrate.

Further experiments demonstrate that the cathodic deposition current density is crucial to the formation of hollow sphere structure. When the current density is increased to 1.0 mA/cm^2, the bowl-like ordered porous arrays can still be formed after a short deposition time (e.g., 12 minutes), while hollow sphere arrays cannot be obtained, even by increasing deposition time. A middle current density of 0.4 mA/cm^2 leads to a transition. Hollow spherical structure is still formed after a long deposition time, while the spherical shells are composed of coarse particles and thus forming a much rougher surface. These Ni hollow sphere array shows important morphology-related magnetic properties [61]. This could be importance both in fundamental magnetic researches and in device applications.

8.3.3.1.2 Preferential Deposition on PS Spheres

Formation of the hollow sphere arrays can be easily understood. The PS spheres' surface is negatively charged, which induces the Ni^{2+} ions (existing in the form of $[Ni(NH_3)_2]^{2+}$) to be adsorbed on the PS spheres' surface easily. Thus, PS spheres' surface should be of lower barrier for Ni nucleation and growth compared with ITO substrate, which causes preferential nucleation and growth on the PS spheres. When the cathodic current density is low (e.g., 0.25 mA/cm^2), quasi-equilibrium process takes place and Ni nuclei are preferentially formed at the bottom interstitial sites between the PS spheres and the substrate, and then the bowl-like ordered porous arrays or hollow sphere arrays are obtained, depending on the deposition time since the deposition along PS spheres' surface took place. When the cathodic current density is high enough (e.g., 1.0 mA/cm^2), however, deposition rate should be large; homogeneous (or unselective) nucleation occurs on the substrate or the preferential growth along PS spheres will not happen. In this case, when the film thickness is smaller than the PS sphere's radius, corresponding to shorter deposition time, bowl-like pore morphology are formed due to the PSs' geometry. For films with thickness larger than PS sphere's radius, corresponding to longer deposition time, however, hollow spherical shells are not formed.

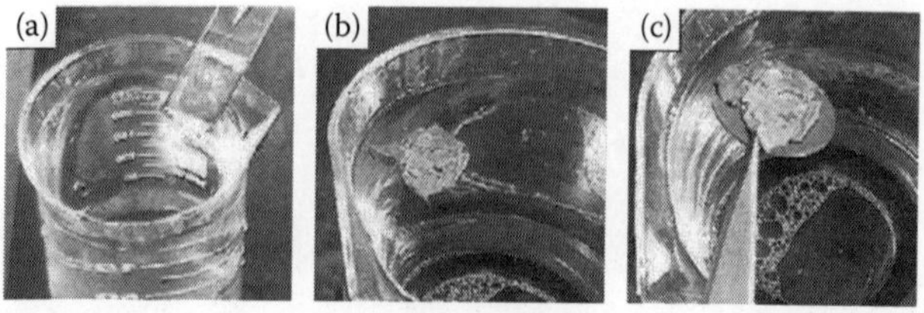

FIGURE 8.36 Photos taken by camera depicting transferring of Ni hollow sphere arrays on indium tin oxide (ITO) glass to a silicon wafer. (a) A Ni hollow sphere array on ITO substrate is dipped in water. (b) The array film is lifted off and floats on the water surface. (c) The floating film is picked up by a silicon wafer. (Reprinted with permission from Duan et al., 2006, 7184–8. Copyright 2006 American Chemical Society.)

8.3.3.1.3 Transferability of Hollow Sphere Arrays

Furthermore, as mentioned in Section 8.3.1.2 for the Au ordered pore array films shown in Figure 8.24, such Ni hollow sphere arrays on a conductive substrate (ITO glass) can also be transferred, in water, to any other desired substrates including the insulated one. Figure 8.36 shows a transferring process of the Ni hollow sphere array. However, the transferable mechanism is different from that of the gold porous arrays [63]. Here, Ni hollow sphere array is formed due to preferential nucleation and growth along PSs' surface. There thus exist a large amount of interstitials between the deposited film and the substrate or only partial contact between the array film and the substrate, which would lead to a weak adherence force, between the substrate and the Ni film, and hence the transferability due to water surface tension.

8.3.3.2 Hierarchically Micro/Nanostructured Monolayer Hollow Sphere Arrays

Some groups have fabricated, by a mild hydrothermal process, various hierarchically structured hollow spheres with nanoparticles, nanorods, or nanosheets as building blocks [33,34]. If such hollow spheres are periodically arranged on a substrate, monolayer hierarchically structured hollow sphere arrays are obtained. Such arrays combine the features of hierarchical structure, hollow spheres, and patterned arrays and possess new properties that the building blocks do not have [11]. Here, we introduce $Ni(OH)_2$ hierarchically micro/nanostructured hollow sphere arrays based on the strategy shown in Figure 8.22. Electrodeposition experiment is similar to that mentioned in the preceding discussions. Briefly, the electrolyte is composed of 1 M $Ni(NO_3)_2$ aqueous solution and its pH value adjusted to 1.7 with nitric acid. A cleaned polycrystalline nickel sheet is used as the auxiliary electrode. The distance between the working electrode and the auxiliary electrode is about 6 cm. Electrodeposition was carried out at 60°C with cathodic current density of 1.2 mA/cm^2 [11].

8.3.3.2.1 Morphology and Structure

After deposition for a certain time and removal of the PS monolayer by dissolution in CH_2Cl_2, the hollow sphere array film is obtained. Figure 8.37 gives some typical results. Clearly, the microspheres are uniformly packed into an array with hexagonal

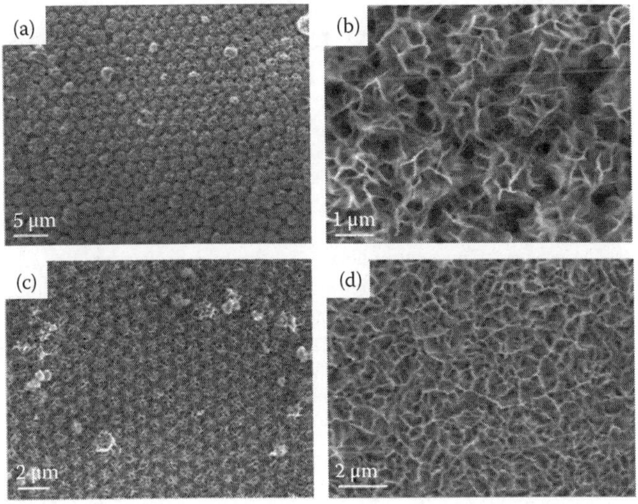

FIGURE 8.37 FESEM images of the as-deposited $Ni(OH)_2$ films. The polystyrene (PS) spheres' size and deposition time are (a and b) 2 μm and 100 minutes, (c) 1 μm and 60 minutes, and (d) no PS used and 60 minutes, respectively. (Duan et al., *Adv. Funct. Mater.* 2007. 17. 644–50. Copyright Wiley-VCH Verlag GmbH & Co. KGaA. Reproduced with permission.)

symmetry. The periodicity, that is, the central distance between the adjacent spheres in the film, can be controlled by PS's size (Figure 8.37a and c). The microspheres in the array are of hierarchical structure and composed of massive ultrathin sheets or nanoflakelets (or nanowall) nearly vertical to the spherical surface, as shown in Figure 8.37b. Contrarily, when pure ITO glass without PS template is used as substrate in the electrodeposition experiment, the film consists of netlike (or wall-like) arranged nanoflakelets and the periodical structure cannot be observed (Figure 8.37d). Furthermore, the corresponding XRD confirms that all of as-deposited samples are α-nickel hydroxide with lattice parameters a = 3.09 Å and c = 22.11 Å [11]. SEM observation in the edge of the film shows that the microspheres in the array are hollow in structure, as demonstrated in Figure 8.38a, in which some broken hollow spheres are clearly seen. TEM examination has also confirmed such hollow structure, as shown in Figure 8.38b. This indicates that the shell layer of the spheres consists of incompact-arranged nanoflakelets. The corresponding SAED pattern reveals that the flakelet is not an intact single crystal because of weak diffraction rings penetrating the diffraction spots. However, both the diffraction spots and the rings show that the flakelet has an orientation with the planar surface perpendicular to *c*-axis (see the inset in Figure 8.38c).

From Figure 8.38a corresponding to the edge region of the sample, one can see that $Ni(OH)_2$ grows both in the place of the substrate without PS template and along the PS spheres in the existence of the PS colloidal monolayer. More importantly, such monolayer hollow sphere film can be detached partially from the substrate by blade, as illustrated in Figure 8.39a. After detachment of the hollow sphere layer, a layer of $Ni(OH)_2$ with bowl-like pore array is left due to the PSs' geometry (Figure 8.39b).

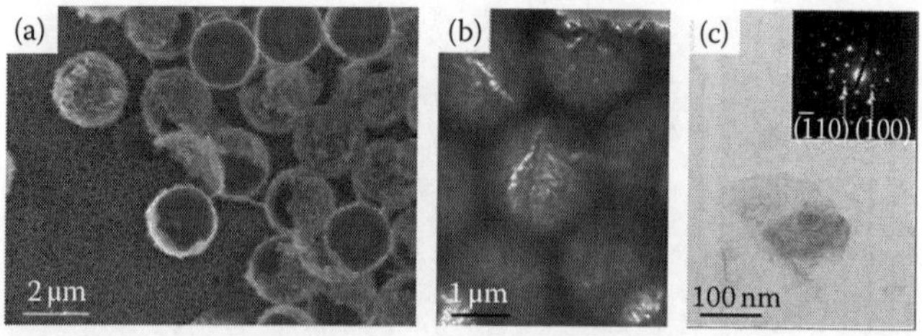

FIGURE 8.38 Microstructural examination of the α-nickel hydroxide microsphere array. (a) FESEM observation in the edge region of the sample shown in Figure 8.37a. (b) TEM image of the detached layer of the microsphere array. (c) TEM image of single nanoflakelet. The inset: selected-area electron diffraction pattern of the nanoflakelet. (Duan et al., *Adv. Funct. Mater.* 2007. 17. 644–50. Copyright Wiley-VCH Verlag GmbH & Co. KGaA. Reproduced with permission.)

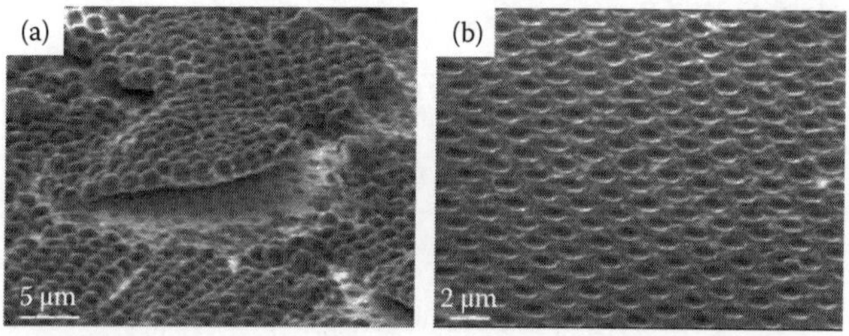

FIGURE 8.39 FESEM images of the monolayer — hollow sphere array shown in Figure 8.37a with partial detachment (a) and after detachment of the hollow sphere array (b). (Duan et al., *Adv. Funct. Mater.* 2007. 17. 644–50. Copyright Wiley-VCH Verlag GmbH & Co. KGaA. Reproduced with permission.)

Obviously, the thickness of the pore array is much smaller than the radius of PS, indicating that the growth on substrate will be depressed after formation of a layer of netlike film in the existence of the PS colloidal monolayer.

8.3.3.2.2 Structural Formula of α-Nickel Hydroxide

For determination of structural formula of α-nickel hydroxide, infrared (IR) spectral measurement and thermogravimetric analysis (TGA) were conducted for the as-deposited monolayer hollow sphere array, as shown in Figure 8.37a. The corresponding results are shown in Figure 8.40. Apparently, the IR spectrum includes (1) a broad peak centered around 3420 cm^{-1} corresponding to the interlamellar water and OH-bond vibration, (2) absorption at 2424 and 2356 cm^{-1} due to CO_2 in air, (3) absorptions in the range of 1000–1500 cm^{-1} due to intercalated anions NO_3^-, and (4) absorptions at 648 and 485 cm^{-1} due to the Ni–O–H bending and Ni–O stretching vibrations, respectively. Thus, the composition of hollow sphere can be marked as $Ni(OH)_x(NO_3)_{2-x}{\cdot}yH_2O$ [74], in which the values of x and y can be determined by the TGA curve, with $x = 1.693$ and $y = 0.117$ [74], or formula $Ni(OH)_{1.693}(NO_3)_{0.307}{\cdot}0.117H_2O$.

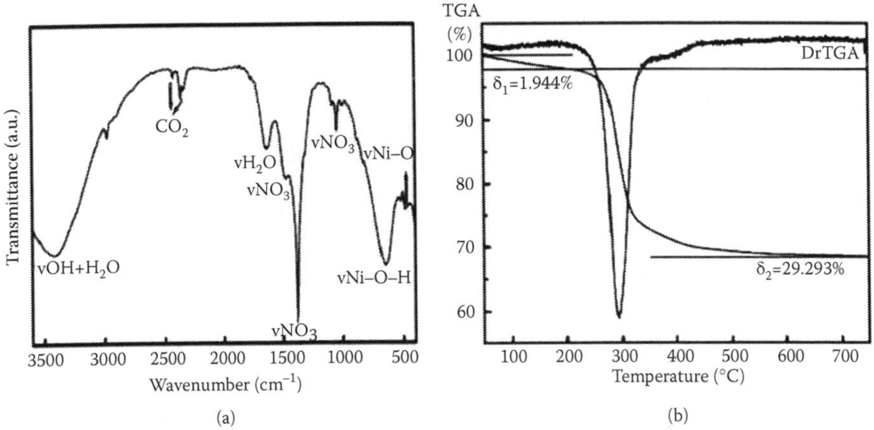

FIGURE 8.40 Infrared spectrum (a) and thermogravimetric analysis (TGA) curve (b) of the as-deposited $Ni(OH)_2$ hollow sphere array shown in Figure 8.37a. (Duan et al., *Adv. Funct. Mater.* 2007. 17. 644–50. Copyright Wiley-VCH Verlag GmbH & Co. KGaA. Reproduced with permission.)

8.3.3.2.3 Influence Factors

For growth of the hierarchical hollow sphere array film, deposition time and cathodic deposition current density are two key parameters. A short deposition time (e.g., 30 minutes) only can form a ringlike array film, as shown in Figure 8.41a. It means that nucleation and growth all start at the sites between PSs and the substrate. When the deposition time is up to 50 minutes, the ordered bowl-like pore array film can be fabricated (Figure 8.41b). At the same time, the netlike film can also be seen at the interstitial sites among PSs, which demonstrates that the film growth on the ITO substrate was completed before the full formation of hollow spheres. After the formation of hollow sphere array, the shell of the hollow spheres becomes denser with the increasing deposition time. On the other hand, if increasing the cathodic current density (J) from 1.2 to 2.0 mA/cm^2, homogeneous nucleation takes place close to the electrode and the $Ni(OH)_2$ grows into the aggregation of nanoflakelets without spherical hollows, as shown in Figure 8.41c and d. Obviously, the high J value is nonbeneficial to the preferential deposition due to fast deposition rate. On the contrary, too smaller value of J is also not appropriate due to the low deposition efficiency.

8.3.3.2.4 Formation of Hierarchically Structured Hollow Spheres

The formation of α-$Ni(OH)_2$ by electrochemical reaction can be described briefly as follows [60,74]:

$$NO_3^- + H_2O + 2e^- \rightarrow NO_2^- + 2OH^- \quad (8.1)$$

$$Ni^{2+} + 2OH^- \rightarrow Ni(OH)_2 \quad (8.2)$$

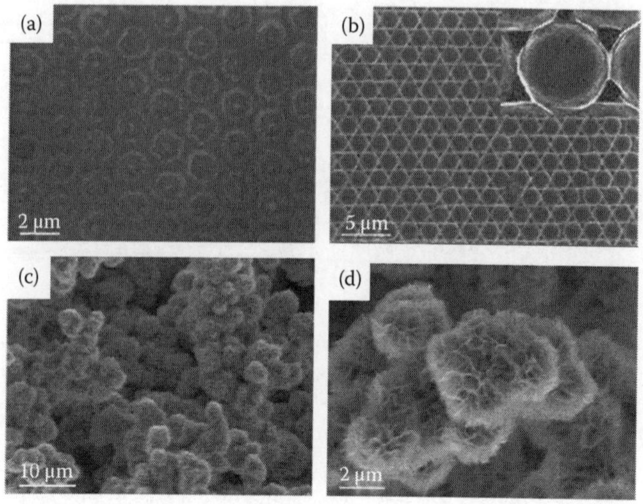

FIGURE 8.41 Influence of the deposition time and current density on the morphology of $Ni(OH)_2$. (a) and (b) FESEM images of ordered $Ni(OH)_2$ arrays after deposition for 30 and 50 minutes, respectively ($J = 1.2$ mA/cm^2). (c) FESEM image of $Ni(OH)_2$ hierarchical structures electrodeposited with current density 2 mA/cm^2 and for 100 minutes without polystyrene (PS) template. (d) Local magnification of (c). The existence of PS colloidal monolayer also leads to similar morphology. Also, hollow sphere arrays cannot be obtained at this current density by decreasing the deposition time. (Duan et al., *Adv. Funct. Mater.* 2007. 17. 644–50. Copyright Wiley-VCH Verlag GmbH & Co. KGaA. Reproduced with permission.)

Initially, OH^- ions were produced by the reduction of NO_3^- close to cathode, then Ni^{2+} ions in the solution react with OH^- and $Ni(OH)_2$ is formed. Here, the PS spheres are negatively charged and thus they can be easily covered with a smooth inorganic layer by the hydrolysis of metal ions. $Ni(OH)_2$ subsequently deposits or nucleates on both the ITO substrate and the interstitial sites between the PS spheres and the substrate. For the latter, deposition along the PS sphere's surface takes place, leading to the final spherical shell, as schematically illustrated in Figure 8.42. Obviously, Ni cannot be formed in this process due to the higher reaction potential of Ni^{2+} to Ni compared with that of NO_3^- to NO_2^- (−0.23 V vs. 0.01 V) [60].

Since the $Ni(OH)_2$ crystal is of layered structure with CdI_2 type, which shows weak interaction between layers and strong binding in the layered planes, surface energy of the layered plane (001) is the lowest. Nickel hydroxide preferentially grows along the layered plane after formation of its nuclei. Also, due to the directional deposition under electric potential, only the oriented nuclei with the (001) plane vertical to substrate and PS sphere's surface preferentially grows, leading to the final formation of wall-like fine structure of $Ni(OH)_2$ nanoflakelets nearly vertical to the substrate and PS sphere's surface (see Figure 8.42). It should be mentioned that partial NO_3^- anions and H_2O molecules will be inserted into the $Ni(OH)_2$ layers during electrodeposition, forming α-type structure.

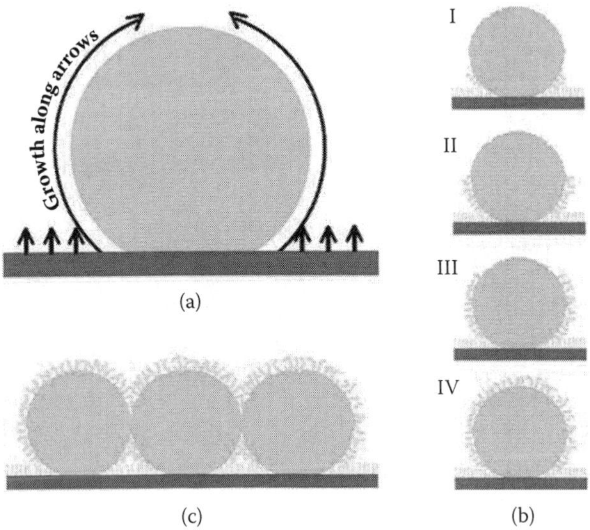

FIGURE 8.42 Schematic illustration for the formation of hierarchical $Ni(OH)_2$ hollow microsphere array. (a) and (c) The schematic initial and final states. (b) The schematic growth process. Steps I through V show formation of Ni(OH) nanoflakelets, nearly vertical to the PS surface, along with the PS surface from bottom to top. (Duan et al., *Adv. Funct. Mater.* 2007. 17. 644–50. Copyright Wiley-VCH Verlag GmbH & Co. KGaA. Reproduced with permission.)

8.3.3.3 Standing Ag Nanoplate–Built Hollow Microsphere Arrays

It is reported that the Ag nanoplates vertically standing and cross-linked on a substrate with nanointerspacings would be of high structural stability, high specific surface area, and high number density of "hot spots" for SERS performances [75]. Sun et al. produced Ag nanoplate arrays standing on a semiconductor substrate by a galvanic reaction approach and showed good SERS performance [76,77]; Liu et al. fabricated Ag nanoplate arrays standing on Au-coated Si substrate and ITO, based on electrochemical deposition under a low current density, exhibiting high enhancement factor of SERS effect (higher than 2×10^5) [78,79]. Here, we introduce Ag nanoplate–built hollow microsphere arrays with controllable structural parameters and centimeter-squared dimension based on electrodeposition on the Au-coated monolayer PS colloidal crystal template under low current density [80].

8.3.3.3.1 Fabrication Route

Figure 8.43 shows the corresponding strategy. Briefly, a preformed hexagonally close-packed PS colloidal monolayer is transferred to a silicon substrate (Figure 8.43a), followed by argon plasma etching to obtain a nonclose-packed monolayer (Figure 8.43b). A thin gold layer is then coated on the PS spheres' surfaces in the monolayer (Figure 8.43c) before electrochemical deposition onto the monolayer. Under a low current density, nanoplates grow vertically on the PS spheres

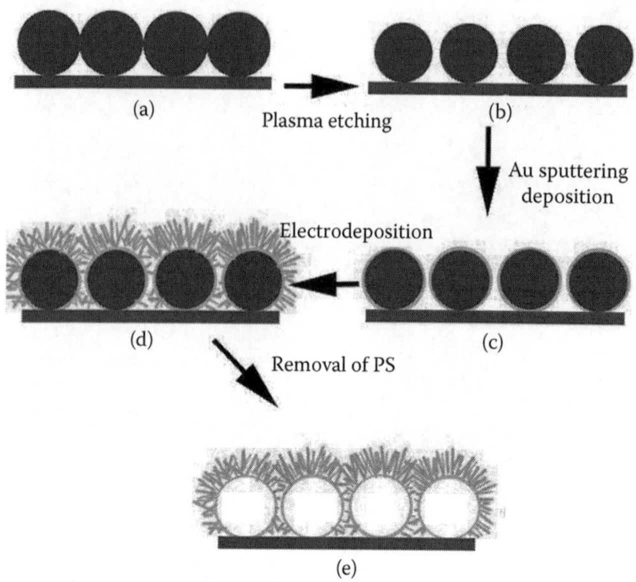

FIGURE 8.43 Schematic illustration of the strategy based on electrodeposition onto the Au-coated monolayer colloidal crystal template under low current density. (a) A close-packed polystyrene (PS) monolayer template on a silicon substrate; (b) argon plasma etching–induced nonclose-packed PS monolayer on the silicon substrate; (c) gold-coated nonclose-packed PS monolayer by ion-sputtering deposition; (d) growth of Ag nanoplates on the gold-coated PSs by electrodeposition; (e) Ag nanoplate–built hollow sphere array after removal of PS. (Liu et al., *J. Mater. Chem.*, 22, 3177–84, 2012. Reproduced by permission of The Royal Society of Chemistry.)

(Figure 8.43d). A Ag nanoplate–built hollow microsphere array can thus be obtained by subsequent removal of PS (Figure 8.43e). The obtained arrays are composed of periodically arranged microsized hollow spheres, which are built by vertically standing and cross-linking Ag nanoplates. The structure of the arrays, including the nanoplates' number density, the microspheres' size, and spacing, is controllable depending on the deposition conditions and template parameters.

Based on the route shown in Figure 8.43, the close-packed monolayer PS colloidal crystal on the silicon substrate was initially etched by argon plasma on a machine (at a pressure of 0.2 mbar and an input power of 100 mW) for 20 minutes, as previously described [81]. The PS colloidal crystal was thus changed to a nonclose-packed template with 50-nm spacing between adjacent PSs, as seen in Figure 8.44a. A thin gold layer (20 nm in thickness) was then coated on the etched PSs' surface by plasma sputtering, as previously reported [82]. A graphite flake was used as anode and Au-coated PS monolayer as cathode. The electrodeposition was conducted in the aqueous solution, containing 1 g/L $AgNO_3$, 5 g/L polyvinylpyrrolidone (PVP), and 1 g/L sodium citrate, under constant current mode with a low current density (5 $\mu A/cm^2$), for about 10 hours at room temperature.

FIGURE 8.44 Typical FESEM images of polystyrene monolayer colloidal crystal after plasma etching (a) and the as-prepared sample (b). The insets correspond to the local magnified images. (Liu et al., *J. Mater. Chem.*, 22, 3177–84, 2012. Reproduced by permission of The Royal Society of Chemistry.)

8.3.3.3.2 Morphology and Structure

After electrochemical deposition at 5 μA/cm^2 for 10 hours and removal of PS, the final products were confirmed by XRD to be of the fcc structure of silver with space group *Fm3m* (JCPDS). The intensity ratio of diffraction peaks in XRD spectrum {111} to {200} is much higher than that of bulk (about 2) [80]. This indicates that the as-prepared sample was abundant in the {111} planes.

Figure 8.44b shows the corresponding morphology. Clearly, the micro-sized spheres are arranged in an array with hexagonal packed pattern with very rough surface. Local magnification reveals that the rough surface is built of vertically standing and cross-linking nanoplates (inset of Figure 8.44b). The nanoplates are several hundreds of nanometers in planar dimension and about 30 nm in thickness. It is 6.4×10^9 nanoplates/cm^2, estimated from the FESEM image. Interestingly, almost all nanoplates are standing on the surface of microspheres, forming cross-linking structure and hence high number density of "nanogaps." The size of these gaps depends on the number density of the standing nanoplates, which can be controlled by citrate concentration in the electrolyte. For the PS colloidal monolayer with 2-μm PS sphere diameter, similar morphology and structure could be obtained with almost the same number density of nanoplates [80].

Correspondingly, TEM examination reveals that all microspheres are hollow in structure (Figure 8.45a). The nanoplate is irregular in profile and rough on surface with many grooves, showing columnar or directional crystals, as typically shown in Figure 8.45b. As for the few black dots on the surface of Ag nanoplates, they should be attributed to the Ag nanoparticles attached on the nanoplates [83]. Figure 8.45c shows the SAED pattern, by directing the electron beam perpendicular to the planar surface of an individual nanoplate. The inner set of spots in the SAED pattern originates from the $\frac{1}{3}\{422\}$ planes normally forbidden by an fcc lattice. Such $\frac{1}{3}\{422\}$ forbidden reflections observed on the platelike structures of silver or gold could be attributed to {111} stacking faults parallel to the {111} surface and extending across the entire nanoplates, as previously reported [84–86]. The second spots correspond to Ag {220} Bragg reflections, indicating that the Ag nanoplates are single crystal in structure and the planar surface of nanoplates is parallel to {111}. A high-resolution

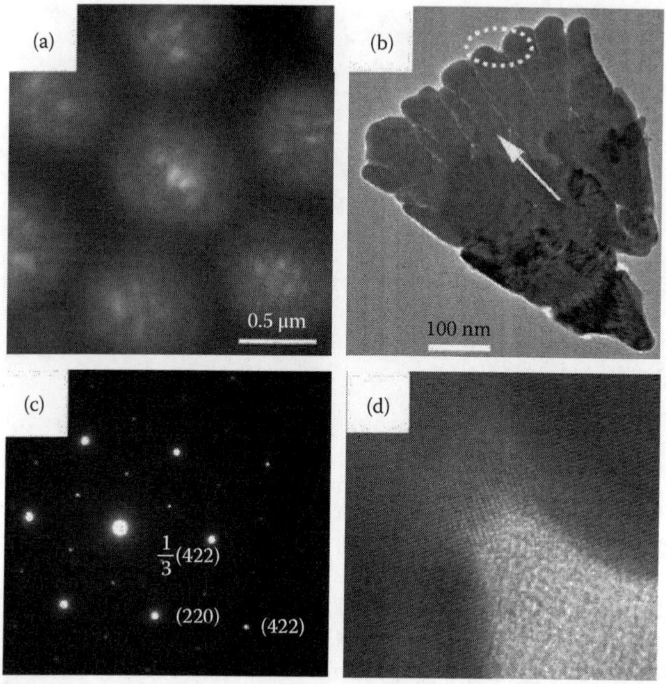

FIGURE 8.45 Microstructural examination of the products shown in Figure 8.44b. (a) TEM image of the sample; (b) morphology of a single nanoplate. The arrow corresponds to the growth direction of the nanoplate. (c) The electron diffraction pattern of the Ag nanoplate shown in (b); (d) high-resolution TEM image corresponding to the area marked in (b). (Liu et al., *J. Mater. Chem.*, 22, 3177–84, 2012. Reproduced by permission of The Royal Society of Chemistry.)

TEM observation of the area across a groove (marked in Figure 8.45b) confirms coherent lattice fringes across two adjacent columnar crystals and further reveals the single crystalline structure of the nanoplates, as demonstrated in Figure 8.45d.

In addition, morphological evolution of the Ag nanoplate–built hollow microsphere arrays with deposition time was examined. During the initial stage of the deposition (e.g., within 1 hour), only precoated Au seeds are observed on the surface of microspheres. No obvious deposition products are found. When the deposition time was up to 3 hours, the platelike products were formed, nearly vertically standing on the spheres' surface, as shown in Figure 8.46a. Further deposition leads to more and ever-growing standing nanoplates (see Figures 8.46b and 8.44b).

8.3.3.3.3 Influencing Factors

Further experiments have revealed that the low deposition current density is crucial to formation of the standing nanoplate-built microspheres. Presence of sodium citrate and surfactant PVP is also important for the morphologies, sizes, and coverage density of the final products.

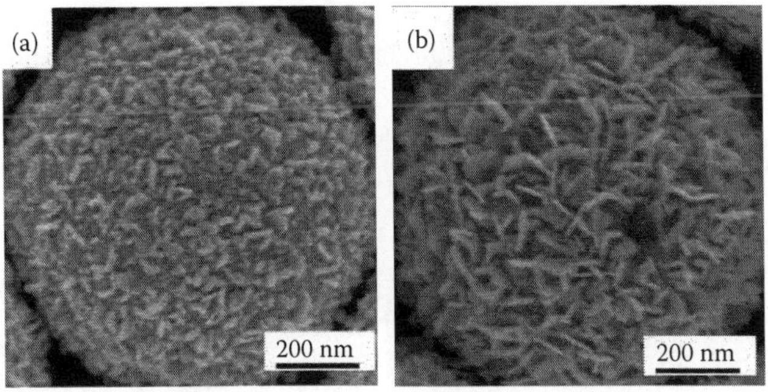

FIGURE 8.46 FESEM images of a single microsphere corresponding to deposition time for (a) 3 hours and (b) 5 hours. (Note: Other conditions are the same as the sample shown in Figure 8.44b). (Liu et al., *J. Mater. Chem.*, 22, 3177–84, 2012. Reproduced by permission of The Royal Society of Chemistry.)

1. Deposition current density: Increasing the deposition current would result in formation of thicker plates but with smaller planar dimension. If deposition was carried out at a current density up to 50μA/cm^2 for the same time (10 hours), a mixture of particles and nanoplates was formed on the microspheres' surface (see Figure 8.47a). Only few are nanoplates similar, in size, to those shown in Figure 8.44b. Most are irregular and thicker (up to 50 nm in thickness) but smaller platelike particles (<300 nm in planar dimension), compared with those shown in Figure 8.44b. Furthermore, when the current density was increased to a very high value (e.g., 100 μA/cm^2), all products are standing rough platelike particles, with 70 nm in thickness, small planar dimension (<250 nm), and slightly higher number density (about 8×10^9/nanoplates/cm^2) (Figure 8.47b).
2. Citrate content in electrolyte: The influence of citrate concentration was inspected keeping other conditions unchanged. Without citrate in the electrolyte, there were only sparse nanoplates standing on the surface of microspheres inhomogeneously with planar size of about 300 nm (see Figure 8.47c). When the citrate was increased to 0.5 g/L, more nanoplates were formed, uniformly standing on the microspheres' surface with number density of about 4.3×10^9 nanoplates/cm^2 (see Figure 8.47d). Further increasing citrate led to higher number density of nanoplates, as shown in Figure 8.44b corresponding to 1 g/L in concentration. So, it can be concluded that citrate concentration determines the number density of nanoplates on microspheres and the formation process. More citrate concentration can result in higher density.
3. PVP content: Furthermore, the amount of PVP in the electrolyte is also crucial to formation of nanoplates. With too less or without PVP but keeping the other conditions unchanged, thicker nanoplates with smaller planar dimension and higher number density, compared with the sample shown in Figure 8.44b, were formed, as shown in Figure 8.47e. The

more PVP content gives rise to formation of the nanoplates with high quality, but increases planar dimension of the nanoplates and decreases the number density, as illustrated in Figures 8.44b and 8.47f (2.5×10^9 nanoplates/cm^2) corresponding to PVP concentrations 5 and 10 g/L, respectively.

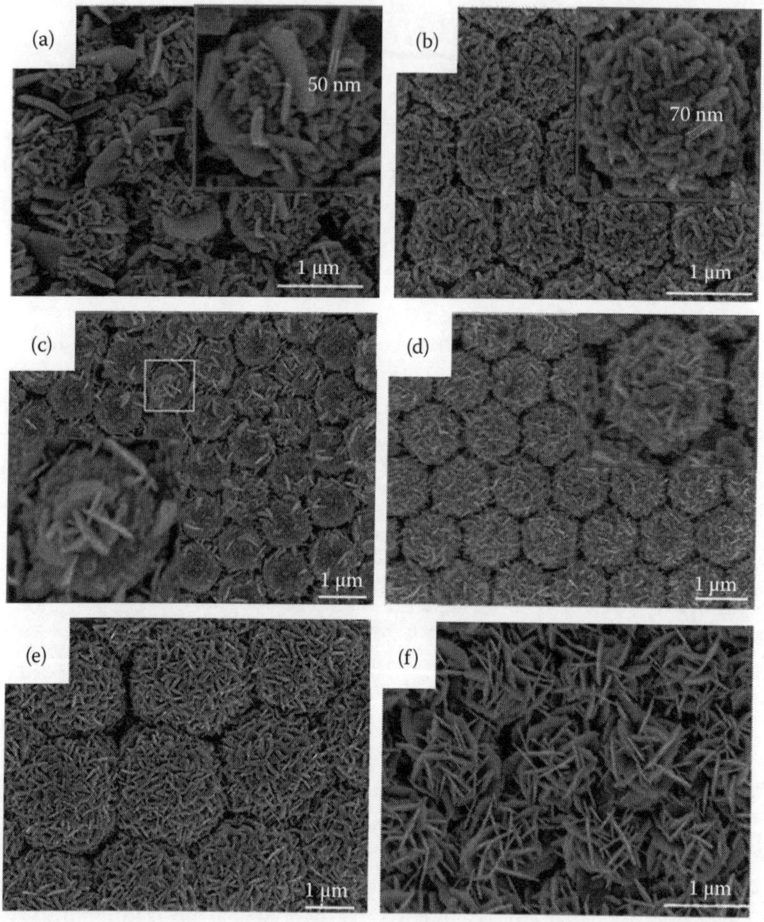

FIGURE 8.47 Influence of deposition conditions and electrolyte composition on the products' morphology. (a) and (b) FESEM images of the products prepared under the deposition current densities 50 and 100 μA/cm^2, respectively; the insets are the local magnified images. (c) and (d) FESEM images of the samples obtained without and with 0.5 g/L trisodium citrate, respectively. The inset in (c) is the magnified image corresponding to the area marked in the smaller frame; the inset in (d) is the local magnified image. (e) and (f) FESEM images of the samples obtained in the electrolyte solution without PVP and with PVP of 10 g/L, respectively. (Liu et al., *J. Mater. Chem.*, 22, 3177–84, 2012. Reproduced by permission of The Royal Society of Chemistry.)

8.3.3.3.4 Nanoplates' Growth and Oriented Connection

Under electric field, Ag^+ ions in the electrolyte move to the surface of cathode and the reduction reaction

$$Ag^+ + e \rightarrow Ag^0 \tag{8.3}$$

takes place on the substrate. On the other hand, citrate in the electrolytic solution is a weak reducing agent, which can reduce Ag^+ to Ag^0 at a mild rate [87,88] and form Ag colloids in solution [80]. So, the formation of Ag nanoplates originates from normal electrodeposition and the Ag colloidal electrophoretic deposition.

1. Normal electrodeposition-induced nanoplates' growth: In the initial stage (e.g., within 3 hours), normal electrodeposition is dominant. The reduced Ag^0 atoms on the cathode's surface nucleate on the random-oriented Au seeds–coated PSs because of low nucleation energy. Ag nucleation could preferentially take place on some Au seeds with surface of {110} plane (or {110}-oriented Au seeds) because of high surface energy [$\gamma_{(110)} > \gamma_{(100)} > \gamma_{(111)}$] [89,90]. Orientation of the formed Ag nuclei can have the relation Ag {110}//Au {110} due to their very close lattice constants, as seen in step (i) of Figure 8.48a. So, the Ag nuclei, formed on the {110}-oriented Au seeds, grow with the terminal planes {111}, based on a quasi-equilibrium growth mode due to the low current density [91], forming nanoplates (see step (ii) of Figure 8.48a). Obviously, when a nanoplate is oriented in out-PS sphere's surface, and especially, perpendicular to the surface, it would grow up more easily. Otherwise, if its orientation is in-PS sphere's surface, it cannot grow sufficiently because of space limitation. Also, if the nanoplates are very dense, they will jostle each other and form cross-linking structure nearly vertically standing on the PS sphere's surface.
2. Electrophoretic deposition (EPD)–induced oriented connection of Ag nanoparticles: By increasing the deposition time (e.g., >3 hours), enough Ag colloids are formed in the solution due to the reduction role of citrate. The formed Ag colloids are stable in the electrolyte solution since the citrate is also a stabilizer and can prevent Ag colloids from aggregation [76]. So, the electrophoretic deposition of the Ag colloids should take place under an external electric field because of their charged property [92], and would be dominant in addition to the normal electrodeposition. For Ag colloidal electrophoretic deposition under a low current density, oriented connection growth occurs, that is, two conjoined single-crystal particles merge into one larger single-crystal particle by rotating or fine-tuning to the same orientation so that the interface disappears and the free energy decreases, as previously reported [92–98]. In this case, when Ag colloids deposit or attach on the ends of preformed nanoplates (see step (i) in Figure 8.48b), there should be a longer time period to complete such oriented connection, due to the low current

density, leading to the nanoplates' growth (see step (ii) in Figure 8.48b). The grooves on the nanoplates (Figure 8.45b) originate from such connection of Ag nanoparticles but both sides of the grooves are of the same orientation (see Figure 8.45d). Meanwhile, the normal Ag^0 atomic deposition will preferentially take place on the grooves due to the energy (see step (ii) in Figure 8.48b). Thus, such grooves tend to disappear after longer time period (see step (iii) in Figure 8.48b).

As for the influence factors, when the current density is high, the electrophoretic deposition rate is high and hence there is no enough time to finish orientation modulation, which is nonbeneficial to formation of high-quality nanoplates, leading to smaller and thicker platelike particles (see Figure 8.47a and b). For additives such as citrate and PVP, the former acts as a stabilizer, which prevents the colloidal Ag, formed in the solution, from aggregation, as previously reported [76], and the latter adsorbs on the planar surface {111} of the standing nanoplates [98], which is why additives are favorable for formation of high-quality Ag nanoplates.

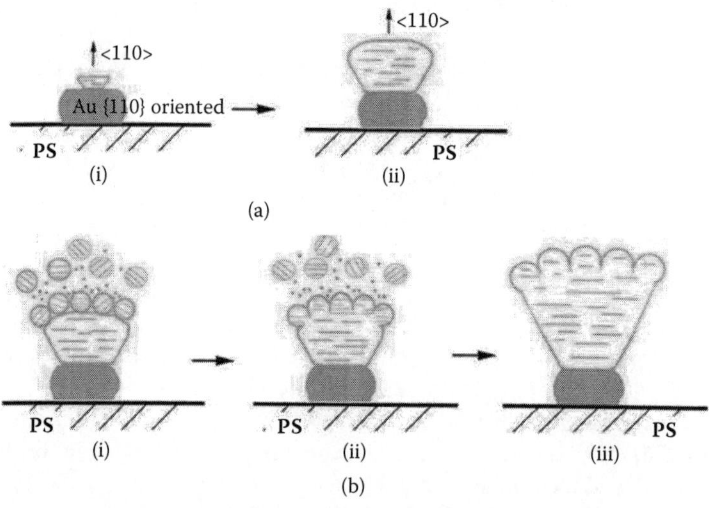

FIGURE 8.48 Schematic illustration of Ag nanoplates' growth and oriented connection. (a) The preferential nucleation and oriented growth for a standing Ag nanoplate during initial electrodeposition. (i) A Ag nucleus is preferentially formed on a {110}-oriented Au seed on polystyrene (PS); (ii) oriented growth of the nucleus along the fastest <110> within {111} plane under a low deposition current density. (b) Subsequent electrophoretic deposition–induced oriented connection of Ag nanoplates. (i) Ag colloidal attachment on the end of a preformed standing nanoplate. (ii) Self-orientation modulation and merging into a single crystal, together with preferential attachment of Ag^0 atoms on the grooves. (iii) Formation of Ag nanoplate with irregular profile and grooves on the planar surface. (Liu et al., *J. Mater. Chem.*, 22, 3177–84, 2012. Reproduced by permission of The Royal Society of Chemistry.)

8.4 MICRO/NANOSTRUCTURED ORDERED PORE ARRAYS BASED ON OTHER ROUTES

In addition to the solution routes and electrodeposition strategies mentioned in the preceding discussion, there are many other techniques for fabrication of micro/nanostructured ordered porous arrays or films. In this section, we introduce few ordered porous films fabricated by electrophoretic deposition and gas-phase route based on PS template.

8.4.1 Micro/Nanostructured Porous Array Based on Electrophoretic Deposition

EPD is another very promising technique for the fabrication of the porous films on conductive substrates (such as ITO glass substrate and silicon wafer) because of the low cost, fairly rapid and simple equipment, and formation of uniform layers with controlled thickness and homogeneous microstructure. Combining the PS colloidal monolayers with electrophoretic deposition strategy, various micro/nanostructured ordered porous films can be fabricated.

8.4.1.1 Strategy

In EPD, charged colloidal particles in a stable suspension are deposited onto an oppositely charged substrate by application of a DC electric field [99–101]. If the particles are positively charged, deposition occurs on the cathode. Otherwise, deposition occurs on the anode. This method consists of two experimental processes: the movement of charged particles in a suspension in an electric field between two electrodes (electrophoresis) and particle deposition on one of the electrodes or on a membrane (electrocoagulation). Additionally, it should be mentioned that well-dispersed and stable suspensions are very important for this strategy.

Figure 8.49 demonstrates the corresponding EPD strategy based on PS colloidal monolayer. Initially, the monolayer PS colloidal crystal template, with close-packed arrangement or nonclose-packed arrangement induced by etching, on a conducting substrate is used as an electrode Figure 8.49a). The electrophoresis is then performed in a colloidal solution. Due to the charged property of the colloids in the solution, the particles deposit on the surface of the PS spheres-covered substrate under a certain electric filed (Figure 8.49b). Finally, the hollow sphere array with porous shell structure can be obtained after removal of the PS by burning or dissolution (Figure 8.49c). Obviously, this strategy has advantages: the morphology and porous shell structure of the hollow spheres can be easily controlled by the electrophoresis parameters or the shape and size of the colloids in the solution. Using EPD, we can obtain various ordered porous array films. In the following sections, we mainly discuss the fabrication of Si (a semiconductor) hollow sphere arrays with controlled structure to demonstrate the above-mentioned advantages.

8.4.1.2 Fabrication of Colloidal Solutions by Laser Ablation in Liquids

There are many methods to prepare colloidal solutions. Here, we discuss preparation by laser ablation of solid target in liquids. The colloidal solution with different particle sizes can be obtained by this method [102,103]. The semiconductor Si is used

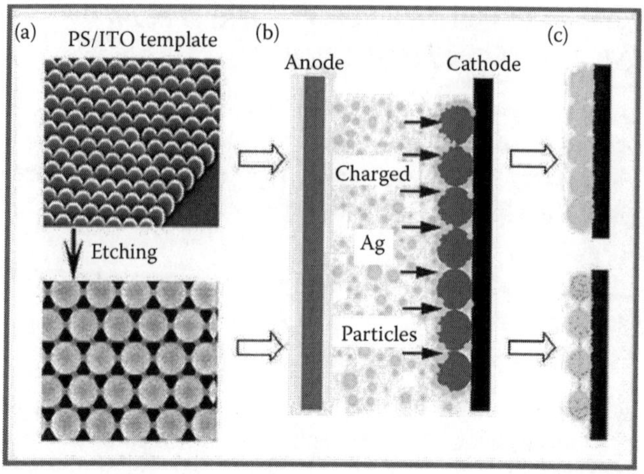

FIGURE 8.49 Schematic illustration for electrophoretic deposition strategy. (a): Polystyrene (PS) monolayer on indium tin oxide (ITO). (b) Electrophoresis of the colloids in colloidal solution. (c) Nanoparticles on PS spheres, which induce hollow sphere array after removal of PS. (Yang et al., *Adv. Funct. Mater.*, 20, 2527–33, 2010. Reproduced by permission of The Royal Society of Chemistry.)

as the typical example. Briefly, for Si colloidal solution, a N-type crystalline silicon wafer of (111) plane is used as the target immersed in ethanol solution and irradiated for 20 minutes by the first harmonic (1064 nm) of Nd:YAG laser, operated at 100 mJ/pulse with a pulse duration of 10 ns and frequency of 10 Hz. It is similar to the preparation of silver colloidal solution [102]. Figure 8.50 shows the as-prepared Si nanoparticles with about 6 nm in size by using the strategy of laser ablation in liquid [103].

8.4.1.3 Morphology and Structure

Using this Si colloidal solution, we obtain the Si hollow sphere arrays based on the strategy given in Figure 8.49. Figure 8.51 illustrates the hollow sphere array after electrophoretic deposition for 10 hours at a deposition current of 10 $\mu A/cm^2$ in fresh Si colloidal ethanol solution. Here, the diameter of PS spheres is 1000 nm. Clearly, the hollow spherical structure is confirmed from the cross-sectional image in the inset in Figure 8.51a. In the locally magnified image (Figure 8.51b), one can observe that the shell layer is porous and consists of Si nanoparticles with about 30 nm in mean size. The bigger Si nanoparticles than the Si colloids in the solution may result from coagulation of the nanoparticles in the solution during electrophoresis. Also, the thickness of the sphere shell layer is comparable to the size of such coagulated nanoparticles. The Si hollow spheres are well adhered to the ITO and do not fall off unless scalped by blade.

After electrophoresis for different time periods, the morphology evolution of these Si hollow sphere array can be observed. After electrophoresis for 2 hours, many relatively small aggregates below 50 nm in size are dispersedly formed on the surface of PS spheres. The coverage area is relatively low, as shown in Figure 8.52a. Electrophoresis for 4 hours leads to an increase of the coverage area (Figure 8.52b). Further electrophoresis induces nearly complete coverage of Si nanoparticles on the PS spheres (see Figure 8.51a).

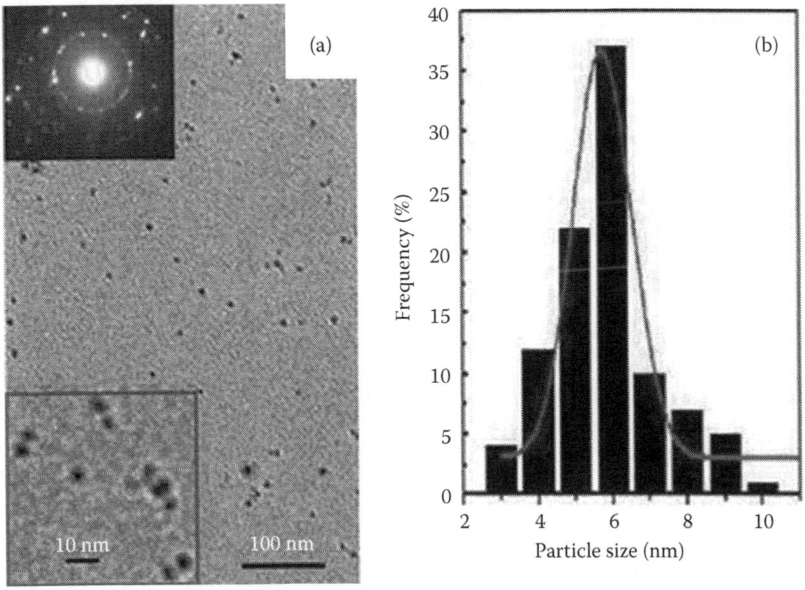

FIGURE 8.50 Morphology and structure of Si colloids induced by laser ablation in ethanol. (a) TEM image of the Si colloids; insets are selected-area electron diffraction (up-left) and local magnification (down-left). (b) Particle size distribution corresponding to (a) (line: fitting result). (Reprinted with permission from Yang et al., 2009, 8287–91. Copyright 2009 American Chemical Society.)

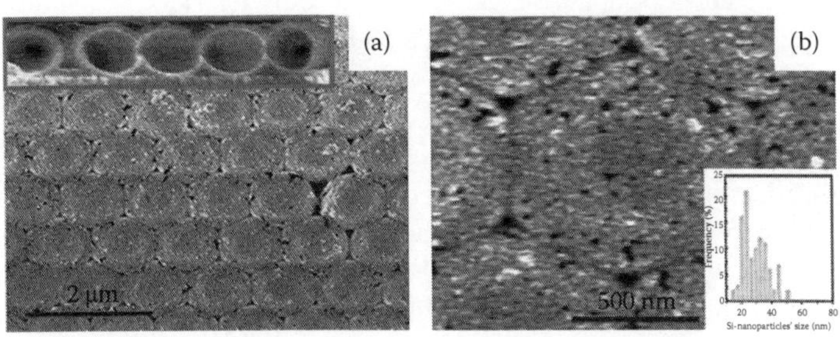

FIGURE 8.51 Morphology of Si hollow sphere array after electrophoretic deposition for 10 hours at 10 μA/cm^2. (a) FESEM image of Si hollow sphere array after removal of polystyrene spheres. The inset: Corresponding local cross-section image. (b) Local magnified image of (a). The inset: Size distribution of Si nanoparticles in the shell layer. (Reprinted with permission from Yang et al., 2009, 8287–91. Copyright 2009 American Chemical Society.)

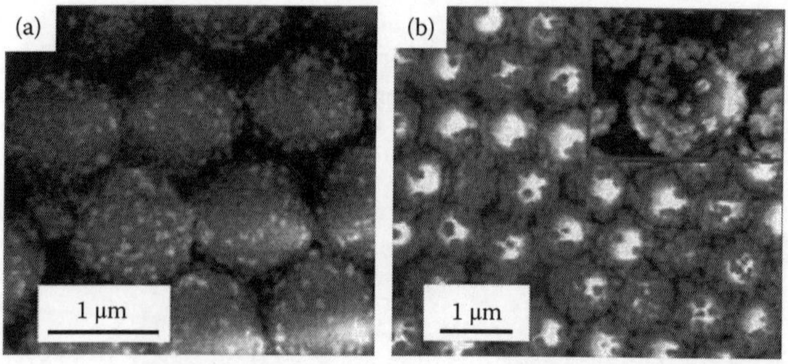

FIGURE 8.52 FESEM images after electrophoretic deposition on the polystyrene (PS)/indium tin oxide at 10 μA/cm^2 for (a) 2 hours and (b) 4 hours (without removal of PSs). The inset in (b) is a locally magnified image of the Si nanoparticles on PS spheres. (Reprinted with permission from Yang et al., 2009, 8287–91. Copyright 2009 American Chemical Society.)

8.4.1.4 Influence of Current and Substrate

Furthermore, deposition current is a very important parameter for the formation of hollow spheres. When deposition current is changed, the morphology is also adjusted. When the current is too low (e.g., 2 μA/cm^2) during electrophoresis, hollow spheres are not formed on the substrate, as shown in Figure 8.53a. Si nanoparticles only fill in the interstitial areas between PS spheres and substrate, resulting in the netlike pore structure from top view after removal of the PS spheres. If current is increased (e.g., 10 μA/cm^2 or higher), the hollow sphere array is obtained. Figure 8.53b shows the morphology of the hollow sphere array on ITO after electrophoresis at 50 μA/cm^2, which is similar to that shown in Figure 8.51. The Si nanoparticles in the shell layer become much bigger (about 100 nm in mean size). In addition, the substrate used is also crucial to the formation of hollow sphere. If substrate PS/Si wafer is used instead of PS/ITO, Si hollow sphere arrays cannot be formed at 10 μA/cm^2. The morphology is netlike pore structure from top view instead of hollow sphere array (see Figure 8.53c). Even electrophoresis at 50 μA/cm^2 only leads to formation of partial Si hollow spheres, as illustrated in Figure 8.53d (rectangular marks).

8.4.1.5 Formation of Hollow Spheres

The formation of Si hollow sphere array on PS/ITO is easily understood. Since the PS spheres are weakly negatively charged, the positively charged Si colloids and coagulated nanoparticles preferentially deposit on the PS sphere's surfaces (see Figure 8.52), leading to the comparatively homogeneous thickness of the shell layer and formation of hollow sphere array after removal of PS. However, low current density (or low deposition rate) may induce nearly equilibrium deposition and preferential filling in the interstitial within the PS monolayer, forming the netlike pore array (Figure 8.53a). For PS/Si wafer, due to homogeneity of Si nanoparticles and Si wafer, Si colloids preferentially deposit on the Si substrate within the interstitials at a low current density. Only when the deposition current is large enough, the Si nanoparticles deposit both on the PS sphere's surfaces and on the interstitials due

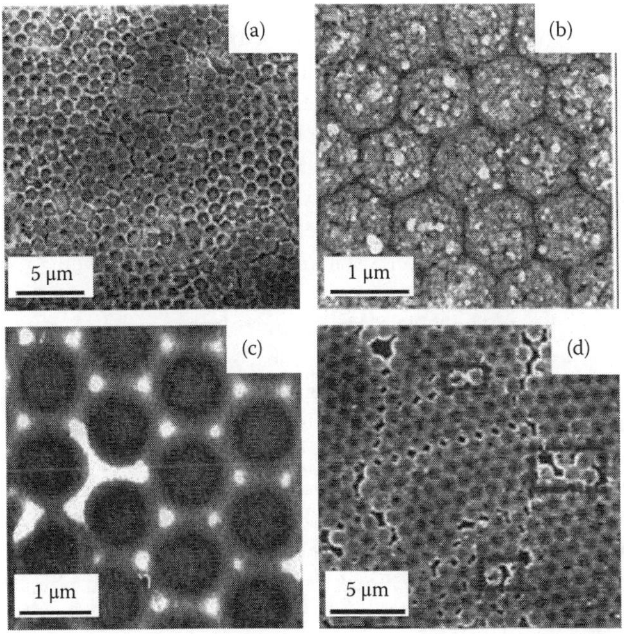

FIGURE 8.53 Influence of the current and substrate on morphologies of the samples after electrophoretic deposition for 10 hours and removal of polystyrene (PS). (a) and (b) FESEM images for the samples after deposition on the PS/ITO at 2 and 50 μA/cm^2, respectively. (c) and (d) FESEM images after deposition on the PS/Si wafer at 10 and 50 μA/cm^2, respectively. (Reprinted with permission from Yang et al., 2009, 8287–91. Copyright 2009 American Chemical Society.)

to the high deposition rate, leading to formation of partial Si hollow spheres. As for the size dependence of Si nanoparticles in the hollow sphere shells on the deposition current, it is associated with the deposition rate. The larger current corresponds to higher deposition rate or faster movement of Si-colloids, leading to easier colloidal aggregation and hence formation of bigger Si nanoparticles in the shells.

Obviously, Si hollow sphere arrays with different periods can also be fabricated by changing the PS size in the template or using nonclose-packed colloidal crystal as the template. Additionally, the size of the pores in the shell layer can be conveniently tuned by electrophoresis parameters and the morphology of the colloidal nanoparticles.

8.4.1.6 Universality of Strategy

Furthermore, for other charged nanoparticles including Ag, ZnO, TiO_2, WO_3, Fe_2O_3, and so on [102,105,106], they also can be fabricated as ordered porous films by the strategy shown in Figure 8.49 and even nonclose-packed and hybrid (by multistep deposition) ordered micro/nanostructured films. Here, we only introduce the ordered porous films from charged Ag nanopaticles based on electrophoretic deposition strategy shown in Figure 8.49.

The electrophoretic deposition was carried out in Ag colloidal solution, which contains nearly spherical Ag particles with mean size of 10 nm obtained by laser ablation of an Ag flake in water. Due to the charged surfaces of Ag nanoparticles in the colloidal solution, Ag nanoparticles were electrophoretically deposited on the conductive electrode (i.e., PS sphere template), forming Ag nanoshells on PS spheres (Figure 8.49b and c). Nonclose-packed PS colloidal monolayer arrays can be synthesized with cross-linked bars between the neighboring PS spheres by using a plasma etching process on as-prepared close-packed PS sphere arrays. The spacing between the nonclose-packed PS spheres is determined by etching time, and the bars can be removed by a subsequent thermal treatment process. Both the close-packed and nonclose-packed PS colloidal monolayers can be used as templates for the electrophoretic growth of Ag nanoshells (Figure 8.49c). The PS spheres under the Ag nanoshells can be further removed by a chemical process, producing ordered arrays of hollow sphere arrays. A typical Ag nanoshell array on a Si wafer is shown in Figure 8.54a, in which highly roughened surfaces of nanoshells are demonstrated. The nanoshells are composed of many Ag nanoparticles, and crevices with a very high density are formed between the nanoparticles (Figure 8.54a and inset). XRD measurement and TEM observation reveal that the nanoshells are fcc crystalline structures of Ag (Figure 8.54b). Here, it should be mentioned that preevaporated Ag thin film on the surfaces of PS spheres is of great importance for the growth of Ag nanoshells. Without the thin Ag film, Ag particles will only deposit in the substrate area where PS spheres are situated and there no nanoshells will be formed on PS spheres.

The diameter of Ag hollow spheres is mainly determined by the diameter of PS spheres. Since, generally, the diameter of PS spheres can be changed from about 200 nm to 4.5 μm, the diameter of Ag hollow spheres can be adjusted in a similar

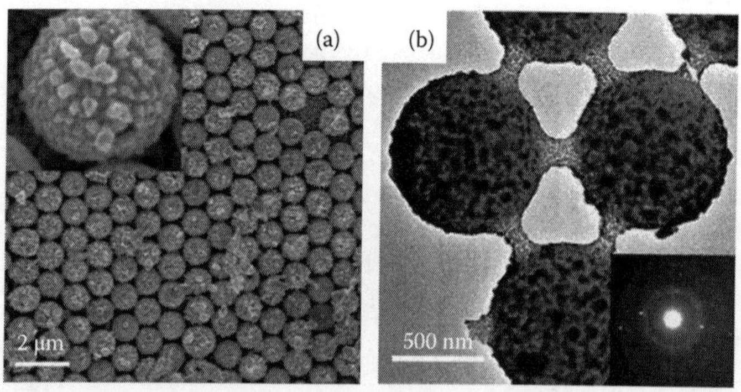

FIGURE 8.54 (a) Field emission scanning electron microscopy image with top view of Ag nanoshell arrays prepared on 1-μm PS spheres under 100 μA/cm^2 electrophoretic current density; the inset is an enlarged view of a nanoshell. (b) Transmission electron microscopy results of Ag nanoshells; the inset is the selected-area electron diffraction pattern that confirms the fcc structure of the Ag nanoshells. (Yang et al., *Adv. Funct. Mater.* 2010. 20. 2527–33. Copyright Wiley-VCH Verlag GmbH & Co. KGaA. Reproduced with permission.)

size range. Ag nanoshells with diameters of about 1000 and 750 nm are shown in Figures 8.54 and 8.55, respectively. Besides the diameters, the spacing between the neighboring nanoshells can also be adjusted. Figure 8.55 reflects nonclose-packed Ag hollow sphere arrays with different intershell spacings. The spacing of the hollow spheres increases with increase of plasma etching time. Correspondingly, the diameter of the Ag hollow spheres decreases with the etching time, from about 750 nm in the beginning to 730, 715, 685, 660, and 640 nm. Therefore, plasma etching time can be used to modify both the intershell spacing and the diameter of the Ag nanoshells.

Interestingly, the size of the Ag nanoparticles in nanoshells can be adjusted by controlling the electrophoretic current density. Figure 8.56a through c shows nonclose-packed nanoshells prepared under current densities of 10, 50, and 100 μA/cm², respectively, indicating that the size of nanoparticles increases with the current density. The size distribution as a function of the electrophoretic current density is shown in Figure 8.56d. The mean size of nanoparticles is about 25, 40, and 60 nm when the current density is 10, 50, and 100 μA/cm², respectively. Therefore, the structural parameters of the Ag hollow sphere arrays can be controlled triply. First, the diameter of the Ag nanoshells can be tuned by changing the diameter of PS spheres. Moreover, the nanoshell diameter can be modified by the plasma etching time of PS templates. Second, the surface roughness of the Ag nanoshells is controllable by selecting appropriate electrophoretic current densities. Third, the

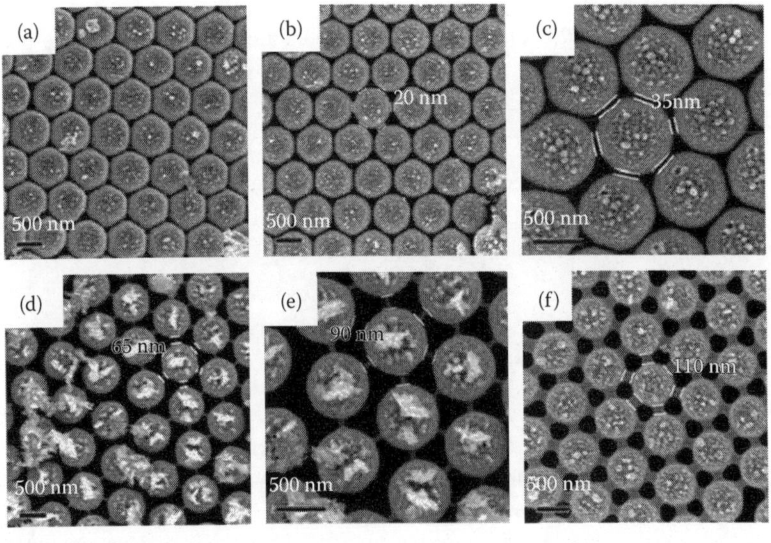

FIGURE 8.55 FESEM images (top view) of a set of six samples showing the spacing control of Ag nanoshells based on the adjustment of the plasma etching time of polystyrene (PS) sphere templates (diameter of PS spheres is about 750 nm before etching). The electrophoretic current and time for fabricating the Ag nanoshells are 50 μA/cm² and 9 hours, respectively. (Yang et al., *Adv. Funct. Mater.* 2010. 20. 2527–33. Copyright Wiley-VCH Verlag GmbH & Co. KGaA. Reproduced with permission.)

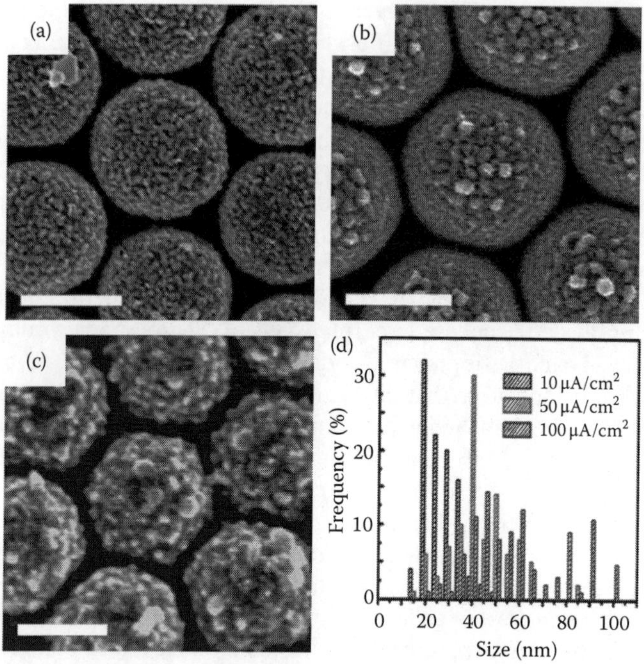

FIGURE 8.56 FESEM images of Ag nanoshell arrays synthesized under different electrophoretic deposition current densities. (a) 10 μA/cm^2, (b) 50 μA/cm^2, (c) 100 μA/cm^2. Scale bars: 500 nm. (d) The size distribution of the Ag nanoparticles in nanoshells. The mean size of nanoparticles is about 25, 40, and 60 nm corresponding to (a), (b), and (c), respectively. (Yang et al., *Adv. Funct. Mater.* 2010. 20. 2527–33. Copyright Wiley-VCH Verlag GmbH & Co. KGaA. Reproduced with permission.)

spacing between the neighboring nanoshells can be controlled by choosing different PS spheres with different intersphere spacings (a range between 20 and 110 nm of the interspacing is shown in Figure 8.55).

8.4.2 Noble Metal 3D Micro/Nanostructured Films Based on Electrophoretic Deposition

Noble metal (Au, Ag, Pt, etc.) 3D micro/nanostructured films have received considerable attention because of their outstanding properties and potential applications in catalysis [107], superhydrophobicity [108], lithium ion batteries [109], SERS sensors [110], and so forth. In fabrication process, it is very important for practical applications with respect to tuning, morphology, and the size of micro/nanostructured unit. The promising ways have been developed to build 3D micro/nanostructured films by the micro/nanostructured porous blocks with tunable structures (such as tubes, spheres, and nanorods). For instance, Au/ZnO micro/nanorod arrays were prepared by electrophoresis in the Au colloidal solution [111], Au nanochain–built 3D netlike porous films were fabricated by electrophoretic deposition [112], the silver porous nanotube–built 3D films with structural tunability were synthesized by the nanofiber template–plasma etching strategy [113], and micro/nanostructured Ag nanoparticle/PS sphere arrays were obtained by eletrophoretic deposition [104]. Such 3D films are controllable and tunable in structure and hence of practical performance.

8.4.2.1 Au/ZnO Micro/Nanorod Arrays by Electrophoresis in the Au Colloidal Solution

A simple and green route has been developed to decorate ZnO nanorod array with Au nanoparticles by EPD in the Au colloidal solution [111]. High-quality arrays of ZnO nanorods with different number densities and diameters were first prepared using a chemical vapor deposition (CVD) method, and the gold colloidal solutions were obtained by laser ablation of a gold target in water, which can produce the Au nanoparticles with fresh surface. The size and monodispersivity of Au spherical nanoparticles in the colloidal solution were controlled by subsequent further laser irradiation. It has been shown that the coating of Au nanoparticles on the ZnO nanorods by EPD is homogeneous and similar in characteristic structure to the conventional 2D nanoparticle films produced by EPD. The Au nanoparticles homogeneously and strongly cover the ZnO nanorods' surfaces.

8.4.2.1.1 Preparation of ZnO Nanorod Arrays

The ZnO nanorod arrays were first prepared on smooth Si wafer by CVD in a horizontal tube furnace. The diameter and length of the nanorods in the arrays were controlled by Zn vapor partial pressure. The higher Zn vapor partial pressure corresponded to thicker and longer ZnO nanorods under the same experimental conditions. ZnO nanoparticle film was introduced as substrate to increase the number density of ZnO nanorods in the array because its surface has dense pits that could offer more nucleation sites than smooth Si wafer, as shown in Figure 8.57. Such ZnO buffer layer is prepared by thermal oxidation of ion beam–sputtered Zn film.

8.4.2.1.2 Preparation of Au Colloidal Solutions

Au colloidal solution is then prepared by laser ablation of an Au target in deionized water. The metal plate is fixed on a bracket in a glass vessel filled with 20 mL of water. Then it is irradiated for 20 minutes by the first harmonic of a Nd:YAG pulsed laser with a power of 90 mJ/pulse and a spot size of about 2 mm. The liquid phase is vigorously stirred with a magnetic stirrer during irradiation. The yield is estimated to be about 0.1 μg/min by the laser ablation–induced mass loss of the gold target. Au colloidal solution is thus obtained with Au content of 100 μg/L. The obtained Au colloidal solution, after removal of the gold plate, is subsequently irradiated again for 60 minutes with stirring, for smaller sized and monodispersed Au colloids. Figure 8.58 presents the TEM images of Au nanoparticles formed by laser ablation. After laser ablation of Au target in water for 20 minutes, Au colloidal solution can be obtained. Au nanoparticles in the colloidal solution are nearly spherical with sizes in a wide range of 20–80 nm, exhibiting a big dimensional dispersion (see Figure 8.58a). However, after additional laser irradiation of the preformed colloidal solution for 60 minutes, Au nanoparticles are much smaller in size and have a mean value of 10 nm with near monodispersion.

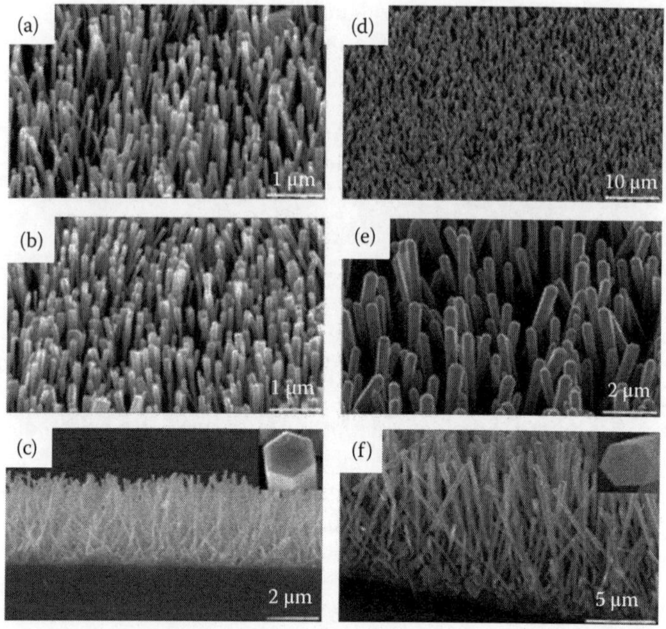

FIGURE 8.57 ZnO nanorod arrays with different number densities and dimension of nanorods. (a) and (b) ZnO nanorods grown on a Si wafer and a ZnO film, respectively. (c) Cross-sectional view of (b); inset: magnified image of an individual nanorod. (d) The array on a Si wafer, with much thicker and longer ZnO nanorods. (e) and (f) A local magnified image and cross-sectional view of (d), respectively. Inset in (f): Magnified image of an individual nanorod. (Reprinted with permission from He et al., 2010, 8925–32. Copyright 2010 American Chemical Society.)

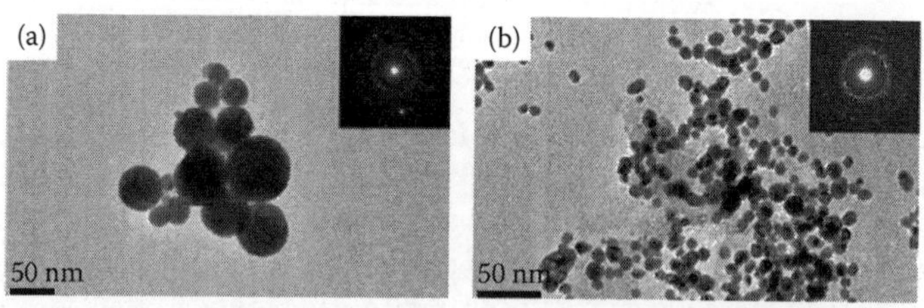

FIGURE 8.58 TEM images of Au nanoparticles in the colloidal solution. (a) For the colloidal solution formed by laser ablation of Au target in water. (b) After additional laser irradiation of the colloidal solution in (a). Insets: Corresponding selected-area electron diffraction patterns. (Reprinted with permission from He et al., 2010, 8925–32. Copyright 2010 American Chemical Society.)

8.4.2.1.3 Electrophoretic Deposition

The EPD cell used is cubic and two electrodes are spaced 2 cm apart. The ZnO nanorod array on Si substrate is used as the electrode and immersed in the 20mL Au colloidal solution. The array is used as the anode. EPD is performed for 1 hour at a DC voltage of 40 V, as illustrated in Figure 8.59.

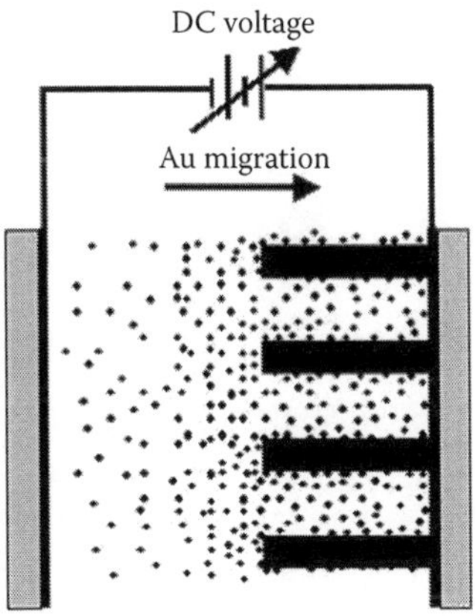

FIGURE 8.59 Schematic illustration of surface decoration strategy of semiconductor nanorod array based on electrophoresis deposition in the Au colloidal solution. (Reprinted with permission from He et al., 2010, 8925–32. Copyright 2010 American Chemical Society.)

8.4.2.1.4 Morphology and Structure

After EPD on the ZnO nanorod arrays, Au nanoparticles were deposited on the nanorods in the arrays. Figure 8.60 shows the morphology of the Au nanoparticle–coated thick ZnO nanorod array shown in Figure 8.57d through f. The morphology of sample is uniform (see Figure 8.60a). It reveals that the whole nanorods are coated with Au nanoparticles homogeneously, as shown in Figure 8.60b and c. Further examination shows that the diameter of the coated nanorods is bigger than that shown in Figure 8.57d through f due to the coating. However, the standing nanorods are still well separated from each other though the spacings between neighboring nanorods decrease. The coating thickness can be controlled by the EPD time and/or concentration of the colloidal solution. Au coating layer on the nanorod's surface is rough or porous at nanoscale due to the packing or aggregation of nanoparticles, as illustrated in Figure 8.60c. These nanosized pores in the coating layer are of nanometers to tens of nanometers in scale. This characteristic structure is similar to the conventional 2D nanoparticle films produced by EPD, as shown in Figure 8.60d, corresponding to the 2D Au nanoparticle film coated on a ZnO substrate prepared under the same EPD conditions. Such Au nanoparticle–decorated nanorod array is beneficial in comparison to SERS performance. Obviously, the EPD technique can be applied to the surface modification of 1D nanostructures using nanoparticles as building blocks for novel electronic and optoelectronic device applications.

The related experiments indicate that electrophoretic potential, the size of Au nanoparticles in the colloidal solution, and the interrod's spacings in the arrays are the key influential factors determining the decoration morphology.

8.4.2.1.5 Formation of Homogeneous and Strong Coating Layer

Before EPD, the Au colloidal solutions are stabilized electrostatically against agglomeration. The formation of the nanoparticle film on the substrate with flat surface (Figure 8.60d) is easily understood. Upon applying voltage, electrophoresis

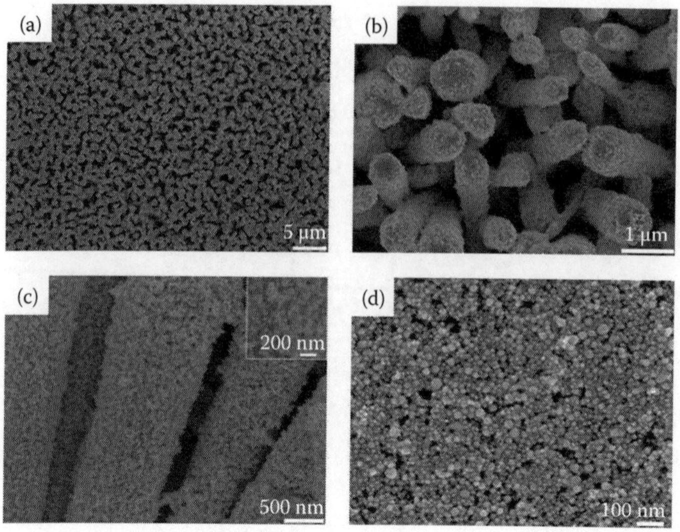

FIGURE 8.60 FESEM images of (a) through (c): Au-coated thick ZnO nanorod array shown in Figure 8.57d at different magnifications. (d) The corresponding two-dimensional Au nanoparticle film coated on a planar ZnO substrate (for comparison). The inset in (c): Local magnification. (Reprinted with permission from He et al., 2010, 8925–32. Copyright 2010 American Chemical Society.)

occurs in the solution. According to Derjaguin–Landau–Verwey–Overbeek (DLVO) theory [114], the negatively charged Au nanoparticles deposit on the anode surface, forming a homogeneous film. For the electrode of semiconductor nanorod array, however, the surface is uneven and 3D structured. The electric field distribution on the electrode surface is uneven. Electric field around the edges or corners of each nanorod, especially at its top part, is stronger than other parts of the nanorod. The concentration of Au colloidal solution is different at various sites near the nanorod surface during EPD due to the different accessibility of the charged nanoparticles to different places. The motion velocity of nanoparticles in the colloidal solution depends on the applied voltage. Under a low electric potential, Au nanoparticles move slowly in the colloidal solution and preferentially deposit on top parts of nanorods due to relatively strong electric field around nanorod tips. Only very few charged nanoparticles can access the lower parts of the channels, leading to a lower concentration of colloidal solution within the interstitials and hence uneven deposition (Figure 8.61a). Contrarily, if the applied potential is too high, the charged nanoparticles in the colloidal solution are fast in motion in response to the electric field. The fast-moving nanoparticles should more easily overcome relatively strong attraction from the nanorod tips and enter the interstitials between the nanorods. However, such moving colloids are easily aggregated together, before reaching the array surface, due to the collision induced by the difference in motion velocity of the particles with different sizes, leading to the formation of uneven film on the array surface (Figure 8.61b). So, compared with the case on a substrate with flat surface, formation of homogeneous nanoparticle coating layer on the nanorod array is difficult. Only

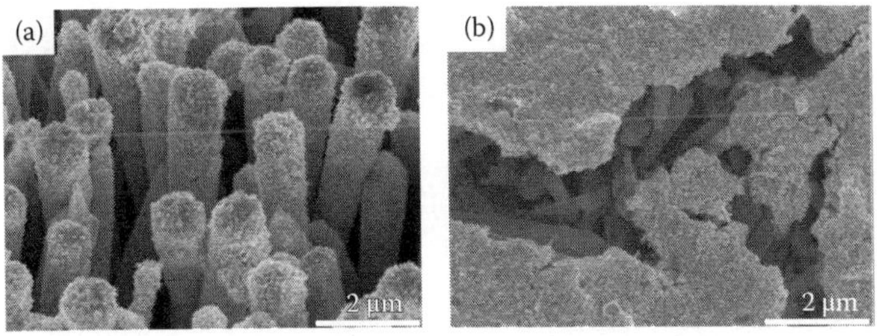

FIGURE 8.61 FESEM images of Au-decorated thick ZnO nanorod arrays after electrophoretic deposition at different potentials for 1 hour: (a) 20 V and (b) 60 V. (Reprinted with permission from He et al., 2010, 8925–32. Copyright 2010 American Chemical Society.)

at a medium or a suitable potential can we obtain the homogeneous decoration on the nanorod arrays. In our case, when a medium potential (e.g., 40 V) is applied, nanoparticles in the colloidal solution move faster to enter the interstitials between the nanorods but is not sufficient to aggregate, inducing an even concentration distribution around the nanorods and hence forming comparatively homogeneous coating (see Figure 8.60). As for the strong interfacial binding between the Au nanoparticles and ZnO nanorods, it could be attributed to the Au colloidal solution prepared by laser ablation in deionized water. Au nanoparticles are formed in water without exposure to air or some additives and retain the fresh surfaces. Obviously, such fresh surfaces are beneficial to the good interfacial connection, leading to strong binding. Furthermore, the surface decoration of ZnO nanorod arrays by electrophoresis has two advantages. First, the entire fabrication process is free of any additives and the fresh surfaces of Au nanoparticles are thus retained, which are not only favorable to the interfacial connection but also advantageous for keeping high surface activity and adsorbing detected molecules. Second, the sizes and even composition of such nanoparticles on the nanorods can be easily controlled by laser treatment for further functional optimization.

8.4.2.2 Au Nanochain–Built 3D Netlike Porous Films by Electrophoretic Deposition

As already mentioned, Au nanoparticle suspension can be prepared by laser ablation of an Au target in deionized water. Very interestingly, Au nanochain colloidal solution can be obtained by laser ablation process for a longer time period [112]. Based on this fact, a route to fabricate 3D structured Au porous films was developed by further electrophoretic deposition in the Au nanochain colloidal solution [112]. The obtained film is built of Au nanochains and homogeneous in macrosize but rough and porous in nanoscale. The fabrication strategy is inexpensive, easy to control, and entirely away from chemical reagents, consequently making them have potential applications in SERS to detect organic molecules.

Figure 8.62 presents the TEM images of colloidal nanoparticles formed by laser ablation in water for different time periods. If the laser ablation lasts only 20 minutes, Au nanoparticles in the colloidal solution are early spherical with sizes in the range

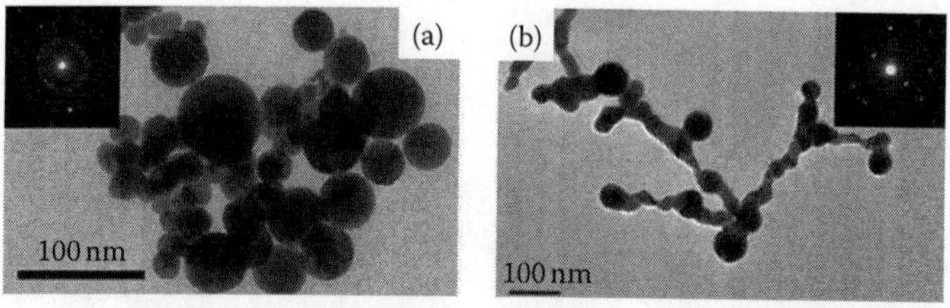

FIGURE 8.62 TEM images of Au nanoparticles in the colloidal solution formed by laser ablation of Au target in water with a Nd:YAG pulsed laser (wavelength 1064 nm, frequency 10 Hz, pulse duration 10 ns, and energy 90 mJ/pulse), for different time periods. (a) 20 minutes and (b) 60 minutes. Insets: Corresponding selected-area electron diffraction patterns. (He et al., *Chem. Commun.*, 46, 7223–5, 2010. Reproduced by permission of The Royal Society of Chemistry.)

of 20–80 nm, exhibiting a big dimensional dispersion (Figure 8.62a). However, when laser ablation lasts 60 minutes, most nanoparticles are well sintered, forming elongated nanochain structures with branches (Figure 8.62b). These nanochains are composed of small Au nanoparticles, having 20–40 nm in width and 100–3000 nm in length by statistics in TEM fields. As for formation of the nanochains during laser ablation in water, there have been some reports recently [115,116]. In our case, Au nanoparticles are first formed in water during initial laser ablation. During subsequent laser ablation, the preformed big nanoparticles are fragmented into smaller ones due to absorption of photon energy. The formed Au nanoparticles keep fresh surfaces in water. When the colloidal concentration is high enough during the laser ablation, the nanoparticles in the solution form chainlike structures due to collision-induced aggregation and laser-induced sintering. Furthermore, it has been shown that the zeta potential of as-prepared colloidal solution is –30 mV (neutral pH). The surface charge of Au colloids is because of the adsorption of dissociated OH^- in water.

After subsequent EPD in the Au nanochain colloidal solution for 3 hours at applied voltage of 30 V, a homogeneous film over a large scale on ITO glass (anode) is obtained (see the photo in the inset of Figure 8.63a). XRD shows that the obtained film is gold with fcc structure, as shown in Figure 8.63a. FESEM images indicate that the film is porous and netlike in structure (Figure 8.63b). The local magnification reveals that the netlike film is 3D and stacked by nanochains (Figure 8.63c). The nanochains are composed of spherical nanoparticles. The pore sizes in the film are from few nanometers to tens of nanometers in scale. The film thickness can be controlled by the EPD time and/or concentration of the colloidal solution. Compared with the original colloidal solution, the sizes of particles in Au nanochains in the prepared film are almost unchanged during the EPD process. TEM examination indicates that the netlike structure is strong enough in spite of the ultrasonic dispersion during TEM sample preparation, as seen in Figure 8.63d. Hence, the EPD technique can be applied to the synthesis of netlike porous films using nanochains as building blocks for novel electronic and optoelectronic device applications.

The formation of 3D netlike Au porous film has been investigated depending on the applied potential and the Au colloids used. The applied potential determines

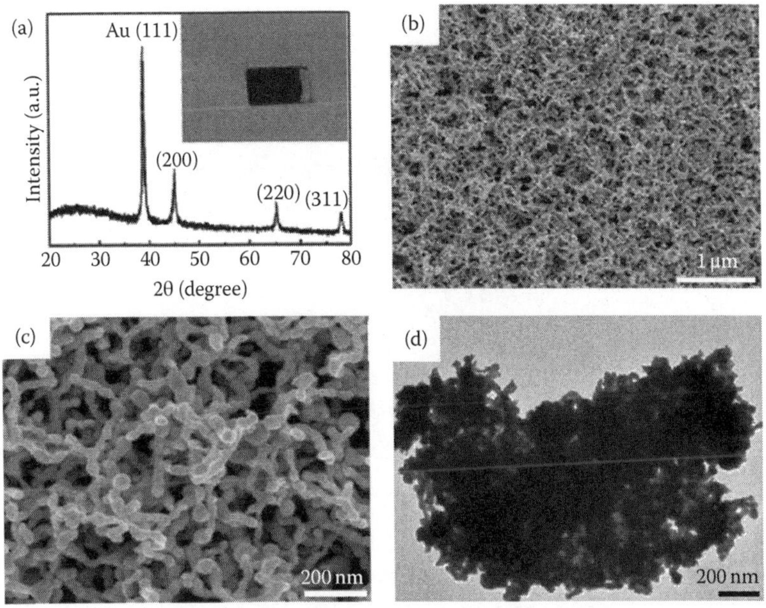

FIGURE 8.63 Structure and morphology of the sample prepared by electrophoretic deposition. (a) X-ray diffraction result. Inset: Photo of the sample on indium tin oxide glass. (b and c) FESEM images with different magnifications. (d) TEM image of the product scraped from the sample. (He et al., *Chem. Commun.*, 46, 7223–5, 2010. Reproduced by permission of The Royal Society of Chemistry.)

uniformity of the film. When the applied potential is low (lower than 30 V, e.g., 5 V) in the Au colloidal solution shown in Figure 8.62b, the obtained film is composed of loose nanochains and a few isolated nanoparticles. Contrarily, very high potential (>100 V) is also nonbeneficial for formation of the homogeneous porous film due to excess gas bubbles from water electrolysis. Also, EPD in the solution with different colloidal configurations leads to different film morphologies. If EPD is carried out in the Au colloidal solution as shown in Figure 8.61a (isolated nanoparticles suspended in water), at 30 V, we obtain spherical nanoparticles–packed film, without netlike structure, as shown in Figure 8.64a. Furthermore, laser irradiation to the Au colloidal solution decreases the size of Au nanoparticles in the colloidal solution [117,118], and hence, the morphology or configuration of the EPD film can also be controlled by laser irradiation of colloidal solution. Figure 8.64b shows the result from EPD in the colloidal solution after additional laser irradiation, for 90 minutes, of the colloidal solution shown in Figure 8.62a. The nanoparticle size in the film is much smaller than that shown in Figure 8.64a. Similarly, after EPD was carried out in the colloidal solution which was additionally laser-irradiated, we obtain the 3D netlike porous film consisting of much thinner nanochains than that shown in Figure 8.63c, as shown in Figure 8.64c. The easy control of the film polymorphism, based on laser ablation and/or subsequent irradiation, is important in practical applications.

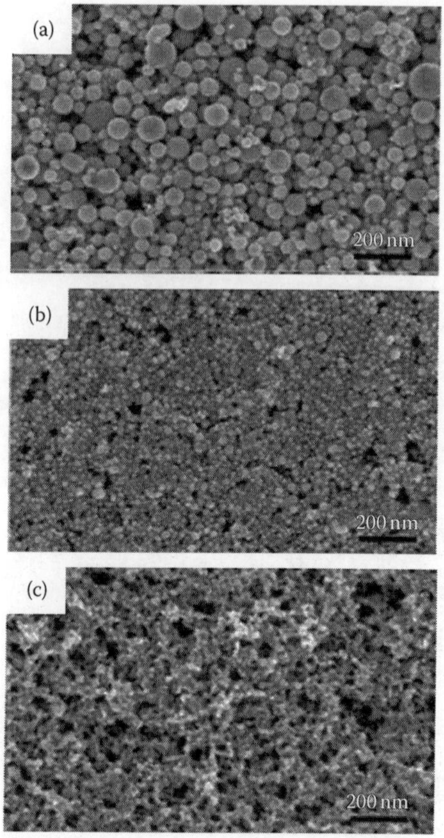

FIGURE 8.64 FESEM images of the films after electrophoretic deposition (EPD) in the Au colloidal solutions with different colloidal configurations (the same EPD conditions as in Figure 8.63). (a) In the colloidal solution shown in Figure 8.62a. (b) and (c) In the colloidal solutions after additional laser irradiation, for 90 minutes, of the colloidal solutions shown in Figure 8.62a and b, respectively. (He et al., *Chem. Commun.*, 46, 7223–5, 2010. Reproduced by permission of The Royal Society of Chemistry.)

The formation of the 3D netlike porous film can be easily understood. Before EPD, the Au colloidal solution is stabilized electrostatically against agglomeration. Upon applying potential, electrophoresis occurs in the solution. According to DLVO theory [114], the negatively charged Au nanochains deposit on the anode surface, and a homogeneous film is finally formed. In addition, the nanochains tend to aggregate and settle in the solution, due to gravity, during electrophoresis, similar to the EPD of large particles (>1 μm) in traditional ceramic areas. Low deposition potential corresponds to slow colloidal movement or low deposition rate. It means that many nanochains in the solution, for the EPD at a low potential, settle in the solution before moving to the anode and hence only form an incomplete film on the substrate. So, a high potential is needed for EPD to obtain a homogeneous film. This strategy presented here can be also extended to fabricate other 3D netlike porous films, from simple metals or semiconductors to multicomponent hybrids. We have obtained many kinds of 3D porous films, such as those of C, Ag, Au_xAg_{1-x}, and mixtures of Au and Ag. We can also get the configuration similar to that shown in Figure 8.63.

8.4.3 Micro/Nanostructured Hollow Sphere Arrays Based on Gas-Phase Surface Sol-Gel Process

In addition to the solution routes and electrochemical techniques, as mentioned in the preceding discussions, gas-phase surface sol-gel approach based on colloidal crystal template is also a facile and efficient strategy for fabrication of hollow sphere arrays. Here, we discuss the fabrication of titania hollow sphere arrays. Li et al. [43] placed net titanium (IV) butoxide in a small glass bottle. By heating, the bottle is saturated with titanium (IV) butoxide vapor. Then, colloidal template is placed in this atmosphere for a certain time period. Next, colloidal template is transferred to an oven with stationary humidity and temperature to hydrolyze the adsorbed titanium (IV) butoxide. Finally, a titania surface is generated. The shell thickness increases by repeating the process a number of times.

Figure 8.65 shows the SEM images of hollow titania sphere array. It should be mentioned that the process is repeated 10 times here. Clearly, the hollow titania spheres are arranged in hexagonal way. From the higher-magnification SEM image, it can be seen that there are broken titania spheres with smooth inner surface and a relatively rough outer surface (Figure 8.65b). The titania shell has a uniform thickness of 37 nm, which infers that the thickness of the shell grows to about 3.7 nm in an individual process. Furthermore, the shell thickness of the hollow spheres can be

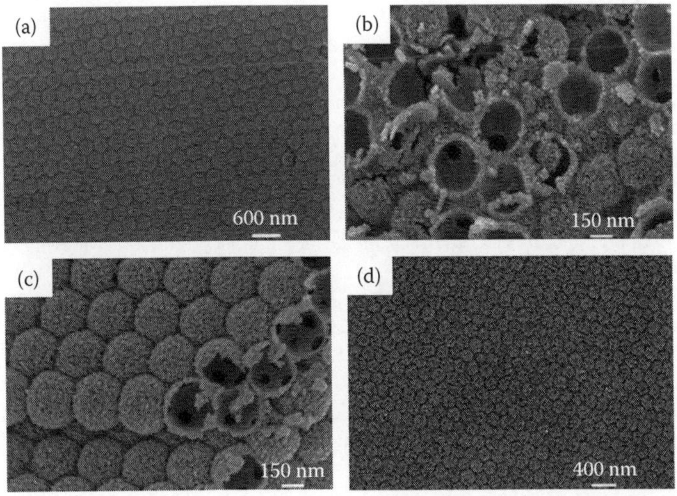

FIGURE 8.65 FESEM images of the ordered titania hollow sphere array films. (a) through (c) Diameter = 328 nm. (a) Top view. (b) and (c) Images of intentionally broken spheres with thicknesses 37 and 26 nm of the shell layer, respectively. (d) Diameter = 198 nm. (Reprinted with permission from Li et al., 2006, 13000–4. Copyright 2006 American Chemical Society.)

adjusted by changing the casting cycles of the titania precursor. When the casting cycle was decreased to 7, the shell thickness reduced to about 26 nm (Figure 8.65c). Similarly, other hollow spheres with different sizes could also be prepared, as displayed in Figure 8.65d.

8.5 BRIEF SUMMARY

A series of micro/nanostructured porous films have been introduced, including monolayer and multilayer porous arrays with homopore size or heteropore size and monolayer hollow sphere arrays. These ordered pore arrays can be fabricated by solution routes, electrochemical strategies, physical vapor deposition, and their combination based on PS colloidal templates, which have been demonstrated to be flexible, simple, low cost, and universal. Using these methods, we can obtain porous arrays of metals, semiconductors, and compounds. The morphology and structural parameters for the arrays can be easily controlled by the template and fabrication conditions.

REFERENCES

1. Jia, F. L.; Yu, C. F.; Ai, Z. H.; and Zhang, L. Z. 2007. Fabrication of nanoporous gold film electrodes with ultrahigh surface area and electrochemical activity. *Chem. Mater.* 19: 3648–53.
2. Scott, R. W. J.; Yang, S. M.; Coombs, N.; Ozin, G. A.; and Williams, D. E. 2003. Engineered sensitivity of structured tin dioxide chemical sensors: Opaline architectures with controlled necking. *Adv. Funct. Mater.* 13: 225–31.
3. Scott, R. W. J.; Yang, S. M.; Chabanis, G.; Coombs, N.; Williams, D. E.; and Ozin, G. A. 2001. Tin dioxide opals and inverted opals: Near-ideal microstructures for gas sensors. *Adv. Mater.* 13: 1468–72.
4. Sun, F. Q.; Cai, W. P.; Li, Y.; Jia, L. C.; and Lu, F. 2005. Direct growth of mono- and multilayer nanostructured porous films on curved surfaces and their application as gas sensors. *Adv. Mater.* 17: 2872–7.
5. Lorenz, B.; Mey, I.; Steltenkamp, S.; Fine, T.; Rommel, C.; Mueller, M. M.; Maiwald, A.; Wegener, J.; Steinem, C.; and Janshoff, A. 2009. Elasticity mapping of pore-suspending native cell membranes. *Small* 5: 832–8.
6. Duan, G.; Cai, W.; Luo, Y.; Li, Z.; and Li, Y. 2006. Electrochemically induced flowerlike gold nanoarchitectures and their strong surface-enhanced Raman scattering effect. *Appl. Phys. Lett.* 89: 211905.
7. Duan, G.; Cai, W.; Luo, Y.; Li, Y.; and Lei, Y. 2006. Hierarchical surface rough ordered Au particle arrays and their surface enhanced Raman scattering. *Appl. Phys. Lett.* 89: 181918.
8. Li, Y.; Lee, E. J.; and Cho, S. O. 2007. Superhydrophobic coatings on curved surfaces featuring remarkable supporting force. *J. Phys. Chem. C* 111: 14813–7.
9. Li, Y.; Huang, X. J.; Heo, S. H.; Li, C. C.; Choi, Y. K.; Cai, W. P.; and Cho, S. O. 2007. Superhydrophobic bionic surfaces with hierarchical microsphere/SWCNT composite arrays. *Langmuir* 23: 2169–74.
10. Liu, Z. Q.; Feng, T. H.; Dai, Q. F.; Wu, L. J.; and Lan, S. 2009. Fabrication of high-quality three-dimensional photonic crystal heterostructures. *Chinese Phys. B* 18: 2383–8.
11. Duan, G.; Cai, W.; Luo, Y.; and Sun, F. 2007. A hierarchically structured $Ni(OH)_2$ monolayer hollow-sphere array and its tunable optical properties over a large region. *Adv. Funct. Mater.* 17: 644–50.

12. Farcau, C.; Canpean, V.; Gabor, M.; Petrisor, T., Jr.; and Astilean, S. 2008. Periodically nanostructured noble-metal thin films with enhanced optical properties. *J. Optoelectron. Adv. Mater.* 10: 809–12.
13. Lyon, L. A.; Musick, M. D.; and Natan, M. J. 1998. Morphology-dependent optical and electrical properties of nanostructured metal films. *Abstr. Pap. Am. Chem. S.* 216: U44–U44.
14. Howard, R. E.; Liao, P. F.; Skocpol, W. J.; Jackel, L. D.; and Craighead, H. G. 1983. Microfabrication as a scientific tool. *Science* 221: 117–21.
15. Totzeck, M.; Ulrich, W.; Goehnermeier, A.; and Kaiser, W. 2007. Semiconductor fabrication: Pushing deep ultraviolet lithography to its limits. *Nat. Photonics* 1: 629–31.
16. Ito, T. and Okazaki, S. 2000. Pushing the limits of lithography. *Nature* 406: 1027–31.
17. Guan, Y.; Fowlkes, J. D.; Retterer, S. T.; Simpson, M. L.; and Rack, P. D. 2008. Nanoscale lithography via electron beam induced deposition. *Nanotechnology* 19: 505302.
18. Ruchhoeft, P. and Wolfe, J. C. 2001. Ion beam aperture-array lithography. *J. Vac. Sci. Technol. B* 19: 2529–32.
19. Lu, G.; Chen, Y.; Li, B.; Zhou, X.; Xue, C.; Ma, J.; Boey, F. Y. C.; and Zhang, H. 2009. Dip-pen nanolithography-generated patterns used as gold etch resists: A comparison study of 16-mercaptohexadecanioc acid and 1-octadecanethiol. *J. Phys. Chem. C* 113: 4184–7.
20. Falcaro, P.; Costacurta, S.; Malfatti, L.; Takahashi, M.; Kidchob, T.; Casula, M. F.; Piccinini, M. et al. 2008. Fabrication of mesoporous functionalized arrays by integrating deep X-ray lithography with dip-pen writing. *Adv. Mater.* 20: 1864–9.
21. Fang, T. H.; Weng, C. I.; and Chang, J. G. 2000. Machining characterization of the nanolithography process using atomic force microscopy. *Nanotechnology* 11: 181–7.
22. Li, F.; Wang, Z.; Ergang, N. S.; Fyfe, C. A.; and Stein, A. 2007. Controlling the shape and alignment of mesopores by confinement in colloidal crystals: Designer pathways to silica monoliths with hierarchical porosity. *Langmuir* 23: 3996–4004.
23. Lee, J. H.; Leu, I. C.; Chung, Y. W.; and Hon, M. H. 2008. Morphology-controlled 2D ordered microstructure arrays by surface modification of colloidal template. *J. Nanosci. Nanotechnol.* 8: 4436–40.
24. Chung, Y. W.; Leu, I. C.; Lee, J. H.; and Hon, M. H. 2007. Effect of heating rate on the fabrication of tunable ordered macroporous structures mediated by colloidal crystal template. *J. Alloy. Compd.* 433: 345–51.
25. Matsushita, S. I.; Miwa, T.; Tryk, D. A.; and Fujishima, A. 1998. New mesostructured porous TiO_2 surface prepared using a two-dimensional array-based template of silica particles. *Langmuir* 14: 6441–7.
26. Zhou, H.; Fallert, J.; Sartor, J.; Dietz, R. J. B.; Klingshirn, C.; Kalt, H.; Weissenberger, D.; Gerthsen, D.; Zeng, H.; and Cai, W. 2008. Ordered n-type ZnO nanorod arrays. *Appl. Phys. Lett.* 92: 132112.
27. Sun, F. Q.; Cai, W. P.; Li, Y.; Cao, B. Q.; Lei, Y.; and Zhang, L. D. 2004. Morphology-controlled growth of large-area two-dimensional ordered pore arrays. *Adv. Funct. Mater.* 14: 283–8.
28. Li, Y.; Cai, W.; and Duan, G. 2008. Ordered micro/nanostructured arrays based on the monolayer colloidal crystals. *Chem. Mater.* 20: 615–24.
29. Duan, G.; Cai, W.; Luo, Y.; Li, Z.; and Lei, Y. 2006. Hierarchical structured Ni nanoring and hollow sphere arrays by morphology inheritance based on ordered through-pore template and electrodeposition. *J. Phys. Chem. B* 110: 15729–33.
30. Yang, S. M.; Jang, S. G.; Choi, D. G.; Kim, S.; and Yu, H. K. 2006. Nanomachining by colloidal lithography. *Small* 2: 458–75.
31. Dai, Z. F.; Li, Y.; Duan, G. T.; Jia, L. C.; and Cai, W. P. 2012. Phase diagram, design of monolayer binary colloidal crystals, and their fabrication based on ethanol-assisted self-assembly at the air/water interface. *ACS Nano* 6: 6706–16.

32. Li, Y.; Cai, W. P.; Duan, G. T.; Cao, B. Q.; and Sun, F. Q. 2005. Two-dimensional ordered polymer hollow sphere and convex structure arrays based on monolayer pore films. *J. Mater. Res.* 20: 338–43.
33. Sun, F. Q.; Cai, W. P.; Li, Y.; Cao, B. Q.; Lei, Y.; and Zhang, L. D. 2005. Morphology controlled growth of large area ordered porous film. *Mater. Sci. Technol.* 21: 500–4.
34. Li, Y.; Duan, G.; and Cai, W. 2007. Controllable superhydrophobic and lipophobic properties of ordered pore indium oxide array films. *J. Colloid Interf. Sci.* 314: 615–20.
35. Li, Y.; Cai, W.; Duan, G.; Sun, F.; Cao, B.; Lu, F.; Fang, Q.; and Boyd, I. W. 2005. Large-area In_2O_3 ordered pore arrays and their photoluminescence properties. *Appl. Phys. A-Mater.* 81: 269–73.
36. Berger, S.; Jakubka, F.; and Schmuki, P. 2009. Self-ordered hexagonal nanoporous hafnium oxide and transition to aligned HfO_2 nanotube layers. *Electrochem. Solid-State Lett.* 12: 45–8.
37. Li, Z.; Cai, W.; Duan, G.; Zeng, H.; and Liu, P. 2009. Morphology dependent magnetic properties of two-dimensional alpha-Fe_2O_3 ordered nanostructured arrays. *J. Nanosci. Nanotechnol.* 9: 2970–5.
38. Liu, H. J.; Cui, W. J.; Jin, L. H.; Wang, C. X.; and Xia, Y. Y. 2009. Preparation of three-dimensional ordered mesoporous carbon sphere arrays by a two-step templating route and their application for supercapacitors. *J. Mater. Chem.* 19: 3661–7.
39. Wang, K.; Birjukovs, P.; Erts, D.; Phelan, R.; Morris, M. A.; Zhou, H.; and Holmes, J. D. 2009. Synthesis and characterisation of ordered arrays of mesoporous carbon nanofibres. *J. Mater. Chem.* 19: 1331–8.
40. Zhang, T.; Qian, J.; Tuo, X.; Yuan, J.; and Wang, X. 2009. Fabricating ordered porous monolayers from colloidal monolayer and multilayer. *Colloid. Surface. A* 335: 202–6.
41. Sun, F.Q.; Yu, J. C.; and Wang, X. C. 2006. Construction of size-controllable hierarchical nanoporous TiO_2 ring arrays and their modifications. *Chem. Mater.* 18: 3774–9.
42. Chen, J.; Liao, W. S.; Chen, X.; Yang, T.; Wark, S. E.; Son, D. H.; Batteas, J. D.; and Cremer, P. S. 2009. Evaporation-induced assembly of quantum dots into nanorings. *ACS Nano* 3: 173–80.
43. Li, Y. Z.; Kunitake, T.; and Fujikawa, S. 2006. Efficient fabrication and enhanced photocatalytic activities of 3D-ordered films of titania hollow spheres. *J. Phys. Chem. B* 110: 13000–4.
44. Wang, C. C.; Kei, C. C.; Wang, C. L.; and Perng, T. P. 2008. Preparation and optical property of TiO_2 nanohoneycomb. *Jap. J. Appl. Phys.* 47: 757–9.
45. Wang, Y. F.; Zhang, J. H.; Chen, X. L.; Li, X.; Sun, Z. Q.; Zhang, K.; Wang, D. Y.; and Yang, B. 2008. Morphology-controlled fabrication of polygonal ZnO nanobowls templated from spherical polymeric nanowell arrays. *J. Colloid Interf. Sci.* 322: 327–32.
46. Li, Y.; Cai, W. P.; Cao, B. Q.; Duan, G. T.; Sun, F. Q.; Li, C. C.; and Jia, L. C. 2006. Two-dimensional hierarchical porous silica film and its tunable superhydrophobicity. *Nanotechnology* 17: 238–43.
47. Cai, W. P.; Zhang, L. D.; Zhong, H. C.; and He, G. L. 1998. Annealing of mesoporous silica loaded with silver nanoparticles within its pores from isothermal sorption. *J. Mater. Res.* 13: 2888–95.
48. Koyama, H.; Fujimoto, M.; Ohno, T.; Suzuki, H.; and Tanaka, J. 2006. Effects of thermal annealing on formation of micro porous titanium oxide by the sol-gel method. *J. Am. Ceram. Soc.* 89: 3536–40.
49. Wang, H. Q.; Li, G. H.; Jia, L. C.; Wang, G. Z.; and Li, L. 2008. General in situ chemical etching synthesis of ZnO nanotips array. *Appl. Phys. Lett.* 93: 153110-1–3.
50. Li, Y.; Cai, W. P.; Cao, B. Q.; Duan, G. T.; and Sun, F. Q. 2005. Fabrication of the periodic nanopillar arrays by heat-induced deformation of 2D polymer colloidal monolayer. *Polymer* 46: 12033–6.

51. Li, Y.; Cai, W. P.; Cao, B. Q.; Duan, G. T.; Li, C. C.; Sun, F. Q.; and Zeng, H. B. 2006. Morphology-controlled 2D ordered arrays by heating-induced deformation of 2D colloidal monolayer. *J. Mater. Chem.* 16: 609–12.
52. Ng, H. T.; Li, J.; Smith, M. K.; Nguyen, P.; Cassell, A.; Han, J.; and Meyyappan, M. 2003. Growth of epitaxial nanowires at the junctions of nanowalls. *Science* 300: 1249–9.
53. Tripathy, D. and Adeyeye, A. O. 2009. Magnetic properties of exchange biased Co/CoO elongated nanoring arrays. *J. Appl. Phys.* 105: 07C110–3.
54. Zhou, L.; Fu, X. F.; Yu, L.; Zhang, X.; Yu, X. F.; and Hao, Z. H. 2009. Crystal structure and optical properties of silver nanorings. *Appl. Phys. Lett.* 94: 153102.
55. Burmeister, F.; Schafle, C.; Matthes, T.; Bohmisch, M.; Boneberg, J.; and Leiderer, P. 1997. Colloid monolayers as versatile lithographic masks. *Langmuir* 13: 2983–7.
56. Wang, H. Q.; Li, G. H.; Jia, L. C.; Li, L.; and Wang, G. Z. 2009. High-temperature anisotropic silicon-etching steered synthesis of horizontally aligned silicon-based Zn_2SiO_4 nanowires. *Chem. Commun.* 3786–8.
57. Duan, G.; Cai, W.; Luo, Y.; Lv, F.; Yang, J.; and Li, Y. 2009. Design and electrochemical fabrication of gold binary ordered micro/nanostructured porous arrays via step-by-step colloidal lithography. *Langmuir* 25: 2558–62.
58. Jia, L.; Cai, W.; and Wang, H. 2009. Layer-by-layer strategy for the general synthesis of 2D ordered micro/nanostructured porous arrays: Structural, morphological and compositional controllability. *J. Mater. Chem.* 19: 7301–7.
59. Jia, L.; Cai, W.; Wang, H.; Sun, F.; and Li, Y. 2009. Hetero-apertured micro/nanostructured ordered porous array: Layer-by-layered construction and structure-induced sensing parameter controllability. *ACS Nano* 3: 2697–705.
60. Therese, G. H. A. and Kamath, P. V. 2000. Electrochemical synthesis of metal oxides and hydroxides. *Chem. Mater.* 12: 1195–204.
61. Duan, G. T.; Cai, W. P.; Li, Y.; Li, Z. G.; Cao, B. Q.; and Luo, Y. Y. 2006. Transferable ordered Ni hollow sphere arrays induced by electrodeposition on colloidal monolayer. *J. Phys. Chem. B* 110: 7184–8.
62. Bartlett, P. N.; Baumberg, J. J.; Coyle, S.; and Abdelsalam, M. E. 2004. Optical properties of nanostructured metal films. *Faraday Discuss.* 125: 117–32.
63. Sun, F. Q.; Cai, W. P.; Li, Y.; Cao, B. Q.; Lu, F.; Duan, G. T.; and Zhang, L. D. 2004. Morphology control and transferability of ordered through-pore arrays based on electrodeposition and colloidal monolayers. *Adv. Mater.* 16: 1116–21.
64. Jiang, P.; Bertone, J. F.; and Colvin, V. L. 2001. A lost-wax approach to monodisperse colloids and their crystals. *Science* 291: 453–7.
65. He, A. Q.; Djurfors, B.; Akhlaghi, S.; and Ivey, D. G. 2002. Pulse plating of gold-tin alloys for microelectronic and optoelectronic applications. *Plat. Surf. Finish.* 89: 48–53.
66. Sun, W. and Ivey, D. G. 1999. Development of an electroplating solution for codepositing Au-Sn alloys. *Mater. Sci. Eng. B* 65: 111–22.
67. Winand, R. 1994. Electrodeposition of metals and alloys: New results and perspectives. *Electrochim. Acta* 39: 1091–105.
68. Cao, B. Q.; Cai, W. P.; Sun, F. Q.; Li, Y.; Lei, Y.; and Zhang, L. D. 2004. Fabrication of large-scale zinc oxide ordered pore arrays with controllable morphology. *Chem. Commun.* July 21: 1604–5.
69. Cao, B. Q.; Sun, F. Q.; and Cai, W. P. 2005. Electrodeposition-induced highly oriented zinc oxide ordered pore arrays and their ultraviolet emissions. *Electrochem. Solid-State Lett.* 8: 237–40.
70. Wang, Z. L. 2004. Zinc oxide nanostructures: Growth, properties and applications. *J. Phys: Condens. Mat.* 16: R829–58.
71. Choi, B. B.; Myung, N.; and Rajeshwar, K. 2007. Double template electrosynthesis of ZnO nanodot array. *Electrochem. Commun.* 9: 1592–5.

72. An, X.; Meng, G.; Zhang, M.; Tian, Y.; Sun, S.; and Zhang, L. 2006. Synthesis and optical absorption property of ordered macroporous titania film doped with Ag nanoparticles. *Mater. Lett.* 60: 2586–9.
73. Yeo, S. H.; Teh, L. K.; and Wong, C. C. 2006. Fabrication and characterization of macroporous CdSe nanostructure via colloidal crystal templating with electrodeposition method. *J. Porous Mater.* 13: 281–5.
74. Jayashree, R. S. and Kamath, P. V. 1999. Factors governing the electrochemical synthesis of alpha-nickel (II) hydroxide. *J. Appl. Electrochem.* 29: 449–54.
75. Sau, T. K. and Murphy, C. J. 2004. Room temperature, high-yield synthesis of multiple shapes of gold nanoparticles in aqueous solution. *J. Am. Chem. Soc.* 126: 8648–9.
76. Sun, Y. and Wiederrecht, G. P. 2007. Surfactantless synthesis of silver nanoplates and their application in SERS. *Small* 3: 1964–75.
77. Sun, Y. G. 2007. Direct growth of dense, pristine metal nanoplates with well-controlled dimensions on semiconductor substrates. *Chem. Mater.* 19: 5845–7.
78. Liu, G. Q.; Cai, W. P.; and Liang, C. H. 2008. Trapeziform Ag nanosheet arrays induced by electrochemical deposition on Au-coated substrate. *Cryst. Growth Des.* 8: 2748–52.
79. Liu, G.; Cai, W. P.; Kong, L. C.; Duan, G. T.; and Lv, F. 2010. Vertical cross-linking silver nanoplate arrays with controllable density based on seed-assisted electrochemical growth and their structurally enhanced SERS activity. *J. Mater. Chem.* 20: 767–72.
80. Liu, G. Q.; Cai, W. P.; Kong, L. C.; Duan, G. T.; Li, Y.; Wang, J. J.; Zuo, G. M.; and Cheng, Z. X. 2012. Standing Ag nanoplate-built hollow microsphere arrays: Controllable structural parameters and strong SERS performances. *J. Mater. Chem.* 22: 3177–84.
81. Yang, J. L.; Duan, G. T.; and Cai, W. P. 2009. Controllable fabrication and tunable magnetism of nickel nanostructured ordered porous arrays. *J. Phys. Chem. C* 113: 3973–7.
82. Duan, G.; Lv, F.; Cai, W.; Luo, Y.; Li, Y.; and Liu, G. 2010. General synthesis of 2D ordered hollow sphere arrays based on nonshadow deposition dominated colloidal lithography. *Langmuir* 26: 6295–302.
83. Hirsch, P.; Howie, A.; Nicholson, R. B.; Pashley, D. W.; and Whelan, M. J. 1977. *Electron Microscopy of Thin Crystals*, Robert E. Krieger, Huntington, NY, Chapter 7.
84. Kirkland, A. I.; Jefferson, D. A.; Duff, D. G.; Edwards, P. P.; Gameson, I.; Johnson, B. F. G.; and Smith, D. J. 1993. Structural studies of trigonal lamellar particles of gold and silver. *J. Proc. R. Soc. London., Ser. A* 440: 589–609.
85. Jin, R. C.; Cao, Y. W.; Mirkin, C. A.; Kelly, K. L.; Schatz, G. C.; and Zheng, J. G. 2001. Photoinduced conversion of silver nanospheres to nanoprisms. *Science* 294: 1901–3.
86. Germain, V.; Li, J.; Ingert, D.; Wang, Z. L.; and Pileni, M. P. 2003. Stacking faults in formation of silver nanodisks. *J. Phys. Chem. B* 107: 8717–20.
87. Maillard, M.; Huang, P. R.; and Brus, L. 2003. Silver nanodisk growth by surface plasmon enhanced photoreduction of adsorbed Ag+. *Nano Lett.* 3: 1611–5.
88. An, J.; Tang, B.; Ning, X. H.; Zhou, J.; Xu, S. P.; Zhao, B.; Xu, W. Q.; Corredor, C.; and Lombardi, J. R. 2007. Photoinduced shape evolution: From triangular to hexagonal silver nanoplates. *J. Phys. Chem. C* 111: 18055–9.
89. Xu, C. W.; Wang, H.; Shen, P. K.; and Jiang, S. P. 2007. Highly ordered Pd nanowire arrays as effective electrocatalysts for ethanol oxidation in direct alcohol fuel cells. *Adv. Mater.* 19: 4256–9.
90. Chang, G.; Zhang, J. D.; Oyama, M.; and Hirao, K. 2005. Silver-nanoparticle-attached indium tin oxide surfaces fabricated by a seed-mediated growth approach. *J. Phys. Chem. B* 109: 1204–9.
91. Liu, G.; Cai, W.; Kong, L.; Duan, G.; and Lu, F. 2010. Vertically cross-linking silver nanoplate arrays with controllable density based on seed-assisted electrochemical growth and their structurally enhanced SERS activity. *J. Mater. Chem.* 20: 767–72.

92. Yang, S. K.; Cai, W. P.; Liu, G. Q.; and Zeng, H. B. 2009. From nanoparticles to nanoplates: Preferential oriented connection of Ag colloids during electrophoretic deposition. *J. Phys. Chem. C* 113: 7692–6.
93. Tang, Z. Y.; Kotov, N. A.; and Giersig, M. 2002. Spontaneous organization of single CdTe nanoparticles into luminescent nanowires. *Science* 297: 237–40.
94. Cho, K. S.; Talapin, D. V.; Gaschler, W.; and Murray, C. B. 2005. Designing PbSe nanowires and nanorings through oriented attachment of nanoparticles. *J. Am. Chem. Soc.* 127: 7140–7.
95. Pradhan, N.; Xu, H. F.; and Peng, X. G. 2006. Colloidal CdSe quantum wires by oriented attachment. *Nano Lett.* 6: 720–4.
96. Ethayaraja, M. and Bandyopadhyaya, R. 2007. Mechanism and modeling of nanorod formation from nanodots. *Langmuir* 23: 6418–23.
97. Zeng, H. B.; Liu, P. S.; Cai, W. P.; Cao, X. L.; and Yang, S. K. 2007. Aging-induced self-assembly of Zn/ZnO treelike nanostructures from nanoparticles and enhanced visible emission. *Cryst. Growth Des.* 7: 1092–7.
98. Sun, Y. G.; Mayers, B.; and Xia, Y. N. 2003. Transformation of silver nanospheres into nanobelts and triangular nanoplates through a thermal process. *Nano Lett.* 3: 675–9.
99. Formento, A.; Montanaro, L.; and Swain, M. V. 1999. Micromechanical characterization of electrophoretic-deposited green films. *J. Am. Ceram. Soc.* 82: 3521–8.
100. Anne, G.; Neirinck, B.; Vanmeensel, K.; Van der Biest, O.; and Vleugels, J. 2006. Origin of the potential drop over the deposit during electrophoretic deposition. *J. Am. Ceram. Soc.* 89: 823–8.
101. Limmer, S. J. and Cao, G. Z. 2003. Sol-gel electrophoretic deposition for the growth of oxide nanorods. *Adv. Mater.* 15: 427–31.
102. Yang, S. K.; Cai, W. P.; Kong, L. C.; and Lei, Y. 2010. Surface nanometer-scale patterning in realizing large-scale ordered arrays of metallic nanoshells with well-defined structures and controllable properties. *Adv. Funct. Mater.* 20: 2527–33.
103. Yang, S.; Cai, W.; Zeng, H.; and Li, Z. 2008. Polycrystalline Si nanoparticles and their strong aging enhancement of blue photoluminescence. *J. Appl. Phys.* 104: 023516.
104. Yang, S.; Cai, W.; Yang, J.; and Zeng, H. 2009. General and simple route to micro/nanostructured hollow-sphere arrays based on electrophoresis of colloids induced by laser ablation in liquid. *Langmuir* 25: 8287–91.
105. Matsushita, S. I.; Fukuda, N.; and Shimomura, M. 2005. Photochemically functional photonic crystals prepared by using a two-dimensional particle-array template. *Colloid Surface Physicochem. Eng. Aspect.* 257–58: 15–7.
106. Grinis, L.; Dor, S.; Ofir, A.; and Zaban, A. 2008. Electrophoretic deposition and compression of titania nanoparticle films for dye-sensitized solar cells. *J. Photoch. Photobio. A* 198: 52–9.
107. Huang, J.; Vongehr, S.; Tang, S.; Lu, H.; Shen, J.; and Meng, X. 2009. Ag dendrite-based Au/Ag bimetallic nanostructures with strongly enhanced catalytic activity. *Langmuir* 25: 11890–6.
108. Li, Y.; Li, C.; Cho, S. O.; Duan, G.; and Cai, W. 2007. Silver hierarchical bowl-like array: Synthesis, superhydrophobicity, and optical properties. *Langmuir* 23: 9802–7.
109. Lee, Y. J.; Lee, Y.; Oh, D.; Chen, T.; Ceder, G.; and Belcher, A. M. 2010. Biologically activated noble metal alloys at the nanoscale: For lithium ion battery anodes. *Nano Lett.* 10: 2433–40.
110. Banholzer, M. J.; Millstone, J. E.; Qin, L.; and Mirkin, C. A. 2008. Rationally designed nanostructures for surface-enhanced Raman spectroscopy. *Chem. Soc. Rev.* 37: 885–97.
111. He, H.; Cai, W.; Lin, Y.; and Chen, B. 2010. Surface decoration of ZnO nanorod arrays by electrophoresis in the Au colloidal solution prepared by laser ablation in water. *Langmuir* 26: 8925–32.

112. He, H.; Cai, W.; Lin, Y.; and Chen, B. 2010. Au nanochain-built 3D netlike porous films based on laser ablation in water and electrophoretic deposition. *Chem. Commun.* 46: 7223–5.
113. He, H.; Cai, W.; Lin, Y.; and Dai, Z. 2011. Silver porous nanotube built three-dimensional films with structural tunability based on the nanofiber template-plasma etching strategy. *Langmuir* 27: 1551–5.
114. Sarkar, P. and Nicholson, P. S. 1996. Electrophoretic deposition (EPD): Mechanisms, kinetics, and application to ceramics. *J. Am. Ceram. Soc.* 79: 1987–2002.
115. Matsuo, N.; Muto, H.; Miyajima, K.; and Mafuné, F. 2007. Single laser pulse induced aggregation of gold nanoparticles. *Phys. Chem. Chem. Phys.* 9: 6027–31.
116. Mafuné, F.; Kohno, J. Y.; Takeda, Y.; and Kondow, T. 2003. Formation of gold nanonetworks and small gold nanoparticles by irradiation of intense pulsed laser onto gold nanoparticles. *J. Phys. Chem. B* 107: 12589–96.
117. Kurita, H.; Takami, A.; and Koda, S. 1998. Size reduction of gold particles in aqueous solution by pulsed laser irradiation. *Appl. Phys. Lett.* 72: 789–91.
118. Inasawa, S.; Sugiyama, M.; Noda, S.; and Yamaguchi, Y. 2006. Spectroscopic study of laser-induced phase transition of gold nanoparticles on nanosecond time scales and longer. *J. Phys. Chem. B* 110: 3114–9.

9 Surface Wettability and Self-Cleaning Properties

9.1 INTRODUCTION

The wettability of solid surfaces is an important performance, which depends on both chemical composition and micro/nanostructures on the surfaces [1]. It can be evaluated by water contact angle (CA), which is an angle conventionally measured through the water droplet where a liquid–vapor interface meets a solid surface. Superhydrophobic surfaces (CA > 150°) and superhydrophilic surfaces (CA < 10°) have attracted much attention for their practical applications, such as the prevention of adhesion of snow, fog, and raindrops to antennas and windows; reduction of friction drag; and creation of self-cleaning, antioxidation, and microfluid devices [2]. A self-cleaning surface is usually defined as a surface that has the ability to remove dirt or contaminants from it when water droplets slide along the surface. Self-cleaning is closely related to surface wettability. The self-cleaning effect is normally attributed to superhydrophobicity with a sliding angle (SA) less than 10° or superhydrophilicity of the surface [3,4]. For superhydrophobicity with a self-cleaning effect, contaminants adhere to the water droplet surface and are removed after the water droplet slides off the solid surface with a small tilted angle, due to large water CA and low surface free energy. For superhydrophilic surfaces, contaminants can easily be swept away by adding water droplets on them, due to very low water CA. Generally, when a water droplet dips into the pores or grooves of a rough surface, wettability can be enhanced by increasing surface roughness, according to Wenzel's equation [5]:

$$\cos \theta_r = r \cos \theta \tag{9.1}$$

where r is the roughness factor, defined as the ratio of total surface area to projected area on a horizontal plane; θ_r is the CA of the film with a rough surface; and θ is the CA of the film with a smooth surface. Obviously, increased roughness can enhance the hydrophobicity and/or hydrophilicity of hydrophobic and/or hydrophilic surfaces.

When a droplet contact with a rough surface and the liquid droplet is completely lifted up by the roughness features and it cannot dip into pores or grooves on the rough surfaces, another model was presented by Cassie as the following equation [6]:

$$\cos \theta_r = f_1 \cos \theta - f_2 \tag{9.2}$$

where f_1 and f_2 (= 1 − f_1) are the area fractions of a water droplet in contact with a surface and with air on the surface, respectively. Obviously, increasing f_2 can lead to a larger θ_r. Equations 9.1 and 9.2 mean that the roughness of solid surfaces is important to achieve superhydrophilic or superhydrophobic surfaces with a self-cleaning effect. If the surfaces are rough enough, superhydrophilicity will be achieved when native flat solid surfaces are very hydrophilic, whereas superhydrophobicity will be obtained when the native solid surfaces are very hydrophobic. According to these two models, one can see that for the formation of superhydrophobic surfaces it is necessary to meet two conditions: one is enough roughness on the surface and the other is low surface free energy on the surface, which can produce a hydrophobic property on the native flat surface.

As mentioned in Chapter 7, the synthesis of micro/nanostructured arrays based on colloidal crystals has been well developed [3–8]. Plentiful ordered micro-/nanostructured array films could be prepared by colloidal monolayer templates, including pore arrays, pillar arrays, and hierarchical micro/nanostructured arrays [7–17]. These ordered arrays and the colloidal crystals are rough in the microscale or nanoscale. It is expected that such ordered structured arrays show superhydrophilicity or superhydrophobicity. This means that nanodevices built from these micro/nanostructured arrays could be waterproof or self-cleaning, in addition to having special device functions if they possess superhydrophilicity or superhydrophobicity.

In this chapter, we introduce recent developments regarding the wettability of periodic structured arrays based on PS colloidal crystal templates, including metal (Ag), semiconductor (ZnO and In_2O_3), and compound (SiO_2 and TiO_2) micro/nanostructured arrays, as well as composite arrays (such as the hierarchical PS microsphere/silver nanoparticle composite arrays and hierarchical PS microsphere/carbon nanotube [CNT] composite arrays).

9.2 TUNABLE WETTABILITY OF PERIODIC ZnO PORE ARRAY FILMS

Periodic pore array films can be created by PS colloidal monolayer template techniques, and their morphologies are closely dependent on the experimental conditions, for example, precursor concentrations. Using the solution-dipping template strategy based on the colloidal monolayer (see Chapter 8), ZnO ordered pore arrays can be fabricated [11]. The morphologies of ZnO ordered pore arrays can be controlled well by increasing the precursor concentration. The surface roughness increases with an increase in precursor concentration. Therefore, it is expected to use this phenomenon in controlling surface wettability. Figure 9.1 shows ZnO ordered pore arrays with various morphologies prepared by different precursor concentrations. These three kinds of surface microstructures correspond to different precursor (zinc acetate) concentrations. At a low precursor concentration (0.3 M) (Figure 9.1a), the pores in the film demonstrate truncated hollow spheres and the pore sizes are smaller than the diameter of the colloidal sphere of the template. The depth is also smaller than the radius of the template. With an increase in precursor concentration to 0.5 M, each pore looks like a hollow hemisphere and the pore size increases to about the diameter of the PS (Figure 9.1b). If the concentration is further increased (1.0 M), the pores show a noncircular shape from the top view. The surface morphology exhibits a hierarchical structure, which is composed of close-packed rough wreaths and some small particles 30 nm in size, as shown in Figure 9.1c and d.

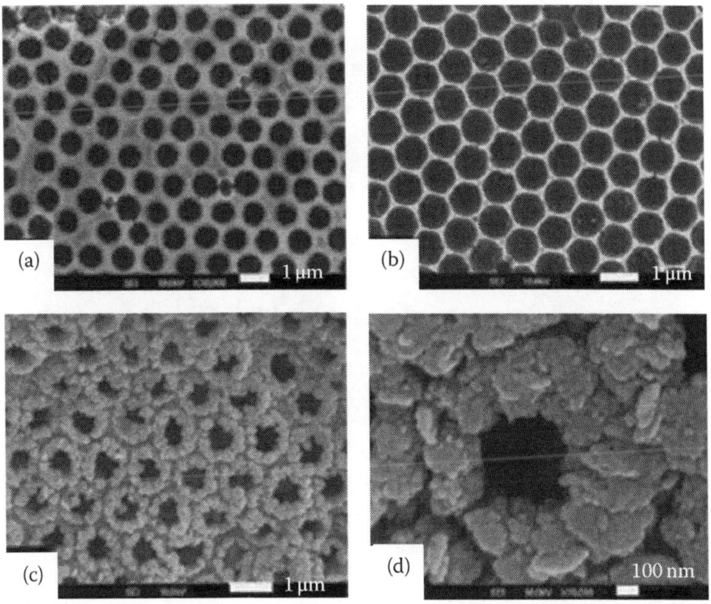

FIGURE 9.1 Field emission scanning electron microscopy (FESEM) images of ZnO ordered pore array films prepared by solution-dipping method using different precursor concentrations: (a) 0.3, (b) 0.5, and (c) and (d) 1.0 M. (Reprinted from Li et al., *J. Colloid. Interf. Sci.*, 287, 634–9, Copyright 2005, with permission from Elsevier.)

The wettability or the water CA of as-prepared ZnO porous arrays with different surface morphologies was examined by carefully dropping water droplets on their surfaces in a dark chamber. Figure 9.2a through c gives the photographs of the water droplets on different films, and the corresponding water CAs are 125°, 131°, and 143°, respectively. These indicate that such ordered porous structures can effectively increase the hydrophobicity in comparison with relatively flat ZnO films (CA, 109°). Moreover, wettability shows a clear dependence on surface microstructures, which are determined by the precursor concentration. Moreover, after chemical modification with low-surface-energy materials, fluoroalkylsilanes, the aforementioned three samples demonstrate superhydrophobicity, and the corresponding water CAs increase to 152°, 156°, and 165°, respectively, as shown in Figure 9.2d through f, where the water droplets display a rather spherical shape.

Interestingly, the water droplets on the surfaces of the modified samples shown in Figure 9.1a and b did not slide even when the surfaces were almost tilted vertically. The water droplets on the modified surface shown in Figure 9.1c, however, rolled off quickly even when the surface was slightly tilted (<5°). This phenomenon can be described more accurately by dynamic CAs, advancing CA (θ_A) and receding CA (θ_R). The dynamic CAs of the modified ordered pore array films fabricated by the 1.0 M precursor are too small to be measured (SA < 5°). Nevertheless, its CA hysteresis, $\theta_H = \theta_A - \theta_R$, is the smallest due to its largest water CA and smallest SA. Our results indicate that with the rise in precursor concentration the CA hysteresis

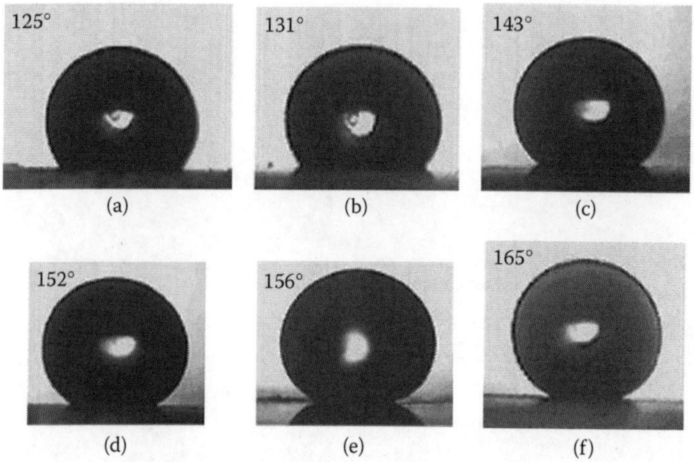

FIGURE 9.2 (a) Through (c) water droplet shapes on ZnO pore arrays and the corresponding static CAs on the as-prepared samples, corresponding to Figure 9.1a through c; (d) through (f) the water droplets and CAs on the same samples after chemical modifications, corresponding to (a) through (c). (Reprinted from Li et al., *J. Colloid. Interf. Sci.*, 287, 634–9, Copyright 2005, with permission from Elsevier.)

of both as-prepared and modified nanostructured films decreases. Moreover, the as-prepared samples have a relatively larger CA hysteresis (>21°), whereas the modified films have a smaller one (<9°).

As we know, dewetting behavior of the rough surface that can be described with Wenzel model show larger θ_H than surfaces that can be described with Cassie model due to the bigger adhesive force between the water droplet and film [18]. In this case, the as-prepared sample has a large θ_H (>21°). Consequently, this dewetting behavior principally prefers the Wenzel model. From Figure 9.1, it can be found that the roughness of the porous array film increases with the increase of precursor concentration. Correspondingly, according to the Wenzel model, θ_r should increase, which agrees well with these results. After modification with fluoroalkylsilane, these ordered pore array films exhibit superhydrophobicity (CA > 150°) with a small θ_H (<9°), which corresponds to the Cassie surface. After modification with lower surface free energy materials, air can be trapped into this ordered porous film. For the modified samples, since the parameter f_s decreases with precursor concentration the corresponding water CA also increases.

With the increase of precursor concentration, CA hysteresis decreases for both as-prepared and modified samples, which is possible due to the gradual reduction of the adhesive forces between water droplets and films induced by the concentration-dependent roughness. When the precursor concentration is not high (say, 0.3 or 0.5 M), although the modified samples have superhydrophobicity, water droplets on their surface do not roll off when the films are tilted. This is mainly attributed to the continuous, stable three-phase (air–liquid–solid) contact line for such netlike ordered pore array films. However, when the precursor concentration is high enough (say, 1.0 M), the surface of the ordered porous film is much rougher and shows a hierarchical structure (Figure 9.1c and d), which is similar to that of the well-known lotus leaves [19,20], leading to superhydrophobicity with large water CA (165°) and small SA (<5°). This film could be expected to show self-cleaning effect.

9.3 CONTROLLABLE SUPERHYDROPHOBICITY OF In_2O_3 ORDERED PORE ARRAY FILMS

Ordered indium oxide pore array films with different morphologies were prepared by the sol-dipping method using PS colloidal monolayers as templates [21]. These porous films took on hydrophilicity. After modification with fluoroalkylsilanes, however, all of these pore array films displayed superhydrophobicity due to rough surface and low surface free energy materials modified on their surfaces. Interestingly, with the increase of pore size in the films, the superhydrophobicity could be controlled and was gradually enhanced due to the corresponding increase of roughness caused by nanogaps produced by the thermal stress in the annealing process with increase of film thickness. The ordered pore array films could be fabricated on the substrate after removal of the PSs and annealing at 400°C in air for 1 hour [21]. Figure 9.3 shows the morphology of the ordered pore array films prepared using the colloidal monolayer with different PS sphere sizes (Figure 9.3a, 1 μm; Figure 9.3c, 2 μm; Figure 9.3e, 4.5 μm). Macroporous pores exhibit an orderly hexagonal arrangement, which corresponds well to the colloidal monolayer template.

When the PS sphere size is 1 μm in the colloidal template, honeycomb-structured ordered indium oxide macropore array films with hexagonal alignment are fabricated by the aforementioned method, as shown in Figure 9.3a and b. If the size of the PS sphere increases to 2 μm, the pore array film can also be formed and pore shapes are close to hollow hemispheres (Figure 9.3c and d). However, most of the walls between two neighboring pores cracked and many nanogaps were produced, as clearly shown in the magnified image (Figure 9.3d). Further increasing the PS sphere size to say 4.5 μm, the macropore shapes were seriously deformed and changed from hollow hemisphere shape to irregular shape. Additionally, many nanogaps with big size also were produced on the walls between the two neighboring pores, as shown in Figure 9.3e and f.

In the annealing process of film materials, some thermal stress will generally be produced between materials and their supporting substrates. With increasing film thickness, the influence of thermal stress on film morphology becomes more serious, especially for the rupture behavior of film. Here, corresponding to an increase of PS sphere size in the colloidal monolayer template the thickness of the array film fabricated using this template will increase. When the PS sphere size is relatively small (1 μm), the height or thickness of the pore array film is about one-half of the sphere size and the thermal stress has nearly no influence on pore array films in the annealing process at 400°C for 1 hour, so the morphology of the pore array film corresponds perfectly to that of the colloidal monolayer template and exhibits hexagonally arranged regular pores. However, when a colloidal monolayer with a PS sphere diameter of 2 μm is used as a template and the thickness of the pore array

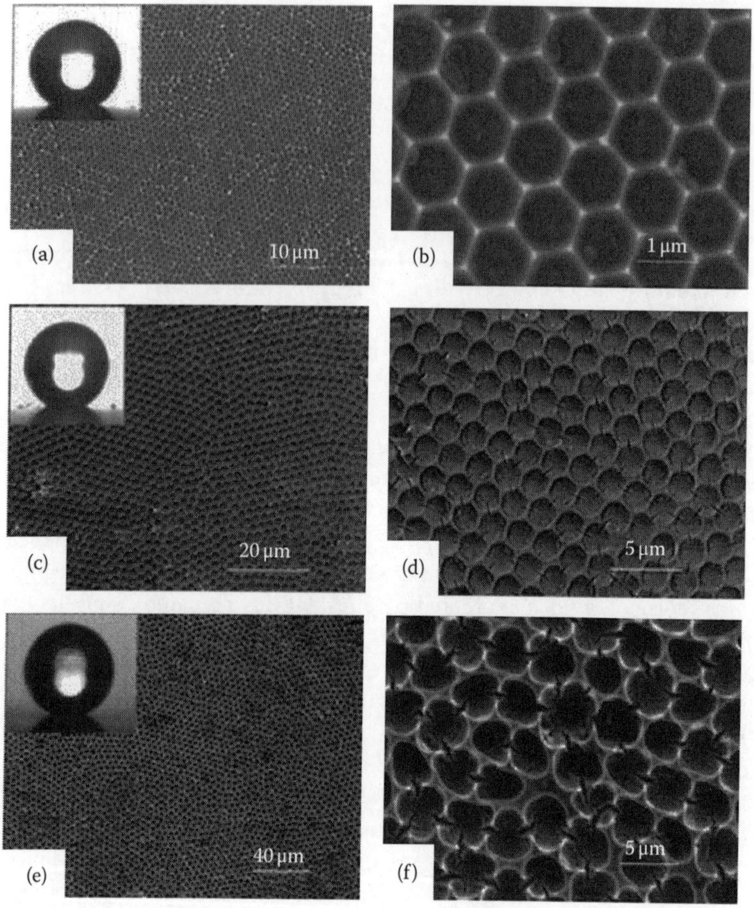

FIGURE 9.3 FESEM images of as-prepared indium oxide pore array films. The diameters of the polystyrene spheres in the colloidal monolayer are as follows: (a) and (b) 1 μm, (c) and (d) 2 μm, and (e) and (f) 4.5 μm. The images of (b), (d), and (f) are the magnifications of (a), (c), and (e), respectively. The precursor concentrations are as follows: (a) 0.2, (c) 0.4, and (e) 0.5 M. The insets in (a), (c), and (e) are the photographs of the water droplets on the corresponding pore arrays after the chemical modifications with fluoroalkylsilane, and the corresponding water contact angles are 155°, 158°, and 163°, respectively. (Reprinted from Li et al., *J. Colloid. Interf. Sci.*, 314, 615–20, Copyright 2007, with permission from Elsevier.)

film correspondingly increases, the thermal stress begins to play an important role in the formation of the pore array film and causes many nanogaps to be produced on the pore walls. If the thickness of the pore array film is increased further by using the colloidal template with a sphere size of 4.5 μm, the influence of thermal stress becomes more serious, resulting in deformed pore shapes and nanogaps on the pore wall. In addition, for the pore array film, it belongs to a kind of rough surface essentially. According to these results, one can clearly see that roughness of the pore array film increased with an increase in PS sphere size in the colloidal monolayer templates due to the effect of thermal stress.

Before being modified with fluoroalkylsilane, all samples took on hydrophilicity (water CA < 90°) and water CA decreased with the increase in pore size, as shown in curve (a) in Figure 9.4. After modification with fluoroalkylsilane, the morphologies of such ordered pore arrays remained nearly unchanged. However, the wettability was changed from the original hydrophilicity to superhydrophobicity. Moreover, with an increase in the pore size from 1 to 2 μm and finally to 4.5 μm, the water CA correspondingly increased from 155° to 158° and 163°, as shown in curve (b) in Figure 9.4. Additionally, the water droplet on them came gradually close to a spherical shape with the increase of pore size in the films, as shown in the insets in Figure 9.3a, c, and e, reflecting that the superhydrophobicity was enhanced with the increase of pore size. Our results indicate that the superhydrophobicity can be controlled well by changing the periodicities of colloidal templates.

The Wenzel model can explain well why such pore array films exhibit hydrophilicity and the hydrophilicity was enhanced with an increase in pore size. For a Wenzel-type surface, obviously, high roughness can enhance both hydrophobicity of hydrophobic surfaces and hydrophilicity of hydrophilic surfaces. The θ value of a relative flat indium oxide film prepared by the dip-coating method without using a colloidal monolayer template was 85°, indicating that the wettability of a flat film is hydrophilic. Here, indium oxide pore array films are much rougher than the relative flat surface, and the roughness of pore array films increases with the increase

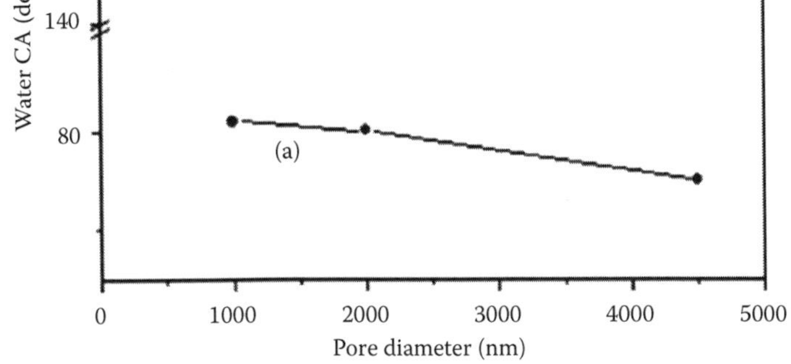

FIGURE 9.4 Curves of water contact angles on In_2O_3 ordered pore array film with different pore sizes before and after modification with fluoroalkylsilane. Curves (a) and (b) show pore sizes before and after modification, respectively. (Reprinted from Li et al., *J. Colloid. Interf. Sci.*, 314, 615–20, Copyright 2007, with permission from Elsevier.)

of pore size due to the many nanogaps produced on the pore walls as mentioned earlier. The relative flat indium oxide film displayed hydrophobicity with a water CA of 115° after chemical modification. According to the Wenzel model, for the same reason, the indium oxide pore films are so rough that the hydrophobicity is enhanced to water superrepellence and all such films displayed superhydrophobicity. The superhydrophobicity increases with an increase in pore size due to the corresponding increase in roughness.

9.4 IRRADIATION-INDUCED REVERSIBLE WETTABILITY OF ZnO PORE ARRAY FILMS

Very interestingly, ultraviolet (UV) light irradiation of ZnO pore array films would lead to transition of wettability from hydrophobicity to hydrophilicity for the as-prepared samples [11]. Figure 9.5 demonstrates the results for as-prepared ZnO ordered pore arrays kept in a dark chamber for 7 days before and after irradiation for 2 hours by UV light from a 500 W Hg lamp with a 400 nm filter. The differences between hydrophobic and hydrophilic CAs increase from 89° to 138° with rise in precursor concentration. If we alternate between keeping the samples in dark for 7 days and then exposing them to UV light, this wettability transition becomes reversible.

As we know, the electron–hole pairs generated by UV irradiation will move to the surface and the holes will react with lattice oxygen to form surface oxygen vacancies. Meanwhile, water and oxygen may compete to adsorb on them. Defective sites are kinetically more favorable for hydroxyl adsorption than oxygen adsorption. This reason and the rough surface can lead to hydrophilicity. Due to the instability of the hydroxyl absorption and the thermodynamical favorite of the oxygen adsorption,

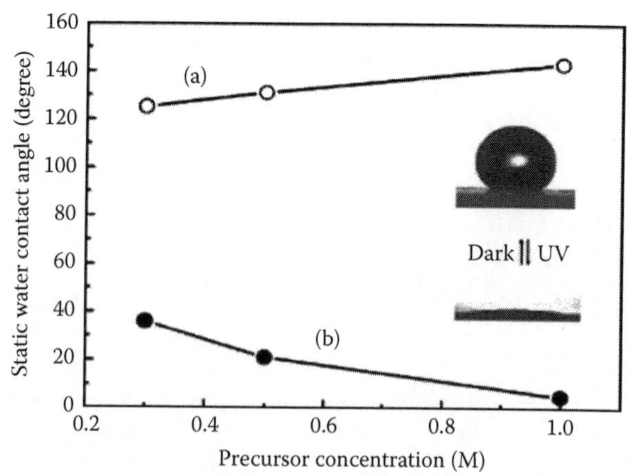

FIGURE 9.5 Transition between hydrophobicity and hydrophilicity induced by ultraviolet (UV) irradiation for the as-prepared samples. Curve (a): stored in dark chamber. Curve (b): irradiated by UV. (Reprinted from Li et al., *J. Colloid. Interf. Sci.*, 287, 634–9, Copyright 2005, with permission from Elsevier.)

oxygen atoms will replace the hydroxyl groups adsorbed on the defective sites gradually when the UV-irradiated ordered porous films were placed into the dark chamber. So when the original state of the surface is recovered, the wettability is reconverted from hydrophilicity to hydrophobicity. According to the Wenzel model, surface roughness enhances both hydrophobicity of hydrophobic surfaces and hydrophilicity of hydrophilic ones. The increase of the porous film roughness factor induced by the increase of precursor concentration can enhance both hydrophobicity and hydrophilicity of the surface with two contrary states, leading to ever-increasing CA differences in reversible wettability transition with rise in precursor concentration, as shown in Figure 9.5. Such a surface is very important in microfluidic devices applications [22,23].

9.5 SUPERHYDROPHOBICITY OF SILVER HIERARCHICAL BOWL-LIKE ARRAYS

Using a colloidal monolayer with PS sphere of 5 μm as the template and silver acetate as the precursor solution, and heating at 360°C for 3 hours, a silver ordered micro/nanostructured array was fabricated on the substrate, as shown in Chapter 7, Figure 7.9 [24]. This ordered array with a periodicity of 5 μm exhibits a hexagonal alignment. Each unit takes on a bowl-like structure (Chapter 7, Figure 7.9a and b) in the array, and the whole bowl-like array has rough inner walls composed of nanoparticles with an average size of approximately 135 nm, as shown in the inset of Figure 7.9b. The most silver nanoparticles are welded with neighbors because of the surface melt during the heating process. The heating process lead to the formation of highly durable micro/nanostructured arrays: these structures were not destroyed and the whole hierarchical arrays were not detached from a substrate even when the substrate was ultrasonically washed in water for 30 minutes [24].

Study of the wettability of the as-prepared ordered array shown in Chapter 7, Figure 7.9 was performed [24]. Before modification with thiol, when a water droplet with a small volume was added on the silver hierarchically ordered structured array, it spread out rapidly and the surface exhibited hydrophilicity with a water CA of 23°. After modification, however, the shape of the water droplet was nearly spherical and the array film showed superhydrophobicity with a water CA of 169°, as shown in Figure 9.6. Additionally, the SA for the water droplet was only 3°, indicating a self-cleaning effect.

The origin of the superhydrophobic property of the as-prepared micro/nanostructured film was investigated. The water CAs of various silver films, including the relative flat silver film, silver nanoparticle film, and ordered pore array film with smooth inner walls, were systematically examined before and after the modification of surfaces with thiol, as demonstrated in Figure 9.7. The water CA of the flat silver film was 69°, and the film showed hydrophobicity with a water CA of 108° after the surface modification (Figure 9.7a). For the silver

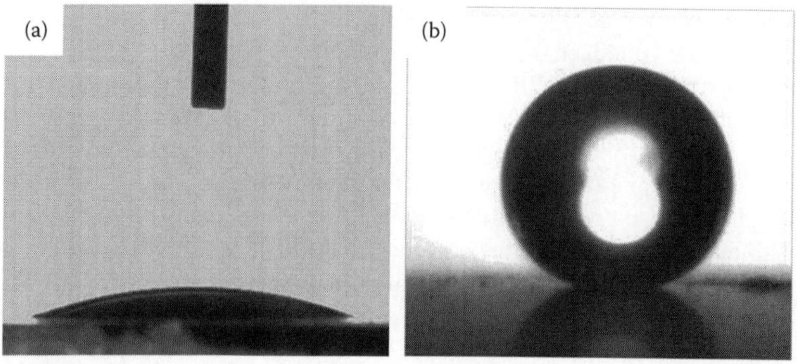

FIGURE 9.6 The water droplets on the silver bowl-like micro/nanostructured array film: (a) before modification with thiol, contact angle (CA) = 23°; (b) after modification, CA = 169°. (Reprinted with permission from Li et al., 2007, 9802–7. Copyright 2007 American Chemical Society.)

nanoparticle film, the water CA was 35° and 134° before and after modification, respectively. These experimental data indicate that the silver flat film and the nanoparticle film were unable to induce superhydrophobicity. The silver pore array film had water CAs of 33°, which was smaller than that of the flat silver film, before and 151° after the surface modification (Figure 9.7b). In addition, the water droplet of 3 mg could not roll off from its surface when the film was tilted to any angle, or even when it was turned upside down, which was mainly due to the stable, continuous three-phase contact line (air–liquid–solid) formed on this netlike pore array structure when the water drop was added on the surface. Although the silver pore array exhibited superhydrophobicity after modification, it was far away from the requirement of the self-cleaning effect. Therefore, the results demonstrated that the self-cleaning effect of as-prepared ordered bowl-like array film was resulted from the special combination of micro- and nano structures of hierarchical array, like lotus leaves.

The Wenzel model can explain well why these surfaces are more hydrophilic and more hydrophobic before and after modification. In the aforementioned cases, the hierarchical bowl-like array has the smallest water CA and the biggest CA before and after modification, respectively, which indicates that this hierarchical structure has the highest roughness. When a water droplet is added to the chemically modified ordered bowl-like array film with a hierarchical structure, air can be trapped in the interstices or corrugations that are produced between the microstructure and the nanostructure. In this case, according to Cassie's equation, f_2 for the as-prepared surface is calculated to be 0.98. For the pore array film obtained by electrodeposition, the water CA after modification is 151° and the corresponding f_2 is only 0.90. This analysis reveals that the hierarchical micro/nanostructured surface produces a large amount of air traps between the microstructure and the nanostructure and that the strong superhydrophobicity of the bionic surface is mainly caused by the unique hierarchical micro/nanostructures and the subsequent surface chemical treatment.

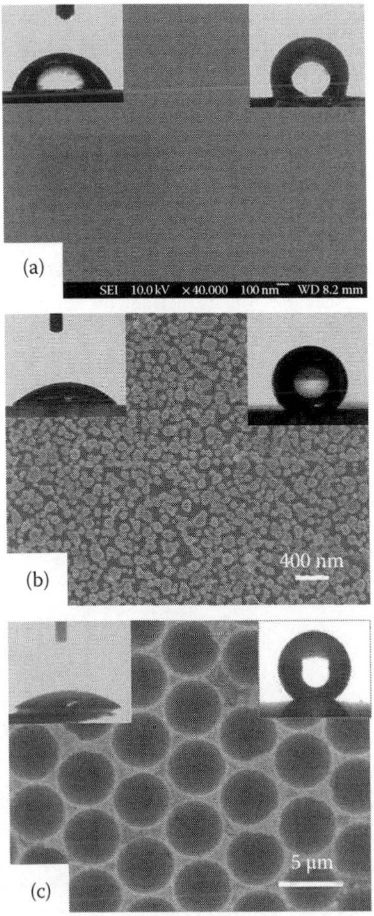

FIGURE 9.7 Morphology of silver films and water droplet shapes: (a) a relative flat silver film on the glass substrate obtained by thermal evaporation deposition, (b) a silver particle film on the glass substrate synthesized by the decomposition of AgAc coating at 360°C for 3 hours, and (c) a silver pore array film with smooth pore walls synthesized by the electrodeposition using a colloidal monolayer with a sphere size of 5 μm. The insets on top-left and top-right are water droplet shapes before and after modification with thiol, respectively. (Reprinted with permission from Li et al., 2007, 9802–7. Copyright 2007 American Chemical Society.)

9.6 WETTABILITY OF SILICA ORDERED MICRO/NANOSTRUCTURED ARRAYS

9.6.1 Micro/Nanostructured Ordered Pore Array

Using the colloidal monolayer as template, dropping the precursor sol on it, subsequent gelling, removing the template, and finally heating treatment, two-dimensional (2D) ordered pore array films of SiO_2, can be fabricated with hierarchically porous structure [25]. The as-prepared 2D SiO_2 ordered pore array is shown in Chapter 8,

Figure 8.8. This array has a hierarchical structure composed of ordered micropores and disordered mesopores in its skeleton (see Chapter 8, Figure 8.9).

Such structured films show different wettability from normal silica films. Figure 9.8 shows the shape of a water droplet and corresponding CAs on a silica nanostructured porous film (shown in Chapter 8, Figure 8.8) before and after surface modification. When a water droplet was added on the fresh silica ordered pore array, it rapidly spread out on the surface of the sample, which shows superhydrophilicity with a CA of about 5° (Figure 9.8a). After surface modification with fluoroalkylsilane, however, the wettability of this porous film was changed to superhydrophobicity. The shape of the water droplet on the film was spherical after inclining and the corresponding CA was 154°, as displayed in Figures 9.8b and 9.9. As we mentioned earlier, this silica ordered structure is actually a kind of hierarchical structure composed of ordered micropores and disordered mesopores in its skeleton. For comparison, the wettability of a flat silica film (about 1 μm in thickness) on a glass substrate, prepared without using a colloidal monolayer template, was also investigated; the water CA of the film was about 10° (Figure 9.8c), which is mainly caused by rich

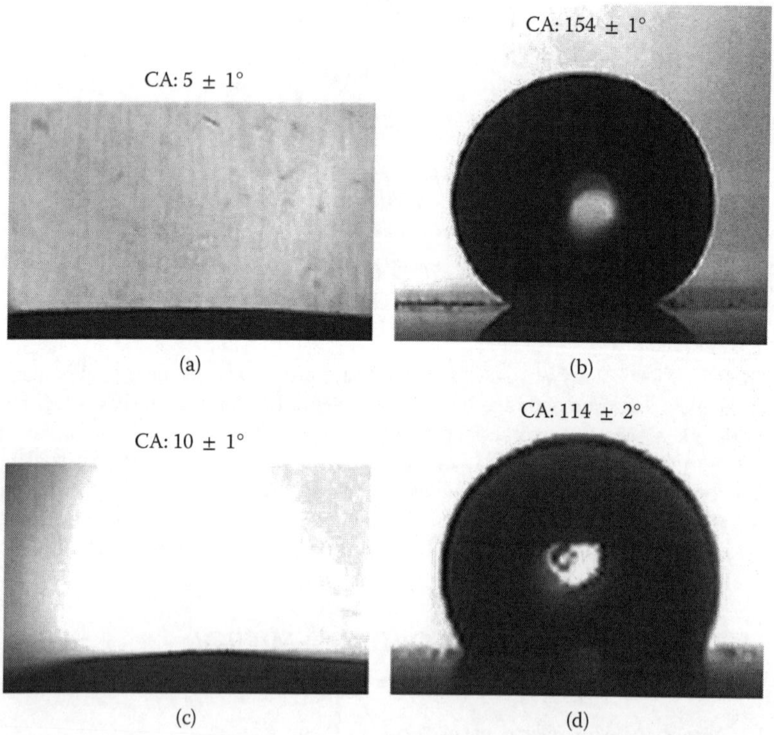

FIGURE 9.8 Water droplet shapes on different silica film surfaces: micropore arrays (shown in Chapter 8, Figure 8.8) (a) before and (b) after surface modification are shown. Flat silica film without micropores (c) before and (d) after modification is shown. (The source of the material including Li et al., *Nanotechnology*, 2006 and Institute of Physics is acknowledged.)

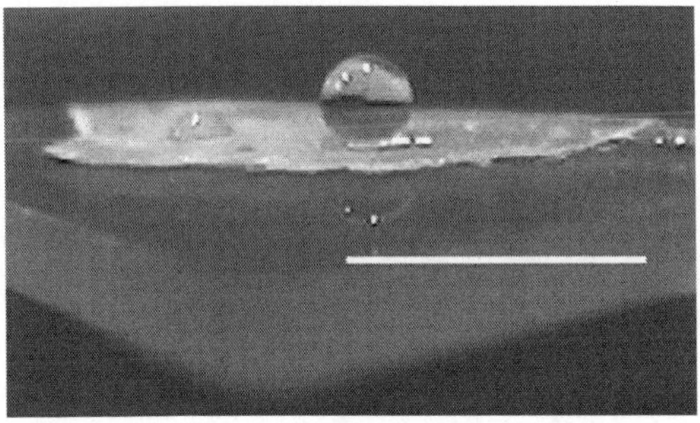

FIGURE 9.9 The photograph of a water droplet on the sample, shown in Figure 9.8b, taken by a charge-coupled device camera. Bar: 0.5 cm. (The source of the material including Li et al., *Nanotechnology*, 2006 and Institute of Physics is acknowledged.)

hydroxyl, possessing good affinity to water vapor on such high specific area surface materials. After modification, the water CA was only 114° (Figure 9.8d). Our results indicate that before and after modification the hierarchical porous silica film can effectively improve its wettability to superhydrophilicity and superhydrophobicity, respectively, compared to the sample without micropores.

According to Wenzel's equation, obviously, high roughness can enhance the hydrophilicity of hydrophilic native surfaces. In our study, such porous films were much rougher than the relatively flat surfaces obtained without using templates, so this model can explain why our as-synthesized films showed superhydrophilicity with a CA of about 5°. After modification with a lower surface free energy material, air can be trapped in this ordered pore array and, hence, a composite surface composed of air and pore array is formed. Measurements of wettability show that the surface is superhydrophobic with a water CA of 154°. In this case, Cassie's equation is available. Based on a simplified model, as schematically illustrated in Figure 9.10a, the relationship between f_2 and radii of the micropores at the film surface (r) and the PS (R) can be described by $f_2 = (\pi\sqrt{3}/6)(r^2/R^2)$. Here, $r = 470$ nm and $R = 500$ nm. We can thus obtain the value $f_2 = 0.8013$. The θ value for the flat silica surface without micropores after modification is experimentally about 114°. So the CA for our modified microporous sample can be estimated to be about 152.5° by Cassie's equation, which is in good agreement with our result.

Furthermore, measurements of SA show that although the modified samples have superhydrophobicity, a small water droplet (5 mg) on their surface cannot roll off when the films are tilted to any angle; even when they are turned upside down, the droplet still remains firmly pinned on the surface. Generally, the three-phase (air–liquid–solid) contact line plays an important role in the sliding behavior of water droplets; proper design of this contact line can lower the energy barrier for droplet motion and improve the droplet's sliding on the surface. According to the morphology of this netlike ordered pore array, it is easy to form the continuous, stable three-phase contact line (see Figure 9.10b), which leads to a larger energy barrier for water droplet motion and makes difficult the droplet's sliding on the surface. Such special properties have great significance on liquid microtransport without loss in microfluidic devices.

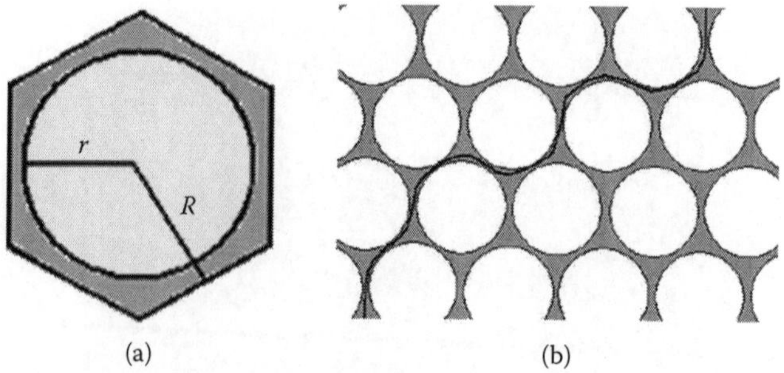

FIGURE 9.10 (a) A schematic illustration of unit with a micropore (top view). *R* and *r* are the radii of polystyrenes in the template and the micropores at the film surface, respectively. (b) Two-dimensional (*x*–*y*) presentation of the silica micropore array surface and probable three-phase contact line (dark line). (The source of the material including Li et al., *Nanotechnology*, 2006 and Institute of Physics is acknowledged.)

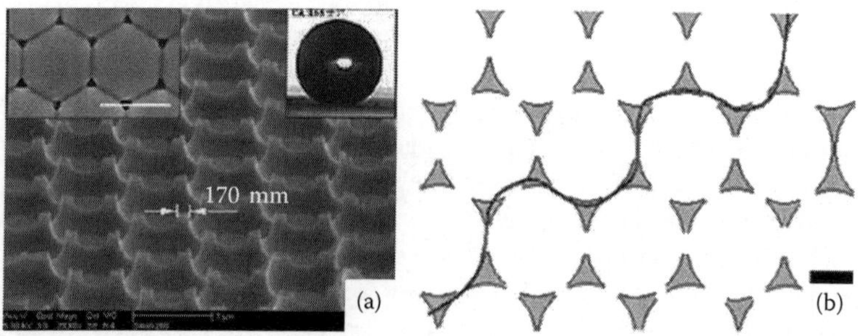

FIGURE 9.11 (a) FESEM image of silica porous film with nanopillar array. Inset on left top in (a) is a monolayer colloid crystal sintered at 120°C for 10 minutes; the bar is 1 μm. Inset on top-right in (a) is the water droplet shape and corresponding contact angle. (b) Probable three-phase contact line (dark line) on the nanopillar array film. (Li et al., *J. Mater. Chem.*, 16, 609–12, 2006. Reproduced by permission of The Royal Society of Chemistry.)

9.6.2 Ordered Nanopillar Arrays

Most recently, it was found that ordered structures with fairly high roughness could be prepared by using heat-deformed colloidal crystals as templates [14,26]. For example, an ordered silica network with triangular prism–shaped pillars on almost each node could be fabricated by a deformed colloidal monolayer, which sintered at 120°C for 10 minutes (see Figure 9.11a). Before modification with fluoroalkylsilane, this film also shows superhydrophilicity (CA < 5°); however, after modification its water CA was increased to 165°. From scanning electron microscopy (SEM) image, we know that the height of regular triangular section of the prism is about 170 nm, according to the geometric relationship, f_2, f_1 can be easily calculated with the values of 0.9615, 0.0385, respectively. By Cassie's equation, the water CA of this nanostructured film is evaluated to be 167.7°, which is in accordance with our results.

Corresponding result of the SA, 36°, further indicates such silica film with nanopillar array largely enhanced its superhydrophobicity in comparison with ordered pore array film obtained by template without sintering, which suggests that we can control the superhydrophobicity using the heat-deformed template with different degrees. Additionally, the existence of the 36° SA indicates that the continuous, stable three-phase (air–liquid–solid) contact line for the ordered pore film (Figure 9.10b) can be broken and the energy barrier for water droplet motion can be lowered by the pillar array on it (Figure 9.11b) [27,28], which is very useful for improving water droplet motion and enhancing superhydrophobicity.

9.7 WETTABILITY OF HIERARCHICAL MICRO/NANOCOMPOSITE ARRAYS

As mentioned in Chapter 7, fabrication techniques of monolayer colloidal crystals have been well developed because of the promising applications of these crystals in optical gratings, optical filters, and antireflective surface coatings. Monolayer colloidal crystals with large areas could be easily fabricated by the self-assembling process, for example, spin coating and dip coating. Because colloidal crystals consisting of hexagonally close-packed microspheres provide surfaces with a regularly ordered and well-defined roughness, they may lead to an enhancement of the surfaces' hydrophobicity. Unfortunately, such surface roughness is not sufficient to induce superhydrophobicity. Therefore, to achieve superhydrophobicity using monolayer colloidal crystals a much rougher surface texture should be provided on the colloidal crystals. For instance, if we decorate zero-dimensional (0D) structures (i.e., nanoparticles) or one-dimensional (1D) ones (i.e., nanotubes or nanowires) on a colloidal monolayer with microsized PS spheres, hierarchical micro/nanostructures like lotus leaves will easily be obtained, which gives us a big opportunity to get a surface with self-cleaning ability.

9.7.1 Zero-Dimensional Nanostructures on Microsized Polystyrene Spheres

Uniform silver nanoparticles could be formed on a substrate by the thermal decomposition of silver acetate (AgAc) at a low temperature [29], as introduced in Chapter 7. The silver nanoparticles on the colloidal monolayer to get bionic hierarchical micro/nanostructured arrays could be successfully prepared, as shown in Chapter 7, Figure 7.6. Briefly, AgAc aqueous solution was dripped onto the colloidal monolayer, forming a thin AgAc coating on it. The colloidal monolayer with the AgAc coating was heated in an oven, leading to the formation of a comparatively uniform decoration of silver nanoparticles on the surfaces of the PS spheres. Finally, hierarchical structures consisting of ordered PS microspheres and silver nanoparticles were created [29].

Figure 9.12 shows the field emission scanning electron microscopy (FESEM) image of the synthesized Ag hierarchical micro/nanostructured array. The monolayer colloidal crystal has a periodicity of 5 μm, and the nanoparticles on the colloidal crystal have an average size of 180 nm. The hierarchical structure was fabricated with a precursor solution of 0.5 M and at a heating temperature of 200°C for 3 hours. The synthesized hierarchical structure mimicked well the surface of a lotus leaf. The water CA of the as-prepared bionic structure was measured to be 29°, showing hydrophilicity (the top-left inset in Figure 9.12). However, the wettability of the structure was changed into superhydrophobicity after the chemical modification of the surface with 1-dodecanethiol, a low-surface-energy material; the water CA was dramatically increased to 168° (the top-right inset in Figure 9.12). In addition, the modified surface exhibited a small SA of about 2°. These results indicated that the fabricated surface with the hierarchical micro/nanostructure had a typical self-cleaning property. In the process of CA measurement, we found it difficult to add a water droplet on the fabricated bionic surface, demonstrating that the surface has a very low adhesive force and very small CA hysteresis. The fact of superhydrophobicity with a very low SA and the difficulty in dropping water on the surface provide strong evidence for the *lotus effect* for the synthesized bionic surface with the hierarchical micro/nanostructure.

To identify the origin of the superhydrophobic properties of the synthesized structures, we fabricated a flat and uniform silver film, film of silver nanoparticles, and monolayer PS colloidal crystal coated with an uniform silver film and then systematically investigated the CAs of the films before and after the modification of the surfaces with thiol (Figure 9.13). The uniform silver films were prepared by plasma sputtering, and the film thickness was 30 nm. The film of silver nanoparticles was provided on a flat substrate by the thermal decomposition of AgAc. The nanoparticles

FIGURE 9.12 FESEM image of a large-scaled bionic surface fabricated using the colloidal monolayer (the diameter of polystyrene sphere: 5 μm) coated with AgAc after heating at 200°C for 3 hours. The insets on top-left and top-right are the photographs of water droplets before and after modification, respectively. The water contact angles were 29° (left) and 168° (right). The inset on the bottom-left is the local magnification. (Reprinted with permission from Li et al., 2007, 14813–7. Copyright 2007 American Chemical Society.)

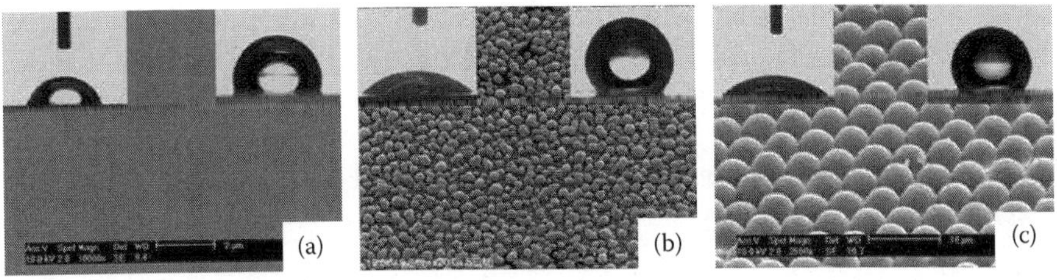

FIGURE 9.13 FESEM images and water droplets on (a) a flat silver film surface, (b) silver nanoparticle film, and (c) colloid monolayer with a thin silver coating. Insets at top-left and top-right in the figures are the water droplets on the films before and after the chemical modification, respectively. The water contact angles before and after the modification were (a) 68° and 110°, (b) 41° and 135°, and (c) 43° and 129°, respectively. (Reprinted with permission from Li et al., 14813–7. Copyright 2007 American Chemical Society.)

had morphologies similar to those in Figure 9.13b. The water CA of the flat silver film was 68°, and the wettability of the film was changed into hydrophobicity with a water CA of 110° after the surface modification (Figure 9.13a). The silver nanoparticle film had a water CA of 41°, which was lower than that of the flat silver film; however, the CA of the film was enhanced to 135° after the surface modification (Figure 9.13b). The silver-coated monolayer PS colloidal crystal showed a water dewetting behavior similar to the silver nanoparticle film: the CA was 43° and increased to 129° after the surface modification (Figure 9.13c). These results show that both the silver nanoparticle film and the PS colloidal crystal did not induce superhydrophobicity and suggest that the strong superhydrophobicity of the synthesized bionic surface originates from the hierarchically combined micro/nanostructure of the surface.

The periodicity of the microstructures can be tuned by changing the PS sphere size in the colloidal monolayer. For example, the bionic structure synthesized from the PS colloidal monolayer of 1.3 μm sphere size induced superhydrophobicity with a water CA of 163° and a SA of about 6° after the modification with thiol.

We present a facile synthetic route to bionic surfaces with remarkable superhydrophobic and self-cleaning properties by preparing 0D nanostructures on PS sphere surfaces. The bionic surfaces are synthesized by decorating silver nanocrystals generated from thermally decomposed AgAc on PS colloidal crystals and subsequent surface modification. Since the bionic surfaces consist of regularly ordered rough structures, one of their features is that the surfaces have uniform superhydrophobicity on the whole surface. Additionally, the wettability of the bionic surfaces can be easily tuned by changing the AgAc precursor concentration, heating temperature of AgAc, and PS sphere size, which is very helpful in microfluidic devices.

9.7.2 One-Dimensional Nanostructures on Microsized Polystyrene Spheres

In addition to creating hierarchical micro/nanostructured arrays by decorating 0D nanostructures on microsized PS spheres, in Chapter 7 we also introduced a facile and alternative method to load 1D nanostructures (CNTs) on microsized PS spheres, which would form hierarchically micro/nanocombined structures. Such arrays would be of superhydrophobic bionic surfaces [30]. The microstructure was supplied by PS colloidal monolayer on a glass or silicon substrate, and the nanostructure was supplied by single-walled carbon nanotubes (SWCNTs) decorated on the microstructures by wet chemical self-assembly, which was introduced in Chapter 7. The morphology and the distribution density of the nanostructure are seen in Chapter 7, Figure 7.17 or Figure 9.14a and can be easily controlled by the concentration of the SWCNT solution [30]. The presented route exhibits well the concept of bionic fabrication. The morphology of the resultant product bears a high resemblance to the natural lotus leaf (Figure 9.14b) and consequently shows strong superhydrophobicity.

The wettability of the synthesized hierarchical structures was investigated by both static water CAs and SAs. The water CA of the as-prepared samples with the hierarchical structures was measured to be 33° (Figure 9.14c), exhibiting hydrophilicity. To reduce the surface energy on the structures, the surface of the as-prepared samples were chemically modified with fluoroalkylsilane: the samples were immersed

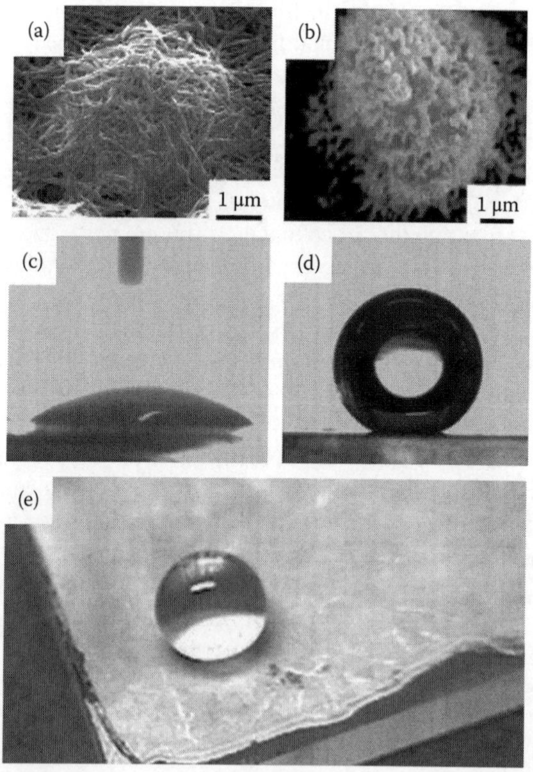

FIGURE 9.14 (a) FESEM image of a unit in a hierarchical microsphere/single-walled carbon nanotube (SWCNT) composite array obtained with a tilting angle of 40°. (b) Surface microstructure of natural lotus leaf [3]. Shapes of water droplets on the fabricated bionic surface (c) before and (d) after the surface modification are shown. The corresponding contact angles are 33° and 165°, respectively. (e) A photograph of a water droplet on the as-prepared bionic surface. (Reprinted with permission from Li et al., 2007, 2169–74. Copyright 2007 American Chemical Society.)

in a hexane solution of 20 mM 1H, 1H, 2H, 2H–perfluorodecyltrichlorosilane for 30 minutes and subsequently dried in an oven at 50°C for 30 minutes. After the chemical modification, the CA of the sample was dramatically increased to 165° and the water droplet was nearly spherical (Figure 9.14d and e). In addition, the surface exhibited a small SA of about 5°. The presence of superhydrophobicity with a very low SA provides strong evidence of the lotus effect for the synthesized hierarchical micro/nanostructures.

The CAs of a monolayer PS colloidal crystal coated with a gold film and an SWCNT film on a flat substrate were systematically investigated [30]. An SWCNT film was prepared on a gold-coated silicon substrate by wet chemical self-assembly, as described earlier. The CA of the gold-coated colloidal monolayer was 94°, and it increased to 138° after the surface modification (Figure 9.15a). Moreover, the SWCNT film on the flat substrate exhibited hydrophilicity with a water CA of 63° and showed hydrophobicity with a CA of 132° after the surface modification (Figure 9.15b). These results indicate that the strong superhydrophobicity of the fabricated bionic surface originates from its unique hierarchical structure combined with the nanoscaled net structure of SWCNTs and the microscaled PS spheres.

Before the surface modification, the wettability of the SWCNT film on the flat surface was hydrophilic (CA: 63°): the hydrophilicity of the CNT film was attributed from –COOH groups on the CNT surfaces. Evidently, SWCNTs on the colloidal monolayers increase the surface roughness. According to the Wenzel model, a rough surface is more hydrophilic than a flat surface and, accordingly, hydrophilicity increases with increasing roughness, which can explain why the water CA (33°) of the SWCNTs on the colloidal monolayer is smaller than the CA (63°) of the SWCNT film on the flat substrate. When the bionic surfaces are modified with the low-surface-energy material, air can be trapped in grooves or interstices on these surfaces. In this case, Cassie's equation can be used to explain the phenomenon. Since the given water CAs of the flat graphite surface and the bionic surface modified with fluoroalkylsilane are 108° and 165°, respectively, f_2 is calculated to be 0.95. This

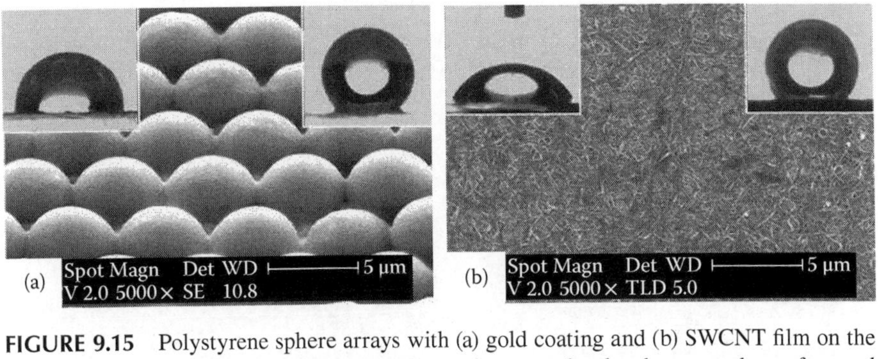

FIGURE 9.15 Polystyrene sphere arrays with (a) gold coating and (b) SWCNT film on the substrate. The insets on the left in (a) and (b) are the water droplet shapes on the surface and the contact angles (CAs) are 94° and 63°, respectively. The insets on the right in (a) and (b) are water droplet shapes after the surface modification and the CAs are 138° and 132°, respectively. (Reprinted with permission from Li et al., 2007, 2169–74. Copyright 2007 American Chemical Society.)

indicates that the strong superhydrophobicity of the bionic surface is mainly due to the air trapped in the rough surface by the combining microstructure of the ordered PS sphere arrays and the nanostructure of the SWCNTs adsorbed on the surface.

In this route, the SWCNT distribution can be controlled on the microsphere surfaces by changing the concentration of the SWCNT solution and the corresponding wettability also can be tuned by changes in morphology. For a typical example, by decreasing the SWCNT concentration from 2.0 to 1.0 $mg{\cdot}L^{-1}$, the corresponding distribution density of SWCNTs was reduced and the microsphere surfaces were not completely coated by SWCNTs (Figure 9.16). This reflects that at lower SWCNT concentrations a smaller quantity of SWCNTs was assembled during wet chemical self-assembly process on the PS spheres. Consequently, the surface roughness becomes less, resulting in a lower CA of 156° after the surface modification. The experimental results showed that the CAs decreased further by decreasing the SWCNT concentration and that superhydrophobic bionic surfaces with CAs higher than 150° were obtained at a concentration range of 2.5–0.7 $mg{\cdot}L^{-1}$.

As we know, SWCNTs are more expensive that multiwalled carbon nanotubes (MWCNTs); if we can use MWCNTs to replace SWCNTs to fabricate bionic surfaces, the costs will be largely reduced. Therefore, we also tried to fabricate MWCNTs on microsized PS spheres using the same route as that for SWCNTs. Figure 9.17 shows

FIGURE 9.16 A FESEM image of the microsphere/SWCNT composition array at SWCNT concentrations of 1.0 $mg{\cdot}L^{-1}$. The left inset is the magnified FESEM image of the structure, and the right inset shows the water droplet on the surface. (Reprinted with permission from Li et al., 2007, 2169–74. Copyright 2007 American Chemical Society.)

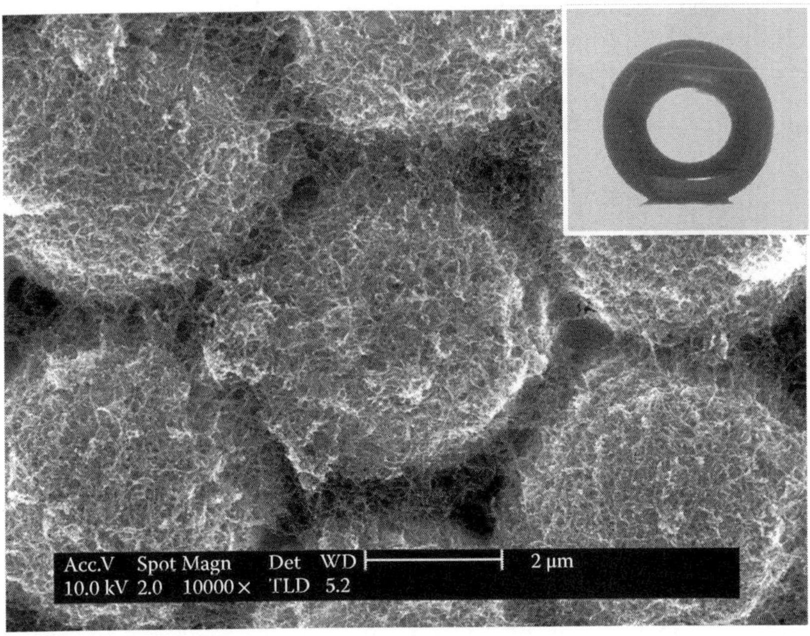

FIGURE 9.17 FESEM image of the polystyrene microsphere/multiwalled carbon nanotube (MWCNT) composite arrays and the water droplet on the surface. (Reprinted with permission from Li et al., 2007, 2169–74. Copyright 2007 American Chemical Society.)

a FESEM image of the hierarchical structured microsphere/MWCNT composite arrays and the water droplet shapes on the surface of these arrays. MWCNTs combined with the monolayer colloidal crystals also exhibited superhydrophobicity. The hierarchical structure was obtained from the MWCNT solution of concentration 2.0 $mg{\cdot}L^{-1}$ and the PS spheres of diameter 5.0 μm. After surface treatment, the surface also displayed superhydrophobicity with a CA of 166° and a SA of 5°.

It should be noted that compared to lotus leaves with randomly distributed microstructures with nonuniform sizes, the prepared bionic surfaces consist of regularly ordered microsphere/CNT composite arrays. Therefore, the synthesized bionic surfaces had a very uniform wettability on the whole surface, which can be proved by the variation of the measured water CA was around 1° on the whole surface.

This work has the following advantages: (1) the synthesized bionic surfaces were very similar to the natural lotus leaves that consist of rugged, hierarchical micro/nanostructures and, thus, the surfaces exhibited strong superhydrophobicity with a low SA after the surface treatment, like lotus leaves. (2) The wettability of the bionic surface can be controlled well by changing the distribution density of CNTs on the colloidal monolayers as well as changing the periodicity of the colloidal monolayers, which allows them to have many potential applications in the fields of microfluidic devices, bioseparation devices, and liquid transportation without loss. (3) CNTs with a continuous and homogeneous distribution on PS spheres have unique electrical and electrochemical properties, and such rough surfaces with hierarchical microsphere/CNT structures have a large specific area. Thus, the fabricated bionic structures can also be used for other devices such as gas sensors with good selectivity and high sensitivity.

9.8 SUPERHYDROPHOBIC SURFACES ON CURVED SUBSTRATES

It is well known that a water strider can float and move quickly on water surfaces. Recently, biomimetic research has revealed that this interesting feature is mainly the result of the superhydrophobicity of the water strider's legs [31]. A water strider's legs have a special hierarchical structure consisting of microsetae with nanogrooves, leading to the superhydrophobicity. A leg of a water strider gives a supporting force corresponding to 15 times the water strider's weight against the water surface. This phenomenon has inspired scientists to fabricate similar objects, which have important applications in the fields of miniaturized aquatic devices operating on water or under water and lossless liquid transportation channels. Although various techniques to synthesize superhydrophobic surfaces have been developed, most of these techniques are restricted on flat substrates. However, in order to allow the striking water-supporting force to an object with curved surface, all the outer surface of the object should be enclosed with superhydrophobic materials. So far, only a few methods to fabricate superhydrophobic surfaces that can mimic the legs of water striders have been investigated, such as superhydrophobic surfaces on metal wires by the electrochemical deposition utilizing the property of physical chemistry of metals, which can mimic well the legs of a water strider. It has been found that a polymer colloidal monolayer on a substrate can be transferred onto another substrate, while retaining its integrality. Inspired by this, bionic superhydrophobic surfaces were fabricated on a curved surface. The fabricated superhydrophobic coating on a convex tube exhibited a strong water-repellent property and supplied a high supporting force when it was floated on water surfaces [29].

As introduced in Chapter 7, the transferring technique of the PS colloidal monolayer was used for the fabrication of a superhydrophobic coating on a curved surface using a precursor solution (silver acetate) as a medium and subsequently performing heating decomposition. Utilizing this strategy, superhydrophobic Ag nanoparticle/PS sphere composite arrays could be successfully created on a convex glass substrate, including the outer surface of a glass tube and the inner surface of the glass tube (outer diameter: 1.4 mm), as illustrated in Chapter 7, Figures 7.11 and 7.12.

To demonstrate the water-repellent property of the superhydrophobic coating, such a coating was prepared on the glass tube and the supporting force of the tube was measured, as demonstrated in Figure 9.18. Both ends of the tube were sealed to prevent water from permeating into the inside of the tube. The outer diameter, inner diameter, and length of the tube were 4.87, 3.05, and 13.15 mm, respectively. The weight of the tube was 332 dyn, which is about 300 times heavier than a water strider (an adult *Gerris remigis*) [32].

Surprisingly, when a glass tube with a superhydrophobic coating was dipped in water and then taken out, no water droplets were found on the tube, exhibiting that the coating has excellent water-repellent superhydrophobic property. As demonstrated

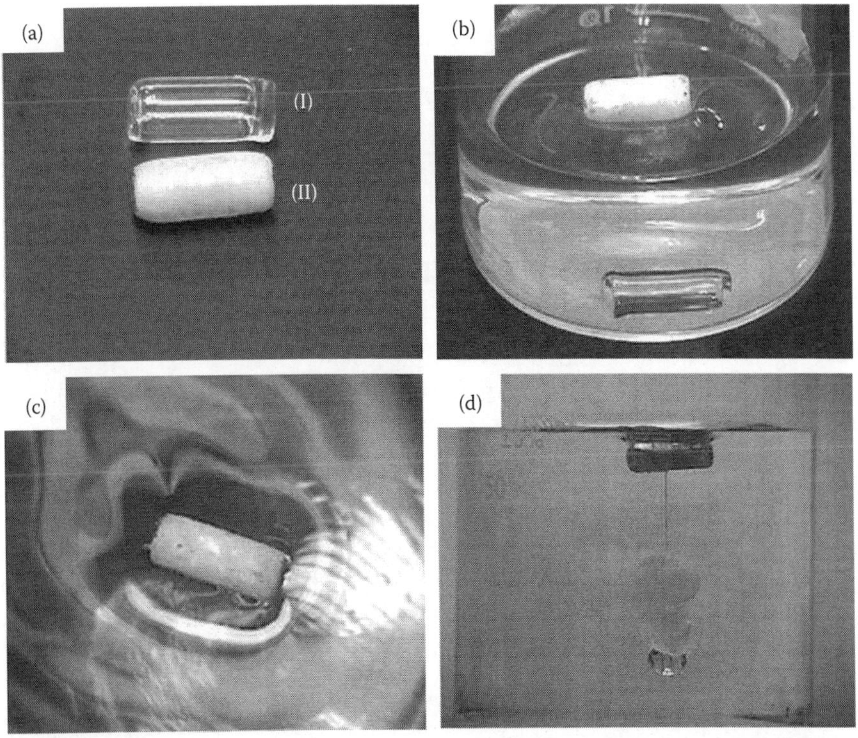

FIGURE 9.18 (a) Photographs of a bare glass tube (I) and a glass tube with a superhydrophobic coating (II). (b) The behavior of the two tubes when they are slowly dropped on water surface. (c) The glass tube with the superhydrophobic coating on water surface. Deformed meniscus is clearly seen on the water surface near the tube. (d) The glass tube with the superhydrophobic coating is weighted with plastic beads. (Reprinted with permission from Li et al., 2007, 14813–7. Copyright 2007 American Chemical Society.)

in Figure 9.18b, a bare glass tube sank to the bottom of vessel. However, the glass tube covered with the hierarchical structure could easily float on the water surface. Moreover, a deformed meniscus is clearly seen on the water surface near the tube in Figure 9.18c, showing that a strong supporting force was produced by the superhydrophobic coating on the tube. For measuring the maximum supporting force, the glass tube was weighted with plastic beads through a thin plastic wire that was fixed at the middle of the glass tube until the glass tube sank to the bottom of the vessel (Figure 9.18d).

The maximum supporting force is obtained when $mg = F_b + F_c$. Here, m is the total mass of the glass tube, plastic beads, and thin wire that makes the weighted tube sink to the water bottom; g is the gravitation acceleration speed. The total weight, mg, is supported by the combination of two forces, the buoyancy force (F_b) and the curvature force (F_c) [32]. F_b can be determined by integrating the hydrostatic pressure over the body area in contact with water. F_c is associated with surface tension and, thus, it is the more important factor characterizing the nonwetting property of the superhydrophobic coating than F_b. The total weight, *mg*, was measured to be 1420 dyn, which is around 4.3 times the weight of the glass tube. In addition, the maximal curvature force was calculated to be 386 dyn, which is 4.5 times larger than that of a leg of a water strider (an adult *G. remigis*), as reported by Lu and his coworkers [32]. The results show that a very high supporting force was produced by the fabricated superhydrophobic coating, which could effectively capture the air in the interstices among the micro/nanohierarchical structure.

9.9 SUPERAMPHIPHILICITY OF TiO_2 HIERARCHICAL MICRO/NANOROD ARRAYS WITHOUT ULTRAVIOLET IRRADIATION

In Chapter 7, we have introduced the fabrication of amorphous, porous TiO_2 micro/nanorod arrays by combining pulsed laser deposition (PLD) using colloidal monolayers [33,34]. Such micro/nanostructured periodic arrays based on colloidal monolayers are actually rough films, as illustrated in Chapter 7, Figure 7.18 or Figure 9.19a and b. It is expected that these micro-/nanostructured arrays can induce surface superhydrophilicity or superhydrophobicity with a self-cleaning effect, due to their high roughness. These arrays exhibited strong superhydrophilicity. When a small water droplet was dropped on a nanorod array, the droplet spread out rapidly on the surface and displayed a water CA of 0° in 0.225 second, as shown in Figure 9.19. Additionally, this nanorod film exhibited superoleophilicity when a small oil droplet was placed on the nanorod surface and the oil CA became 0° in 0.5 second, as seen in Figure 9.20. These results suggest that this amorphous 1D nanorod array had superamphiphilicity with 0° of both water CA and oil CA.

A TiO_2 film with superamphiphilicity can generally be obtained by UV irradiation, due to hydroxyls generated by oxygen defects or dangling bonds on its surface, induced by photochemical processes [35]. However, the TiO_2 nanorod array film mentioned here possessed superamphiphilicity without further UV irradiation. The ions (e.g., Ti^{4+} and O^{2-}) and electrons are released into the PLD chamber, and some oxygen species are lost in the vacuum environment in PLD after a TiO_2 target absorbs energy from laser irradiation by exceeding its threshold. Oxygen vacancies are produced in the deposited TiO_2 during PLD, converting relevant Ti^{4+} sites to Ti^{3+} sites that are favorable for dissociative water adsorption. Therefore, these defect sites microscopically form hydrophilic domains on the TiO_2 surface. However, the other parts surrounding the hydrophilic domain remain oleophilic on the surface. A composite TiO_2 surface having hydrophilic and oleophilic domains on a microscopically distinguishable scale demonstrates macroscopic amphiphilicity on the TiO_2 surface [35]. Additionally, a TiO_2 nanoparticle film prepared by PLD on a substrate without a PS colloidal monolayer exhibited a water CA of 15° and an oil CA of 27°, as illustrated in Figure 9.21. The roughness of the hexagonal-close-packed (hcp) TiO_2 micro/nanorod array film was much higher than that of the nanoparticle TiO_2 film produced by PLD without using a colloidal monolayer. According to Wenzel's equation, wettability is enhanced from amphiphilicity to superamphiphilicity. Therefore, the superamphiphilicity of the amorphous 1D nanorod array originates from the combination of the amphiphilicity produced by PLD and the special rough structures of hcp hierarchical nanorod arrays.

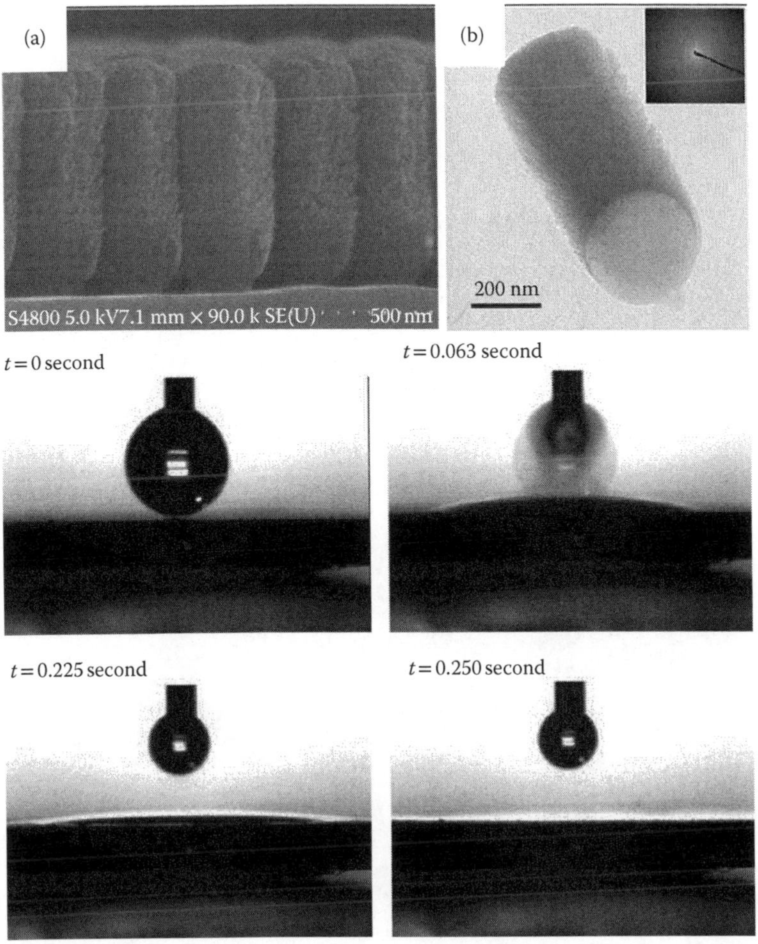

FIGURE 9.19 (a) and (b) The morphology of amorphous TiO_2 hierarchical micro/nanorod arrays (also see Chapter 7, Figures 7.18 and 7.19). The lower four frames show the time course of water-contacting behavior on the hexagonal-close-packed one-dimensional amorphous TiO_2 nanorod array film. (Reprinted with permission from Li et al., 2008, 14755–62. Copyright 2008 American Chemical Society.)

More importantly, this amorphous TiO_2 nanorod array demonstrated very good photocatalytic activity for organic molecular degradation (e.g., effective decomposition of stearic acid under UV illumination) [33]. Compared with the other TiO_2 films (e.g., an amorphous TiO_2 film produced by PLD without using a colloidal monolayer and an anatase TiO_2 nanorod array), the hcp amorphous TiO_2 micro/nanorod array on the colloidal monolayer demonstrated the best performance for degradation of organic materials, which is attributed to its porous structures and a much higher specific surface area than that of the anatase micro/nanorod array. Besides the higher specific surface area, special hierarchical structures composed of radiation-shaped nanobranches emanating from a center point on the PS sphere of

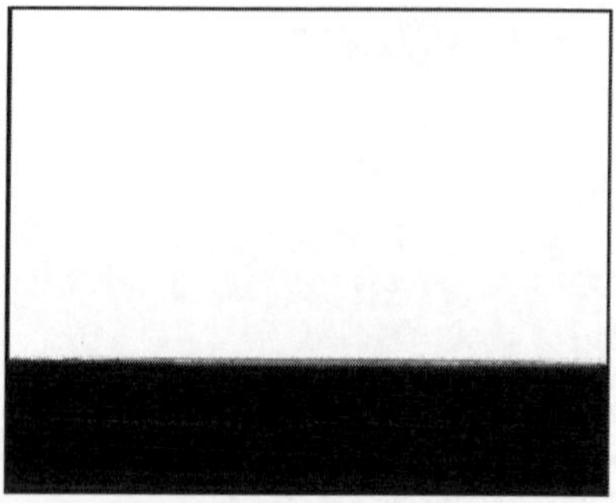

FIGURE 9.20 Oil (rapeseed) droplet shape on the TiO_2 nanocolumn array: the oil contact angle becomes 0° 0.5 second after it was dropped onto the surface. (Reprinted with permission from Li et al., 2008, 14755–62. Copyright 2008 American Chemical Society.)

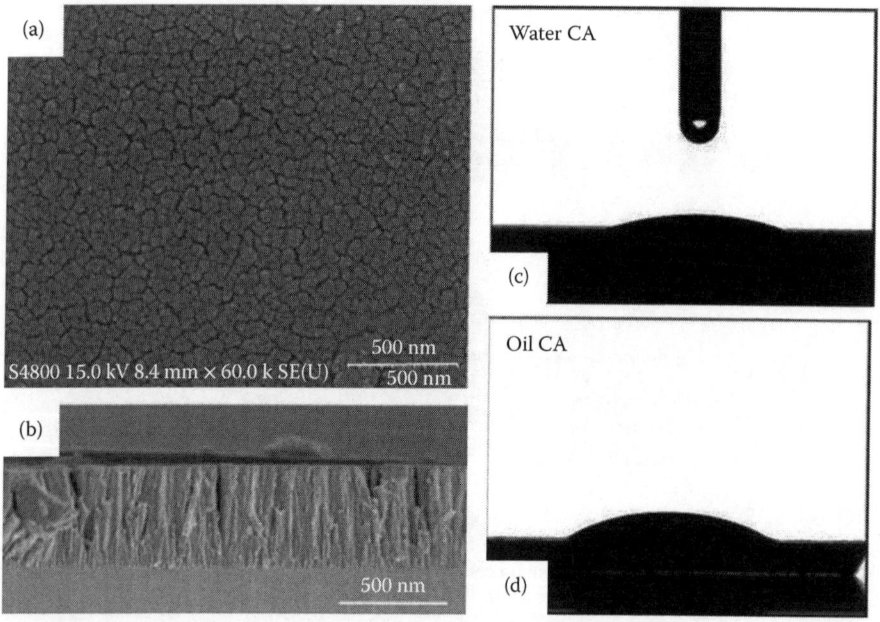

FIGURE 9.21 Morphology and wettability of the TiO_2 film obtained by pulsed laser deposition directly on a bare substrate without using polystyrene colloidal spheres: the FESEM images of the film from (a) top and (b) cross-sectional views are shown. The photographs of (c) water and (d) oil droplets on the film are also shown. The water contact angle (CA) is 15°, and the oil (rapeseed) CA is 27°. (Reprinted with permission from Li et al., 2008, 14755–62. Copyright 2008 American Chemical Society.)

a periodic structured nanorod array of amorphous TiO_2 can enhance photocatalytic activity, compared to an amorphous TiO_2 thin film produced by PLD without using a colloidal monolayer.

A combination of superamphiphilicity and photocatalytic activity can yield a self-cleaning surface. For instance, an oily liquid contaminant spreads out on a surface due to superoleophilicity, which is helpful in improving the photocatalytic efficiency under light illumination. An organic contaminant including oil gradually degrades under sunlight irradiation (sunlight contains some UV light). The self-cleaning effect can be realized after washing away contamination from the superhydrophilic surface.

9.10 BRIEF SUMMARY AND REMARKS

Biomimetic research reveals that the self-cleaning effect of a lotus leaf is ascribed to the combination of both a hierarchical micro/nanostructure on the surface and a low-surface-energy material covering the surface. According to this phenomenon, one can see that it is necessary to meet two conditions to create superhydrophobic surfaces: one is a large enough roughness factor on the surface, and the other is a coating of low surface free energy on the surface. Inspired by the lotus effect, a lot of techniques to prepare bionic superhydrophobic surfaces have been developed. As there are well-developed synthesis methods of colloidal crystals and micro/nanostructured arrays based on the colloidal crystals, a lot of ordered micro/nanostructured arrays could be fabricated. These ordered arrays and the colloidal crystals are rough in the microscale or nanoscale on the surfaces, which gives a good chance to create superhydrophobicity on the array surfaces. In this chapter, we introduce superhydrophobic surfaces based on colloidal crystal techniques. From these results, one can clearly see that hierarchical micro/nanostructures are very important in realizing superhydrophobicity with a self-cleaning effect. This can be proved well by the examples of silver hierarchical bowl-like arrays, hierarchical PS microsphere/silver nanoparticle composite arrays, and hierarchical PS microsphere/CNT composite arrays, which exhibited the very large water CAs and very small SAs. Importantly, the technique to fabricate superhydrophobic surfaces on curved substrates has important applications in miniaturized aquatic devices operating on water or under water and lossless liquid transportation channels. It is well known that the micro/nanostructured arrays fabricated by PS colloidal monolayer templates have important applications in photonics, photoelectronic devices, and so on. This suggests that nanodevices built from these micro/nanostructured arrays could be waterproof and self-cleaning, in addition to their special device functions after possessing superhydrophobicity. However, the as-prepared superhydrophobic surfaces with self-cleaning effect cannot be used in the practical applications (e.g., the coating on the building) due to the high costs, easy contamination. Besides these facts, another big problem, easily being damaged because of the decrease in mechanical stability with increase in surface roughness, also prevents them from being used in normal applications. Therefore, the design and preparation of bionic self-cleaning superhydrophobic surfaces with low cost and high mechanical stability will be an important and challenging task based on colloidal crystal techniques. More attention should be paid to these problems in the near future.

REFERENCES

1. Gau, H.; Herminghaus, S.; Lenz, P.; and Lipowsky, R. 1999. Liquid morphologies on structured surfaces: From microchannels to microchips. *Science* 283: 46–9.
2. Nakajima, A.; Hashimoto, K.; and Watanabe, T. 2001. Recent studies on super-hydrophobic films. *Monatsh. Chem.* 132: 31–41.
3. Sun, T.; Feng, L.; Gao, X.; and Jiang, L. 2005. Bioinspired surfaces with special wettability. *Acc. Chem. Res.* 38: 644–52.
4. Li, X. M.; Reinhoudt, D.; and Crego-Calama, M. 2007. What do we need for a superhydrophobic surface? A review on the recent progress in the preparation of super-hydrophobic surfaces. *Chem. Soc. Rev.* 36: 1350–68.
5. Wenzel, R. N. 1949. Surface roughness and contact angle. *J. Phys. Colloid Chem.* 53: 1446–67.
6. Cassie, A. B. D. 1948. Permeability to water and water vapour of tetiles and other fibrous material. *Diss. Faraday Soc.* 44: 239–43.
7. Sun, F.; Cai, W.; Li, Y.; Cao, B.; Lei, Y.; and Zhang, L. 2004. Morphology-controlled growth of large-area two-dimensional ordered pore arrays. *Adv. Funct. Mater.* 14: 283–8.
8. Sun, F. Q.; Cai, W. P.; Li, Y.; Cao, B. Q.; Lu, F.; Duan, G. T.; and Zhang, L. D. 2004. Morphology control and transferability of ordered through-pore arrays based on electrodeposition and colloidal monolayers. *Adv. Mater.* 16: 1116–21.
9. Cao, B. Q.; Cai, W. P.; Sun, F. Q.; Li, Y.; Lei, Y.; and Zhang, L. D. 2004. Fabrication of large-scale zinc oxide ordered pore arrays with controllable morphology. *Chem. Commun.* 1604–5.
10. Cao, B. Q.; Sun, F. Q.; and Cai, W. P. 2005. Electrodeposition-induced highly oriented zinc oxide ordered pore arrays and its ultraviolet emissions. *Electrochem. Solid-State Lett.* 8: G237–40.
11. Li, Y.; Cai, W. P.; Duan, G. T.; Cao, B. Q;. Sun, F. Q.; and Lu, F. 2005. Superhydrophobicity of 2D ZnO ordered pore arrays formed. *J. Colloid. Interf. Sci.* 287: 634–9.
12. Duan, G. T.; Cai, W. P.; Luo, Y. Y.; Li, Z. G.; and Lei, Y. 2006. Hierarchical structured Ni nanoring and hollow sphere arrays by morphology inheritance based on ordered through-pore template and electrodeposition. *J. Phys. Chem. B* 110: 15729–33.
13. Duan, G. T.; Cai, W. P.; Li, Y.; Li, Z. G.; Cao, B. Q.; and Luo, Y. Y. 2006. Transferable ordered Ni hollow sphere arrays induced by electrodeposition on colloidal monolayer. *J. Phys. Chem. B* 110: 7184–8.
14. Li, Y.; Cai, W. P.; Cao, B. Q.; Duan, G. T.; and Sun, F. Q. 2005. Fabrication of the periodic nanopillar arrays by heat-induced deformation of 2D polymer colloidal monolayer. *Polymer*. 46: 12033–6.
15. Li, Y.; Cai, W. P.; Duan, G. T.; Cao, B. Q.; and Sun, F. Q. 2005. 2D ordered polymer hollow sphere and convex structure arrays based on monolayer pore films. *J. Mater. Res.* 20: 338–43.
16. Li, Y.; Cai, W. P.; Duan, G. T.; Sun, F. Q.; and Lu, F. Large area In_2O_3 ordered pore arrays and their photoluminescence properties. *Appl. Phys. A: Mater. Sci. Process.* 81: 269.
17. Li, Y.; Cai, W. P.; Duan, G. T.; Sun, F. Q.; Cao, B. Q.; and Lu, F. 2005. Synthesis and optical absorption property of ordered macroporous titania film doped with Ag nanoparticles. *Mater. Lett.* 59: 276–9.
18. Lafuma, A. and Quéré, D. 2003. Superhydrophobic states. *Nature Mater.* 2: 457–60.
19. Barthlott, W. and Neinhuis, C. 1997. Purity of the sacred lotus, or escape from contamination in biological surfaces. *Planta* 202: 1–8.
20. Neinhuis, C. and Barthlott, W. 1997. Characterization and distribution of water-repellent, self-cleaning plant surfaces. *Ann. Bot.* 79: 667–77.

21. Li, Y.; Cai, W. P.; and Duan, G. T. 2007. Controllable superhydrophobic and lipophobic properties of ordered pore indium oxide array films. *J. Colloid. Interf. Sci.* 314: 615–20.
22. Liu, H.; Feng, L.; Zhai, J.; Jiang, L.; and Zhu, D. 2004. Reversible wettability of a chemical vapor deposition prepared ZnO film between superhydrophobicity and superhydrophilicity. *Langmuir* 20: 5659–61.
23. Feng, J.; Feng, L.; Jin, M.; Zhai, J.; Jiang, L.; and Zhu, D. 2004. Reversible super-hydrophobicity to super-hydrophilicity transition of aligned ZnO nanorod films. *J. Am. Chem. Soc.* 126: 62–3.
24. Li, Y.; Li, C. C.; Cho, S. O.; Duan, G. T.; and Cai, W. P. 2007. Silver hierarchical bowl-like array: Synthesis, superhydrophobicity, and optical properties. *Langmuir* 23: 9802–7.
25. Li, Y.; Cai, W. P.; Cao, B. Q.; Duan, G. T.; Sun, F. Q.; Li, C. C.; and Jia, L. C. 2006. Two-dimensional hierarchical porous silica film and its tunable superhydrophobicity. *Nanotechnology* 17: 238.
26. Li, Y.; Cai, W. P.; Cao, B. Q.; Duan, G. T.; Li, C. C.; Sun, F. Q.; and Zeng, H. B. 2006. Morphology-controlled 2D ordered arrays by heating-induced deformation of 2D colloidal monolayer. *J. Mater. Chem.* 16: 609–12.
27. Yoshimitsu, Z.; Nakajima, A.; Watanabe, T.; and Hashimoto, K. 2002. Effects of surface structure on the hydrophobicity and sliding behavior of water droplets. *Langmuir* 18: 5818–22.
28. Oner, D. and McCarthy, T. J. 2000. Ultrahydrophobic surfaces. Effects of topography length scales on wettability. *Langmuir* 16: 7777–82.
29. Li, Y.; Lee, E. J.; and Cho, S. O. 2007. Superhydrophobic coatings on curved surfaces featuring remarkable supporting force. *J. Phys. Chem. C* 111: 14813–7.
30. Li, Y.; Huang, X. J.; Heo, S. H.; Li, C. C.; Choi, Y. K.; Cai, W. P.; and Cho, S. O. 2007. Superhydrophobic bionic surfaces with hierarchical microsphere/SWCNT composite arrays. *Langmuir* 23: 2169–74.
31. Gao, X. F. and Jiang, L. 2004. Biophysics: Water-repellent legs of water striders. *Nature*. 432: 36.
32. Hu, D. L.; John, B. C.; and Bush, W. M. 2003. The hydrodynamics of water strider locomotion. *Nature* 24: 663–6.
33. Li, Y.; Sasaki, T.; Shimizu, Y.; and Koshizaki, N. 2008. Hexagonal-close-packed, hierarchical amorphous TiO_2 nanocolumn arrays: Transferability, enhanced photocatalytic activity, and superamphiphilicity without UV irradiation. *J. Am. Chem. Soc.* 130: 14755–62.
34. Li, Y.; Sasaki, T.; Shimizu, Y.; and Koshizaki, N. 2008. A hierarchically ordered TiO_2 hemispherical particle array with hexagonal-non-close-packed tops: Synthesis and stable superhydrophilicity without UV irradiation. *Small* 4: 2286–91.
35. Wang, R.; Hashimoto, K.; Fujishima, A.; Chikuni, M.; Kojima, E.; Kitamura, A.; Shimohigoshi, M. et al. 1998. Photogeneration of highly amphiphilic TiO_2 surfaces. *Adv. Mater.* 10: 135–8.

10 Optical Properties and Devices

10.1 INTRODUCTION

In Chapter 9, we have introduced wettability of the periodic micro/nanostructured arrays based on polystyrene (PS) colloidal crystal templates. In fact, in addition to the special surface wettability, these periodic micro/nanostructured arrays also exhibit some special structure-dependent optical properties including tunable optical properties, such as controlled visible light absorption, terahertz (THz) absorption, diffraction, photoluminescence (PL) properties, incident angle–independent stop band, optical gas sensing, and simple optical frequency splitting, which have important applications in surface-enhanced Raman scattering (SERS), sensors, optoelectronic device, wavelength meters, and so on. In this chapter, we introduce few progresses made in optical properties of the micro/nanostructured arrays, their structural dependences, and the array-based optical devices.

10.2 CONTROLLED OPTICAL PROPERTY OF GOLD NANOPARTICLE ARRAYS

In Chapter 7, we introduced fabrication of gold nanoparticle arrays based on the thermal evaporation on the PS colloidal monolayer template [1]. The Au nanoparticles on the substrate are hexagonally arranged with a P6mm symmetry and have triangular shape from top view due to the template geometry, as shown in Figure 7.2. It has been found that the optical properties of gold nanoparticle arrays are sensitive to the shape of the nanoparticles and the structural parameter of the arrays [2,3]. The morphology of the arrays could be directly and selectively manipulated by laser beam irradiation. As described in Chapter 7, the morphology of the gold nanoparticles in the array could gradually evolve from the triangular shape to the spherical shape with laser irradiation, as illustrated in Figure 7.3 or in the inset of Figure 10.1.

The optical absorption spectral evolution with the particle shape changing by laser irradiation is demonstrated in Figure 10.1. The as-prepared samples show a very broad absorbance peak centered around 680 nm together with a shoulder extending well into the near-infrared (NIR) region, which indicates that the peak is composed of at least two peaks. The peak at 680 nm decreases and disappears as the laser irradiation is increased up to 100 pulses. In addition, after irradiation by about 60 pulses, another peak emerges around 550 nm. After about 100 pulses, the peak shifts to 530 nm.

Jin et al.'s work [4] and Mie theory [5] indicate that triangular gold particles should exhibit surface plasmon resonance (SPR) containing one out-plane resonance

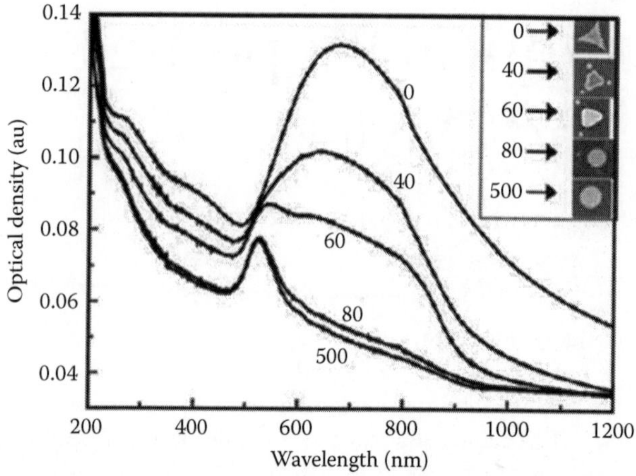

FIGURE 10.1 Evolution of the optical absorption spectra of gold nanoparticle arrays on quartz versus laser irradiation with the indicated number of laser pulses. Inset: the corresponding morphology of individual dots at each pulse number. (With kind permission from Springer Science+Business Media: *Appl. Phys. B*, 81, 2005, 765–8, Sun et al.)

and two in-plane resonances. The out-plane resonance should be at a shorter wavelength that is too weak to be discerned. One of the in-plane resonances should be around 530 nm and the other at a longer wavelength (depending on a morphology factor). Generally, spherical particles have only a single SPR band at about 530 nm. In the spectra shown in Figure 10.1, the broad absorption band around 680 nm for the array before irradiation can be attributed to the superposition of the two bands of the triangular-shaped particles. Laser irradiation makes particles spheroidize (see Figure 7.3e), leading to a decrease and eventual disappearance of the shape-dependent band around 680 nm, and leaving the single SPR band at 530 nm corresponding to a spherical particle. The variations of spectra can reflect the morphological changes of the particle arrays, which show the potential application in the optical data storage devices.

10.3 TERAHERTZ ABSORPTION BANDS IN AU/POLYSTYRENE SPHERE ARRAYS

Au was deposited on the PS colloidal monolayer arrays by means of ion-beam sputtering and the Au shell/PS sphere arrays were then formed [6]. The Au/PS spheres with gold shell thicknesses of 5, 10, and 15 nm were fabricated on Si substrate in the present study. The THz time-domain spectroscopy (TDS) was carried out for such micro/nanostructured periodic Au/dielectric sphere arrays. The metal–insulator transition can be achieved in THz bandwidth by varying the structural parameters, such as the thickness of the Au shell and the diameter of the Au/dielectric spheres. The Au/PS sphere arrays do not show metallic THz response when the Au shell thickness is larger than 10 nm and the PS sphere diameter is smaller than 500 nm. This effect is in sharp contrast to the observations in flat Au films on Si substrate.

Interestingly, the Au/PS sphere arrays with 5-nm Au shell thickness show extraordinary THz absorption bands or metallic optical conductance when the diameter of the sphere is larger than 200 nm. This effect is related to the quantum confinement effect in which the electrons in the structure are trapped in the sphere potential well of the gold shell.

THz TDS was performed using the standard configuration for the measurement of the THz transmission spectrum (see Figure 10.2). Here, (1) femtosecond laser pulses are employed as pumping light source, (2) pumping laser pulses are focused on GaAs photoconductive antenna that provides the broadband and pulsed THz radiation source, (3) the free-space electro-optic sampling through ZnTe crystal is applied for detection of light transmission in time domain, and (4) Fourier transformation of the measured data is carried out to obtain the THz transmission spectrum. In the present study, the THz TDS measurements were undertaken under a dry nitrogen purge at room temperature.

One of the advantages of using Au/PS sphere array as an optoelectronic device is that the optoelectronic properties of the structure can be tuned and modulated by varying the structural parameters of the arrays. For the presentation of the experimental results from this study, the transmission spectra of the Au/PS sphere arrays on Si substrate are normalized with the transmittance with respect to the Si substrate. In Figure 10.3, we show THz transmission spectra for different-sized Au/PS sphere arrays at a fixed Au shell thickness of 10 nm. The transmission spectrum for a 10-nm-thick Au film on Si substrate is shown for reference. The significant THz absorption occurs for the flat Au/Si film. This implies that a relatively high optical conductance is achieved in the flat film sample. For a normal conducting thin film on an insulating substrate, the electromagnetic energy extinction in the THz region is mainly

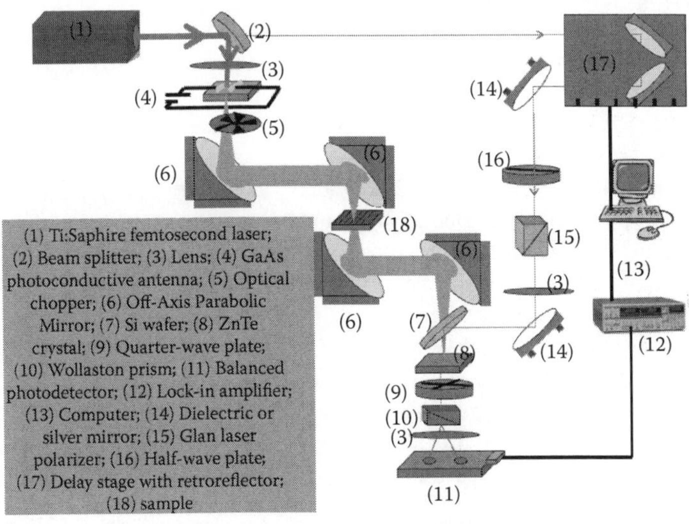

FIGURE 10.2 The experimental setup for the terahertz time-domain spectroscopy measurement (details are given in the text). (With kind permission from Springer Science+Business Media: *Nanoscale Res. Lett.*, 7, 2012, 657, Duan et al.)

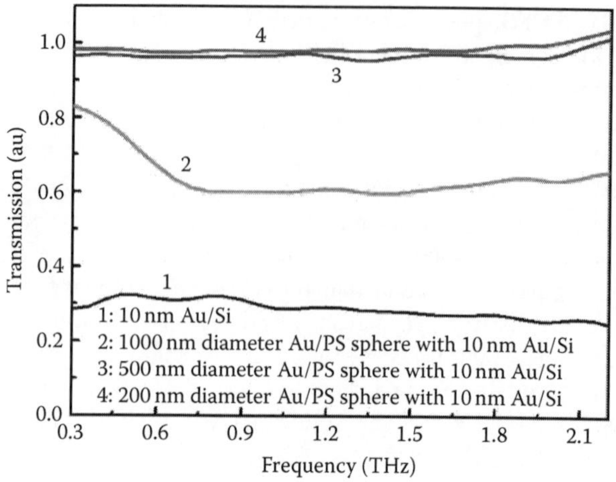

FIGURE 10.3 Normalized transmission spectra for different sized Au/polystyrene (PS) sphere array with 10-nm-thick Au shell. (With kind permission from Springer Science+Business Media: *Nanoscale Res. Lett.*, 7, 2012, 657, Duan et al.)

dominated by free-carrier motion driven by electric field component of the THz wave when neglecting the magnetic interaction in the film. Therefore, the THz transmission is directly associated with the complex conductivity of the film. It is well known that the Au film on dielectric substrate experiences a metal–insulator transition in the thickness range of 6–7 nm, which results in abrupt increase in THz absorption (or decrease in THz transmission) with increasing Au film thickness [7]. The THz transmittance of the thin Au/Si film is attributed to reduced electronic conduction, which arises due to backscattering of the carriers localized in disconnected gold islands [7]. The experimental results for flat Au/Si films are in line with these experimental findings. However, it should be noted that in the work by Walther et al. [7], the light transmittance for 10-nm-thick Au/Si films was observed at about 0.1, which is smaller than our results. The samples used in the measurements in the work by Walther et al. [7] were produced by thermal evaporation. Such a technique differs significantly from the ion-beam sputtering employed in the sample preparation in the present study. It is known that different techniques to grow Au films on Si wafer can result in different crystallizations of the Au nanoclusters. The better crystallization of the Au nanofilms prepared by thermal evaporation is the main reason why the light transmittance observed in the work by Walther et al. [7] is lower than that measured in this study.

The transmission spectra normalized with respect to the Si substrate for different diameters of Au/PS sphere in the array structures are shown in Figure 10.3, where the Au shell thickness is fixed to be 10 nm. The results for 10-nm-thick Au film on Si substrate are also shown for reference. It can be seen that the transmission spectra of the Au/PS sphere arrays with 10-nm Au shell thickness do not indicate effects similar to the metallic light response. Instead, the intensity of the transmittance for the Au/PS sphere arrays shows a strong dependence on the size of the Au/PS sphere. The THz transmission in a sphere array with 1000-nm sphere diameter and 10-nm Au shell thickness is pronouncedly larger than that for a flat Au film with the same thickness of the Au layer. In particular, the 200- and 500-nm diameter sphere arrays are almost transparent within the investigated THz bandwidth. This implies that the array samples are optically insulating in the THz region. Similar results can be observed for the array samples with 15-nm Au shell thickness. It is known that in a metal sphere array structure, the electron–electron interaction at the interface between adjacent spheres plays a crucial role in affecting optical conductance and transmittance. The arrangement of the Au/PS spheres in the manner of packing closely to each other can introduce relatively high interface carrier density. Normally, such a density is larger in arrays with smaller Au/PS spheres than in those with larger spheres. A larger interface carrier density corresponds to a stronger interface electronic scattering, which can reduce macroscopic conductivity of the sample. Thus, the optical conductance is smaller in arrays with smaller Au/PS spheres. Hence, the THz transmission increases with decreasing size of the Au/PS spheres in the array structures, as shown in Figure 10.3.

Figure 10.4 gives the normalized THz transmission spectra for Au/PS sphere arrays with 5-nm thickness in the gold shell and different diameters of spheres. Interestingly, two absorption peaks appeared at about 1.2 and 1.7 THz, which can be clearly observed from the transmission spectra in the samples with 1000- and 500-nm-diameter spheres. The transmission spectra with 500-nm-diameter Au/PS sphere in the arrays are shown in Figure 10.5 for different thicknesses of gold shells. The THz absorption peaks can only be observed clearly for the sample with 5-nm Au shell thickness. The results shown in Figures 10.4 and 10.5 indicate that there exists the critical thickness (about 5 nm) of Au shell for the metal–insulator transition in Au/PS sphere arrays. There are no absorption peaks on PS spheres without depositing gold layer and flat gold film on Si substrate. This effect suggests that the

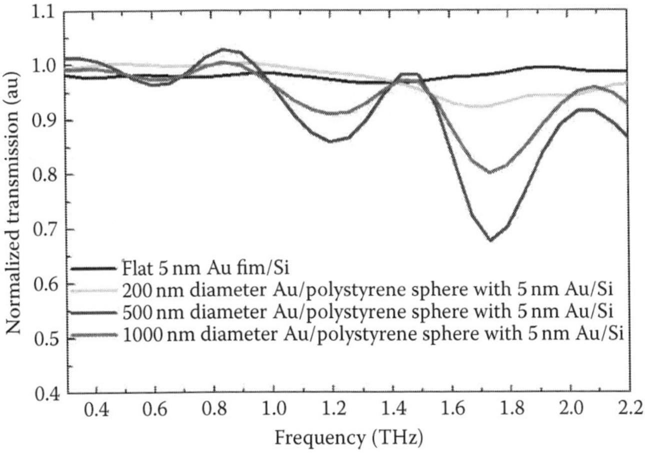

FIGURE 10.4 Normalized transmission spectra for Au/ polystyrene sphere array with different diameters and fixed 5-nm-thick Au shell. (With kind permission from Springer Science+Business Media: *Nanoscale Res. Lett.*, 7, 2012, 657, Duan et al.)

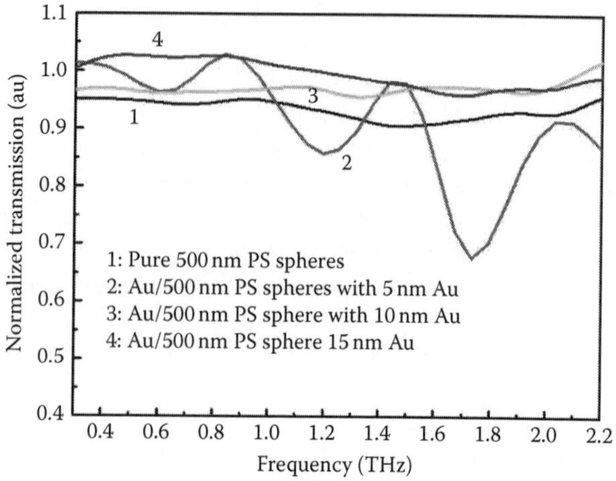

FIGURE 10.5 Normalized transmission spectra for Au/polystyrene (PS) sphere arrays with 500 nm diameter and different Au shell thicknesses. (With kind permission from Springer Science+Business Media: *Nanoscale Res. Lett.*, 7, 2012, 657, Duan et al.)

gold shell on the PS spheres affects the THz response in the array structure. The Surface plasmon polariton (SPP) is known to give rise to the extraordinary high transmission of light radiation through arrays of two-dimensional (2D) subwavelength holes. The observation of the THz absorption bands in Au/PS sphere arrays differs from the reported extraordinary THz transmission enhanced in 2D periodical metallic-dielectric structures. The lattice constant (about 500 nm) of the Au/PS sphere arrays is far smaller than that in common SPP THz devices (a few tens of microns). The first-order SPP resonant frequency is estimated to be around 170 THz, which is far beyond the spectroscopy range in this measurement.

Different from traditional planar metallic structural arrays on dielectric substrates, the Au/PS microsphere arrays introduce the quantum confinement of electron motion in radial direction between the barriers from dielectric spheres and air, when the thickness of the gold shell is thin. The highly confined electrons in the shell structure's potential well can form quantized electronic states with energy spacing in the THz range. This can result in the resonant absorption in THz bandwidth due to inter-subband electronic transition accompanied by the absorption of THz photons. In addition, the periodical structure of the Au shell arrays can also modulate the electronic states and the corresponding electron wave functions. For the closely packed microsphere arrays with thin Au shell, the electron wave function in one sphere can penetrate to the other neighboring spheres and form mini-band structures. This can result in a broadened THz absorption spectrum in the samples. Furthermore, the SPP modes in Au/PS sphere arrays differ significantly from those in Au/substrate film structures. Thus, the coupling between THz electromagnetic field and electrons trapped in the Au shell arrays has some unique features. The results here indicate that the strength of coupling between THz light radiation and electrons in the Au/PS sphere array can be efficiently tuned and modulated by varying the structural parameters such as the diameter of the Au/PS sphere and the thickness of the Au shell. Consequently, the Au/PS sphere array is a good electronic device in examining photon-induced metal–insulator transition in THz bandwidth.

It should be noted that different curvatures of the microspheres can normally lead to different cluster distributions on the spheres. This can also affect the THz transmission in the array structure. The investigation into different cluster modes in different arrays with different diameters needs considerable scanning electron microscopy study, which needs further work. As shown in Figure 10.4, the THz absorption bands occur at about 1.2 and 1.7 THz at a fixed shell thickness of about 5 nm for sphere diameters from 200 to 1000 nm. Our very recent theoretical calculations indicate that for hollow spherical structures and for diameters of the sphere larger than 100 nm, the electronic subband energies E_n degenerate and depend only on the shell thickness d_s through

$$E_n = \frac{\hbar^2 \pi^2 n^2}{2m^* d_s^2} \tag{10.1}$$

where n is the quantum number and m^* is the effective mass for an electron in gold shell [6]. We know that the THz absorption in the Au/PS microsphere array is mainly induced by the SPR and the absorption frequency is determined mainly by the surface plasmon frequency. In a quantum structure such as Au microsphere array, the surface plasmon frequency depends mainly on the energy difference between different electronic subbands. Thus, when the diameter of the Au nanosphere array is larger than 100 nm, the THz absorption frequency depends mainly on the Au shell thickness, instead of the sphere's diameter.

10.4 TUNABLE SURFACE PLASMON RESONANCE OF AU OPENING-NANOSHELL ORDERED ARRAYS

Nanoshells of noble metals, especially for Au, have attracted much attention due to their controllable local SPR properties by changing their structural parameters (curvature and thickness). In addition, Au nanoshell possesses a good biological compatibility and has a potential application in thermotherapy [8–12]. Au opening nanoshell can exhibit enhanced optical absorption and scattering cross sections at higher wavelengths, compared with an Au hollow sphere with nanosized shell [13–16]. If using the array consisting of Au opening nanoshell as a SERS substrate, it will have an advantage of good structural uniformity, which is highly valuable to design devices detecting organic molecules with stable performance, in addition to the strong electromagnetic coupling between two neighboring nanoshelled objects in the array.

As described in Chapters 7 and 8, the template technology based on a PS monolayer colloidal crystal gives us a flexible tool to fabricate periodically structured arrays, which can ensure the structural uniformity on a whole substrate [17–26]. However, tuning the local SPR of noble metals in a large range, especially to the NIR region, is still expected. Local SPR in NIR region for an array is very important for fiber-optic SERS sensors [27,28]. The reason is mainly as follows: (1) transmission of the SERS signal has little loss at the NIR region due to the absorption window; (2) excitation at a wavelength in the NIR region can reduce the effects of native autofluorescence and absorption on the SERS signal output; and (3) the array can also be used for biomedical application based on SERS effect because blood and tissue are most transparent in this region.

10.4.1 Arrays' Structure and Tunable SPR

In this section, we introduce tunable SPR of Au opening-nanoshell ordered arrays [29]. The Au opening-shell arrays could be fabricated based on sputtering deposition of Au onto monolayer PS colloidal crystal [29]. Briefly, a hexagonally close-packed PS colloidal monolayer on a silicon substrate is first bombarded or etched by argon plasma to obtain nonclose-packed colloidal monolayer [30–32]. A thin gold layer is then coated on the etched PS sphere surfaces through a plasma sputtering deposition. Because of the area contact of the PS spheres with the substrate, which can be controlled by heating, Au opening-nanoshell ordered arrays can be thus obtained after removal of PS colloidal template. The structural parameters of the array (shell thickness, curvature, opening size, and spacings between adjacent nanounits) can be controlled by deposition and templated conditions. So, the SPR and hence SERS performances can be tunable.

Figure 10.6a illustrates a typical morphology of the as-prepared Au nanoshell array, from the PS spheres with initial size of 200 and deposition thickness of 2.4 nm. It exhibits a hexagonal nonclose-packed array. The gaps or spacings between two

FIGURE 10.6 The morphology of the Au opening-nanoshell array obtained from the polystyrene (PS) colloidal template with 200-nm-diameter PS and deposition thickness of 2.4 nm. (a) Typical field emission scanning electron microscopy (FESEM) image. (b) The local magnified image of (a). (c) Transmission electron microscopy (TEM) image of the sample shown in (a), and the inset is the corresponding fast Fourier transform. (d) TEM image observed with a slantwise angle for the sample shown in (a). (Reprinted with permission from Liu et al., 2011, 1–5. Copyright 2011 American Chemical Society.)

adjacent units in the array are about 15 nm, and there are six symmetrical gaps around each unit, as shown in Figure 10.6b. Figure 10.6c shows the corresponding transmission electron microscopy (TEM) image of the array after scraping from the substrate. It still keeps the hexagonal nonclose-packed arrangement. The inset is the corresponding fast Fourier transform, and the diffraction pattern originates from the periodical structure of the Au array, which indicates the hexagonal-packed structure. In addition, Figure 10.6d presents the TEM image of the slightly slantwise array, from which the shell structure of each unit can be identified, according to the thick brim and the thin midst. And each unit in the array is an incomplete hollow sphere with a "placket" at bottom. So, it is called opening-shell structure. This is attributed to nonshadow deposition on PS spheres under plasma sputtering [33]. In addition, the spacings between neighboring two opening-nanoshells can be tuned by the sputtering deposition time, as shown in Figure 10.7, corresponding to different spacings from ~5 to ~20 nm depending on the sputtering deposition time.

Correspondingly, the spacing-dependent SPR could be measured. Figure 10.8a shows the optical absorbance spectra for the Au opening-shell arrays with spacings from 16 to 2 nm. There exists a local SPR peak around 760 nm for the array with 16-nm spacing (or 2-nm shell thickness). The SPR redshifts to 1530 nm (in NIR region) with decrease of the spacing from 16 to 2 nm or increase of the shell thickness from 2 to 9 nm. The SPR position shows near linear dependence on the spacings between two adjacent units in such arrays, as illustrated in Figure 10.8b (data from Figure 10.8a).

Such spacing or shell thickness–dependent SPR is different from that of the previous reports on the isolated Au shells (or hollow particles), which showed the SPR shifting to short wavelength by increasing the shell thickness [9,11]. Therefore, the

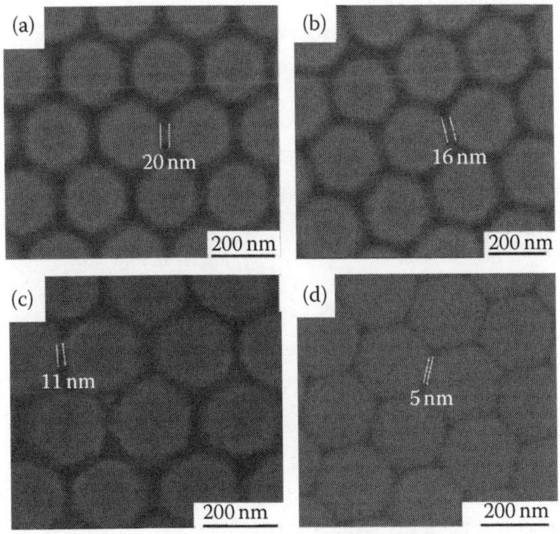

FIGURE 10.7 FESEM images of the Au opening-nanoshell arrays obtained by sputtering deposition for different time periods. (a) 5 seconds. (b) 20 seconds. (c) 45 seconds. (d) 75 seconds. (Reprinted with permission from Liu et al., 2011, 1–5. Copyright 2011 American Chemical Society.)

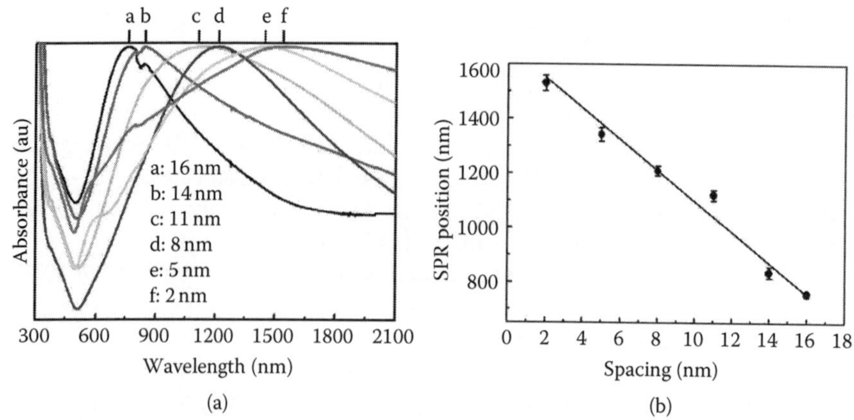

FIGURE 10.8 Optical absorption properties for the Au opening-nanoshell arrays with different spacings or shell thicknesses. (a) Optical absorbance spectra. (b) The surface plasmon resonance (SPR) position as a function of the spacing. (Data from [a], the line: guide to aid the eye). (Reprinted with permission from Liu et al., 2011, 1–5. Copyright 2011 American Chemical Society.)

SPR in this case should not mainly originate from a single Au opening nanoshell. Actually, increase of shell thickness corresponds to reduction of the spacing between the adjacent nanounits in the arrays. Such reduced spacing would induce the electromagnetic plasmonic coupling of the adjacent Au opening nanoshells. The coupling effect could enhance electromagnetic fields at the nanospaced sites [34–36]. Generally, the SPR peak will redshift with decrease of the spacing between adjacent nanounits as described by [25,37]

$$\frac{\Delta\lambda}{\lambda_0} = a\exp\left(-\frac{d}{bD}\right) \tag{10.2}$$

where λ_0 is the SPR wavelength in the uncoupling case, $\Delta\lambda$ is the redshift of SPR in wavelength, a and b are the fitting parameters (>0), and D and d are the size of a nanounit and the spacing between two adjacent ones, respectively. Obviously, $\Delta\lambda$ increases by decreasing the value of d (see Figure 10.8a). If $(d/bD) << 1$, Equation 10.2 can be changed to

$$\Delta\lambda \approx A - Bd \tag{10.3}$$

where A and B are the constants related to a, b, D, and λ_0. In this case, the SPR shift $\Delta\lambda$ should exhibit near linear dependence on the spacings. Obviously, Equation 10.3 can well describe the results shown in Figure 10.8b.

10.4.2 Dependence of SERS Effect on SPR

As we know, when the excitation is carried out at a wavelength near the SPR position, the achieved SERS signal will be the strongest [38,39]. However, the accurate tuning of the SPR around the position of the excitation wavelength is still expected. Here, we take the normal Raman excitation wavelength of 785 nm as an example to demonstrate the feasibility of accurate-tuning the SPR position and to show the dependence of the SERS activity on the SPR wavelength.

The SPR at 785 nm can be easily controlled for the Au opening-nanoshell array just by sputtering deposition time (24 seconds), as shown in curve c of Figure 10.9a, corresponding to the array in Figure 10.6. For comparison, the other Au opening-nanoshell arrays with the SPR near the excitation wavelength (785 nm) was also fabricated by sputtering deposition for different time periods, as illustrated in curves a and b and d–f of Figure 10.9a. Figure 10.9b shows the Raman spectra of 4-ATP molecules absorbed on these arrays with the SPR positions shown in Figure 10.9a correspondingly. The SERS signal is the most intense when the SPR is located at the site of the excitation wavelength, as shown in curve c of Figure 10.9b. This should be easily understood. When exciting at the position of SPR peak, the strongest electromagnetic plasmonic coupling will take place and induce the strongest SERS signal.

Such arrays, if used for SERS substrate, have the following advantages: (1) It has an even surface, which can ensure the uniformity of SERS signal across whole substrate, and stability for the device application based on SERS effect. (2) There exist many gaps, in this array, as "hot spots," leading to high SERS activity. (3) The local SPR property of this array can be tuned from visible to NIR region easily by changing the spacing between two adjacent nanounits. Therefore, the appropriate arrays can be controllably fabricated with the SPR position in accord with the excitation wavelength.

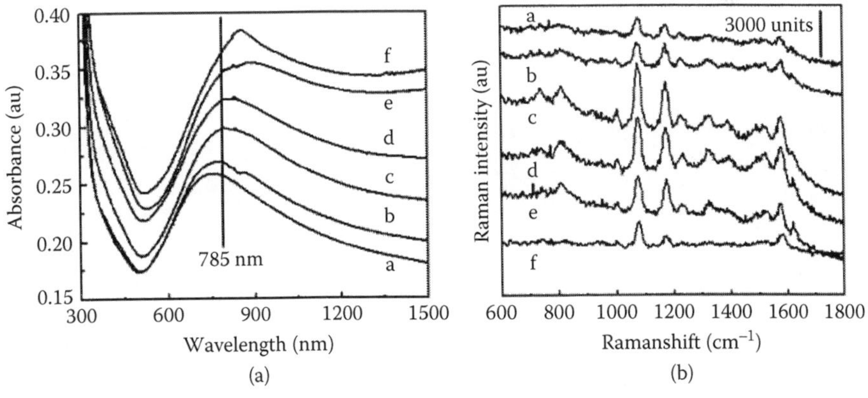

FIGURE 10.9 Comparison between (a) optical absorption spectra, for Au opening-nanoshell arrays with different fine-tuned shell thicknesses, and (b) Raman spectra of 4-ATP on the arrays. Curves a–f: the arrays by sputtering deposition for 22, 23, 24, 25, 26, 27 seconds, respectively. Excitation wavelength for Raman spectra: 785 nm. (Reprinted with permission from Liu et al., 2011, 1–5. Copyright 2011 American Chemical Society.)

10.5 PHOTOLUMINESCENCE PROPERTIES OF SEMICONDUCTOR PERIODIC ARRAYS

For some semiconductor ordered pore array films based on the PS colloidal monolayer, the PL properties were also investigated for such arrays.

10.5.1 PL of Oriented ZnO Ordered Pore Arrays

As described in Chapter 8, the oriented ZnO ordered pore array (along *c*-axis) induced by Au(111)/Si substrates can be fabricated based on the colloidal monolayer and electrodeposition technique (see Figure 8.31). Such oriented ordered pore array exhibited a strong UV emission without visible light, which was attributed to good crystal quality [40]. This ordered pore array could thus be of potential applications in future functional optical micro/nano devices.

Figure 10.10a shows the PL spectra of the oriented ordered porous ZnO films electrodeposited on Au(111)/Si substrate by the deposition currents of 0.7, 0.9, and 1.1 mA, shown in Figure 8.31. For all the arrays on Au/Si substrates, a UV light emission peak centered at about 380 nm (3.26 eV) was observed near the band edge of ZnO and the peak position did not show obvious shift with change of the electrodeposition conditions. Except the array electrodeposited at current of 0.9 mA, another emission peak was also observed at about 550 nm. This visible emission is usually attributed to the oxygen vacancies (Ov) in ZnO and originates from the recombination of a photogenerated hole with a single ionized charge state of this defect [40]. The intensity ratios of the UV to the green emission are about 3 and 20 for the samples deposited by currents of 0.7 and 1.1 mA, respectively. Additionally, the array prepared by current of 0.9 mA

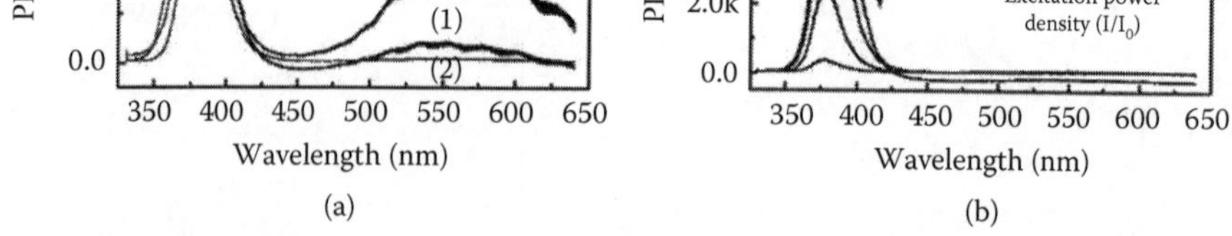

FIGURE 10.10 Room temperature photoluminescence (PL) spectra of ZnO ordered pore arrays on Au/Si substrates at the excitation wavelength of 325-nm He-Cd laser. (a) Different electrodeposition currents and time: 0.7 mA, 150 minutes; 0.9 mA, 120 minutes; 1.1 mA, 120 minutes at the same exciting intensity of 0.5 I_0. (b) Electrodeposition current: 0.9 mA, 120 minutes. The curves from bottom to top correspond to the excitation intensities: 0.10 I_0, 0.25 I_0, 0.50 I_0, and 1.00 I_0, respectively, and the inset shows plot of PL integrated intensity versus excitation intensity (R: the linear correlated coefficient). The excitation light source was 325-nm He-Cd laser and the full exciting power intensity (I_0) was about 2 kW/cm^2. (From Cao et al., *Electrochem. Solid State Lett.*, 8, G237–40, 2005.)

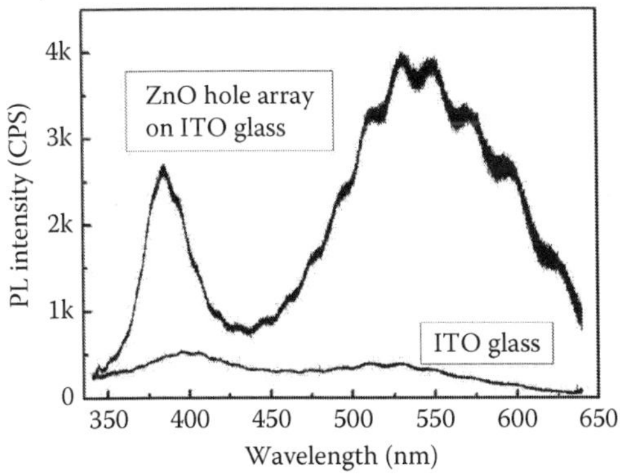

FIGURE 10.11 Room temperature photoluminescence (PL) spectra of ZnO ordered pore arrays (1.0 V, 120 minutes) on ITO-coated glass substrate and void substrate at the excitation wavelength of 325-nm He-Cd laser. (From Cao et al., *Electrochem. Solid State Lett.*, 8, G237–40, 2005.)

shows only UV PL without defect-related visible emission. Since the crystal quality of ZnO film is strongly related to the stability of the UV emission and the ratio of UV to visible emission, the oriented ZnO pore arrays are of high crystal quality, especially for that of 0.9 mA. Figure 10.10b shows the PL spectra of the samples deposited by current of 0.9 mA under different excitation power densities; only UV emission, no green emission, was detected. The integrated PL intensity increases linearly with the increase of excitation power density (see the inset of Figure 10.10b). The full width at half maximum of the UV emission is about 20 nm (~170 meV) and the peak does not show obvious shift for different excitation power densities. All these indicate that the UV luminescence originates from free or bound exciton recombination. The slight redshift of the UV emission at 3.26 eV, with respect to the theoretical position ($h\varpi = E_g - E_0 = 3.31\,\text{eV}$, E_g is the bandgap [3.37 eV]; E_0 is the ZnO exciton binding energy [60 meV]), can be attributed to the heating effect of excitation laser used in the experiment [40].

For comparison, the PL spectrum for the ZnO porous array electrodeposited under the potential of 1.0 V for 120 minutes on indium tin oxide (ITO)-coated glass substrate is shown in Figure 10.11, in which PL for ITO-coated glass substrate under the same excitation is also included. Besides the typical UV emission band, a broad and strong visible emission band is also detected, indicating the poor crystal quality and existence of many defects, such as Ov, in the random oriented ZnO pore array.

10.5.2 PL of In_2O_3 Ordered Pore Arrays

The In_2O_3 ordered pore arrays on silicon substrate could be fabricated by the sol-dipping and PS template method [41]. The morphologies of these ordered arrays are all similar to that of SiO_2 shown in Figure 8.8, as typically illustrated in Figure 10.12a. When excited at 250 nm, such In_2O_3 ordered pore arrays exhibit a PL emission peak mainly located in the blue-green region with maximum intensity centered at around 465 nm. Figure 10.12b shows the PL results of the In_2O_3 ordered porous array films by templates of 1000- and 350-nm PS spheres with 0.35 and 0.20 M precursor

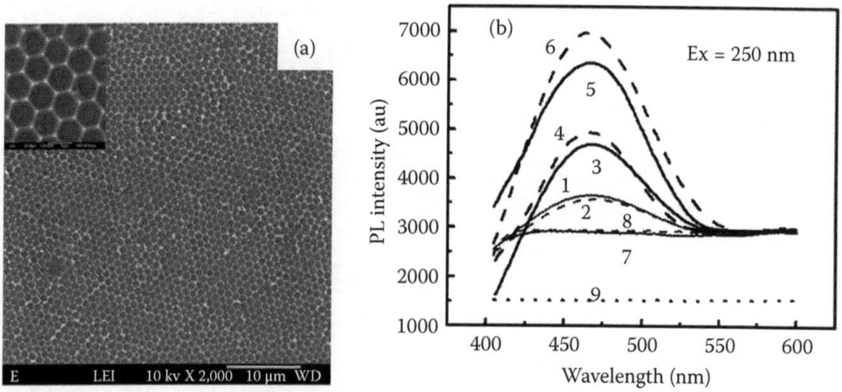

FIGURE 10.12 A typical FESEM image of In_2O_3 ordered pore arrays (a) and their photoluminescence (PL) spectra with excitation of 250 nm (b). Curves 1 and 2: as-prepared sample (original treatment at 400°C in air); curves 3 and 4: subsequent treatment in H_2 at 160°C for 1 hour for the as-prepared samples; curves 5 and 6: subsequent treatment in H_2 at 200°C for 1 hour for the as-prepared samples; curves 7 and 8: subsequent treatment in air at 700°C for 1 hour for the as-prepared samples; curve 9: Si wafer. Dashed line: samples prepared by PSs of 350 nm; solid line: samples prepared by PSs of 1000 nm. (With kind permission from Springer Science+Business Media: *Appl. Phys. A*, 81, 2005, 269–73, Li et al.)

concentration, respectively (see curves 1 and 2 in Figure 10.12b). The PL should come from the In_2O_3 skeleton since no peak was found in this region for silicon substrate. Also, the bulk In_2O_3 cannot emit light at room temperature [42]. Furthermore, if the as-prepared porous films were subsequently heat-treated in H_2 at 160°C or 200°C, the corresponding PL enhances significantly, especially for films with pore spacing of 350 nm (see curves 3–6 in Figure 10.12b). However, subsequent heat-treatment in air at a high temperature for the samples leads to disappearance of this peak (see curves 7 and 8 in Figure 10.12b). Obviously, this peak can mainly be attributed to oxygen deficiencies in In_2O_3. Treatment in H_2 induces oxygen deficiencies in In_2O_3 and hence increases the PL peak. After subsequent heat-treatment in air at high temperature, the oxygen deficiencies disappear, resulting in no PL peak. In addition, since the smaller pore size corresponds to higher specific surface of the arrays, subsequent treatment in H_2 for the array with smaller pore size should induce higher concentration of Ov and thus a more significant increase on PL (see curves 4 and 6 in Figure 10.12b).

10.6 QUASI-PHOTONIC CRYSTAL PROPERTIES OF MONOLAYER HOLLOW MICROSPHERE ARRAY

10.6.1 Tunable Optical Transmission Stop Band

The hierarchically structured $Ni(OH)_2$ monolayer hollow sphere array can be synthesized by electrodeposition, as typically shown in Figure 8.37. Such monolayer hollow sphere array could demonstrate tunable optical properties in a large region [43]. Figure 10.13 illustrates the optical transmission spectra with the incident light aligned perpendicularly to the as-prepared $Ni(OH)_2$ monolayer hollow sphere arrays on ITO. A

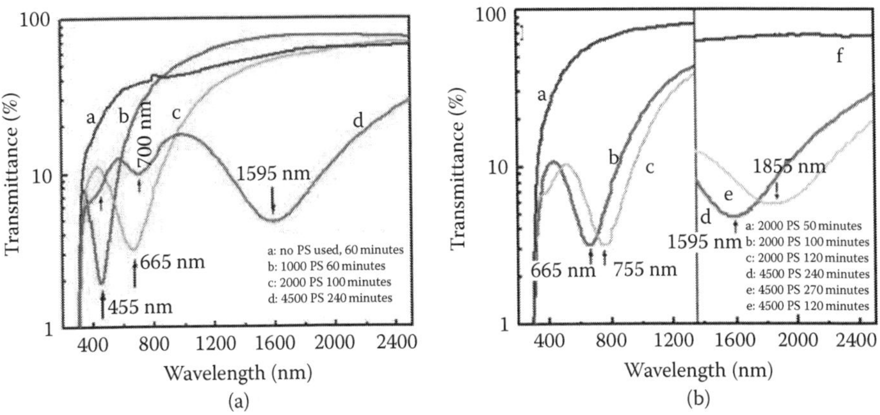

FIGURE 10.13 Optical transmission spectra of $Ni(OH)_2$ monolayer hollow sphere arrays on ITO-coated glass with the incident light perpendicular to the substrate. (a) and (b) Optical spectra of the sample by electrodeposition for different time using PS colloidal monolayer with different PS diameters. (From Duan et al., *Adv. Funct. Mater.*, 17, 644–50, 2007.)

size-dependent optical transmission stop band exists in this monolayer hollow sphere array, which redshifts in a large range from 455 to 1595 nm with the increase of the hollow spheres in size from 1000 to 4500 nm. For the array with hollow spheres from 4500-nm PSs, in addition to the main stop band around 1595 nm, there is another weaker band located around 700 nm together with a shoulder at 450 nm. Since the periodicity of the array is controlled by the PS sphere's size, the optical transmission stop band can thus be easily adjusted in a large region from the visible to NIR region by PSs' size. In contrary, no such stop band is detected for the netlike film (see curve a in Figure 10.13a, corresponding to the sample shown in Figure 8.37d), or for hierarchically structured pore array (see curves a and f in Figure 10.13b). This demonstrates that the monolayer hollow sphere structured array is crucial to produce such tunable optical bands.

In addition, the position of the optical transmission stop band can also be fine-adjusted by electrodeposition deposition time without change of periodicity. Increase of the deposition time leads to redshift of the band for the hollow sphere arrays, as shown in curves b–e of Figure 10.13b, which could be attributed to the denser shell of the hollow spheres. It means that the optical transmission stop band could be controlled flexibly in a large region, by PS sphere's size for coarse-tuning and deposition time for fine adjustment, which is undoubtedly important both in applications and in fundamental researches.

Importantly, further experiments have revealed that the position of the transmission stop band is almost independent of the incident angle θ (the angle between the incident light beam and the normal to the array plane). For the monolayer hollow sphere array induced by 2000-nm PS monolayer, the center of transmission stop band redshifts less than 30 nm with change of the incident angle from 0° to 75°, as demonstrated in Figure 10.14. This is quite small and almost negligible compared with the three-dimensional (3D) photonic crystals, which is very sensitive to θ according to [44] the following:

$$k\lambda_{\min} = 2d_1(n^2 - \sin^2\theta)^{1/2} \qquad (10.4)$$

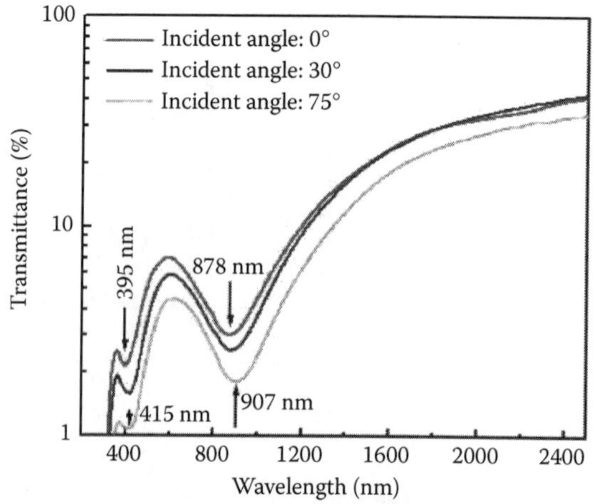

FIGURE 10.14 Optical transmission spectra of $Ni(OH)_2$ monolayer hollow sphere array on ITO-coated glass with different incident angles (the angle between incident light and the normal to the sample plane, θ). The array was prepared by electrodeposition with $J = 1.2$ mA/cm^2 for 140 minutes based on 2000-nm polystyrene monolayer. It has two clear transmission stop bands due to first and second diffractions, which is beneficial to the understanding of the shift of stop band. (Duan et al., *Adv. Funct. Mater.* 2007. 17. 644–50. Copyright Wiley-VCH Verlag GmbH & Co. KGaA. Reproduced with permission.)

where k is the order of the diffraction, $\lambda_{\min}$ is the wavelength of the stop band, n is the mean refraction index of the 3D crystal, and d_l is the periodical constant along the normal of the sample plane. It is well known that θ-dependent position of the stop band is a disadvantage of 3D photonic crystal in application. If one wants to prevent transmission of light from any incident angle, the fully photonic crystal with a θ-independent stop band is needed. This is a challenge for 3D photonic crystals. The monolayer hollow sphere array here could be a good candidate with θ-independent stop band.

10.6.2 Equivalent Double-Layer Photonic Crystal Approximation

Generally, an optical transmission stop band exists in 3D colloidal crystal (or inverse opal structure) but not in a 2D monolayer colloidal crystal if the incident light is perpendicular to the planar surface. The monolayer hollow sphere array here, however, can equivalently be considered as symmetrical double layers (top layer and bottom layer) with interspacing d_l, as shown in Figure 10.15, indicating a photonic crystal with double layers. Under perpendicular incidence of a light beam (or $\theta = 0$, see Figure 10.15a). Equation 10.4 can written as

$$k\lambda_{\min} = 2nd_l \tag{10.5}$$

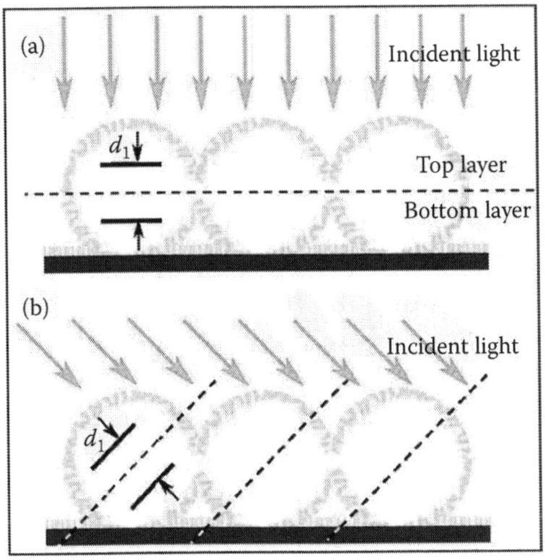

FIGURE 10.15 Schematic illustration of the equivalent double-layer photonic crystal approximation for the monolayer hollow sphere arrays. (a) Incidence of light perpendicular to the array plane ($\theta = 0$). (b) A slant incidence of light ($\theta > 0$). (Duan et al., *Adv. Funct. Mater.* 2007. 17. 644–50. Copyright Wiley-VCH Verlag GmbH & Co. KGaA. Reproduced with permission.)

where n corresponds to the mean refractive index of the layers (consisting of the sphere shell and interstice). Although the exact values of n and d_1 are unknown, the d_1 value should increase with rise in PS sphere's size, and n depends on material species and structure of the shell. Obviously, the denser the sphere shell (or the longer the deposition time), the larger the n value. Thus, for the first-order diffraction ($k = 1$), the transmission stop band should redshift with increase of the sphere size or deposition time (see Figure 10.13). Also, there should be a multiple relation between different diffraction orders ($k = 1, 2, 3$), which is in rough agreement with the result (see curve d in Figure 10.13). The slight deviation is because of the wavelength-dependent optical absorption of the $Ni(OH)_2$, which superimposes on the transmission spectra and hence changes the measured band position.

As for θ-independent stop band, it can be attributed to the special structure of monolayer hollow sphere. Due to the symmetry of the hollow spheres, for the incident light with different θ, d_1 value should be similar (see Figure 10.15b) and thus the position of stop band is almost θ-independent.

10.6.3 Optical Gas-Sensing Devices

As mentioned in Section 10.6.1 (Equation 10.4), the position of the transmission stop band is associated with n. So, the factors influencing n value will lead to the position change of the stop band. Here, the hollow spheres with the shells composed of massive nanoflakelets possess high specific surface area. Obviously, adsorption of

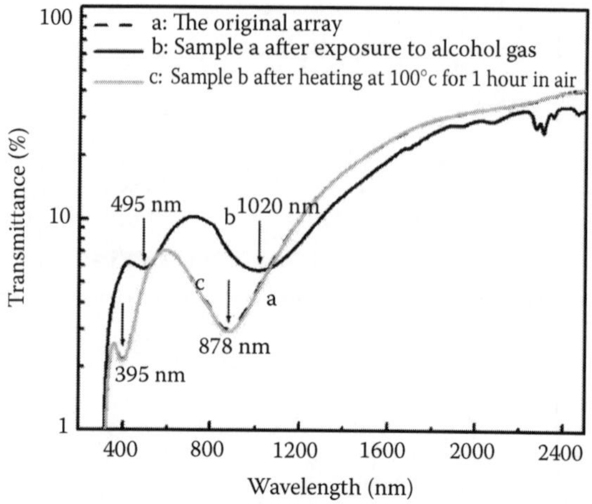

FIGURE 10.16 Optical transmission spectra for the array corresponding to that shown in Figure 10.14, with the incident light perpendicular to the substrate. After exposure of the original sample to alcohol gas (in a sealed cup with alcohol) for 30 minutes, the first and second stop bands redshifts more than 140 and 100 nm, respectively (curve b). By subsequent heating at 100°C for 1 hour in air (curve c), the bands are recovered completely. (Duan et al., *Adv. Funct. Mater.* 2007. 17. 644–50. Copyright Wiley-VCH Verlag GmbH & Co. KGaA. Reproduced with permission.)

environmental gases on the surface of the array will lead to change of n value and hence shift of the stop band. It means that such monolayer hollow sphere array with hierarchical structure could be a good optical gas sensor. Preliminary experiments have confirmed significant shift of the stop band induced by exposure to alcohol atmosphere, and such shift can be recovered completely by subsequent heating at 100°C in air, as demonstrated in Figure 10.16. Alternate exposure and heating would exhibit reversible shift of the band. It is due to $Ni(OH)_2$ that the hollow spheres with nanoflakelets could be produced. The monolayer hollow sphere array with such hierarchical structure provides practical possibility of a new optical sensor for gas detection.

Altogether, such monolayer hollow sphere arrays combine the merits of both patterned array and nanocrystals (or nanoflakelets) and exhibit potential applications in optical devices, photonic crystals, nanoscience, and nanotechnology. Among them could be the candidate of the fully photonic crystal with θ-independent stop band.

10.7 TRANSMISSION MODE WAVEMETER DEVICES BASED ON THE ORDERED PORE ARRAYS

It is well known that the traditional light wavelength measurement is based on Michelson interference principle, which is of high resolution and has been widely used in optical communication. However, the setup is very expensive and big-sized. So, it is expected to develop a simple light wavelength measurement with a small size in design and structure. As we know, when a beam of parallel incident light with the wavelength λ perpendicularly passes through the ordered pore arrays with periodical structure, diffraction will occur and the spots (or ring) will be seen if a screen is put behind the array due to the interference among unit cells of the arrays, which has been observed on ordered colloidal monolayer [45]. The diffraction angle depends on the wavelength of the incident light. On this basis, we can determine the wavelength of incident light by measuring the diffraction angle. So, the 2D micro/nanostructured ordered pore arrays could be used for optical frequency splitter. In this section, we introduce the light wavelength measurement based on the 2D ordered nanostructured porous arrays, which was prepared by solution-dipping on the PS colloidal monolayer [46], as described in Chapter 8. The corresponding setup is simple in design and structure, small in size, and can be applied to light wavelength measurement in some areas [47].

10.7.1 Basic Equations

When an incident light beam passes through a periodic pore array, interference could occur between the neighboring unit cells in the array, and diffraction would be very significant if the periodic parameter of the ordered arrays is comparable to the wavelength of incident light [48]. Usually, the centimeter-scaled 2D nanostructured ordered pore arrays consist of multicrystalline domains. If the beam size Φ of a parallel incident light is smaller than the domain size d_0 of the 2D ordered pore array, diffraction spots will appear on the screen behind the array after the light passes through the array perpendicularly. If $d_0 << \Phi$, however, diffraction rings can be seen on the screen, as schematically shown in Figure 10.17.

So, if the pores in an array are arranged in hexagonal structure, diffraction spots will also be of hexagonal pattern with spacing m between adjacent spots or diffraction ring with radius m, depending on the relative size of d_0 and Φ, where m is also the distance between optical axis and diffraction spots, as illustrated in Figure 10.18.

Generally, $l >> a$, we have

$$y - x = 2a\sin\theta \tag{10.6}$$

in which l is the distance between the screen and the array, $2a$ is the spacing between neighboring unit cells (pores) in the array (or lattice constant), x and y are the distances between pores and the diffraction spot on screen, respectively, and θ is the diffraction angle. Based on the interference condition $y - x = k\lambda$, one obtains

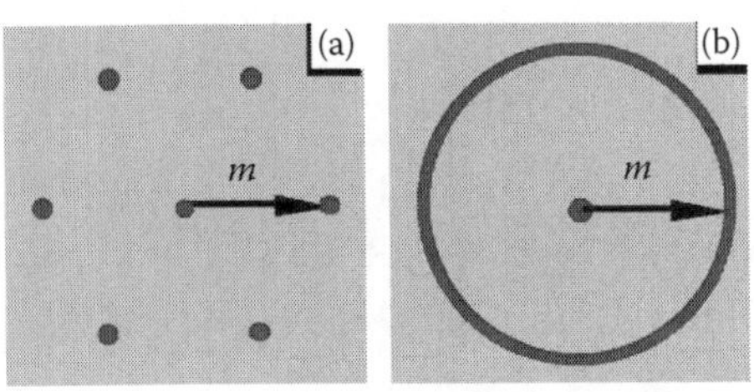

FIGURE 10.17 Schematic illustrations of diffraction patterns when a beam of incident light perpendicularly passes through two-dimensional ordered pore arrays with hexagonal closed packed pores (as shown in Figure 10.19). (a) $\Phi < d_0$, (b) $\Phi >> d_0$. (From Duan et al., *J. Nanosci. Nanotechnol.*, 6, 2474–8, 2006.)

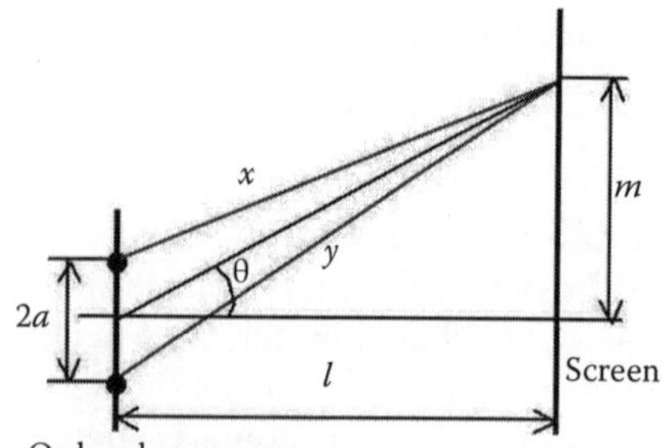

FIGURE 10.18 The sketch of interference between neighboring unit cells. (From Duan et al., *J. Nanosci. Nanotechnol.*, 6, 2474–8, 2006.)

$$\sin\theta = k\frac{\lambda}{2a} \tag{10.7}$$

where k is the interference order with integer 0, 1, 2 …, $\theta \in \left(0, \frac{\pi}{2}\right)$ and $\lambda < 2a$. From Figure 10.18 and letting $k = 1$, we have the relation

$$\lambda = \frac{2am}{\sqrt{m^2 + l^2}} \tag{10.8}$$

Obviously, if the m value is obtained, the wavelength of the incident light can be estimated by Equation 10.8. Such ordered through pore arrays could be used for optical frequency splitter. A new type of wavelength meter can thus be designed completely different from the conventional wavemeter in mode.

10.7.2 Optical Diffraction of the Ordered Pore Arrays

Here, the ordered pore arrays of PVA and Fe_2O_3 are taken as model materials, which were fabricated by the PS template solution-dipping method, as described in Chapter 8. Figure 10.19 shows the typical PVA ordered through-pore array, which is composed of many ordered domains with size d_0 from ten to hundreds of micron.

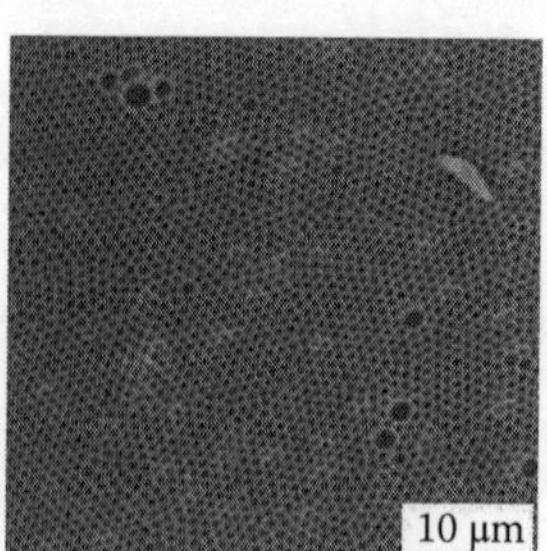

FIGURE 10.19 FESEM image of PVA ordered pore array with lower magnification. The periodic parameter of the ordered arrays is 1000 nm. (From Duan et al., *J. Nanosci. Nanotechnol.*, 6, 2474–8, 2006.)

A beam of coherent light and a beam of natural (white) light were used for incident light perpendicular to the ordered pore arrays. The former comes from second and third harmonic Nd:YAG lasers with a defined wavelength (532 and 355 nm) and about 4 mm in beam size or $\Phi >> d_0$. The screen covered with a piece of fluorescent paper, which can emit in visible region after excitation by ultraviolet light, was located behind and parallel to the sample. To eliminate the refraction influence from the substrate, the samples were put in such a way that the substrate surfaces with the ordered arrays face the screen.

Figure 10.20 gives the typical photos taken by a digital camera corresponding to the 532-nm laser light (Figure 10.20a) and natural light (Figure 10.20b) passing through the ordered pore arrays. There are a very clear and sharp diffraction ring for the laser incidence with single wavelength and colored spectra for the white incident light, indicating that such arrays can split wavelength of white incident light from blue to red continuously [47] (Lack of circularity of the diffraction ring in Figure 10.20b is because the incident white light is not a strict parallel beam.) Such diffraction depends only on the wavelength of incident light and the pore spacing, but independent of the material species, as shown in Table 10.1, which gives the ring radii (corresponding to the distance from center of the ring to the middle of line in the diffraction ring) and the corresponding wavelength calculated from

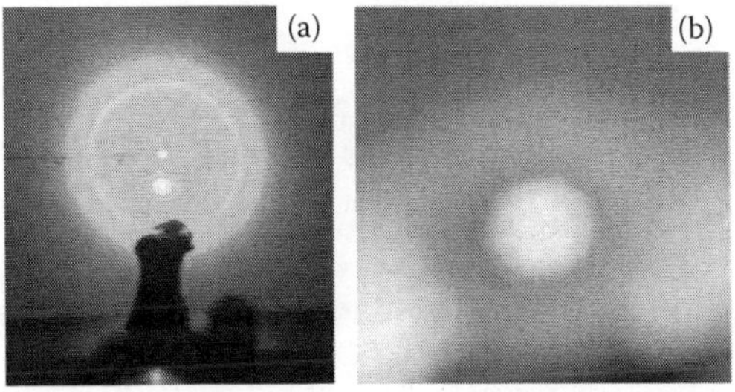

FIGURE 10.20 Diffraction patterns corresponding to the incident light (a) 532-nm laser beam and (b) natural light passing through ferric oxide ordered pore arrays. The periodic parameter is 1000 nm. (From Duan et al., *J. Nanosci. Nanotechnol.*, 6, 2474–8, 2006.)

TABLE 10.1
Diffraction Ring Sizes and Corresponding Wavelength Calculated from Equation 10.8

Incident Light (nm)	Samples	Ring Radius m (cm)	λ from Equation 10.8 (nm)
532	PVA pore array	7.6 ± 0.2	535 ± 10
532	α-Fe_2O_3 pore array	7.5 ± 0.2	530 ± 10
355	PVA pore array	4.6 ± 0.2	357 ± 14
355	α-Fe_2O_3 pore array	4.5 ± 0.2	351 ± 14

Source: Duan et al., *J. Nanosci. Nanotechnol.*, 6, 2474–8, 2006.
Note: $l = 12$ cm and $2a = 1000$ nm.

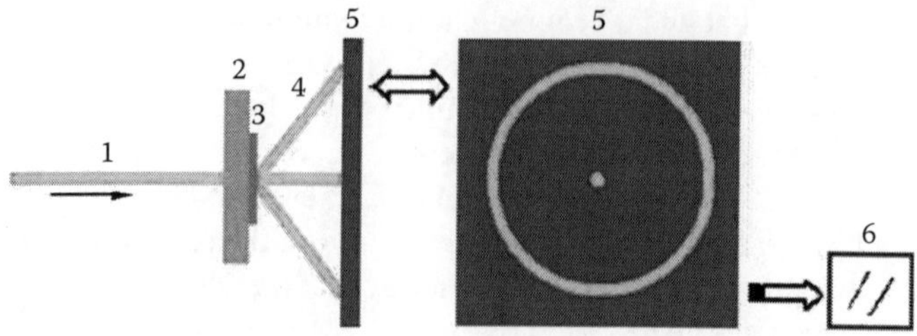

FIGURE 10.21 The sketch of a wavemeter in transmission mode. 1: A parallel incident beam obtained by an optical component; 2: glass substrate; 3: ordered pore arrays on the substrate; 4: diffraction light beam; 5: screen; and 6: controller. (From Duan et al., *J. Nanosci. Nanotechnol.*, 6, 2474–8, 2006.)

Equation 10.8 for all ordered through-pore arrays when $l = 12$ cm and $2a = 1000$ nm using laser incidence. The measured wavelength is in good agreement with the real one. In addition, the ring thickness and the size of zero-order diffraction spot were found always nearly equal to the laser beam size Φ, independent of the l value in this study due to the highly parallel light.

Based on the preceding results, a new type of setup for wavelength measurement could be designed in transmission mode, as schematically illustrated in Figure 10.21. The ordered pore array is a key component. A beam of light passes through a special optical component (e.g., waveguide and lens) to form a beam of parallel light with a certain beam size, which makes incidence on the arrays perpendicularly, leading to diffraction ring pattern on the screen. Then, the ring size can be measured and the wavelength of the incident light can thus be obtained from Equation 10.8.

10.7.3 Resolution and Precision of Wavelength Measurement

The diffraction ring has a thickness, as seen in Figure 10.20a. This should mainly be attributed to the monochromatic quality and beam size of the incident light. Both of them affect the resolution and precision in determination of the wavelength.

The monochromatic quality of a beam of incident light will influence the thickness of diffraction ring. Obviously, the higher resolution corresponds to the thicker ring at a given monochromatic quality. From Equation 10.8, we have

$$\frac{\partial m}{\partial \lambda} = K \tag{10.9}$$

where

$$K = \frac{l}{2a\left[1-(\lambda/2a)^2\right]^{3/2}} \tag{10.10}$$

or the differential form $\delta m = K\delta\lambda$. K determines the effect of monochromatic quality on the ring's thickness, depending on the parameters λ, l, and a. The higher K value corresponds to larger effect. The K versus λ at different l values is illustrated in Figure 10.22. In the long wavelength region (especially close to the value $2a$), the

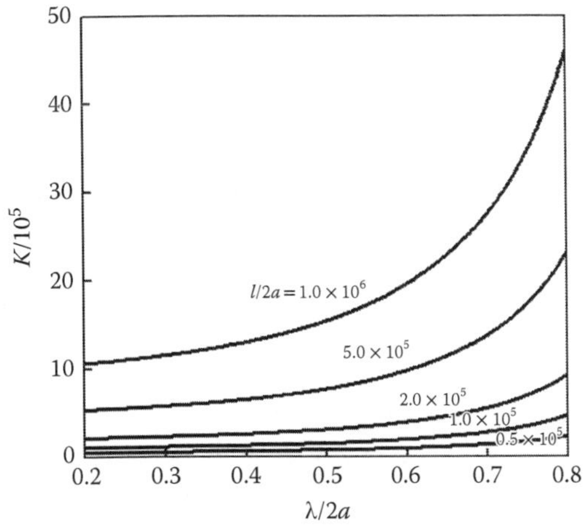

FIGURE 10.22 Curves of K versus λ for different $l/2a$ values. (From Duan et al., *J. Nanosci. Nanotechnol.*, 6, 2474–8, 2006.)

K value is large. When $\lambda/2a <$ ~0.6, the K is kept in a lower value. Furthermore, the larger l value corresponds to the larger K value. These indicate that the wavelength $\lambda/2a <$ ~0.6 and smaller l value correspond to the sharper ring at a given monochromatic quality. On the other hand, the larger K value should correspond to the higher resolution to determine the wavelength of incident light.

For a quantitative understanding, letting minimum discernable spacing on the screen $\delta m = 0.5$ mm, $2a = 1000$ nm, $\lambda/2a = 0.8$–0.6, and $l = 20$ cm, $K = 0.93$–0.39 mm/nm and resolution for incident light wavelength $\delta\lambda = 0.6$–1.3 nm are obtained. It means that the wavelength resolution could reach <1 nm in the wavelength region close to $2a$. It should be mentioned that this is a theoretic resolution without consideration of the effect from the beam size.

A beam with size Φ will induce the ring thickness or $\delta m \approx \Phi$. From Equation 10.8, the corresponding uncertainty of the wavelength for a monochromatic incident light $\delta\lambda = \Phi/K$. Evidently, the wavelength $\lambda/2a > -0.6$ and large l value should lead to the low uncertainty or high precision for determining λ. If considering values of λ/2a, l, $2a$, the same as preceding, $\delta\lambda = (1.1$–$2.6)\Phi$, in which Φ and $\delta\lambda$ are in mm and nm units, respectively. If using $\Phi = 0.5$ mm, $\delta\lambda = 0.6$–1.3 nm. It means that precision of the wavelength determined by this wavelength meter could reach <1 nm in the wavelength region close to $2a$ and the beam size of incident light smaller than 0.5 mm.

In addition, it should be pointed out that there inevitably exist some defects or disorderly arranged areas in the 2D ordered pore arrays with multidomains, such as boundaries among domains, vacancies, and dislocations. Fortunately, the percentage of such defects is low compared with the regular unit cells in the arrays, although it is difficult to measure it accurately. The diffraction rings are produced only by domains with periodic arrangement. The disorderly arranged areas or defects cannot form the diffraction rings or spots on the screen, it cannot significantly change the diffraction ring thickness or spot size caused by order structure in the array.

Obviously, if percentage of defects in the 2D ordered pore array is high or the crystallinity is low, the contrast of the diffraction ring produced by the domains would be weak.

10.8 BRIEF SUMMARY

In summary, we have introduced some optical properties and devices for some typical metal, semiconductor, and compound micro/nanostructured arrays; their structural dependences; and the array-based optical devices. Due to the unique structure and morphology, these periodic micro/nanostructured arrays have exhibited structure-dependent optical properties including tunable optical properties, controlled visible light absorption, THz absorption, diffraction, PL properties, incident angle-independent stop band, optical gas sensing, and simple optical frequency splitting. Also, the possibility that the 2D ordered pore arrays could be used for wavemeter with a certain precision has been demonstrated.

REFERENCES

1. Sun, F.; Cai, W.; Li, Y.; Duan, G.; Nichols, W.; Liang, C.; Koshizaki, N.; Fang, Q.; and Boyd, I. 2005. Laser morphological manipulation of gold nanoparticles periodically arranged on solid supports. *Appl. Phys. B* 81: 765–8.
2. Haynes, C. L. and Van Duyne, R. P. 2001. Nanosphere lithography: A versatile nanofabrication tool for studies of size-dependent nanoparticle optics. *J. Phys. Chem. B* 105: 5599–611.
3. Haynes, C. L.; McFarland, A. D.; Smith, M. T.; Hulteen, J. C.; and Van Duyne, R. P. 2002. Angle-resolved nanosphere lithography: Manipulation of nanoparticle size, shape, and interparticle spacing. *J. Phys. Chem. B* 106: 1898–902.
4. Jin, R.; Cao, Y.; Mirkin, C. A.; Kelly, K.; Schatz, G. C.; and Zheng, J. 2001. Photoinduced conversion of silver nanospheres to nanoprisms. *Science* 294: 1901–3.
5. Kreibig, U. and Vollmer, M. 1995. *Introduction, in Optical Properties of Metal Clusters*, Springer, Berlin, Germany, pp. 1–12.
6. Duan, G.; Su, F.; Xu, W.; Zhang, C.; and Cai, W. 2012. Extraordinary terahertz absorption bands observed in micro/nanostructured Au/polystyrene sphere arrays. *Nanoscale Res. Lett.* 7: 657.
7. Walther, M.; Cooke, D.; Sherstan, C.; Hajar, M.; Freeman, M.; and Hegmann, F. 2007. Terahertz conductivity of thin gold films at the metal-insulator percolation transition. *Phys. Rev. B* 76: 125408.
8. Liu, G. L.; Lu, Y.; Kim, J.; Doll, J. C.; and Lee, L. P. 2005. Magnetic nanocrescents as controllable surface-enhanced Raman scattering nanoprobes for biomolecular imaging. *Adv. Mater.* 17: 2683–8.
9. Zimmermann, C.; Feldmann, C.; Wanner, M.; and Gerthsen, D. 2007. Nanoscale gold hollow spheres through a microemulsion approach. *Small* 3: 1347–9.
10. Prevo, B. G.; Esakoff, S. A.; Mikhailovsky, A.; and Zasadzinski, J. A. 2008. Scalable routes to gold nanoshells with tunable sizes and response to near-infrared pulsed-laser irradiation. *Small* 4: 1183–95.
11. Zeng, J.; Huang, J.; Lu, W.; Wang, X.; Wang, B.; Zhang, S.; and Hou, J. 2007. Necklace-like noble-metal hollow nanoparticle chains: Synthesis and tunable optical properties. *Adv. Mater.* 19: 2172–6.
12. Liang, H. P.; Wan, L. J.; Bai, C. L.; and Jiang, L. 2005. Gold hollow nanospheres: tunable surface plasmon resonance controlled by interior-cavity sizes. *J. Phys. Chem. B* 109: 7795–800.

13. Love, J. C.; Gates, B. D.; Wolfe, D. B.; Paul, K. E.; and Whitesides, G. M. 2002. Fabrication and wetting properties of metallic half-shells with submicron diameters. *Nano Lett.* 2: 891–4.
14. Charnay, C.; Lee, A.; Man, S.-Q.; Moran, C. E.; Radloff, C.; Bradley, R. K.; and Halas, N. J. 2003. Reduced symmetry metallodielectric nanoparticles: Chemical synthesis and plasmonic properties. *J. Phys. Chem. B* 107: 7327–33.
15. Ye, J.; Van Dorpe, P.; Van Roy, W.; Lodewijks, K.; De Vlaminck, I.; Maes, G.; and Borghs, G. 2009. Fabrication and optical properties of gold semishells. *J. Phys. Chem. C* 113: 3110–5.
16. Ye, J.; Van Dorpe, P.; Van Roy, W.; Borghs, G.; and Maes, G. 2009. Fabrication, characterization, and optical properties of gold nanobowl submonolayer structures. *Langmuir* 25: 1822–7.
17. Liu, X.; Linn, N. C.; Sun, C. H.; and Jiang, P. 2010. Templated fabrication of metal half-shells for surface-enhanced Raman scattering. *Phys. Chem. Chem. Phys.* 12: 1379–87.
18. Wang, C.; Ruan, W.; Ji, N.; Ji, W.; Lv, S.; Zhao, C.; and Zhao, B. 2010. Preparation of nanoscale Ag semishell array with tunable interparticle distance and its application in surface-enhanced Raman scattering. *J. Phys. Chem. C* 114: 2886–90.
19. Baia, L.; Baia, M.; Popp, J.; and Astilean, S. 2006. Gold films deposited over regular arrays of polystyrene nanospheres as highly effective SERS substrates from visible to NIR. *J. Phys. Chem. B* 110: 23982–6.
20. Farcau, C. and Astilean, S. 2009. Silver half-shell arrays with controlled plasmonic response for fluorescence enhancement optimization. *Appl. Phys. Lett.* 95: 193110.
21. Liu, J.; McBean, K. E.; Harris, N.; and Cortie, M. B. 2007. Optical properties of suspensions of gold half-shells. *Mater. Sci. Eng. B-Solid* 140: 195–8.
22. Li, Y.; Koshizaki, N.; and Cai, W. 2011. Periodic one-dimensional nanostructured arrays based on colloidal templates, applications, and devices. *Coord. Chem. Rev.* 255: 357–73.
23. Li, Y.; Cai, W.; and Duan, G. 2008. Ordered micro/nanostructured arrays based on the monolayer colloidal crystals. *Chem. Mater.* 20: 615–24.
24. Li, L.; Zhai, T.; Zeng, H.; Fang, X.; Bando, Y.; and Golberg, D. 2011. Polystyrene sphere-assisted one-dimensional nanostructure arrays: Synthesis and applications. *J. Mater. Chem.* 21: 40–56.
25. Zhang, J. and Yang, B. 2010. Patterning colloidal crystals and nanostructure arrays by soft lithography. *Adv. Funct. Mater.* 20: 3411–24.
26. Masson, J. F.; Gibson, K. F.; and Provencher-Girard, A. 2010. Surface-enhanced Raman spectroscopy amplification with film over etched nanospheres. *J. Phys. Chem. C* 114: 22406–12.
27. Konorov, S. O.; Addison, C. J.; Schulze, H. G.; Turner, R. F.; and Blades, M. W. 2006. Hollow-core photonic crystal fiber-optic probes for Raman spectroscopy. *Opt. Lett.* 31: 1911–3.
28. Gessner, R.; Rösch, P.; Petry, R.; Schmitt, M.; Strehle, M.; Kiefer, W.; and Popp, J. 2004. The application of a SERS fiber probe for the investigation of sensitive biological samples. *Analyst* 129: 1193–9.
29. Liu, G.; Li, Y.; Duan, G.; Wang, J.; Liang, C.; and Cai, W. 2011. Tunable surface plasmon resonance and strong SERS performances of Au opening-nanoshell ordered arrays. *ACS Appl. Mater. Interfaces* 4: 1–5.
30. Lou, Y.; Lunardi, L. M.; and Muth, J. F. 2010. Fabrication of nanoshell arrays using directed assembly of nanospheres. *IEEE Sens. J.* 10: 617–20.
31. Murray-Méthot, M.-P.; Ratel, M.; and Masson, J.-F. 2010. Optical properties of Au, Ag, and bimetallic Au on Ag nanohole arrays. *J. Phys. Chem. C* 114: 8268–75.
32. Murray-Methot, M. P.; Menegazzo, N.; and Masson, J. F. 2008. Analytical and physical optimization of nanohole-array sensors prepared by modified nanosphere lithography. *Analyst* 133: 1714–21.

33. Duan, G.; Lv, F.; Cai, W.; Luo, Y.; Li, Y.; and Liu, G. 2010. General synthesis of 2D ordered hollow sphere arrays based on nonshadow deposition dominated colloidal lithography. *Langmuir* 26: 6295–302.
34. Olk, P.; Renger, J.; Härtling, T.; Wenzel, M. T.; and Eng, L. M. 2007. Two particle enhanced nano Raman microscopy and spectroscopy. *Nano Lett.* 7: 1736–40.
35. Talley, C. E.; Jackson, J. B.; Oubre, C.; Grady, N. K.; Hollars, C. W.; Lane, S. M.; Huser, T. R.; Nordlander, P.; and Halas, N. J. 2005. Surface-enhanced Raman scattering from individual Au nanoparticles and nanoparticle dimer substrates. *Nano Lett.* 5: 1569–74.
36. Härtling, T.; Alaverdyan, Y.; Hille, A.; Wenzel, M.; Käll, M.; and Eng, L. 2008. Optically controlled interparticle distance tuning and welding of single gold nanoparticle pairs by photochemical metal deposition. *Opt. Express* 16: 12362–71.
37. Jain, P. K. and El-Sayed, M. A. 2010. Plasmonic coupling in noble metal nanostructures. *Chem. Phys. Lett.* 487: 153–64.
38. Lee, S. H.; Bantz, K. C.; Lindquist, N. C.; Oh, S. H.; and Haynes, C. L. 2009. Self-assembled plasmonic nanohole arrays. *Langmuir* 25: 13685–93.
39. McFarland, A. D.; Young, M. A.; Dieringer, J. A.; and Van Duyne, R. P. 2005. Wavelength-scanned surface-enhanced Raman excitation spectroscopy. *J. Phys. Chem. B* 109: 11279–85.
40. Cao, B.; Sun, F.; and Cai, W. 2005. Electrodeposition-induced highly oriented zinc oxide ordered pore arrays and their ultraviolet emissions. *Electrochem. Solid-State Lett.* 8: G237–40.
41. Li, Y.; Cai, W.; Duan, G.; Sun, F.; Cao, B.; Lu, F.; Fang, Q.; and Boyd, I. 2005. Large-area In_2O_3 ordered pore arrays and their photoluminescence properties. *Appl. Phys. A* 81: 269–73.
42. Ohhata, Y.; Shinoki, F.; and Yoshida, S. 1979. Optical properties of rf. reactive sputtered tin-doped In_2O_3 films. *Thin Solid Films* 59: 255–61.
43. Duan, G.; Cai, W.; Luo, Y.; and Sun, F. 2007. A hierarchically structured $Ni(OH)_2$ monolayer hollow-sphere array and its tunable optical properties over a large region. *Adv. Funct. Mater.* 17: 644–50.
44. Richel, A.; Johnson, N.; and McComb, D. 2000. Observation of Bragg reflection in photonic crystals synthesized from air spheres in a titania matrix. *Appl. Phys. Lett.* 277: 1062.
45. Prevo, B. G. and Velev, O. D. 2004. Controlled, rapid deposition of structured coatings from micro-and nanoparticle suspensions. *Langmuir* 20: 2099–107.
46. Sun, F.; Cai, W.; Li, Y.; Cao, B.; Lei, Y.; and Zhang, L. 2004. Morphology-controlled growth of large-area two-dimensional ordered pore arrays. *Adv. Funct. Mater.* 14: 283–8.
47. Duan, G.; Cai, W.; Li, Y.; and Cao, B. 2006. Measurement of light wavelength based on nanostructured ordered pore arrays. *J. Nanosci. Nanotechnol.* 6: 2474–8.
48. Sommerfeld, A. 1964. *Optics*, Academic Press, San Diego, CA.

11 Gas-Sensing Devices and Structurally Enhanced Gas-Sensing Performances

11.1 INTRODUCTION

Semiconductor gas sensors have attracted much attention because of their applications in medical diagnosis, environmental monitor, personal safety, and national security [1–3]. It realizes the detection, analysis, monitor, and alarm based on the change of electrical signal (such as electrical current, resistance, or voltage), which results from the adsorption and/or reaction of gas molecules on the surface of semiconductor films [4–6]. Their performances correlate with the interaction between the testing gas molecules and the adsorbed oxygen molecules on surface of sensing films and are usually valuated by five basic parameters including sensitivity, selectivity, response time, operating temperature, and long-term stability. Herein, the gas sensitivity (S) is defined as a ratio of two electric resistance values, or $S = R_{air}/R_{g0}$, where R_{air} is the resistance value of the sensor in air, and R_{g0} is the steady value of the resistance after exposure to the test gas. Response time (t_R) is defined as the time required for the resistance to reach 90% of the total variation of resistance after exposure to the test gas.

Obviously, the increase of the surface areas and active surface sites of sensing materials can greatly enhance the properties of gas sensors. Nanoscale materials, especially nanocrystals, have received great attention as potential building blocks due to their inherently high surface area [7–13]. However, nanoparticles are usually interconnected and aggregated, which are unfavorable to the effective use of the active surface sites and the gas diffusion inside the film. Therefore, production of the nanostructured films with the homogeneous thickness as well as controlled pore size and porosity is very important both in scientific interest and practical application for development of the next generation of nanostructured gas sensors. Mono- or multilayer-ordered micro/nanostructured porous films have high surface-to-volume ratio, have large pore size, and facilitate high accessibility for the gas molecules [14–16]. More importantly, these films can be constructed directly on any desired substrate with the homogeneous and controlled film thickness and the controllable pore sizes through various techniques as mentioned above. Thus, ordered porous films would be the good candidates of new gas-sensing elements. In this chapter, we will introduce the gas-sensing devices and performances based on homopore- and heteropore-sized porous films.

11.2 GAS SENSORS BASED ON HOMOPORE-SIZED POROUS FILMS

11.2.1 Construction of Sensing Device

As introduced in Section 8.2.2, the ordered porous films can be directly synthesized on curved surface based on the sol–gel strategy. In this way, gas sensors with the porous films on commercially available ceramic tubes can be fabricated easily [14]. As illustrated in Figure 11.1a, a commercial ceramic tube with two preformed gold electrodes at the parts close to its two ends was used as substrate. The length, outer and inner diameters of the tube are about 5, 2, and 1 mm, respectively. Based on the strategy shown in Figure 8.15, the colloidal monolayer floating on the surface of desired solution is directly picked up by a ceramic tube with a relatively rough surface. After subsequent heat treatments, the desired ordered porous film can be fabricated on surface of the ceramic tube. Then, one Pt wire was put inside the tube to control the operating temperature of gas sensors. Finally, all the electrodes and Pt wire were soldered on the pedestal of gas sensor; a gas-sensing device can be constructed as shown in Figure 11.1b.

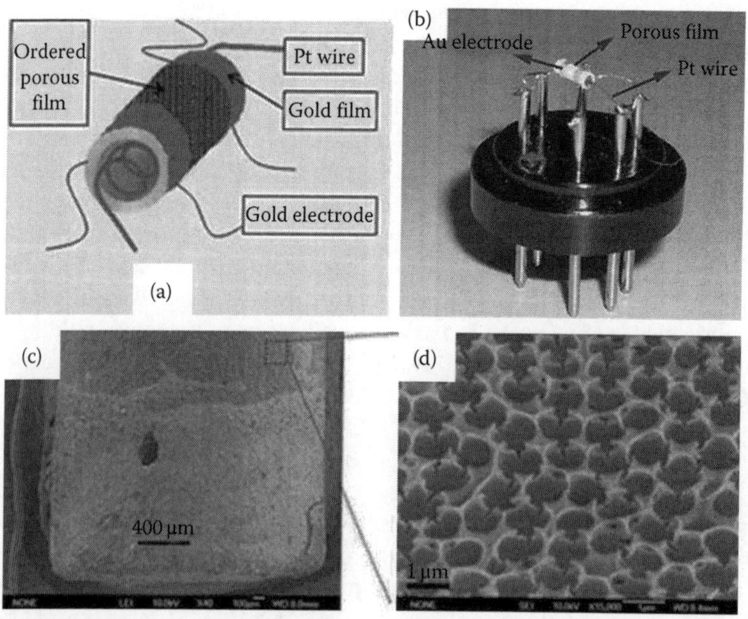

FIGURE 11.1 A gas sensor fabricated with an SnO_2 monolayer–ordered pore array film on a ceramic tube. (a) The structure of the gas sensor. (b) Photograph of a final gas sensor. (c) A low-magnification field emission scanning electron microscopy (FESEM) image of the gas sensor surface. (d) The magnified image of the area marked in (c). (From Sun et al., *Adv. Mater.* 2005. 17. 2872–7. Copyright Wiley-VCH Verlag GmbH & Co. KGaA. Reproduced with permission.)

The monolayer-ordered porous film can wholly cover the surface of the tube and connect the two electrodes, as shown in Figure 11.1c. However, there are many gaps (or cracks) in the film, which derive from mismatched thermal expansion coefficients of semiconductor and the ceramic tube (Figure 11.1d). These breaks can make the resistance of ordered porous film increase greatly. Thus, the sensing signal is very weak, sometimes even no signal. To increase the conductivity, the quality and thickness of porous film should be improved. Further experiments proved that increase of the number of layer is a proper way. By using the ceramic tube which has already been covered with a monolayer porous film as the substrate, and repeating the procedures shown in Figure 8.15, the double or multilayer-ordered porous films can be fabricated on this ceramic tube. It should be noted that the first layer acts as a buffer layer here. The cracks in the lower (or first) layer can be remedied during formation of the second layer due to the presence of the solution. In this way, the microstructures and hence the conductivities of the gas sensors can be improved.

11.2.2 Structure-Dependent Gas-Sensing Performance

Figure 11.2 shows the different morphologies of four-layered SnO_2 porous film–based sensors using 1000, 350, 200 nm polystyrene (PS) spheres as template and their corresponding response curves to 100 ppm ethanol gas at 300°C, respectively. Obviously, the morphologies of all the films are quite homogeneous. There are no breaks or gaps on the surface of porous films compared with the monolayer porous film. Further, for the sensing properties, one can see that the response time decreases and the sensitivity increases with a reduction in the pore size. From this result, one can infer that the response time and sensitivity of these porous films can be controlled by changing the size of PS spheres.

For SnO_2 gas sensors, the changes in resistance mainly originate from the adsorption and desorption of gas molecules from the surfaces of the sensing structures [10,17]. In air, when oxygen molecules are chemisorbed on the surface of SnO_2 as O_2^-, O^-, and O^{2-}, some electrons in SnO_2 are captured, and hence the porous films show a higher resistance. When the film is exposed to reductive gas (such as ethanol), oxygen ions on the surface react with ethanol molecules, which make the conductivity of the film increase. Herein, obviously, the porous films with a higher surface area (or smaller pore size) can adsorb more molecules, and thus possess a higher concentration of oxygen ions. When such films are exposed to the gas, more adsorbed oxygen is desorbed, leading to a higher variation in resistance or higher sensitivity. Contrary, the response time is mainly influenced by the film thickness. When the number of layers is the same for all the films, the film thickness will increase with the increase of the pore size. As we know, the thicker the film, the slower the response time. Thus, the response time decreases with reduction in the pore size.

In fact, except the pore size and film thickness, the other factors can also influence the sensing performance such as the experimental method, the annealing temperature, and the interstice between two particles. Recently, Scott et al. also fabricated SnO_2 opals with controlled neck dimensions between adjacent spheres

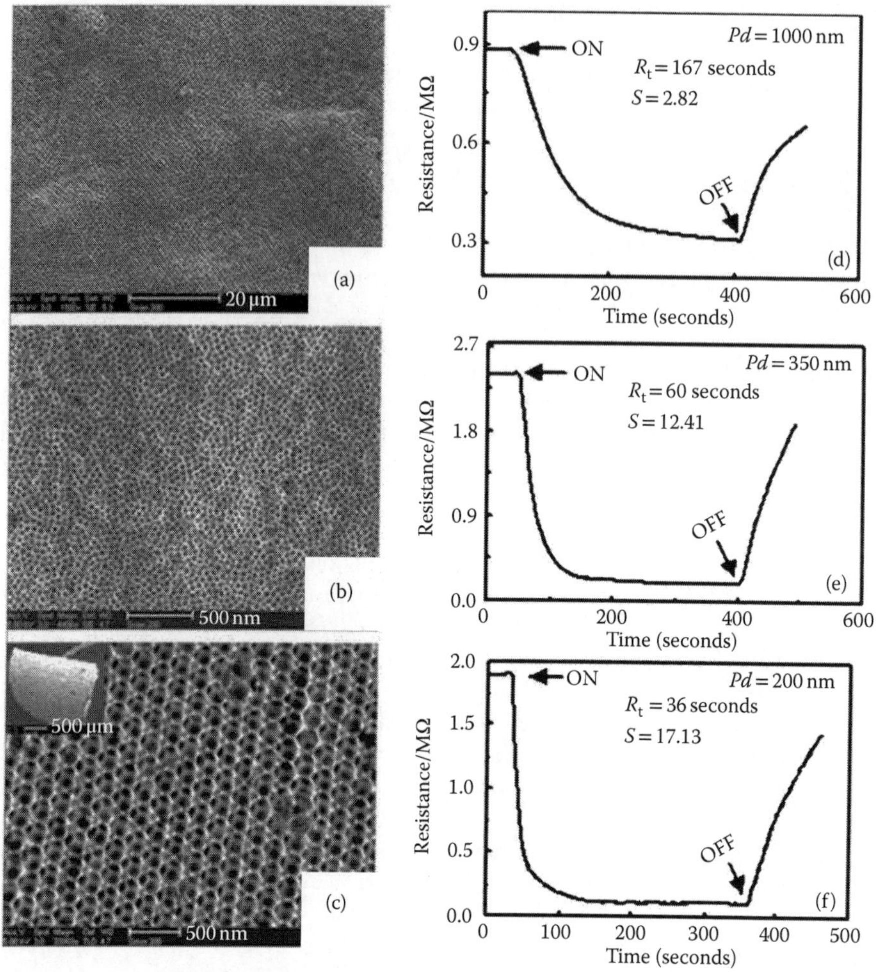

FIGURE 11.2 FESEM images of SnO_2-ordered porous films with four layers fabricated on curved surface of the gas sensors and their corresponding sensing properties in ethanol gas. The sensors have been fabricated with (a) 1000 nm, (b) 350 nm, and (c) 200 nm in diameter of polystyrene (PS) spheres. The inset in (c) is a low-magnification image of a part of the gas sensor. (d) through (f) show the corresponding measured resistance responses to 100 ppm ethanol gas at 300°C. "ON" means injection of gas into the test chamber and "OFF" means venting of the gas from the chamber. *Pd* is the pore diameter, R_t is the response, and *S* is the sensitivity. (From Sun et al., *Adv. Mater.* 2005. 17. 2872–7. Copyright Wiley-VCH Verlag GmbH & Co. KGaA. Reproduced with permission.)

by sintering the parent silica opals at different temperatures [18,19]. They found that the gas-sensing response of these materials to CO greatly depended on the neck diameter and the calcination temperature. When the sintering temperature of the original silica template is increased from 700°C to 1050°C, the average neck diameter of SnO_2 opals between spheres becomes larger. Thus, the potential barrier between grains created by the space-charge layer will decrease, which induce lower sensitivity toward CO. Further, when the sintering temperature of original silica template was fixed, varying the calcination temperature of SnO_2 opals can also lead to the change of sensitivity in CO gas. The sensitivity decreases with the increase of calcination temperature due to the increase in nanocrystallite sizes (which corresponds to smaller surface area).

11.3 GAS-SENSING DEVICES BASED ON HETEROPORE-SIZED POROUS FILMS

Generally, large pore size corresponds to the fast response but low sensitivity due to a low surface area, and vice versa for a small pore size. If heteropore-sized porous films, as illustrated in Figure 8.19, are used in gas sensors, it could combine the advantages of larger and small pores. Higher sensitivity and faster response can also be realized by this way. Further, reasonable combination of the monolayer porous films with different pore sizes would obtain the sensing elements with controllable sensing parameters in a large range. Here, we take In_2O_3 as a model material to demonstrate the sensing performances of double-layered heteropore-sized films [2].

11.3.1 GAS-SENSING DEVICES

Based on the strategy shown in Figure 8.18, we can directly fabricate the double-layered hierarchically micro/nanostructured porous film on the surface of a commercial ceramic tube with two preformed gold electrodes at the parts close to its two ends and one Pt wire inside the tube for heating, as illustrated in Figure 11.3a (the length, outer and inner diameters of the tube are about 5, 2, and 1 mm, respectively). In this way, we can get a series of nanostructured gas sensors with different pore sizes. The morphology of the film on the ceramic tube is representatively illustrated in Figure 11.3b, which is similar to that shown in Figure 8.19a.

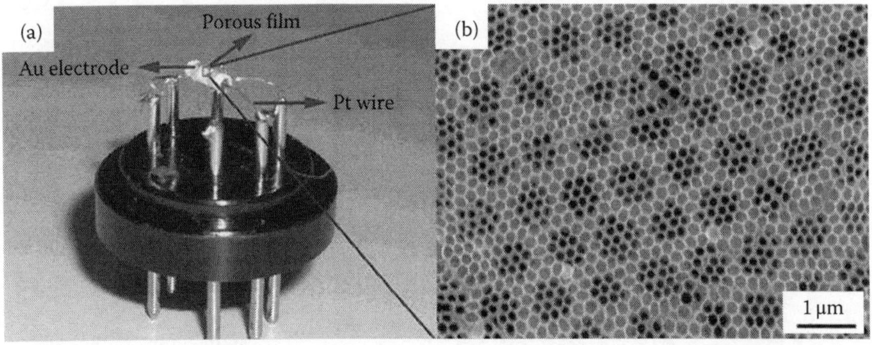

FIGURE 11.3 (a) The photo of the hierarchically structured porous film (1000/200 nm)–based sensor on a ceramic tube and (b) FESEM image of the film marked in (a). (Reprinted with permission from Jia et al., 2009, 2697. Copyright 2009 American Chemical Society.)

11.3.2 Gas-Sensing Performance with Both High Sensitivity and Fast Response

All gas-sensing measurements were made in a static system at a low working temperature (about 60°C) and the experiment environment of 50% relative humidity (RH). Ammonia (NH_3) was used as test gas. The electric response measurements in the NH_3 atmosphere have shown that the gas sensitivity and response time of the sensors are significantly dependent on the structure and morphology of the films. Figure 11.4 presents the response curves to 500 ppm NH_3 gas at the working temperature of about 60°C, for the samples (or sensors) based on the double-layered micro/nanostructured porous films of 1000/200 nm (the pore size of the first layer is 1000 nm, whereas that of the top layer is 200 nm), 200/200 nm, and 1000/1000 nm on the ceramic tubes. For the sensing device with the homopore-sized porous film of 200/200 nm, the gas sensitivity (S) and response time t_R are about 20.3 and 102 seconds, respectively, whereas 3.5 and 10 seconds for the 1000/1000 nm, as shown in Figures 11.4a and b. Obviously, for the devices with the homopore-sized porous films, the big pore size (1000 nm) corresponds to fast response (10 seconds) but a very low sensitivity (3.5), whereas the small pore size (200 nm) leads to the opposite situation or high sensitivity (20.3) but the slow response (102 seconds).

For the sensing device with hierarchically porous-structured film of 1000/200 nm, however, the values of S and t_R are 14.7 and 27 seconds, respectively (see Figure 11.4c). Undoubtedly, the double-layered hierarchically structured porous film–based sensor combines the advantages of both big- and small-sized pores and has the better sensing performance with both high sensitivity and fast response. Furthermore, it is much more sensitive (nine times higher) than the sensing film of the In_2O_3 nanoparticles with 30 nm in average size, which was fabricated on the ceramic tube by a simple printing method, as shown in Figure 11.5, in addition to the comparable t_R.

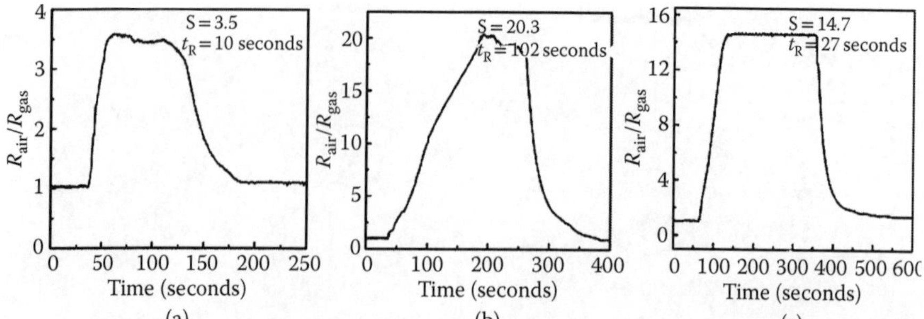

FIGURE 11.4 Response curves to 500 ppm NH_3 test gas at 60°C for different double-layered micro/nanostructured porous films. (a) 1000/1000 nm, (b) 200/200 nm, (c) 1000/200 nm. (Reprinted with permission from Jia et al., 2009, 2697. Copyright 2009 American Chemical Society.)

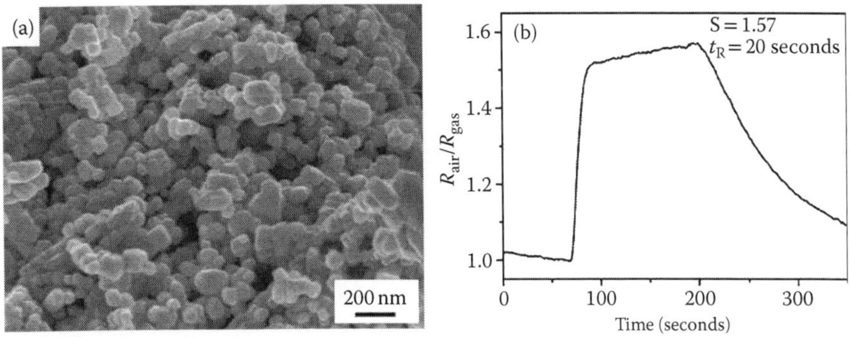

FIGURE 11.5 FESEM image (a) and response curve to 500 ppm NH_3 at 60°C (b) for the In_2O_3 nanoparticle film, which was fabricated on the ceramic tube by a simple printing method. (Reprinted with permission from Jia et al., 2009, 2697. Copyright 2009 American Chemical Society.)

11.3.3 Pore Size–Dependent Gas-Sensing Performance and Diagram of t_R versus S

Further, the double-layered hierarchically nanostructured porous film exhibits the controllability of t_R and S values just by changing the pore sizes. The flexibility of the synthetic strategy, as shown in Figure 8.18, allows for the tunable pore sizes in the hierarchically structured porous films. Figure 8.19a and b shows the morphology of the double-layered hierarchically micro/nanostructured porous films on the ceramic tubes with 1000/350 and 1000/100 nm, respectively. Correspondingly, the response curves to 500 ppm NH_3 test gas at the same experimental conditions as above are illustrated in Figure 11.6, showing a good controllability in sensing performance. The controllability of the hierarchical porous structure is highly expected to modulate the gas-sensing parameters, such as S and t_R in a large range, according to practical application.

Figure 11.7 demonstrates the values of S and t_R, under 500 ppm NH_3 test gas, for all the sensing devices with the double-layered films of different porous structures, or a diagram of t_R versus S. We can see clearly that S value increases mainly with the reduction of the pore size in the top layer when the pore size of the first layer was fixed as 1000 nm (see the dashed line in Figure 11.7). In contrary, t_R value mainly depends on the pore size of the first layer (see points c and d in Figure 11.7). This indicates that the S value can mainly be controlled by the pore size of top layer and the t_R value simply by that of first layer. On basis of such t_R–S diagram in Figure 11.7, the structure of double-layered porous films with desired sensing parameter values can thus be easily designed and fabricated according to the practical requirements. As an example, if we need a sensor with the sensing parameters (S, t_R) located at point e (the black mark) in Figure 11.7, obviously, the pore size of top layer in the film should fall into the range from 1000 to 350 nm (or between points a and b in x-axis), and that of the first layer should be from 1000 to 200 nm (or between points a and d but close to point a in y-axis), which has been confirmed. Position e in Figure 11.7 corresponds to the result of the sensing device with the double-layered hierarchically micro/nanostructured porous film of 750/500 nm.

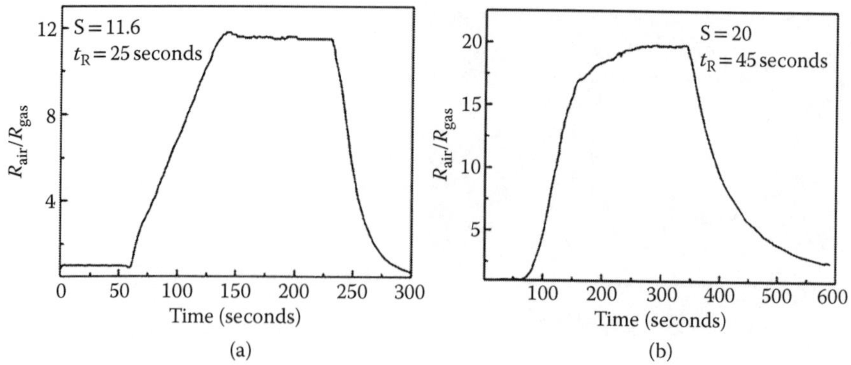

FIGURE 11.6 Response curves in 500 ppm NH_3 for the sensing devices with hierarchically porous structured films of (a) 1000/350 nm and (b) 1000/100 nm. (Reprinted with permission from Jia et al., 2009, 2697. Copyright 2009 American Chemical Society.)

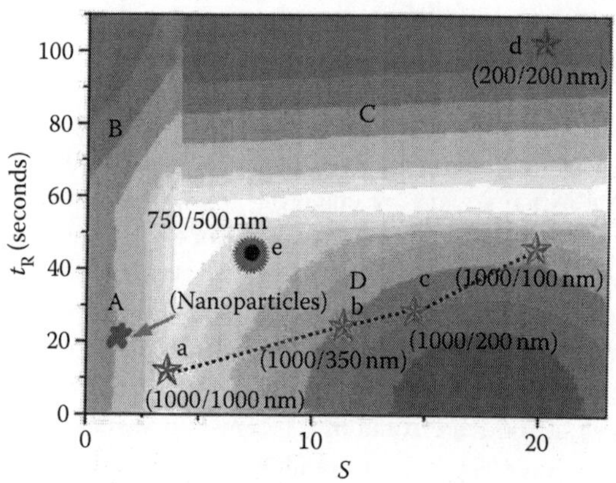

FIGURE 11.7 Diagram of the sensing parameters t_R versus S for In_2O_3 films with different double-layered porous structures in 500 ppm NH_3 atmosphere at 60°C (see text in details). A through D correspond to the different zones in the diagram. (Reprinted with permission from Jia et al., 2009, 2697. Copyright 2009 American Chemical Society.)

In addition, the t_R–S diagram in Figure 11.7 can, in principle, give a measurement of sensing performance for a gas sensor. Different positions in the diagram correspond to different sensing performances. Obviously, in regions A and B, the S value is too low and the response is too slow inregions B and C, meaning that the sensors with sensing parameters in these regions are unfavorable in practical use [4,20]. The parameters in region D correspond to both fast response and high sensitivity, and hence the devices in this region have the practical importance.

The gas-sensing performances of the double-layered film-based devices for the other concentrations (from 20 to 2000 ppm) of the NH_3 test gas were also examined, as demonstrated in Figure 11.8a. Obviously, the sensitivity shows similar increase with the rising concentration of the NH_3 test gas. Furthermore, evolution of the sensing parameters with pore size is also similar to that in Figure 11.7, as illustrated in Figure 11.8b. Further work has revealed that such sensing devices are of the good

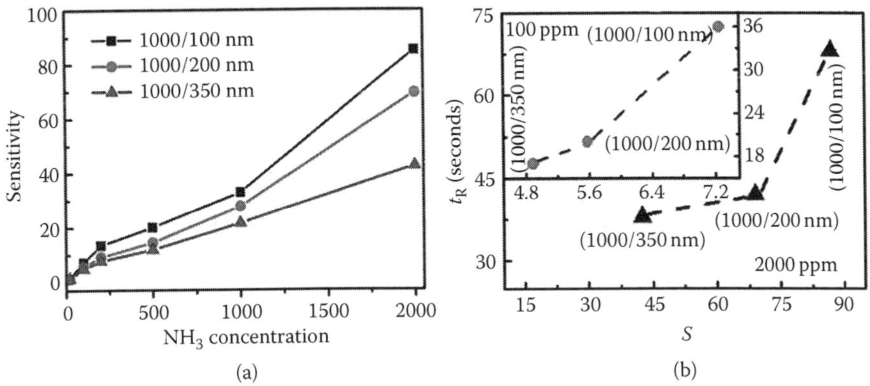

FIGURE 11.8 (a) Sensitivity as a function of concentration of test gas NH_3 at 60°C for different double-layered hierarchically structured porous In_2O_3 films (1000/100 nm, 1000/200 nm, 1000/350 nm). (b) Diagram of t_R versus S in 2000 and 100 ppm (the inset) NH_3 at 60°C for the double-layered hierarchically micro/nanostructured porous In_2O_3 sensors (films) with different pore sizes of top layer. (Reprinted with permission from Jia et al., 2009, 2697. Copyright 2009 American Chemical Society.)

reproducibility of the sensing performances and the good sensing stability [2]. It could be expected that combination of the monolayer porous structure with different pore sizes into multilayer (more than two layers) hierarchically micro/nanostructured porous films would lead to the controllability of S and t_R values of sensors in a large range and meet various practical requirements.

11.3.4 Structurally Induced Controllability of Gas-Sensing Performances

The resistance change of the In_2O_3-based sensing device is induced by the redox reaction on surface of the In_2O_3 film in gas atmosphere. When the sensing film is exposed to air, oxygen in air is chemisorbed as O^{2-} ions on its surface, and some electrons in the film are localized in the film surface, leading to an increase in the resistance [21,22]. When exposure to a reducing gas, such as NH_3, the redox reaction occurs, at a certain temperature, between the reducing gas and adsorbed oxygen molecules, and the amount of the adsorbed oxygen will thus be decreased, inducing release of the surface-trapped electrons back to the In_2O_3 film and hence significant reduction of resistance. Thus, many factors can influence the sensing performance, such as the mobility of conduction electrons and the chemical/thermal stability of metal oxide, the surface area, the thickness, the pore size, and crystallite/grain size of the films. For a given film, the gas-sensing performance should mainly depend on the thickness, crystallite/grain size, and pore size of the film.

Here, the conditions of the film fabrication (precursor solution concentration, and drying and subsequent annealing treatment) are the same for all the devices with different pore sizes. It means that the crystallite/grain size for the films on all sensing devices should be similar (about 12 nm) [2]. So, the controllability of the gas-sensing property in this case should mainly be attributed to the structure of the porous films. Obviously, the higher surface area exposed to the environmental atmosphere should correspond to the higher S value [5]. Meanwhile, the big pore size would lead to a fast transportation of gas molecules and hence fast response or small t_R value. The results in this work are thus easily understood. When the pore size in the top layer is reduced (keeping that of the first layer), the S value will rise due to the increase of the specific surface area, but the response rate slows down because of the small pore-induced difficult transportation of the test gas. Additionally, the peculiar hierarchical micro/nanostructure will produce the much more dangling bonds or the higher surface activity, for the reaction of the reducing gas with the adsorbed oxygen molecules, than the bulk structure [23,24], and thus remarkably decrease the reaction temperature. In this work, the working temperature is only about 60°C, much lower than that (about 300°C) of the other In_2O_3 sensors [21,25], which is very important for the practical application, especially for the detection of some flammable gases.

Obviously, if the monolayer porous film with different pore sizes is combined into multilayered hierarchically micro/nanostructured porous films, both S and t_R values can be modulated separately in a large range, and we can thus obtain the gas sensors with desired sensing performances, or both high sensitivity and fast response, to meet practical requirements. Such sensors are superior to those of the nanoparticle-based ones and can overcome the shortages of the homopore-sized porous films mentioned in Section 11.2, and can improve the S and t_R values to the desirable values.

Similarly, the heteropore-sized porous film devices from some other oxide semiconductors such as SnO_2, Fe_2O_3, TiO_2, ZnO, CuO, and so on, could be fabricated. The structurally induced tunable sensing parameters and the better sensing performances than the conventional nanostructured ones could also be achieved. Representatively, Figure 11.9 gives the results corresponding to Fe_2O_3, which is a widely used material in gas sensor industry. The heteropore-sized porous Fe_2O_3 film device shows both high sensitivity and fast response and exhibits better sensing performance than that of the homopore-sized porous Fe_2O_3 film device, as shown in Figure 11.9e. Besides, it is highly expected that the hetero-multilayer hierarchical micro/nanostructure semiconductor porous film devices could be fabricated by alternately using several different precursor solutions. Such sensing devices could be of the high performances for the detection of multiple gases. Also, the sensing film surface can be decorated with some specific organic and bioorganic receptors to get higher selectivity.

11.3.5 Extension of the Strategy

The strategy shown in Figure 8.18 shows some advantages, such as the controllable pore size in each layer and thus porosity of whole film, the uniform film thickness, the substrate independency, the easy manipulation in layer numbers and the film compositions, and hence the controllability of the gas-sensing parameters in a large region. People can thus design and fabricate the hierarchically micro/nanostructured sensors with desired sensing performances. The main shortage of this strategy is the relatively complicated fabrication.

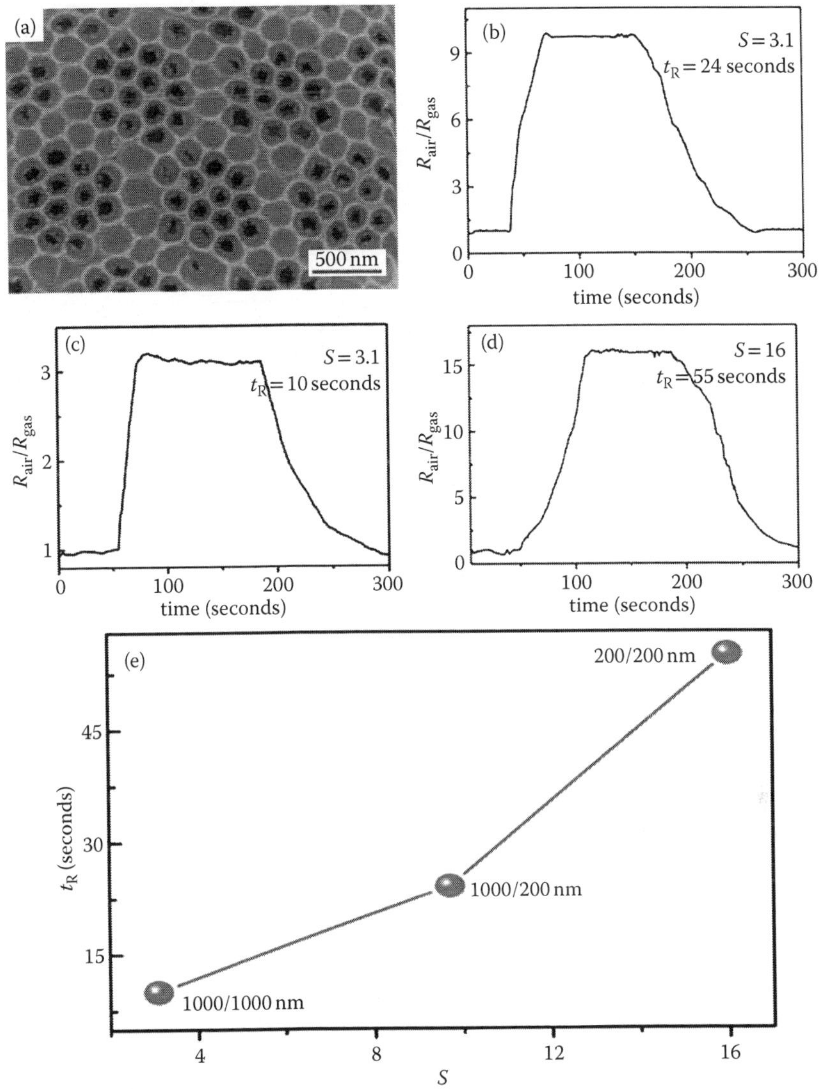

FIGURE 11.9 FESEM image (a) and response curves to 500 ppm NH_3 test gas at 60°C (b) for the heteropore-sized porous Fe_2O_3 film (1000/200 nm). (c) and (d) correspond to the homopore-sized porous Fe_2O_3 films with pore sizes of 1000 and 200 nm, respectively. (e) The diagram of the sensing parameters t_R versus S for Fe_2O_3 micro/nanostructured films with different double-layered porous structures in 500 ppm NH_3 atmosphere at 60°C (the data from [b] through [d]). (Reprinted with permission from Jia et al., 2009, 2697. Copyright 2009 American Chemical Society.)

If the micro/nanostructured porous film is fabricated using the template prepared by simply mixing large and small PSs together (see Figure 11.10a), the heteropore-sized porous films with micro/nanostructure could also be obtained. This could be a good idea since the gas-sensing parameters of such films could also be controlled

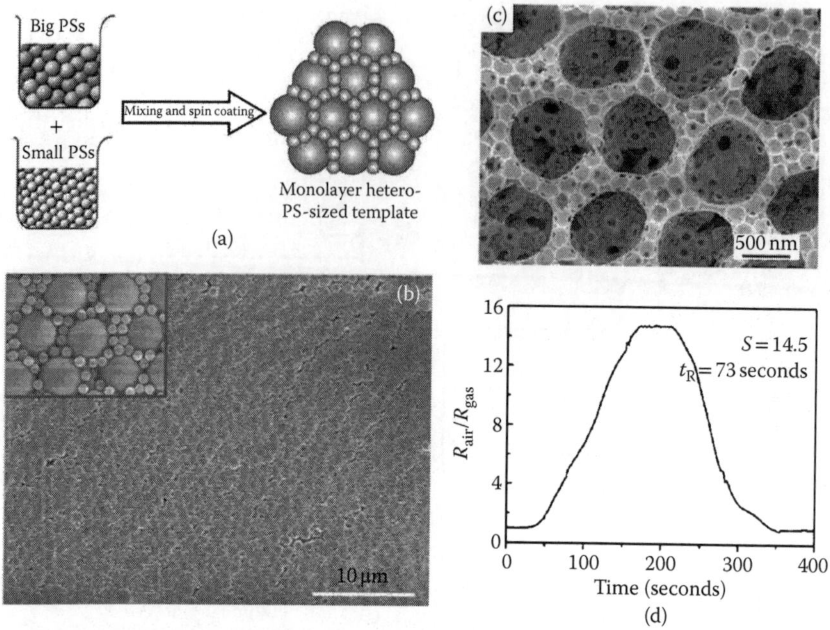

FIGURE 11.10 (a) Schematic strategy of monolayer heterosized PS template by simply mixing large and small PS spheres together. (b) FESEM image of heterosized PS template prepared directly by spin coating the suspension, mixed with 1000 and 200 nm PS spheres, on a glass slide; the insert is the local magnified image. (c) The corresponding mixed micro/nanostructured porous In_2O_3 film on the ceramic tube, from the template shown in (b) and using the strategy shown in Figure 8.18. (d) The response curve of the film shown in (c) in the 500 ppm NH_3 at 60°C. (Jia and Cai, *Adv. Funct. Mater.* 2010. 20. 3765–73. Copyright Wiley-VCH Verlag GmbH & Co. KGaA. Reproduced with permission.)

by pore sizes and the size ratio of large to small PS spheres in the mixture, and the procedure seems to be facile to implement. On this basis, the monolayer mixed-PS-sized templates can be prepared by spin coating the suspension, mixed with large and small PS spheres, on a glass slide. Typically, Figure 11.10b shows the morphology of hetero-PS-sized template (1000/200 nm PS spheres) by one-step spin coating. The larger PS spheres arrange orderly and the smaller PS spheres fill in the interstices among the larger PS spheres. Using such template and the strategy shown in Figure 8.18, the monolayer-mixed porous-structured film sensor can be fabricated. Figure 11.10c shows the corresponding heteropore-sized porous In_2O_3 film, on the ceramic tube, from the mixed template with 1000/200 nm PS spheres. The film consists of orderly arranged microsized pores and the skeleton with smaller pores. Figure 11.10d illustrates the gas-sensing result in the 500 ppm NH_3 at 60°C, showing similar sensitivity to that corresponding to the double-layered heteropore-sized porous films (see Figure 11.4c). Similarly, the sensing performance of such mixed porous film can be controlled in a certain range by pore sizes. Compared with the double-layered heteropore-sized porous structure, for this mixed-porous micro/nanostructured film, fabrication is relatively easier, but controlled range of gas-sensing parameters is comparatively small since the mixed porous films can be formed only when the size ratio of smaller to larger PS spheres keeps in a certain range (if the ratio >1/2, the mixed template cannot be well produced). So which strategy of both should be used depends on practical applications.

11.3.6 Doping-Induced Selectivity of Porous Films

As mentioned above, the separate tunability of the sensitivity and the response time in a large range can be realized, and the sensing performances of both high sensitivity and fast response can also be achieved, just by controlling microstructures of the ordered porous films. Actually, in addition to the controllable high sensitivity and fast response, the strong selectivity to detect a target gas in the atmosphere with multigas molecules are also very much expected for the good gas-sensing devices or the real-time gas-sensing devices. It has been reported that doping metal ions can modify the space charge layer thickness of the sensing films [27–31] and hence alter the gas-sensing performances depending on the target gas molecules, leading to the selectivity. Based on the strategy shown in Figure 8.18, it is easy to obtain the various doped porous films with micro/nanostructures, only by controlling the composition of precursor solution. It has been found that proper doping to the porous films can realize significant selectivity of sensing performance [32]. Here, we take Cr^{3+} or Pd^{2+} ions-doped SnO_2 porous films with double layer and homopore size as example to demonstrate such selectivity.

The precursor solution 0.1 M $SnCl_4{\cdot}5H_2O$ with different concentrations of metal chloride (Cr^{3+} or Pd^{2+}) varying from 0.5 to 5%M was used. The double-layered homo- or heteropore-sized SnO_2 porous films doped with different amounts of Cr^{3+} or Pd^{2+} were thus fabricated based on the strategy shown in Figure 8.18. Figure 11.11 presents the results for Cr^{3+} or Pd^{2+}-doped SnO_2 porous films, which are homogeneous and similar in morphology. X-ray photoelectron spectroscopic measurements have confirmed that the chemical states of Cr and Pd in SnO_2 can be identified as Cr^{+3} and Pd^{+2}, respectively, according to their binding energies [32].

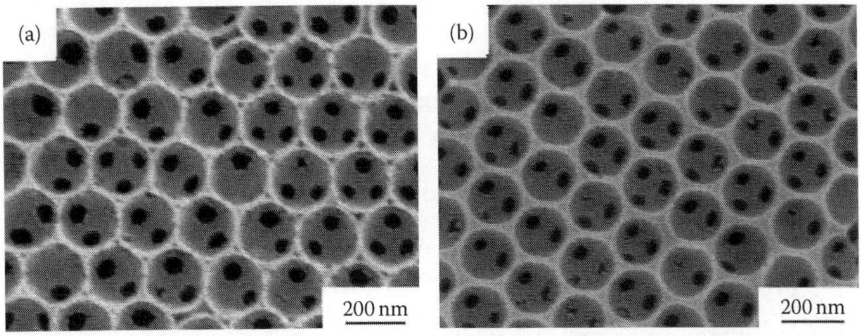

FIGURE 11.11 FESEM images of the double-layered homopore-sized SnO_2 porous films doped with 1%M (a) Cr^{3+} and (b) Pd^{2+}. (Jia and Cai, *Adv. Funct. Mater.* 2010. 20. 3765–73. Copyright Wiley-VCH Verlag GmbH & Co. KGaA. Reproduced with permission.)

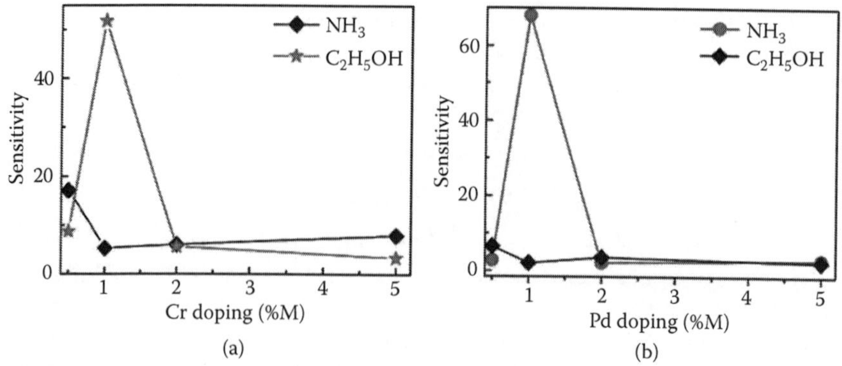

FIGURE 11.12 The relationship of sensitivity versus ion-doping amount under 1000 ppm ethanol or NH_3 gases for the double-layered homopore-sized SnO_2 porous films doped with (a) Cr^{3+} and (b) Pd^{2+}. (Jia et al., *Appl. Phys. Lett.* 2010 IEEE.)

Correspondingly, the gas-sensing performances for all the doped SnO_2 porous film–based sensors have been measured at the working temperature of about 60°C under 1000 ppm ethanol and ammonia gases, respectively. It has been found that the sensitivity to the given test molecules strongly depends on the doping amount and the doping species when the doping amount is not higher than 2%, as illustrated in Figure 11.12. The higher doping amount leads to the sensitivity similar to the undoped samples. For the Cr^{+3}-doped porous films, doping with 1%M induces the highest sensitivity (up to 52) to the ethanol molecules, 10 times higher than that (about 5) to the ammonia gas, indicating highly selective sensing to ethanol, relative to ammonia (see Figure 11.12a). For the Pd^{2+}-doped porous films, however, the opposite is true. 1%M doping amount leads to the highest sensitivity (up to 68) to the ammonia gas, but a very low value (about 2) to the ethanol molecules, exhibiting strongly selective sensing to NH_3, and 34 times higher than that to ethanol molecules (see Figure 11.12b). Obviously, these results have demonstrated that proper doping not only improve the sensing performances but also favor the selective interaction with the target gas and hence the selective sensing to some peculiar gases. This would be an important step toward the practical application of the nanostructured porous film-based sensors. The improved ethanol-sensing properties for the Cr^{3+}-doped case could be attributed to electronic sensitization effect of the additive, and the enhanced ammonia-sensing performance for the Pd^{2+}-doped case could originate from the increased ammonia adsorption. The detailed explanation of doping-induced selectivity is difficult before deep understanding influence of doping on surface properties of the sensing films, which should be further studied.

11.4 HIGH-PERFORMANCE GAS-SENSING DEVICES BASED ON THE POROUS FILMS/MEMS CHIP

As mentioned above, the semiconductor thin-film gas sensors are based on the changes in electric conductance of the thin films induced by interaction with the target gas molecules [5,33], involving diffusion and adsorption of the target gas molecules in the film and their reaction with the sensing film, which need a length of time. So, for the normal conductometric gas sensors, the response is relatively slow (usually in the range of several seconds to minutes) and the sensitivity is comparatively low (the detectable lower limit is above 10 ppm level). In addition, the power consumption is high (above 200 mW) since the sensors would always work at a high temperature (250°C–500°C) to achieve or activate reversible response to gases [34,35]. Thus, the conductometric gas sensors are only used in the areas where the high sensing performances are not needed. Development of the conductometric gas sensors with high performances, such as ultrafast response (second level), high sensitivity (parts per billion [ppb]-level detection limit), low power consumption (10 mW level), high stability, and good selectivity, have been expected and in challenge [36,37]. Although micro/nanostructured porous film–based gas-sensing devices, described in Sections 11.2 and 11.3, could realize the controllable, tunable, and stable sensing parameters and exhibit much better performances than the conventional nano-object-built film-based devices, they are still of high power consumption, relatively low sensitivity (ppm-level detection limit), and slow response (10-second level), and hence far from the high sensing performances.

Actually, the performances of the gas-sensing devices are associated directly with the surface properties, thickness, and microstructures of the sensing films [5,38,39]. Meanwhile, the device system (electrodes-equipped substrate) for supporting sensitive materials as well as their compatibility is also essential for the final performances [38,39]. Generally, the thinner the sensing films are or the larger the surface–volume ratio of the films are, the higher the sensitivity and the faster the response [5,38,40–43]. However, ultrathin films will be highly fissile and lead to the instability of the sensing performance.

As mentioned in Sections 11.2 and 11.3, the template-induced micro/nanostructured metal oxide–ordered pore films are the good candidates of the ideal sensing thin films, which are uniform and controllable in thickness, stable and tunable in microstructure, and reproducible and flexible in fabrication, due to in-situ using template techniques [2,14]. Meanwhile, the sensing film fabricated on a micromachined microhotplate platform allows quickly changing the operation temperature and remarkably decreasing the heating power consumption, which can be used to rapidly regenerate the surface of the metal oxides [38,44,45]. So, if combining the template-induced ordered porous thin films with the micromachined microhotplate platform, we could obtain the conductometric gas-sensing devices with high performances, such as high sensitivity, fast response, and low power consumption. However, the crucial issue is how to fabricate the well-qualified sensing layer efficiently on micro-platforms by a feasible approach. In this section, we will introduce a flexible strategy for such combination. Typically, SnO_2 is taken as the example to demonstrate the validity of this strategy. The fabricated microelectromechanical systems (MEMS)-assisted gas-sensing device could exhibit the real high sensing performances.

11.4.1 Construction Strategy of the High-Performance Sensing Devices

The construction strategy for the high-performance gas-sensing devices could be presented based on the combination of the structure-controllable porous thin film with MEMS-based interdigital electrodes (IDEs)/microheater chip. The supporting system of the device is designed into a sandwich-structured chip. This chip consists of the comb-type electrodes (or the microspaced IDEs), isolating layer, and microheater, which are superposed together on a SiO_2/Si_3N_4 substrate, as demonstrated in Figure 11.13a. The comb-type IDEs (8 Pt fingers, 100 μm long, 10 μm wide, and 10 μm apart) are located on the Si_3N_4 isolating layer with 600 nm in thickness. Underneath is the microheater. Here, the IDEs are used to reduce the resistance value of the sensing film due to the parallel connection effect of the multielelctrodes. Such a chip can be fabricated by a classic MEMS technology. Figure 11.13a presents a typical photo of the as-fabricated chip (ca. 8 mm^2 in area). This chip has four pads, two for detection of the response signal and the other two for the heating source.

This new sensing device can be produced by in situ fabrication of an ordered porous thin film on the MEMS-based IDEs-microheater chip via transferring the solution-dipped organic colloidal template [2,14,26], as schematically presented in Figure 11.13b, which is similar to the strategies shown in Figures 8.15 and 8.18. A glass slide covered with the self-assembled PS colloidal monolayer template is firstly dipped into a precursor solution. The PS template is lifted off the slide due to the surface tension of the solution and floats on the solution surface. The floating monolayer is then transferred to the chip by picking it up. Due to the capillary effect, the PS monolayer on the chip also contains the precursor solution in the interstitials among the PS colloidal spheres and between the chip and the PS monolayer. After subsequent drying and annealing, the PS template would be burnt away and the ordered porous thin film be thus left on the IDEs-microheater chip due

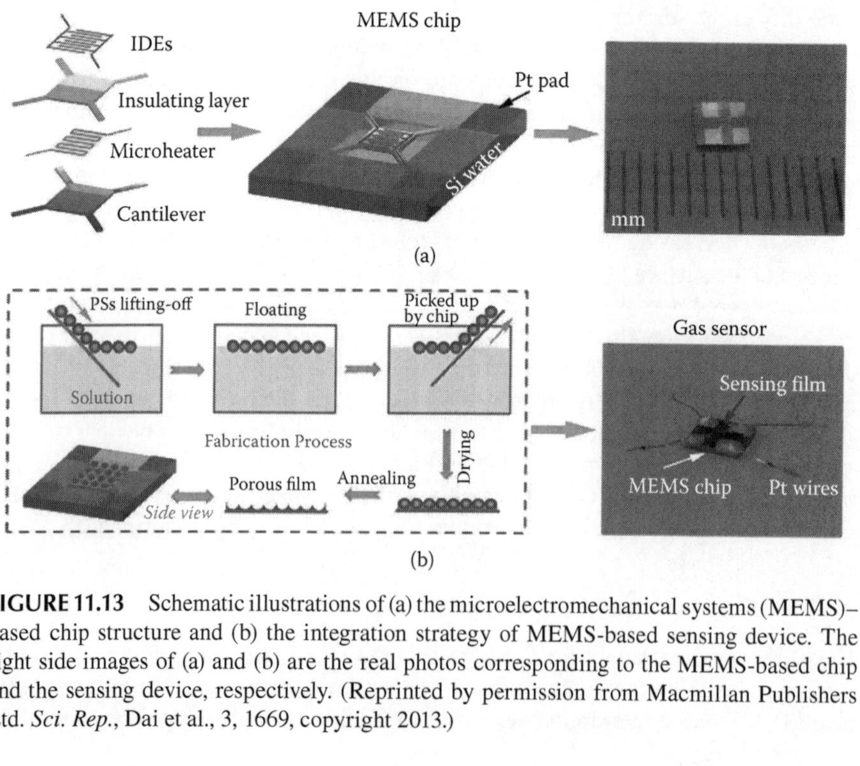

FIGURE 11.13 Schematic illustrations of (a) the microelectromechanical systems (MEMS)–based chip structure and (b) the integration strategy of MEMS-based sensing device. The right side images of (a) and (b) are the real photos corresponding to the MEMS-based chip and the sensing device, respectively. (Reprinted by permission from Macmillan Publishers Ltd. *Sci. Rep.*, Dai et al., 3, 1669, copyright 2013.)

to the template geometry. Obviously, the microstructure and thickness of the thin film are controllable, depending on the size of PS colloidal spheres and concentration of the precursor solution. Further, the doped thin films can also be obtained by controlling the composition of the precursor solution. Such thin film on the chip should be homogenous in structure and reproducible in fabrication due to the template technique. In addition, the multilayer-ordered porous thin films can also be fabricated just by repeating the above procedures. The gas sensor with monolayer- or multilayer-ordered porous thin sensing films can thus be constructed. Such conductometric thin film gas sensor can work independently without any external heating module, owing to the inside microheater.

11.4.2 Integrated Gas Sensors

According to the strategy in Figure 11.13, the monolayer- or multilayer-ordered porous sensing thin films can be in situ fabricated on the IDEs-microheater chip to form a gas-sensing device. Here, the gas-sensing material SnO_2 is taken as an example to demonstrate the microstructure and sensing performances of such sensor. Typically, Figure 11.13b gives a photo of the as-fabricated gas-sensing device, where the IDEs-microheater chip is covered with the SnO_2 thin film from 500 nm PS sphere template and after annealing at 400°C. It exhibits greenish color. By linking the Pt wires on this chip, the resistance of the SnO_2 thin film was measured to be 120 kΩ at 25°C in air, indicating that the gas-sensing device is successfully fabricated based on the above-mentioned strategy for in situ integrating the micro/nanostructured-ordered porous thin film on the MEMS-based chip.

More importantly, based on the strategy in Figure 11.13b, the mass production of such gas-sensing devices can be easily realized. Briefly, the area of the MEMS-based chip is about 8 mm^2 and the monolayer PS colloidal template can be prepared to be of 80 cm^2 on a 4-in. Si wafer (see Figure 11.14a). The IDEs-microheater chips can be manufactured in one Si wafer in batches by MEMS technology. It means that we can produce about 1000 gas-sensing devices in one batch by transferring one solution-dipped wafer-scaled template onto the similarly sized multiple chips-outfitted wafer, as illustrated in Figure 11.14b.

11.4.3 Morphology and Microstructure

Figure 11.15 shows the typical morphology of the as-fabricated SnO_2 porous sensing film on the MEMS-based chip. The working area for the chip is 150 μm × 150 μm (see Figure 11.15a), indicating minimization of the sensing device. Magnified observation has revealed that the IDEs have 8 Pt fingers with 100 μm in length, 10 μm in width, and 10 μm in spacing, and the microheater can also be seen indistinctly, as demonstrated in Figure 11.15b, corresponding to the field emission scanning electron microscopy (FESEM) image. Further, the ordered porous thin film is clearly seen on the IDEs, as shown in Figure 11.15b through d. The pores in the thin film are hexagonally arranged due to the template geometry, as seen in Figure 11.15c, corresponding to a high-magnification FESEM image of the sensing thin film between adjoining fingers. The thin film is highly homogenous in structure and morphology. In

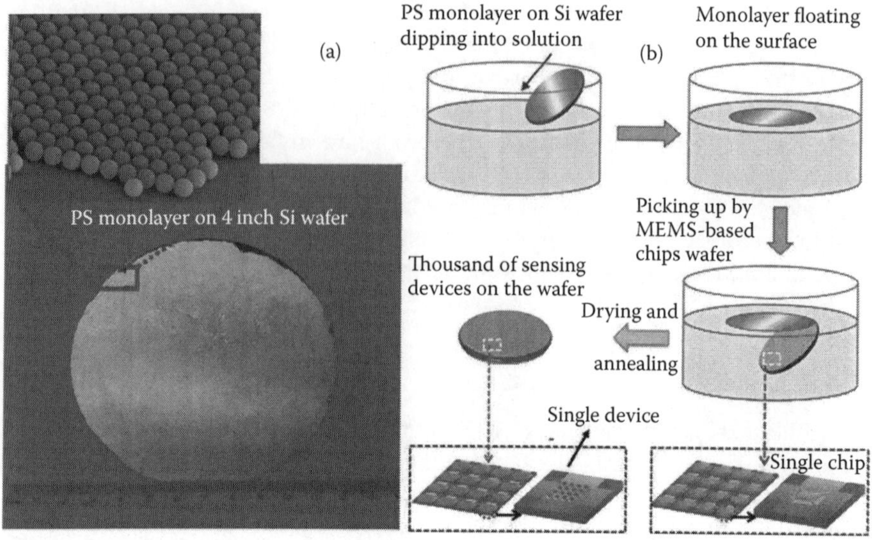

FIGURE 11.14 Illustration for mass production of MEMS-based sensing devices. (a) A real photo of the monolayer PS colloidal template on a 4-in. Si wafer by an air/water interfacial self-assembly. The inset is a local FESEM image. (b) Schematic steps of the mass production of MEMS-based sensing devices on one multiple chips-outfitted wafer. (Reprinted by permission from Macmillan Publishers Ltd. *Sci. Rep.*, Dai et al., 3, 1669, copyright 2013.)

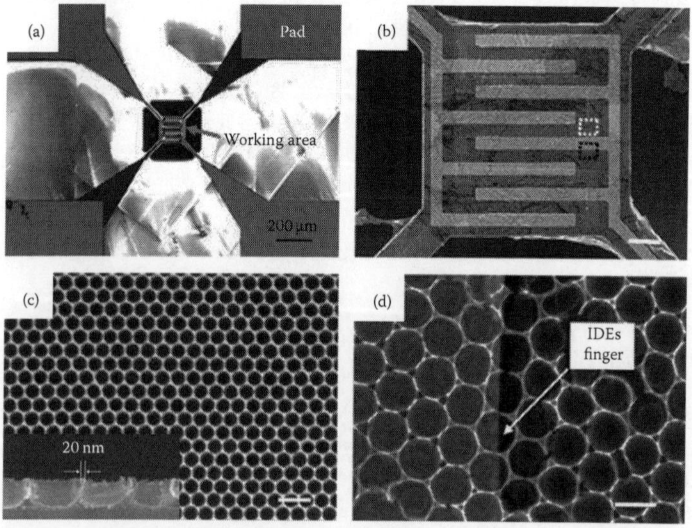

FIGURE 11.15 The top-view FESEM images of the MEMS-based sensing device with different magnifications. (a) Top-view image of the whole device (low magnification). (b) An enlarged image of the working area. (c) The enlarged image corresponding to the area between adjoining fingers [marked by white dot frame in (b)]. The inset is the cross-sectional image. (d) The enlarged image corresponding to the area across a finger edge [marked by black dot frame in (b)]. The scale bars in (b) and (c) are 20 μm and (d) is 500 nm. (Reprinted by permission from Macmillan Publishers Ltd. *Sci. Rep.*, Dai et al., 3, 1669, copyright 2013.)

addition, the cross-sectional observation has shown that the porous thin film and the pore wall are about 200 and 20 nm in thickness, respectively, as illustrated in the inset of Figure 11.15c.

Generally, owing to the thickness of the Pt fingers, the sensing thin films would be easily broken at the edges of IDEs. For instance, screen printing the nanoparticle paste on the chip is difficult to obtain the sensing films with high quality, especially for the very thin films. In addition, using the brush coating or screen printing method, the sensing thin films are uncontrollable in microstructure and thickness, and unreproducible in fabrication, as mentioned in Section 11.3. Here, these problems are well solved due to the solution dipping and template transferring technique. Figure 11.15d clearly exhibits the continuity of the porous sensing thin film at the finger's edges and the compatibility of the micro/nanostructured porous thin film with the MEMS-based chip. Hence, the strategy in Figure 11.13b is a good candidate for the integration of the micro/nanostructured porous sensing thin films on the IDEs-microheater chips.

11.4.4 Power Consumption and Heating Homogeneity

The power consumption of such gas-sensing device was examined, which depends on the working temperature. Since the resistance of the Pt resistor (heater) in the sensing device is temperature dependent, the average temperature T (°C) of the working area in the device can be determined by the resistance R of the heater, or

$$T = T_0 + \frac{R - R_0}{\alpha R_0} \tag{11.1}$$

where α is the temperature coefficient of Pt resistance, R_0 is the original resistance at room temperature T_0 (25°C). Obviously, from Equation 11.1, the average temperature T at the working area in the device can be obtained by measuring the resistance of the heater under a given potential or current. Figure 11.16a shows the heating power consumption as a function of the average temperature of the sensor, showing a nearly linear relation. If the working temperature is at T = 350°C, the power consumption of the microheater is only about 32 mW, which is about seven times lower than an excellent commercialized metal oxide semiconductor (MOS) ethanol sensor (ca. 210 mW; Figaro TGS2620 [Glenview, IL]) [47]. Such low-power consumption is of great significance in the practical application.

In addition, many semiconductor-sensing materials have strong temperature-dependent sensitivity and cross-sensitivity. So, the temperature gradient on the working area should be as small as possible. Based on the structure of the gas-sensing device shown in Figure 11.15 and the finite-element method, the temperature distribution on the device during heating can be obtained by electrothermomechanical simulations using commercial analysis software Coventor (Cary, NC) [48] as typically demonstrated in Figure 11.16b. Under a heating power of 34 mW, the temperature on the working area falls into the range from 350°C to 368°C, indicating a quite homogenous temperature distribution with an average temperature gradient of about 0.17°C·μm^{-1} across the whole working area.

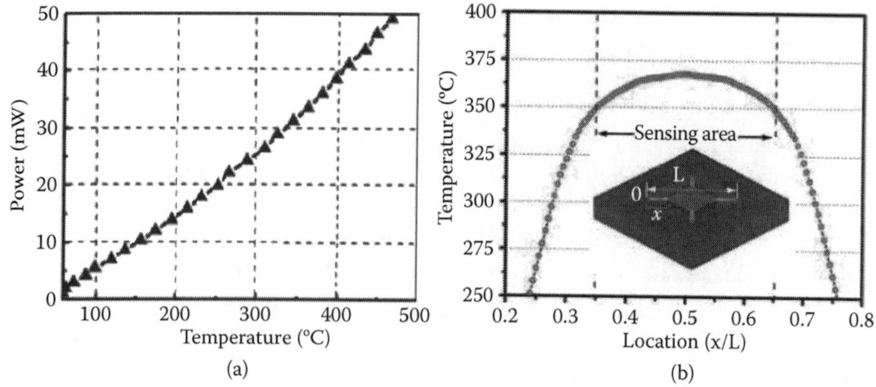

FIGURE 11.16 The power consumption and heating homogeneity. (a) The heating power of the microheater as a function of working temperature for the gas-sensing device. (b) Temperature distribution across the chip by an electrothermomechanical simulations using the analysis software Coventor. The inset shows the temperature distribution illustration. (Reprinted by permission from Macmillan Publishers Ltd. *Sci. Rep.*, Dai et al., 3, 1669, copyright 2013.)

11.4.5 High Gas-Sensing Performances

As mentioned above, this sensing device has low power consumption and relatively homogeneous temperature distribution on the working area, and very thin and uniform sensing film with controllable microstructure. The high gas-sensing performances can thus be expected for such a new device.

11.4.5.1 Response Time and Sensitivity

The average acetone concentration in exhalation from a diabetic patient is higher than 1.8 ppm, as previously reported [49,50]. So, it is necessary to distinguish the diabetic patient from a driver drinking alcohol for avoidance of misdeclaration. Here, the gas-sensing behaviors to ethanol and acetone are examined to demonstrate the high gas-sensing performances for this new sensing device.

Figure 11.17a gives the electric response (R_{air}/R_{gas}) as a function of test time to ethanol under 1–5 ppm at 350°C. The sensing device exhibits a high sensitivity and ultrafast response. With the exposure to 3 ppm ethanol, for instance, the sensitivity is about 3.9 and the response time is less than 1 second (0.73 seconds) (see Figure 11.17b). Similarly, Figure 11.17c and d shows the sensing performance to acetone at 350°C. The sensitivity and response time are 2.5 and 0.62 seconds, respectively, under 3 ppm acetone. In addition, the gas-sensing tests were conducted under 5 ppm ethanol (60% RH) and at 350°C (or 32 mW) for the five parallel MEMS-based sensing devices, as shown in Figure 11.18. The sensing signals have exhibited a good repeatability. The response time for the previously reported MOS gas sensors is usually in the range of several seconds to minutes. For the gas-sensing devices in this case, however, it is within 1 second. Such second-level response is fast enough to be used in some advanced fields, especially in alarming

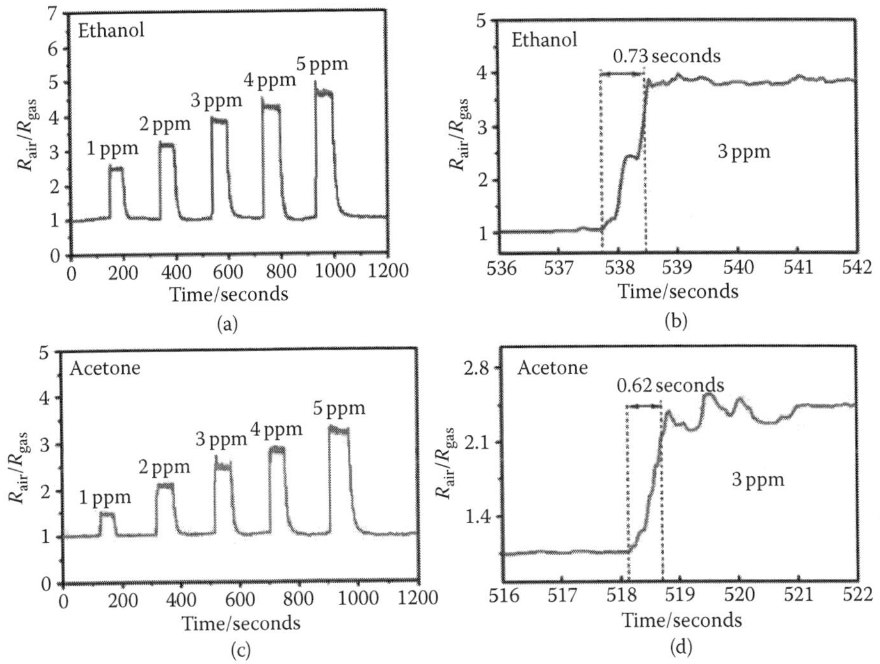

FIGURE 11.17 The electric response (R_{air}/R_{gas}) as functions of test time. (a) 1–5 ppm ethanol at 350°C, (b) a high magnification of (a) at 3 ppm. (c) 1–5 ppm acetone at 350°C and (d) a high magnification of (c) at 3 ppm. (Reprinted by permission from Macmillan Publishers Ltd. *Sci. Rep.*, Dai et al., 3, 1669, copyright 2013.)

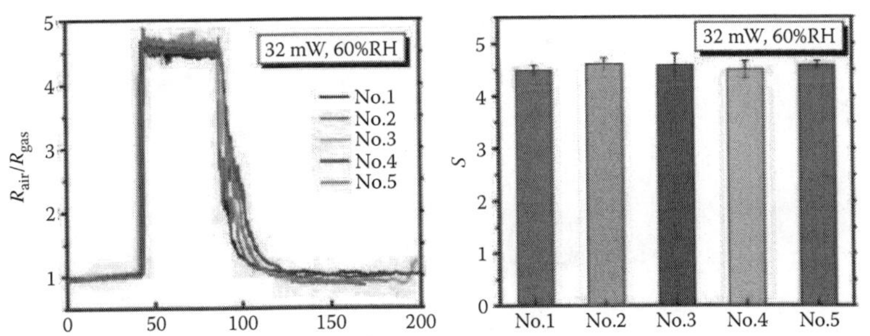

FIGURE 11.18 The sensing responses of five parallel as-fabricated gas-sensing devices under 5 ppm ethanol at 350°C (32 mW) and 60% relative humidity (RH). (a) Response curves. (b) Sensitivity for the five devices (data from [a]). (Reprinted by permission from Macmillan Publishers Ltd. *Sci. Rep.*, Dai et al., 3, 1669, copyright 2013.)

and monitoring system for the high fateful gases such as chemical and biological warfare agents [51].

The sensing mechanism to the ethanol and acetone originates mainly from the target gas-induced oxidation of negatively charged oxygen (O^-or O^{2-}) adsorbed on the SnO_2 film, as previously reported [4,52]. The fast response can be attributed to the porous (microsized) and ultrathin (20 nm) structure of the film (see the inset of Figure 11.15c). Such structure induces easy and fast accessibility to all the film's surface for the target gases after exposure.

11.4.5.2 Lower Limit of Detection

The lower limit of detection is an important parameter for the gas sensors in trace or ultratrace target gas analysis [51]. The sensing performance, for exposure to the sub-ppm ethanol gas, was also examined for the device above. A continuous test without recoveries was taken in the concentration range from 100 to 500 ppb at 350°C, as is shown in Figure 11.19. In this condition, the response time and sensitivity are, respectively, about 1.2 seconds and 1.1 to 100 ppb ethanol. Further, the dependence of the sensitivity on the concentration of the target gases has been obtained, as illustrated in the inset of Figure 11.19, exhibiting a linear relation in the sub-ppm range, or

$$S = 1.0 + 0.9 \times C \tag{11.2}$$

where C is the concentration of the target gas in ppm. Such dependence is in good agreement with the empirical equation for MOS gas sensors [19]. Interestingly, such sensing device displays a strong resolving power to low-concentration ethanol, due to the high signal-to-noise ratio in the testing process. Obviously, if there is no noise, an extremely slight change of the acquired signal can confirm the existence of a gas. Here, the detectable lower limit of the test gas concentration can be reasonably

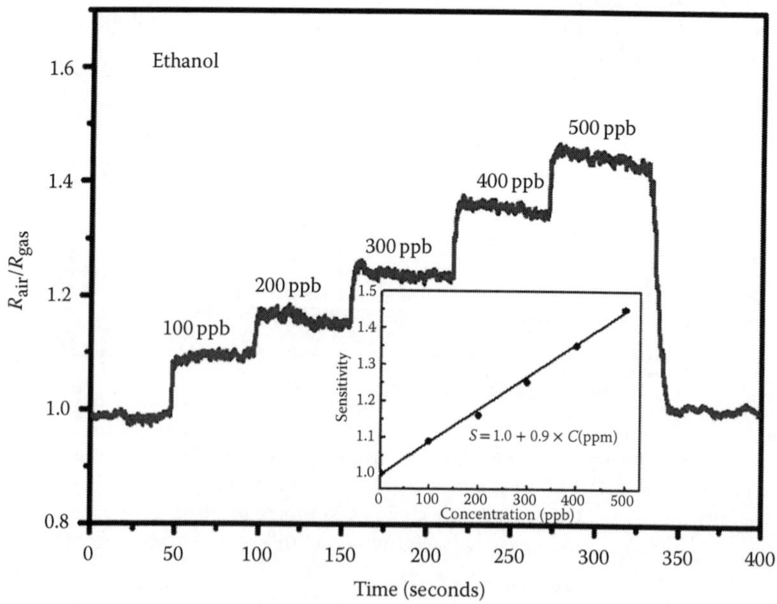

FIGURE 11.19 The electric response (R_{air}/R_{gas}) as functions of time for a continuous test without recoveries in the ethanol concentration ranges from 100 to 500 ppb at 350°C. The inset is the corresponding sensitivity versus gas concentration. (Reprinted by permission from Macmillan Publishers Ltd. *Sci. Rep.*, Dai et al., 3, 1669, copyright 2013.)

predicted on the basis of the present signal-to-noise ratio. In Figure 11.19, the noise signal N is observed to be about 0.01 in scale. The standard requirement of the detection lower limit is $(S - 1)/N > 3$ [53]. So, the response signal should be larger than 0.03 or the value of the sensitivity must be >1.03. From Equation 11.2, the corresponding concentration can be estimated to be about 30 ppb when sensitivity is 1.03. It means that the detectable lower limit of concentration to ethanol is about 30 ppb at 350°C. This is much lower, or three orders of magnitude lower, than that of the normal MOS gas sensors, which is usually higher than 10 ppm [54,55]. This detectable lower limit is close to that of the single nanowire (or nanobelt) sensor [56], which usually need complicated device construction, and possesses weak antijamming, nonuniformity of individual nanowire's size and uncertain contact resistance [57,58].

11.4.5.3 Dependence of Sensitivity on Working Temperature

Further, the dependence of sensitivity on the working temperature was examined. Figure 11.20 shows the results for the as-fabricated sensor exposed to ethanol and acetone, in the concentration range of 1–5 ppm (60%RH). We can see that the optimal working temperatures for ethanol and acetone are 350°C and 400°C, respectively, and such optimal working temperatures are independent of the gas concentrations. The sensitivity to ethanol exhibits different dependence on the working temperature from acetone. Such different dependences for different test gases provide the possibility for design of the devices with sensing selectivity. Here, the sensitivity difference between ethanol and acetone is relatively large around 300°C. On this basis, the sensing device at this working temperature could realize good selectivity to ethanol in despite of the existence of acetone. Further and deep study about this issue is needed.

11.4.5.4 Durability and Stability of the MEMS-Based Sensing Devices

The experiments have revealed that the MEMS-based sensing devices are of good durability and stability in performance. Figure 11.21a shows the response curves

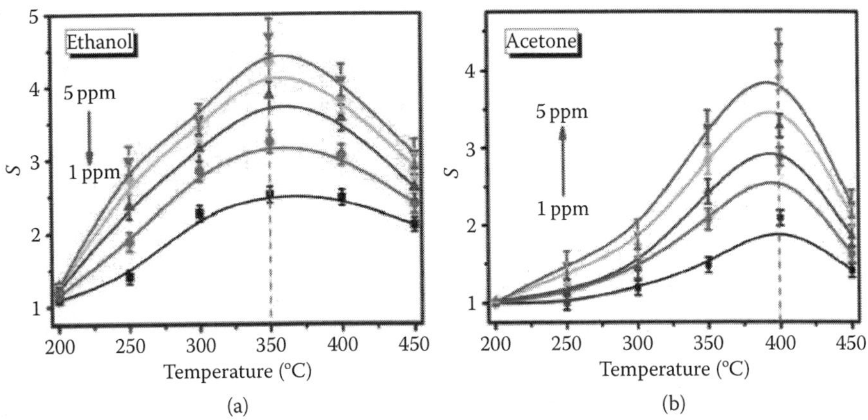

FIGURE 11.20 The curves of sensitivity versus temperature under the gas concentrations of 1, 2, 3, 4, and 5 ppm. (a) Ethanol and (b) acetone. (Reprinted by permission from Macmillan Publishers Ltd. *Sci. Rep.*, Dai et al., 3, 1669, copyright 2013.)

to ethanol at different phases for the as-fabricated device. The response signal only decreases slightly from 4.67 to 4.12 in 6 months under 350°C and 60%RH, which indicates a good durability or long-term stability. As for the effect from ambient, Figure 11.21b shows the ethanol-sensing performance (5 ppm, 350°C) at different ambient relative humidities. The sensitivity (*S*) of the sensing device ranges from 4.2 at 20%RH to 4.8 at 80%RH. The effect of the ambient humidity on response signal is relatively low.

In addition, considering time-weighted average for real applications, the ethanol sensing was measured in the high concentration from 100 to 500 ppm at 350°C and 60%RH, and Figure 11.22 shows the corresponding results. The sensitivity (*S*) falls

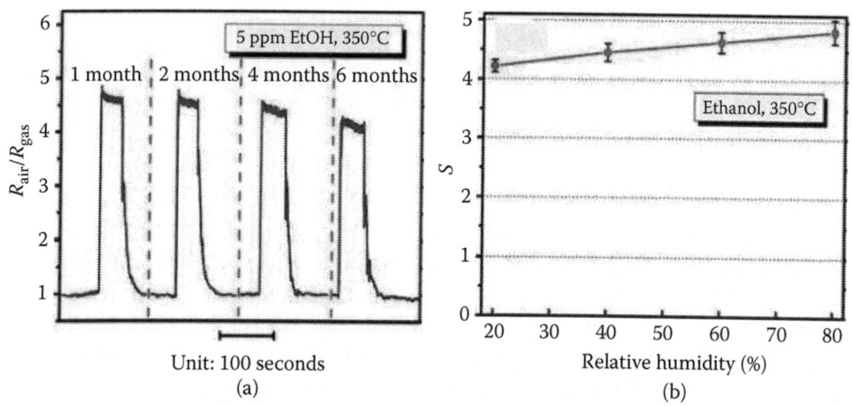

FIGURE 11.21 The durability and stability of the MEMS-based device (SnO_2). (a) The 1, 2, 4, and 6 months-later sensing response of the device to 5 ppm ethanol under 350°C and 60%RH, respectively. (b): The sensitivity of the device to 5 ppm ethanol at 350°C versus the ambient RH. (Reprinted by permission from Macmillan Publishers Ltd. *Sci. Rep.*, Dai et al., 3, 1669, copyright 2013.)

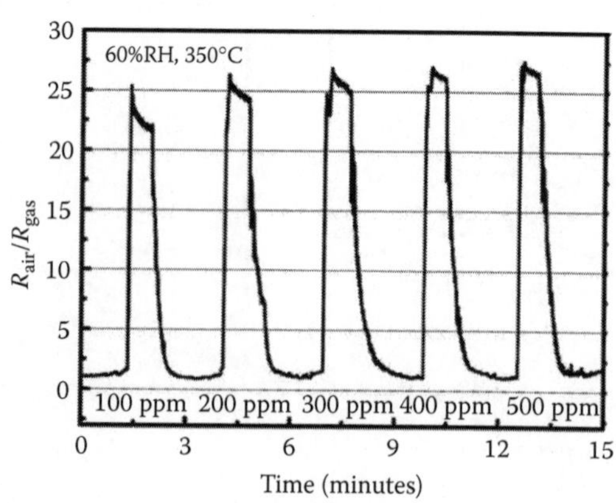

FIGURE 11.22 The real-time sensing response of the MEMS-based device (SnO_2) to 100–500 ppm ethanol under 350°C and 60%RH. (Reprinted by permission from Macmillan Publishers Ltd. *Sci. Rep.*, Dai et al., 3, 1669, copyright 2013.)

into the range from about 23 to 27. It means that the sensing signal reaches saturation in the concentration above 100 ppm. Obviously, such gas sensors are more suitable for detection in low-concentration range. Comparing the results with those in literature, it can found the sensitivity in this case is higher than those from Sr-doped nanostructured $LaFeO_3$ film and sol–gel derived ZnO thin film [59,60], and close to that from ultrathin InN field-effect transistors [60]. However, the response time is smaller than that from InN field-effect transistors [61].

11.5 BRIEF SUMMARY AND REMARKS

Micro/nanostructured ordered porous films as new candidates of gas-sensing element have many unique advantages compared with conventional nanoparticle-based gas-sensing devices, such as well-defined porosity, homogeneous and controlled film thickness, and the controllable pore sizes. In this chapter, we have given an overview of the recent progresses in monolayer-, multilayer-controlled micro/nanostructured porous film-based gas-sensing devices, including the structurally induced control of the gas-sensing performances and selectivity.

Using the transferability of the PS colloidal monolayer in the precursor solution, the various porous films-based sensing devices can be controllably constructed on any desired substrate, including double or multilayer homopore-sized, heteropore-sized, and ion-doped porous films with micro/nanostructures. By controlling the microstructure or the chemical composition of the porous films, the sensing devices with both high sensitivity and fast response or high selectivity can be realized, and further, the sensing parameters (sensitivity, selectivity, and response time) can also be modulated separately, according to need, in a large range. More importantly, from the dependence of sensing parameters on film microstructures, we can easily design and fabricate structure of the porous films with desired sensing parameters according to practical requirements. Further, based on the combination of micro/nanostructured porous thin film and MEMS-based chip, we could achieve the high-performance gas sensors with mW-level power consumption, second-level response, and ppb-level detection limit. Also, based on the solution dipping and PS colloidal template transferring method, we could fabricate such gas-sensing devices with low cost and mass production potential.

It should be noted that deep study in this area is required to obtain the good gas sensors with more practical performances, especially in (1) construction of the other complex structures such as multilayered heteropore-sized porous films or ions-doped films, heteropore-sized and heterochemical composition porous films, which will be advantageous for separately modulating the sensing parameters including sensitivity, response time, and selectivity in large range; (2) surface modification of the porous films with specific organic and bioorganic receptors, which cannot only provide a promising way to enhance the sensing performance but also extend the use of these porous films to the other gases, such as biological agent, toxic organic pollutants, and so on. In addition, systematical study of the influence of the experimental conditions on sensing performances, including final annealing temperature, environmental humidity, and ion-doping way, is also necessary for the more efficient use of these porous films. In a word, the micro/nanostructure porous films are the promising candidates of the next-generation gas sensors with the practical application.

REFERENCES

1. Kohl, D. 2001. Function and applications of gas sensors. *J. Phys. D* 34: R125–49.
2. Jia, L. C.; Cai, W. P.; Wang, H. Q.; Sun, F. Q.; and Li, Y. 2009. Hetero-apertured micro/nanostructured ordered porous array: Layer-by-layered construction and structure-induced sensing parameter controllability. *ACS Nano* 3: 2697–705.
3. Obvintseva, L. A. 2008. Metal oxide semiconductor sensors for determination of reactive gas impurities in air. *Russ. J. Gen. Chem.* 78: 2545–55.
4. Tiemann, M. 2007. Porous metal oxides as gas sensors. *Chem. Eur. J.* 13: 8376–88.
5. Yamazoe, N.; Sakai, G.; and Shimanoe, K. 2003. Oxide semiconductor gas sensors. *Catal. Surv. Asia* 7: 63–75.
6. Franke, M. E.; Koplin, T. J.; and Simon, U. 2006. Metal and metal oxide nanoparticles in chemiresistors: Does the nanoscale matter? *Small* 2: 36–50.
7. Zhang, J.; Wang, S.; Wang, Y.; Xu, M.; Xia, H.; Zhang, S.; Huang, W.; Guo, X.; and Wu, S. 2009. Facile synthesis of highly ethanol-sensitive SnO_2 nanoparticles. *Sens. Actuators B* 139: 369–74.
8. Pokhrel, S.; Simion, C. E.; Teodorescu, V. S.; Barsan, N.; and Weimar, U. 2009. Synthesis, mechanism, and gas-sensing application of surfactant tailored tungsten oxide nanostructures. *Adv. Funct. Mater.* 19: 1767–74.
9. Luyo, C.; Ionescu, R.; Reyes, L. F.; Topalian, Z.; Estrada, W.; Llobet, E.; Granqvist, C. G.; and Heszler, P. 2009. Gas sensing response of NiO nanoparticle films made by reactive gas deposition. *Sens. Actuators B* 138: 14–20.
10. Liu, H.; Gong, S. P.; Hu, Y. X.; Liu, J. Q.; and Zhou, D. X. 2009. Properties and mechanism study of SnO_2 nanocrystals for H_2S thick-film sensors. *Sens. Actuators B* 140: 190–5.
11. Wang, Y. D.; Djerdj, I.; Antonietti, M.; and Smarsly, B. 2008. Polymer-assisted generation of antimony-doped SnO_2 nanoparticles with high crystallinity for application in gas sensors. *Small* 4: 1656–60.
12. Wang, H. Q.; Li, G. H.; Jia, L. C.; Wang, G. Z.; and Tang, C. J. 2008. Controllable preferential-etching synthesis and photocatalytic activity of porous ZnO nanotubes. *J. Phys. Chem. C* 112: 11738–43.
13. Wang, H. Q.; Wang, G. Z.; Jia, L. C.; Tang, C. J.; and Li, G. H. 2007. Polychromatic visible photoluminescence in porous ZnO nanotubes. *J. Phys. D* 40: 6549–53.
14. Sun, F. Q.; Cai, W. P.; Li, Y.; Jia, L. C.; and Lu, F. 2005. Direct growth of mono- and multilayer nanostructured porous films on curved surfaces and their application as gas sensors. *Adv. Mater.* 17: 2872–77.
15. Sun, F. Q.; Cai, W. P.; Li, Y.; Cao, B. Q.; Lei, Y.; and Zhang, L. D. 2004. Morphology-controlled growth of large-area two-dimensional ordered pore arrays. *Adv. Funct. Mater.* 14: 283–88.
16. Li, Y.; Cai, W. P.; and Duan, G. T. 2008. Ordered micro/nanostructured arrays based on the monolayer colloidal crystals. *Chem. Mater.* 20: 615–24.
17. Bukowiecki, S. and Ulli, H. P. 1981. Electrical and chemical behaviour of metal oxide semiconductor gas sensors. *Helv. Phys. Acta* 54: 630–30.
18. Scott, R. W. J.; Yang, S. M.; Coombs, N.; Ozin, G. A.; and Williams, D. E. 2003. Engineered sensitivity of structured tin dioxide chemical sensors: Opaline architectures with controlled necking. *Adv. Funct. Mater.* 13: 225–31.
19. Scott, R. W. J.; Yang, S. M.; Chabanis, G.; Coombs, N.; Williams, D. E.; and Ozin, G. A. 2001. Tin dioxide opals and inverted opals: Near-ideal microstructures for gas sensors. *Adv. Mater.* 13: 1468–72.
20. Gurlo, A. and Riedel, R. 2007. In situ and operando spectroscopy for assessing mechanisms of gas sensing. *Angew. Chem. Int. Ed.* 46: 3826–48.

21. Soulantica, K.; Erades, L.; Sauvan, M.; Senocq, F.; Maisonnat, A.; and Chaudret, B. 2003. Synthesis of indium and indium oxide nanoparticles from indium cyclopentadienyl precursor and their application for gas sensing. *Adv. Funct. Mater.* 13: 553–57.
22. Korotcenkov, G.; Cerneavschi, A.; Brinzari, V.; Vasiliev, A.; Ivanov, M.; Cornet, A.; Morante, J.; Cabot, A.; and Arbiol, J. 2004. In_2O_3 films deposited by spray pyrolysis as a material for ozone gas sensors. *Sens. Actuators B* 99: 297–303.
23. Du, N.; Zhang, H.; Chen, B.; Ma, X.; Liu, Z.; Wu, J.; and Yang, D. 2007. Porous indium oxide nanotubes: Layer-by-layer assembly on carbon-nanotube templates and application for room-temperature NH_3 gas sensors. *Adv. Mater.* 19: 1641–5.
24. Liu, Z. F.; Yamazaki, T.; Shen, Y.; Kikuta, T.; Nakatani, N.; and Kawabata, T. 2007. Room temperature gas sensing of p-type TeO_2 nanowires. *Appl. Phys. Lett.* 90: 3.
25. Pinna, N.; Neri, G.; Antonietti, M.; and Niederberger, M. 2004. Nonaqueous synthesis of nanocrystalline semiconducting metal oxides for gas sensing. *Angew. Chem. Int. Ed.* 43: 4345–9.
26. Jia, L. C. and Cai, W. P. 2010. Micro/nanostructured ordered porous films and their structurally induced control of the gas sensing performances. *Adv. Funct. Mater.* 20: 3765–73.
27. Prim, A.; Pellicer, E.; Rossinyol, E.; Peiro, F.; Cornet, A.; and Morante, J. R. 2007. A novel mesoporous CaO-loaded In_2O_3 material for CO_2 sensing. *Adv. Funct. Mater.* 17: 2957–63.
28. Chen, A.; Bai, S.; Shi, B.; Liu, Z.; Li, D.; and Chung Chiun, L. 2008. Methane gas-sensing and catalytic oxidation activity of SnO_2-In_2O_3 nanocomposites incorporating TiO_2. *Sens. Actuators B* 135: 7–12.
29. McCue, J. T. and Ying, J. Y. 2007. SnO_2-In_2O_3 nanocomposites as semiconductor gas sensors for CO and NOx detection. *Chem. Mater.* 19: 1009–15.
30. Vilaseca, M.; Coronas, J.; Cirera, A.; Cornet, A.; Morante, J. R.; and Santamaria, J. 2008. Development and application of micromachined Pd/SnO_2 gas sensors with zeolite coatings. *Sens. Actuators B* 133: 435–41.
31. Penza, M.; Rossi, R.; Alvisi, M.; Cassano, G.; Signore, M. A.; Serra, E.; and Giorgi, R. 2008. Pt- and Pd-nanoclusters functionalized carbon nanotubes networked films for sub-ppm gas sensors. *Sens. Actuators B* 135: 289–97.
32. Jia, L. C.; Cai, W. P.; and Wang, H. Q. 2010. Metal ion-doped SnO_2 ordered porous films and their strong gas sensing selectivity. *Appl. Phys. Lett.* 96: 103115.
33. Batzill, M. and Diebold, U. 2005. The surface and materials science of tin oxide. *Prog. Surf. Sci.* 79: 47–154.
34. Potje-Kamloth, K. 2008. Semiconductor junction gas sensors. *Chem. Rev.* 108: 367–99.
35. Huebner, M.; Koziej, D.; Bauer, M.; Barsan, N.; Kvashnina, K.; Rossell, M. D.; Weimar, U.; and Grunwaldt, J.-D. 2011. The structure and behavior of platinum in SnO_2-based sensors under working conditions. *Angew. Chem. Int. Ed. Engl.* 50: 2841–4.
36. Korotcenkov, G. 2007. Metal oxides for solid-state gas sensors: What determines our choice? *Mat. Sci. Eng. B* 139: 1–23.
37. Stetter, J. R. and Li, J. 2008. Amperometric gas sensors: A review. *Chem. Rev.* 108: 352–66.
38. Yamazoe, N. and Shimanoe, K. 2009. New perspectives of gas sensor technology. *Sens. Actuators B* 138: 100–07.
39. Tricoli, A.; Righettoni, M.; and Teleki, A. 2010. Semiconductor gas sensors: Dry synthesis and application. *Angew. Chem. Int. Ed. Engl.* 49: 7632–59.
40. Lu, H. L.; Ma, W. C.; Gao, J. H.; and Li, J. M. 2000. Diffusion-reaction theory for conductance response in metal oxide gas sensing thin films. *Sens. Actuators B* 66: 228–31.
41. Sakai, G.; Matsunaga, N.; Shimanoe, K.; and Yamazoe, N. 2001. Theory of gas-diffusion controlled sensitivity for thin film semiconductor gas sensor. *Sens. Actuators B* 80: 125–31.

42. Jimenez-Cadena, G.; Riu, J.; and Rius, F. X. 2007. Gas sensors based on nanostructured materials. *Analyst* 132: 1083–99.
43. Barsan, N.; Koziej, D.; and Weimar, U. 2007. Metal oxide-based gas sensor research: How to? *Sens. Actuators B* 121: 18–35.
44. Muller, G.; Friedberger, A.; Kreisl, P.; Ahlers, S.; Schulz, O.; and Becker, T. 2003. A MEMS toolkit for metal-oxide-based gas sensing systems. *Thin Solid Films* 436: 34–45.
45. Ruiz, A. M.; Illa, X.; Diaz, R.; Romano-Rodriguez, A.; and Morante, J. R. 2006. Analyses of the ammonia response of integrated gas sensors working in pulsed mode. *Sens. Actuators B* 118: 318–22.
46. Dai, Z. F.; Xu, L.; Duan, G. T.; Li, T.; Zhang, H. W.; Li, Y.; Wang, Y.; Wang, Y. L.; and Cai, W. P. 2013. Fast-response, sensitivitive and low-powered chemosensors by fusing nanostructured porous thin film and ides-microheater chip. *Sci. Rep.* 3: 1669.
47. Figaro Engineering Inc: Products, gas sensors. http://www.aliexpress.com/item/FIGARO-TGS-2620-Gas-Sensor-for-the-detection-of-Solvent-Vapors/580243393.html.
48. Xu, L.; Li, T.; and Wang, Y. 2011. A novel three-dimensional microheater. *IEEE Electron Device Lett.* 32: 1284–6.
49. Owen, O. E.; Trapp, V. E.; Skutches, C. L.; Mozzoli, M. A.; Hoeldtke, R. D.; Boden, G.; and Reichard, G. A., Jr. 1982. Acetone metabolism during diabetic ketoacidosis. *Diabetes* 31: 242–8.
50. Cao, W. Q. and Duan, Y. X. 2006. Breath analysis: Potential for clinical diagnosis and exposure assessment. *Clin. Chem.* 52: 800–11.
51. Ishikawa, F. N.; Chang, H.-K.; Curreli, M.; Liao, H.-I.; Olson, C. A.; Chen, P.-C.; Zhang, R. et al. 2009. Label-free, electrical detection of the SARS virus N-protein with nanowire biosensors utilizing antibody mimics as capture probes. *ACS Nano* 3: 1219–24.
52. Xu, C.; Tamaki, J.; Miura, N.; and Yamazoe, N. 1991. Promotion of tin oxide gas sensor by aluminum doping. *Talanta* 38: 1169–75.
53. Kim, I. D.; Rothschild, A.; Hyodo, T.; and Tuller, H. L. 2006. Microsphere templating as means of enhancing surface activity and gas sensitivity of $CaCu_3Ti_4O_{12}$ films. *Nano Lett.* 6: 193–8.
54. Liu, X. H.; Zhang, J.; Guo, X. Z.; Wu, S. H.; and Wang, S. R. 2010. Amino acid-assisted one-pot assembly of Au, Pt nanoparticles onto one-dimensional ZnO microrods. *Nanoscale* 2: 1178–84.
55. Wang, C. X.; Yin, L. W.; Zhang, L. Y.; Qi, Y. X.; Lun, N.; and Liu, N. N. 2010. Large scale synthesis and gas-sensing properties of anatase TiO_2 three-dimensional hierarchical nanostructures. *Langmuir* 26: 12841–8.
56. Mai, L. Q.; Xu, L.; Gao, Q.; Han, C. H.; Hu, B.; and Pi, Y. Q. 2010. Single beta-$AgVO_3$ nanowire H_2S sensor. *Nano Lett.* 10: 2604–8.
57. Sysoev, V. V.; Strelcov, E.; Sommer, M.; Bruns, M.; Kiselev, I.; Habicht, W.; Kar, S.; Gregoratti, L.; Kiskinova, M.; and Kolmakov, A. 2010. Single-nanobelt electronic nose: Engineering and tests of the simplest analytical element. *ACS Nano* 4: 4487–94.
58. Comini, E. 2006. Metal oxide nano-crystals for gas sensing. *Anal. Chim. Acta* 568: 28–40.
59. Murade, P. A.; Sangawar, V. S.; Chaudhari, G. N.; Kapse, V. D.; and Bajpeyee, A. U. 2011. Acetone gas-sensing performance of Sr-doped nanostructured $LaFeO_3$ semiconductor prepared by citrate sol-gel route. *Curr. Appl. Phys.* 11: 451–6.
60. Kakati, N.; Jee, S. H.; Kim, S. H.; Oh, J. Y.; and Yoon, Y. S. 2010. Thickness dependency of sol-gel derived ZnO thin films on gas sensing behaviors. *Thin Solid Films* 519: 494–8.
61. Kao, K.-W.; Hsu, M.-C.; Chang, Y.-H.; Gwo, S.; and Yeh, J. A. 2012. A sub-ppm acetone gas sensor for diabetes detection using 10 nm thick ultrathin InN FETs. *Sensors* 12: 7157–68.

12 Surface-Enhanced Raman Scattering Performances and Detection Applications

12.1 INTRODUCTION

Detection based on surface-enhanced Raman scattering (SERS) effect is a technique that enhances Raman scattering signal of organic molecules adsorbed on rough metal surfaces [1]. The enhancement factor (EF) could be as high as 10^{10} to 10^{11} [2,3], indicating SERS can be used to detect single molecules in environment [4,5]. The periodic micro/nanostructured arrays made by colloidal template techniques possess very rough surfaces and good periodicities, and hence have important applications in SERS devices to detect organic molecules. SERS effect has been proven to be of strong surface-morphology dependence. Such ordered metal arrays with rough surfaces are highly helpful to the SERS applications in the identification and detection of organic molecules because they could be used as the SERS-active substrates. Additionally, the periodicity of such arrays has been proved to be useful to further increase SERS signals [6]. Based on the colloidal template, some strategies to create the noble metal micro/nanostructured arrays have been developed, as described in Chapters 7 and 8. In this chapter, we will introduce the SERS performances of some typical noble metal micro/nanostructured arrays and their detection applications.

12.2 Au HIERARCHICALLY MICRO/NANOSTRUCTURED PARTICLE ARRAYS

Recently, it has been found that Au hierarchically micro/nanostructured arrays present a significantly structurally enhanced SERS effect [7]. Such array could be fabricated by a second templated strategy or electrodeposition on the ordered alumina through-pore template induced by solution dipping on polystyrene (PS) sphere colloidal monolayer, as described in Chapter 7. The Au hierarchical micro/nanoparticle array was shown in Figure 7.14. The individual microsized particle (or building block) in the array shows nanoscaled surface roughness, and the whole array is of periodicity on microscale and hierarchical surface roughness on micro- and nanoscales.

SERS measurements, using Rhodamine 6G (R6G) as a probe molecule, have indicated that the hierarchically rough-structured Au particle array (sample C) exhibits strong SERS effect, as shown in Figure 12.1a. Here, for easy description, let us denote the smooth Au film as sample A (prepared by vacuum physical vapor deposition on indium tin oxide [ITO] substrate), the rough Au film as sample B (prepared by electrodeposition on ITO substrate without template using the same deposition parameters as sample C). The morphology of sample C is shown in Figure 12.1b and exhibits similar surface roughness to sample C. The Raman results for samples A and B are also given in Figure 12.1a.

Before Raman examination, the samples were dipped into a solution with 10^{-6} M R6G with stirring for 10 minutes, rinsed with deionized water, and dried with high-purity flowing nitrogen. Sample A only gives a very weak signal (curve a in Figure 12.1a), and for sample B, the signal remains relatively weak but stronger than that of sample A due to its nanoscaled surface roughness (curve b in Figure 12.1a). For the hierarchically rough-structured Au particle array or sample C, however, it exhibits very strong SERS signal (see curve c in Figure 12.1a), much stronger than that of sample B. These results demonstrate the significantly structurally enhanced SERS effect. The more the roughness in micro/nanoscale for the surface, the stronger the Raman signal. For further confirmation of effect of surface roughness on the SERS effect, laser irradiation, which can modify the morphology of Au particles and decrease the surface roughness [8], as introduced in Chapter 7, was conducted for sample C. The morphology of Au particle shows spherical shape with smooth surface after proper irradiation (see Figure 12.1c). The corresponding SERS intensity was decreased dramatically (curve c' in Figure 12.1a). This indicates that nanosized surface roughness on the microsized particles is also an important factor for the SERS effect. Further, there is

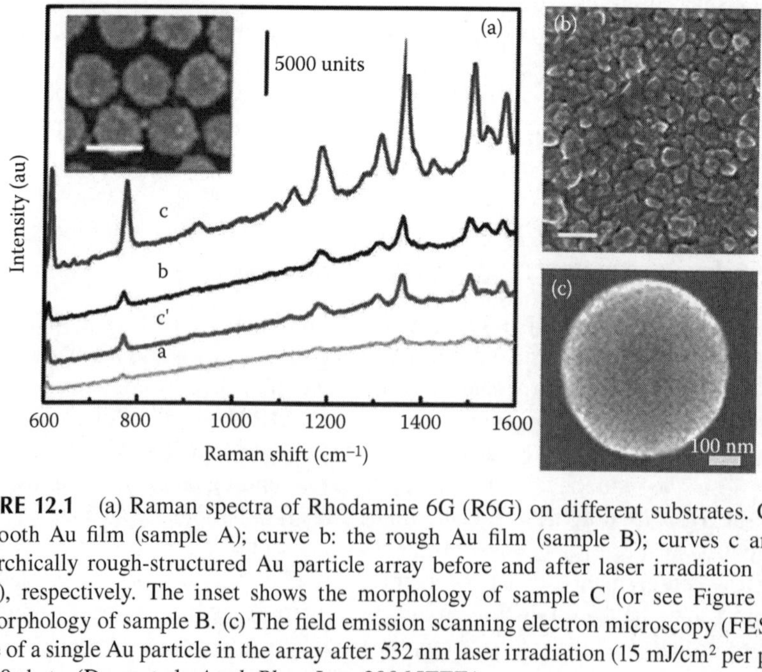

FIGURE 12.1 (a) Raman spectra of Rhodamine 6G (R6G) on different substrates. Curve a: smooth Au film (sample A); curve b: the rough Au film (sample B); curves c and c′: Hierarchically rough-structured Au particle array before and after laser irradiation (sample C), respectively. The inset shows the morphology of sample C (or see Figure 7.14). (b) Morphology of sample B. (c) The field emission scanning electron microscopy (FESEM) image of a single Au particle in the array after 532 nm laser irradiation (15 mJ/cm^2 per pulse) for 800 shots. (Duan et al., *Appl. Phys. Lett.* 2006 IEEE.)

no observable activity loss within 1 month for the SERS from the gold particle array with hierarchical roughness, showing high stability and good reproduction, which could be attributed to the structural stability of such array. Although same dipping conditions do not necessarily mean same numbers of molecules attached and detected, here it could be believed that the possible difference of R6G adsorption amounts on different substrates is not the main reason for the difference of SERS signals. Samples B and C have similar nanoscaled surface roughness, and the surface area should be larger for the former than sample C due to nonclose arrangement of the microparticles. But the opposite is true for the SERS signal (i.e., the Raman signals for sample C is stronger than those for sample B). Also, further experiments have demonstrated that, when the R6G concentration is down to 10^{-8} M, sample C still shows obvious Raman signals while no any signal could be detected for sample B or sample C after laser irradiation [7].

The strong SERS for the micro/nanostructured array (sample C) can be attributed to both the periodic structure and the hierarchical surface roughness. First, according to the previous report [9], the photon density of states may easily redistribute in a periodic structure, which leads to an increase in the density of optical modes and hence the enhancement of the Raman scattering of the detected molecules. Second, the microsized-particles have a roughly spherical-like shape with the diameter on the order of the laser wavelength (514.5 nm). Therefore, the localized plasmon mode can contribute to the Raman scattering enhancement [10,11]. The incident light excites plasmons trapped at the crevices in the long-range-ordered array, which can produce significant SERS effect. Third, the nanoscaled surface roughness and the size of Au particles could be optimal for SERS based on Nie's founding that the most efficient SERS should occur on the nanoparticles in several tens nanometer in size [12].

12.3 Ag NANOPARTICLE–BUILT HOLLOW MICROSPHERE (NANOSHELL) ARRAYS

As described in Chapter 8, Ag nanoparticle–built hollow microsphere (or nanoshell) arrays can be fabricated based on an electrophoretic deposition in the Ag colloidal solutions, induced by laser ablation in liquid, using monolayer PS sphere templates as the electrophoretic electrodes according to the strategy shown in Figure 8.49. The structural parameters of the Ag nanoshell arrays are well controlled, including the diameter, the thickness and the surface roughness of the nanoshells, and the spacing between the neighboring nanoshells (or the microspheres), as shown in Figures 8.54 through 8.56. Correspondingly, the properties of the nanoshell arrays including SERS and localized surface plasmon resonance (LSPR) can also be controlled based on the structural parameters of the nanoshell array. It has been shown that the Ag nanoshell array, consisting of 60-nm-sized nanoparticles and with 20-nm intershell spacing, exhibits the highest SERS activity, which could be used as the SERS substrate with high sensitivity [13]. In this section, we introduce the structural-dependent SERS and surface plasmon resonance of such Ag nanoparticle–built hollow microsphere (nanoshell) arrays.

12.3.1 Tunability of the LSPR and SERS Properties

The Ag nanoparticle–built hollow microsphere (nanoshell) arrays, shown in Figures 8.54 through 8.56, are of high structural controllability and hence show the tunable properties. The dependence of the LSPR and SERS properties of the Ag nanoshell arrays on their structures was investigated systematically [13]. Generally, the LSPR and SERS performance of noble metal (Au and Ag) nanostructured arrays depends on both the morphology of the building blocks and the geometric parameters of the whole assembly. Here, three factors could be used to control the LSPR and SERS properties: (1) the intershell spacings, (2) the diameter of nanoshells, and (3) the mean size of the nanoparticles in nanoshell layers.

Figure 12.2a presents the optical absorption spectra of the nanoshell arrays with different intershell spacings, as shown in Figure 8.55a through e. It is clearly shown that the intershell spacing has prominent influence on the LSPR properties. There are four LSPR peaks located from the visible to the near-infrared range. And these four peaks have the similar evolutions, in the peak position, with the intershell spacing (Figure 12.2b). When the spacing is about 20 nm, the LSPR peaks redshift compared with those of the close-packed nanoshell arrays. Moreover, an additional LSPR peak at 1190 nm appears (the asterisk mark in Figure 12.2a). Further increase of the intershell spacing to about 35 nm leads to no redshifting of the peak positions, and a blueshifting starts, especially for the peaks (2) and (4). As the spacing increases to about 65 nm, the LSPR peaks have prominent blueshifts to high energy levels. Further enlargement of the spacing to about 90 nm results in small redshifts. Similar relations of the LSPR shifting

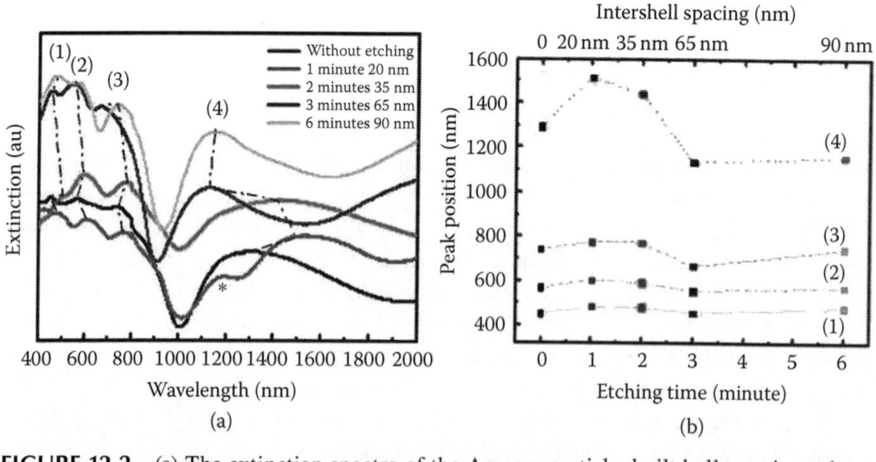

FIGURE 12.2 (a) The extinction spectra of the Ag nanoparticle–built hollow microsphere (nanoshell) arrays with different intershell spacing, as shown in Figure 8.55a through e. (b) Shifting of the surface plasmon resonance (SPR) peak positions for the main four peaks marked in panel (a) with the plasma-etching time and the intershell spacing. (Yang et al., *Adv. Funct. Mater.* 2010. 20. 2527–33. Copyright Wiley-VCH Verlag GmbH & Co. KGaA. Reproduced with permission.)

with the intershell spacing are observed for the nanoshell arrays with different surface roughness.

The size of the Ag nanoparticles in the shell layer or the surface roughness of the microspheres can also be used to adjust the LSPR of the Ag nanoshell array. Figure 12.3a gives the absorption spectra of the Ag nanoshell arrays, corresponding to the three samples shown in Figure 8.56, prepared under different electrophoretic current densities. The LSPR peaks blueshift with the increase in the mean size of nanoparticles (i.e., increase in the deposition current density), as clearly seen in Figure 12.3b.

Correspondingly, Figure 12.4a gives the SERS spectra of R6G molecules adsorbed on Ag nanoshell arrays with different intershell spacings shown in Figure 8.55a through e. It is clearly shown that the nanoshell array with about 20 nm spacing exhibits strong Raman peaks with intensities of five times higher than those of the close-packed one. The intensity of the Raman signal decreases with the increase in the intershell spacing (from 20 to 35, 65, and 90 nm). Further, it has been shown that the diameter of microsphere also influences the SERS intensity. The arrayed microspheres with about 2 μm in diameter show much weaker SERS signal (the bottom line in Figure 12.4a) than those with 660–750 nm in diameter (all the other lines except the bottom line in Figure 12.4a).

Furthermore, Ag nanoshell arrays with different sizes of Ag nanoparticles in the shell layers have different SERS enhancement. Figure 12.4b shows the effect of the nanoparticle size on the SERS signal, which is more prominent than that of the intershell spacing. It is found that the SERS signal of the array with 25 nm Ag nanoparticles in the nanoshells is about three times higher than that with 40-nm-sized particles, whereas the nanoshells with 60-nm-sized particles have the strongest SERS signal (about one magnitude higher than that of the nanoshells with 40-nm-sized particles).

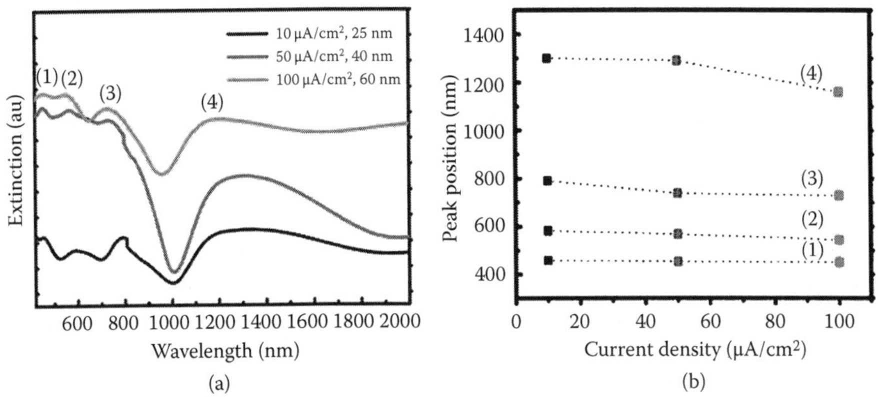

FIGURE 12.3 (a) The extinction spectra of the Ag nanoparticle–built hollow microsphere (nanoshell) arrays with different mean size of the nanoparticles in the nanoshell. (b) Shifting of the SPR peak positions of the main four peaks marked in panel (a) with the deposition current density. (Yang et al., *Adv. Funct. Mater.* 2010. 20. 2527–33. Copyright Wiley-VCH Verlag GmbH & Co. KGaA. Reproduced with permission.)

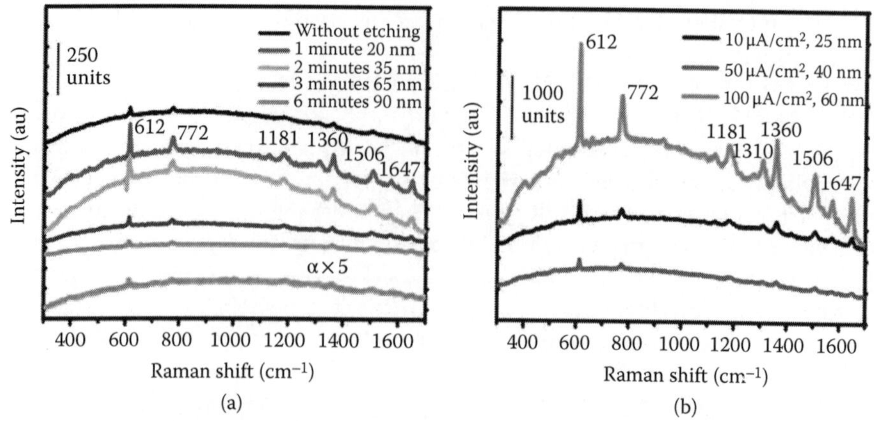

FIGURE 12.4 (a) Surface-enhanced Raman scattering (SERS) spectra of R6G molecules adsorbed on Ag nanoshell arrays with different intershell spacings shown in Figure 8.55. Curve α is the SERS curve (five times enlarged) of Ag nanoshell array with about 3 μm in sphere diameter. (b) SERS spectra of R6G molecules adsorbed on Ag nanoshell arrays with different sizes of nanoparticles in the nanoshell layers shown in Figure 8.56. All the SERS data are averaged by five scans over randomly chosen areas, and standard deviation is less than 10%. (Yang et al., *Adv. Funct. Mater.* 2010. 20. 2527–33. Copyright Wiley-VCH Verlag GmbH & Co. KGaA. Reproduced with permission.)

12.3.2 Structural Parameter-Dependent Coupling Effects

It is well known that the metal nanoparticles, which are much shorter than the wavelength of visible light, have only dipole plasmon resonance [14]. When the size of particles increases, multipole plasmon modes starts. The multipole plasmon modes appear at higher energy range than that of the dipole plasmon resonance in the extinction spectrum, which is known as phase retardation effect. Nanoshells have the similar size-dependent plasmon resonance performance to that of nanoparticles. There are some reports concerning the dipole plasmon properties of noble metal nanoshells with the size smaller than 100 nm, such as single shell and two neighboring shells [15–20]. Here, the Ag microspheres (or nanoshells) with relatively large diameter (500–1000 nm) support higher order multipole plasmon resonance. According to plasmon hybridization theory [15], the interaction of the inner and outer surfaces of nanoshells results in the splitting of plasmon resonances into two new resonances. Therefore, higher order multipole modes are excited in the shell layers and participate in the plasmon hybridization and results in the LSPR features shown in Figure 12.2a. The dipole and quadrupole interaction leads to the appearance of peaks (3) and (4), whereas peaks (1) and (2) are caused by even higher order multipole interactions, such as quadrupole and octupole. The plasmon coupling between arrayed nanoshells with larger shell spacing (e.g., 65 and 90 nm) is weak, which is similar to the LSPR performance of single nanoshells. When the spacing decreases to 35 nm, the LSPR hybridization between the neighboring nanoshells is enhanced, resulting in the redshifting of the plasmon peaks to the lower energy level. When the spacing reaches 20 nm, the plasmon hybridization is further enhanced, giving rise to further redshifted LSPRs. And the strong coupling between the dipole plasmon modes results in the peak at about 1190 nm (the asterisk mark in Figure 12.2a), leading to a complete separation of the dipole hybridization from the quadrupole hybridization. As for the close-packed nanoshells that are connected to each other, electrons flow across the adjacent nanoshells. The consequent blueshifting counteracts the redshifting of LSPR peaks caused by the plasmon coupling, resulting in the LSPR shifting to a high energy level (the black line in Figure 12.2a). Furthermore, the nanoshell thickness increases with the size of the Ag nanoparticles in the nanoshell layers. Thus, the plasmon coupling between the inner and outer surfaces of nanoshells is weakened, leading to the LSPR blueshifting shown in Figure 12.3.

The strong plasmon coupling will produce intense near-field electromagnetic fields (hot spots) between adjacent nanoshells [21–23] and thus generate the periodically located "hot spots" (with large SERS enhancement) across the nanoshell arrays. Therefore, the strong plasmon coupling, existing in the nanoshell arrays with 20 nm intershell spacing, should give rise to a strong SERS enhancement. The SERS enhancement should be weakened when the spacing increases to 35, 65, and 90 nm, which agrees well with the experimental results (Figure 12.4a). As for the arrayed Ag nanoshells with about 2 μm in diameter, a much lower density of "hot spots" in the nanoshell arrays than that of the nanoshells with 750 nm in diameter results in a much lower Raman signal intensity.

Generally, the surface roughness of nanoshells is more dominant for the SERS enhancement than the intershell spacing. As previously reported [4,24], Ag nanoparticles with the size range within 60–100 nm show intense SERS enhancement. Here, Ag nanoshells formed with 60-nm-sized nanoparticles exhibit the best SERS enhancement due to three reasons: (1) the nanoparticles with about 60 nm in mean size have intense SERS enhancement, (2) the large number of crevices (nanogaps) between the nanoparticles induce the SERS enhancement, and (3) the spacings between the neighboring nanoshells also magnify the SERS signal. The arrayed Ag nanoshells, with 60 nm mean-sized nanoparticles and 20 nm intershell spacing, have the strongest SERS enhancement, which could be used for the SERS substrates with high sensitivity. Here, the SERS EFs of this substrate (using 4-aminothiophenol [4-ATP] as probe molecules) can be estimated by the equation:

$$EF = \frac{I_{SERS}/N_{ads}}{I_{bulk}/N_{bulk}} \quad (12.1)$$

where I_{SERS} and I_{bulk} are the Raman signal intensities at 1075 cm^{-1}, which is a representative vibration of 4-ATP molecules (Figure 12.5), for the 4-ATP molecules adsorbed on the nanoshell array and the bulk 4-ATP, respectively; N_{ads} and N_{bulk} are the numbers of 4-ATP molecules absorbed on the nanoshell arrays and the solid 4-ATP exposed to the laser spot, respectively. Based on the measurements in Figure 12.5 and Equation 12.1, the EF value of the array is about 10^6. The Ag nanoshell arrays show much larger SERS enhancement than Ag nanoparticle (with 80 nm mean size) films produced by usual mirror reaction. There are a large number of nanogaps in the Ag nanoshell arrays consisting of 60-nm-sized Ag nanoparticles,

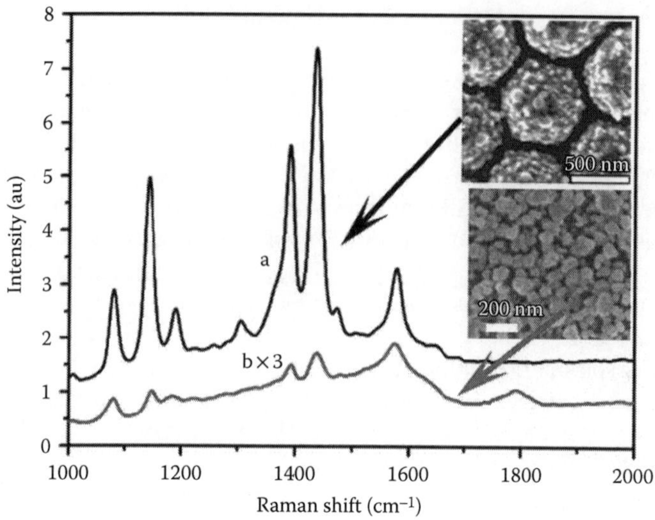

FIGURE 12.5 Raman spectra of 4-ATP molecules on the 60-nm-sized Ag nanoparticle–built Ag nanoshell array (curve a) and Ag nanoparticle film prepared through conventional mirror reaction (curve b). Insets: corresponding scanning electron microscopic images. (Yang et al., *Adv. Funct. Mater.* 2010. 20. 2527–33. Copyright Wiley-VCH Verlag GmbH & Co. KGaA. Reproduced with permission.)

as shown in Figure 8.56c, giving rise to their stronger SERS enhancement than those formed with 40-nm-sized Ag nanoparticles.

12.4 STANDING Ag NANOPLATE–BUILT HOLLOW MICROSPHERE ARRAYS

Based on the electrodeposition onto the Au-coated PS monolayer colloidal crystal template, we developed a simple and flexible strategy to directly fabricate the Ag nanoparticle–built hollow microsphere arrays with centimeter-squared scale [25], as described in Chapter 8. The array consists of periodically arranged microsized hollow spheres, which are built by vertically standing and cross-linking Ag nanoplates, as shown in Figure 8.44b. The nanoplates are of single crystal in structure, several hundred nanometers in the planar dimension, and about 30 nm in the thickness. The number density and size of silver nanoplates can be controlled by deposition conditions. Such micro/nanostructured hollow sphere arrays have shown significant SERS effect associated with their geometry, exhibiting strong SERS performances with high stability and good homogeneity. The minimum detectable concentration of 4-ATP molecules can be lower than 10^{-15} M. Further, such arrays could be reusable based on an argon plasma-cleaning method and used as a very effective SERS substrate for the detection of trace toxicants such as potassium cyanide, whose limit of detection is down to 0.1 parts per billion (ppb) [25].

12.4.1 SERS Activity of Ag Nanoplate–Built Hollow Microsphere Arrays

Silver nanoplates, which are of high specific surface area and rich optical properties but flat surface, are not optimal option for SERS substrate due to its difficulty in the formation of nanointerstitial spots [26–29]. However, when the nanoplates are cross-linked and stand vertically on the substrate, with a high distribution density or high density of nanogaps, the situation should be completely different. This has been confirmed by the standing Ag nanoparticle–built hollow microsphere array. Curve 1 in Figure 12.6 shows the Raman spectra of 4-ATP molecules on the standing nanoplate-built hollow microsphere array (after immersion in 10^{-6} mol/L 4-ATP solution for 30 minutes, integral time: 1 second). For comparison, the corresponding results are also shown in curves 2 and 3 of Figure 12.6 for the substrates Ag nanoparticles film synthesized by usual mirror reaction and Ag nanoplates array growing on the Au-coated flat silicon without using PS template. Obviously, the standing nanoplate-built hollow microsphere array shows much higher SERS activity than the nanoparticle film and also much stronger Raman signal than Ag nanoplates array growing on the Au-coated flat silicon without using PS template. Further, such hollow sphere array exhibits very low limit of detection. Even when the 4-ATP concentration was decreased down to 1×10^{-15} M, the Raman signal is still detectable within 1 second in integrating time, as illustrated in Figure 12.7.

12.4.2 Estimation of Enhancement Factor

The 4-ATP was used as test molecules, and the EF of the standing nanoplate-built hollow microsphere array could be estimated, in order of magnitude, by Equation 12.1. The key issue is the determination of N_{ads} and N_{bulk}. In this experimental condition

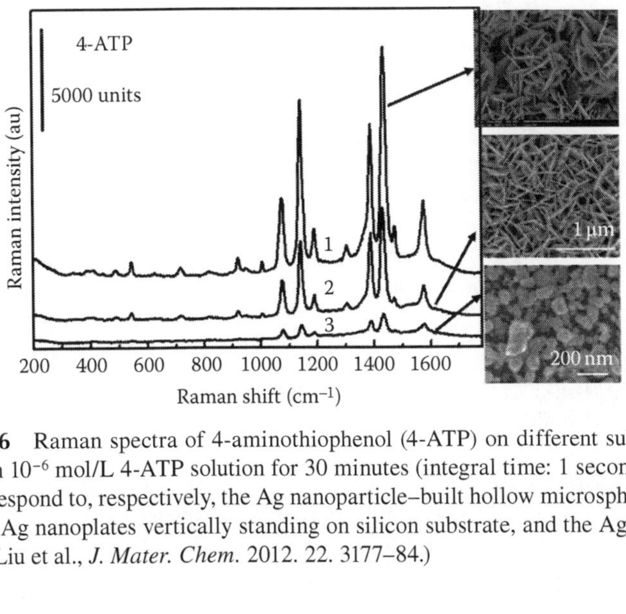

FIGURE 12.6 Raman spectra of 4-aminothiophenol (4-ATP) on different substrates after immersion in 10^{-6} mol/L 4-ATP solution for 30 minutes (integral time: 1 second). Curves 1, 2, and 3 correspond to, respectively, the Ag nanoparticle–built hollow microsphere array, the cross-linked Ag nanoplates vertically standing on silicon substrate, and the Ag nanoparticle film. (From Liu et al., *J. Mater. Chem.* 2012. 22. 3177–84.)

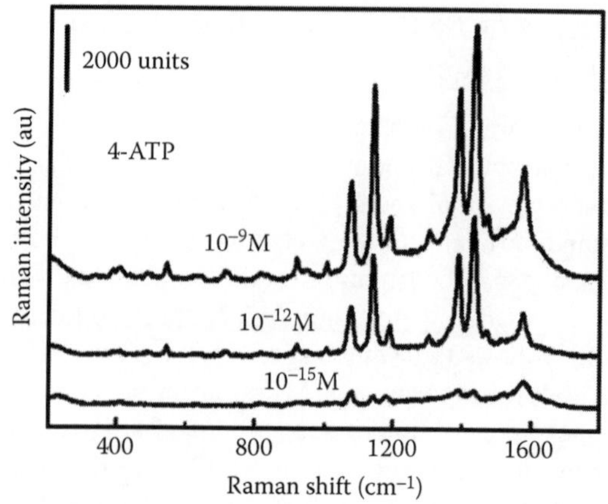

FIGURE 12.7 Raman spectra of the 4-ATP on the Ag nanoparticle–built hollow microsphere array (shown in the upright column of Figure 12.6) after immersion in the 4-ATP solution with different concentrations (integral time: 1 second). (Liu et al., *J. Mater. Chem.* 2012. 22. 3177–84. Reproduced by permission of The Royal Society of Chemistry.)

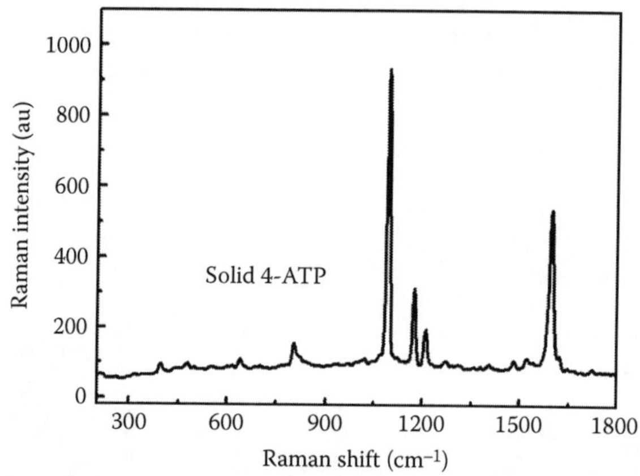

FIGURE 12.8 Raman spectrum of solid 4-ATP (integral time: 10 seconds). (Liu et al., *J. Mater. Chem.* 2012. 22. 3177–84. Reproduced by permission of The Royal Society of Chemistry.)

for solid 4-ATP, the probe volume could be considered to be a tube with a waist diameter of ~1.0 μm and detective depth of ~20 μm (from the instruction of the Raman spectrometer [MiniRam™ II—B&W Tek, Inc., Newark, DE]). So, the N_{bulk} value obtained is to be about 9.41×10^{10} 4-ATP molecules. The Raman spectrum of the solid 4-ATP molecules is shown in Figure 12.8. For the vibration at 1075 cm^{-1}, $I_{bulk}/N_{bulk} = 9.0 \times 10^{-9}$.

For the determination of N_{ads}, a 100 μL 10^{-8} M 4-ATP solution (containing 6×10^{11} molecules) was dropped on the hollow microsphere array with the projective area 1 cm^2 and dispersed uniformly. The molecule adsorption is much less than full monolayer coverage, which needs 5×10^{14} cm^{-2} 4-ATP molecules on the flat surface.

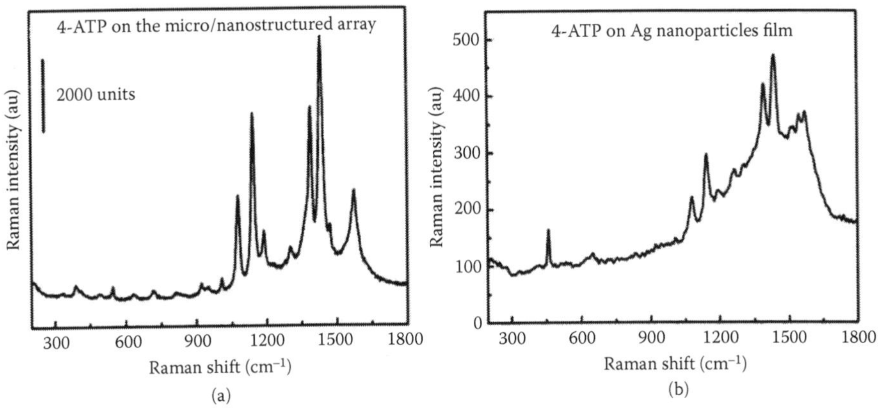

FIGURE 12.9 Raman spectra of 4-ATP on different substrates (integral time: 10 seconds). (a) The Ag nanoparticle–built hollow microsphere array, after dropping a 100 μL × 10^{-8} M on it with the projective area 1 cm^2 and dispersing uniformly; (b) the Ag nanoparticle film after immersion into 10^{-3} M 4-ATP solution. (Liu et al., *J. Mater. Chem.* 2012. 22, 3177–84. Reproduced by permission of The Royal Society of Chemistry.)

So, it could be assumed that all the 4-ATP molecules are adsorbed on the substrate, or all molecules contribute to the measured Raman signals. Correspondingly, the number of the adsorbed molecules within the area of laser spot N_{ads} can be determined to be up to 4.7×10^3. Figure 12.9a presents the corresponding Raman spectrum. The EF value thus obtained, for the vibration at 1075 cm^{-1}, is to be about 1.4×10^8, which is much higher than some previously reported values (up to ~10^5) [30,31]. Here, it should be pointed out that the EF value determined in this way is underestimated. The real value is difficult to determine but should be higher. Similarly, for the Ag nanoparticle film, the EF value can be estimated to be about only 5×10^3 by combining the result in Figure 12.9b. The enhancement effect of the former is much higher (five orders of magnitude) than that of the latter.

12.4.3 Structurally Enhanced Effect

SERS effect is a very local phenomenon occurring at nanogaps or in pores of the rough surface [32,33]. Here, the high SERS activity of the Ag nanoparticle–built hollow microsphere arrays could be attributed to the following: (1) the array contains high number density of the sharp edges on the standing silver nanoplates, which exhibit the strong field enhancement [34,35]; (2) the standing nanoplate-built hollow microsphere array is very uniform in structure, which is important for SERS [36], and importantly, the nanoplates stand nearly vertically on the hollow microspheres' surface with cross-linking structure, forming high density of interstitials or nanogaps that provide large number of the "hot spots" of SERS [23,37,38]; (3) such array is of high specific surface area, which favors the adsorption of more probing molecules; (4) the interstitials among the hollow microspheres could also provide "hot spots" for SERS effect; and (5) the interior of hollow microspheres is a cavity surrounded by Au shell layer, which also contributes to SERS [39].

12.4.4 Practicability of the Array

Further, the practicability for SERS substrate has also been examined for the standing nanoplate-built hollow microsphere arrays.

12.4.4.1 Reproducibility of Measurements

The measurements have revealed that the reproducibility of the SERS signal at different spots on the microsphere arrays is very good. Typically, Figure 12.10 shows the Raman spectra of 4-ATP molecules on the array collected at random 16 spots, with smaller than 8% in standard deviation for the six peaks at 1075, 1143, 1189, 1392, 1442, and 1576 cm^{-1}. Compared with the scientific standards about Raman signal reproducibility which is up to 20% in spot-to-spot variation over 10 mm^2 [40], less than 10% of standard deviations is relatively small. So such array is of the good measurement reproducibility across the whole sample. This should be attributed to the highly homogeneous structure of such micro/nanostructured arrays.

12.4.4.2 Reusability as an SERS Substrate

In practical application, reusability as a SERS substrate is also important. For irreversible adsorption, the substrate should be cleaned before reuse. It has been found that plasma bombardment of the substrate is a good cleaning method. Curve 1 in Figure 12.11a shows the Raman spectrum for the Ag nanoparticle–built hollow sphere array after immersion in 10^{-6} M 4-ATP solution for 30 minutes. After argon plasma bombardment with the input power of 50 W for 5 minutes, the adsorbed 4-ATP molecules can be removed from the substrate completely, as demonstrated in curve 2 of Figure 12.11a. Further, Figure 12.11b presents the values of the 4-ATP Raman peak at 1075 cm^{-1} before and after plasma bombardment for 5 minutes for different cycles, showing the reversible measurements. It means that a good reusability of the SERS substrates can be achieved by the plasma bombardment (or cleaning).

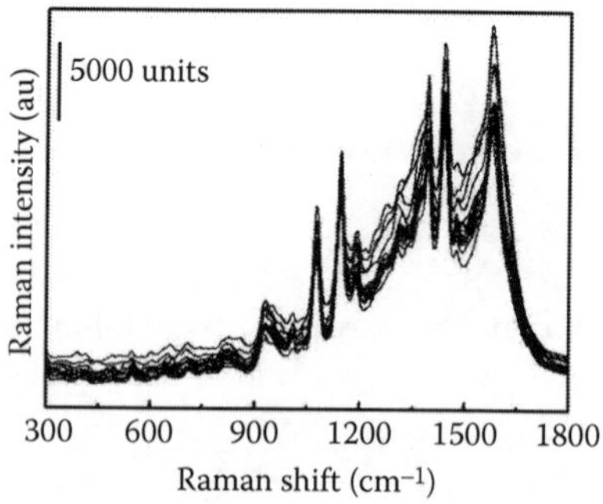

FIGURE 12.10 Raman spectra of 4-ATP from 16 different spots on the Ag nanoparticle–built hollow sphere array. Note: the concentration of 4-ATP is 10^{-6} M, and integral time is 1 second. The background of amorphous carbon is visible in the spectra due to the lower exciting power used. (Liu et al., *J. Mater. Chem.* 2012. 22. 3177–84. Reproduced by permission of The Royal Society of Chemistry.)

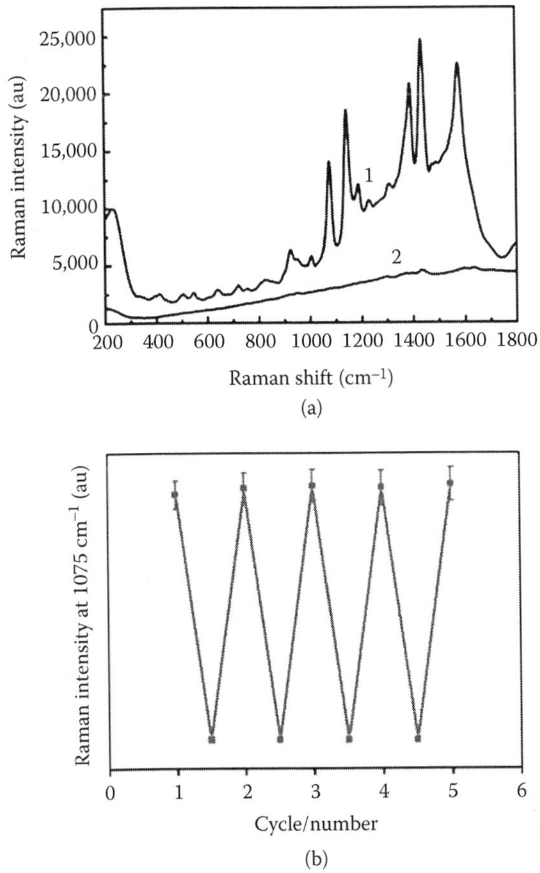

FIGURE 12.11 Reusability of the Ag nanoparticle–built hollow sphere array as a SERS substrate. (a) The Raman spectra for the 4-ATP on the array immersed in 10^{-6} M solution for 30 minutes, before (curve 1) and after (curve 2) plasma-cleaning (power 50 W) for 5 minutes. (b) The Raman intensity of the 4-ATP at 1075 cm^{-1} before (high values) and after (low values) plasma-cleaning (50 W, 5 minutes) for different cycles. (Liu et al., *J. Mater. Chem.* 2012. 22. 3177–84. Reproduced by permission of The Royal Society of Chemistry.)

It is notable that the bombardment power should be moderate. Too small power cannot remove the adsorbed molecules completely and too high bombardment power could destroy the surface morphology. Figure 12.12a gives the reused results corresponding to a high power (the input power of 100 W). The plasma bombardment can induce SERS enhancement within the first 3 to 4 cycles, after which the Raman signal decreased gradually. Surface morphological observation has revealed that after three cycles, there are many "small holes" formed in the surface of nanoplates (see Figure 12.12b), which provide more "hot spots" for the Raman signal enhancement. However, after more cycles, the standing nanoplates are heavily etched and destroyed due to plasma bombardment–induced Ag atom sputtering, leading to the

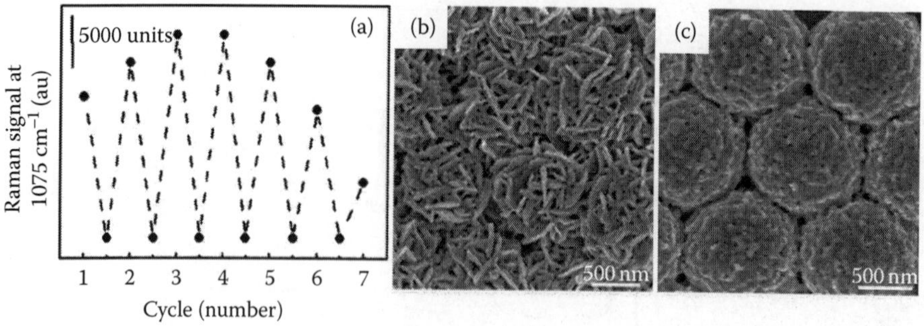

FIGURE 12.12 (a) The Raman signal of the 4-ATP at 1075 cm^{-1} before (high values) and after (low values) plasma-cleaning (100 W, 5 minutes) the Ag nanoparticle–built hollow sphere array for different cycles (integral time: 1 second). (b) and (c) FESEM images of the array corresponding to the plasma-cleaning for three and seven cycles, respectively. (Liu et al., *J. Mater. Chem.* 2012. 22. 3177–84. Reproduced by permission of The Royal Society of Chemistry.)

SERS activity decreasing, as shown in Figure 12.12c corresponding to the array after seven cycles of plasma bombardment with 100 W.

12.4.5 Application in Trace Detection of Cyanide

Cyanide is the lethality compound since the C≡N− in it has strong ability to bind the active site of cytochrome oxidase, which inhibits cellular respiration. It is recognized to be one of the most toxic anions [41–43]. Despite its toxicity, cyanide is widely used industrially in gold mining, electroplating for protecting or decorating, metallurgy, and the syntheses of nylon and other synthetic fibers and resins [44,45]. For instance, the gold mining industry uses more than 100 million pounds of cyanide salt annually in the United States alone. Although the U.S. Environmental Protection Agency has set the Maximum Contaminant Level for cyanide in drinking water to be 0.2 parts per million (ppm) [46], the acceptable level of cyanide in water or soil for people is generally much lower than this calibration. Obviously, it is important to quickly detect the trace amount of cyanide due to its high toxicity. There are many detection methods of cyanide, such as fluorescence or colorimetric sensing methods of free cyanide based on the formation of cyanide–boron complexes, cyanide–hydrogen complexes, and so on [47–55]. However, these methods are easily interfered by other anions, such as fluoride and acetate; that is, it is difficult to realize the fingerprint recognition of cyanide molecules [44,56]. Also, these methods are not suitable for on-site quick detection due to the tedious sample pretreatment or measurement at laboratory. Development of the methods for quick and trace detection of cyanide molecules is highly expected.

The detection based on SERS effect could be a suitable method for quick and trace measurement. There have been some reports about the detection of cyanides based on SERS effect [57–65]. For instance, Bozzini et al. reported that gold nanoparticle–based SERS effect can be used to detect sodium cyanide from water sample at the 110 parts per trillion level [59]; Premasiri et al. [60] presented silver sol–gel substrate to detect cyanide in waste based on SERS effect at ppm level. But, these substrates have lack of structural stability and uniformity in the whole substrate surface, which blocked the practical application. In this section, we will introduce the trace detection of kalium cyanide (KCN) in water by using the standing Ag nanoparticle–built hollow microsphere array as SERS substrate, whose detection limit is down to 0.1 ppb [65].

12.4.5.1 Surface Cleaning of Substrate

In addition to the substrate with high SERS activity, the essentiality of molecules for study is also important. Some molecules can be easily detected because of their strong adsorption on the surface of noble metal or the substrate, such as R6G molecules and 4-ATP molecules, which are thus usually used as target molecules to examine the SERS activity of substrate, as mentioned above. So, some interferential molecules with weak bond, adsorbed on the surface of substrate, have little influence on the detection of R6G or 4-ATP molecules based on SERS effect. But for cyanide, the bonding is not so strong. So the precleaning of substrate is very important, especially for the substrate fabricated in presence of some additives as in this case. As mentioned in Section 12.4.4.2, plasma bombardment of the substrate with appropriate power is a good cleaning method. It has been shown that the initial substrate exhibits the significant Raman peaks, meaning existence of interferential molecules on the surface. After argon plasma bombardment with the input power of 50 W for 2 minutes, the interferential molecules can be removed from the substrate completely [65].

12.4.5.2 Effect of Excited Power

Although there are extensive reports on SERS study, few works are focused on the influence of laser-exciting power on SERS effect. Generally, the intensity of Raman signal is enhanced with increasing the excited laser power when it is not too high. The Raman spectra using the laser powers with 0.01–0.5 mW was measured for the substrate after immersion in the KCN aqueous solution with 1 ppm (integral time: 30 seconds), as shown in Figure 12.13 The power with 0.1–0.2 mW induces the maximal Raman signal of C≡N stretching mode. The lower power (<0.1 mW) leads to the weaker Raman signal. Similarly, the power with >0.2 mW results in reduction of the Raman signal. While using the power of 0.5 mW, almost no signal is detected. The Raman signal reduction induced by the increasing powers should be attributed to the laser irradiation–induced desorption of KCN due to the weak bonding adsorption. Using the power of 0.5 mW induces the nearly complete desorption of the adsorbed KCN during measurement. Obviously, the laser-exciting power plays an important role in detection of cyanide molecules based on SERS effect.

12.4.5.3 Measurement Consistency of Parallel Substrates

In Section 12.4.4.1, we have mentioned the reproducibility of the SERS signal at different spots on the array, as shown in Figure 12.10, which exhibits the good measurement reproducibility across the whole substrate, with smaller than 10%, due to

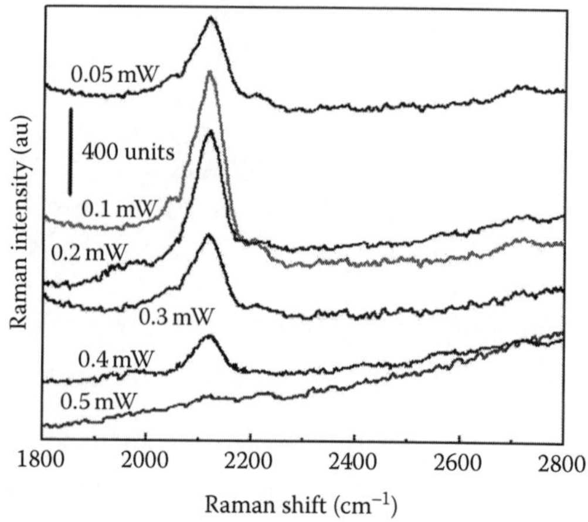

FIGURE 12.13 Raman spectra of kalium cyanide (KCN) on the Ag nanoparticle–built microsphere array after immersion in the KCN aqueous solution with 1 parts per million (ppm) for 10 minutes, excited by different laser powers. (Reprinted from *J. Hazard. Mater.*, 248–249, Liu et al., 435–41, Copyright 2013, with permission from Elsevier.)

the highly homogeneous structure of such arrays induced by the template technique. Further, three parallel arrays (independent samples) were used as substrates to measure the Raman spectra of KCN after immersion into KCN aqueous solution with a concentration of 1 ppm, as shown in Figure 12.14. The intensity of the peak at 2120 cm^{-1} is almost the same among three parallel substrates with the deviation smaller than 10%, showing good reproducibility and consistency of the measurement. This is attributed to the reproducible fabrication of the substrate due to the template technique.

12.4.5.4 Effect of Immersion Time

It is well known that cyanide can dissolve some metal, such as Au and Ag to form metal–cyanide complex, especially at high concentration. Undoubtedly, such dissolution will in turn reduce the accuracy of cyanide detection. So, the effect of immersion time was examined. In this case, the Ag nanoparticle–built microsphere array was immersed into KCN solution with the concentrations up to 1 ppm for 10 minutes and dried with high-purity flowing nitrogen before Raman spectral measurement. It has been demonstrated that the morphology of the substrate is almost unchanged after 10 minute immersion compared with that of the original one shown in Figure 8.44b, as illustrated in Figure 12.15a. However, if immersing in the solution for a long enough time, the surface nanostructure on the array would observably dissolve. Figure 12.15b shows the morphology after immersion in the KCN aqueous solution with 1 ppm for 5 hours. The dissolved surface morphology can be clearly seen on this long-time-immersed substrate. Correspondingly, the Raman spectral measurements have shown the immersion time-dependent intensity at 2120 cm^{-1}, as demonstrated in Figure 12.15c and d. The Raman intensity at 2120 cm^{-1} increases

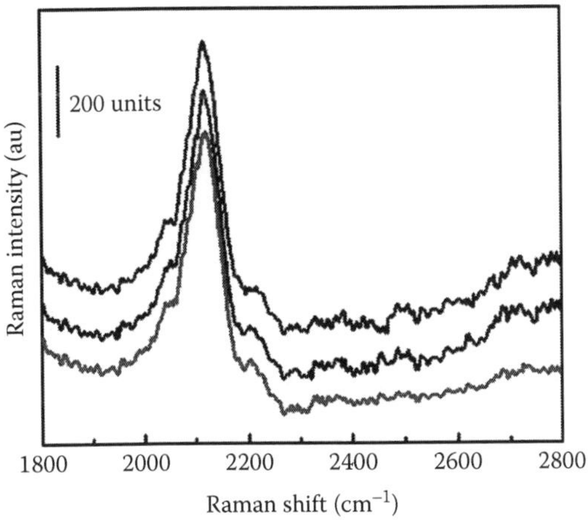

FIGURE 12.14 The Raman spectra of KCN using three parallel (independent) Ag nanoparticle–built microsphere arrays as substrates after immersion into KCN aqueous solution with a concentration of 1 ppm for 10 minutes, showing similar intensity. (Reprinted from *J. Hazard. Mater.*, 248–249, Liu et al., 435–41, Copyright 2013, with permission from Elsevier.)

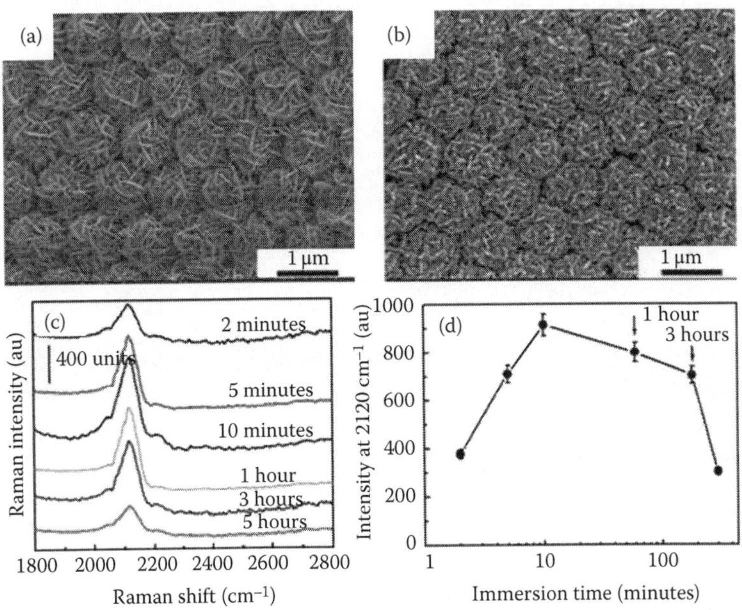

FIGURE 12.15 Influence of the immersion time on the morphology of the substrate and Raman spectra. (a) and (b) The FESEM images of the substrate after immersion in the KCN aqueous solution (1 ppm) for 10 minutes and 5 hours, respectively; (c) the Raman spectra of KCN on the substrate after immersion in the KCN aqueous solution (1 ppm) for different time; (d) Raman intensity at 2120 cm^{-1} versus the immersion time (the data are from [c]). (Reprinted from *J. Hazard. Mater.*, 248–249, Liu et al., 435–41, Copyright 2013, with permission from Elsevier.)

with the immersion time up to 10 minutes due to being insufficiently soaked. Ten minute immersion leads to the maximal Raman signal because of the sufficient soaking. The longer immersion (up to 3 hours) induces insignificant decrease of the Raman signal due to slight dissolution of the surface structure of the substrate. When the immersion time is longer than 3 hours, significant dissolution take places, leading to a large reduction of the Raman signal. Obviously, at low concentration (<1 ppm), the Ag nanoparticle–built microsphere array is quite stable, at least, for the immersion within 3 hours.

12.4.5.5 Trace Detection of Kalium Cyanide

12.4.5.5.1 Concentration Dependence

The cleaned Ag nanoparticle–built microsphere array was used as SERS substrate and immersed in the KCN solution with different concentrations for 10 minutes. Figure 12.16a shows the corresponding Raman spectra, excited at 0.1 mW of laser power (integral time 30 seconds). The peak at 2120 cm^{-1} corresponds to C≡N stretching mode. With increasing the concentration of KCN molecules, the intensity of the peak is increased. The detection limitation can be down to 0.1 ppb, which demonstrates that such array is a very effective SERS substrate for the detection of trace cyanides. Further, if the area under the peak at 2120 cm^{-1} is taken as the Raman signal intensity I, there exists the linear double-logarithm relation between I and the KCN concentration C in the soaking aqueous solution, as demonstrated in Figure 12.16b, or by linear fitting, we have

$$\mathrm{Lg}I = 2.16 + 0.26\mathrm{Lg}C \tag{12.2}$$

Equation 12.2 can be written as a power function:

$$I = 145C^{0.26} \tag{12.3}$$

Such concentration dependence of Raman intensity would be important to the device design based on SERS effect of the Ag nanoparticle–built hollow microsphere arrays.

In addition, the concentration dependence of the sensitivity $\mathrm{d}I/\mathrm{d}C$ can be obtained based on Equation 12.3, or

$$\frac{\mathrm{d}I}{\mathrm{d}C} = 38C^{-0.74} \tag{12.4}$$

The inset in Figure 12.16b demonstrates the evolution of Raman sensitivity with the KCN concentration. Obviously, the lower concentration corresponds to the higher sensitivity. Detection based on SERS effect is thus suitable for trace amount of KCN.

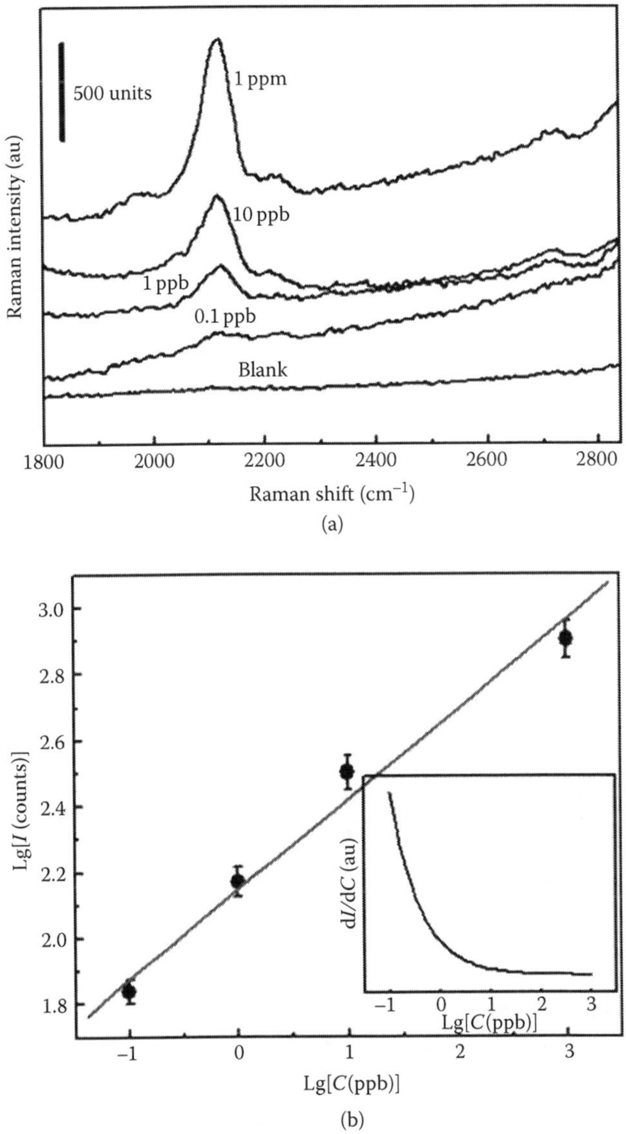

FIGURE 12.16 Raman spectral dependence on KCN concentrations. (a) Raman spectra of KCN on the Ag nanoparticle–built hollow microsphere arrays after immersion in the KCN solution with different concentrations for 10 minutes (laser power: 0.1 mW; integral time: 30 seconds). (b) The double-logarithm relation between the integral area of the Raman peak at 2120 cm^{-1} and the KCN concentration in the soaking solution (the data from [a]). The inset is the sensitivity (dI/dC) of Raman signal as a function of KCN concentration in the KCN aqueous solution. (Reprinted from *J. Hazard. Mater.*, 248–249, Liu et al., 435–41, Copyright 2013, with permission from Elsevier.)

12.4.5.5.2 *Freundlich Adsorption of KCN*

The high sensitivity at low concentrations of KCN could mainly be attributed to the concentration-dependent adsorption. In general, the intensity or integral area of a Raman peak should be proportional to the number of molecules q_e adsorbed on the surface of substrate within the area of a laser spot, or show a linear relationship between them, that is,

$$I = K_1 q_e \tag{12.5}$$

where K_1 is the constant. For adsorption of molecules on a heterogeneous surface, it can be described by Freundlich model [66], or

$$q_e = K_F C^{1/n} \tag{12.6}$$

where the parameters K_F and n are the parameters reflecting the adsorption capacity and adsorption intensity, respectively. If combining Equations 12.5 and 12.6, the relationship between the intensity of Raman peak and the KCN concentration in the soaking solution can be written as

$$I = KC^{1/n} \tag{12.7}$$

where $K = K_1 K_F$, the proportional constant. Obviously, Equation 12.7 is in good agreement with Equation 12.3. Such agreement has also confirmed that KCN adsorption on the array follows Freundlich model. Further, combining Equations 12.3 and 12.7, the parameter value of KCN adsorption can thus be obtained to be n = 3.8. This also provides a simple and convenient method to determine the adsorption parameters, which are usually obtained by the time-consuming measurement of the adsorption isotherm.

12.4.5.5.3 *A Short Remark*

All in all, the standing Ag nanoparticle–built hollow microsized sphere arrays can be used as the excellent SERS substrates to detect trace KCN molecules in water, which have high SERS activity and structural stability, and good measurement reproducibility. There exists a good linear double-logarithm relation between the signal of Raman and the concentration of KCN molecules in water in the range from 0.1 ppb to 1 ppm. In addition, the appropriate immersion in the solution and the suitable laser power for Raman excitation are crucial to trace detection of KCN molecules. Here, it should be pointed out that both the structural stability of the substrate and the selectivity to CN^- are important in the real environments. How to realize the strong selectivity in the real environments is still a big challenge [67,68]. Surface modification of the substrate could be one of the effective ways. The systematic study of the structural stability of the substrate and the selectivity to CN^- in the real environments are required.

12.5 BRIEF SUMMARY

In summary, the noble metal micro/nanostructured periodic arrays with rough surfaces are of high SERS activity and could be used as the SERS-active substrates for the SERS applications in the identification and detection of organic molecules. On the basis of the introduction of the colloidal template strategies for the noble metal micro/nanostructured arrays in Chapters 7 and 8, in this chapter, we have introduced the SERS performances of the some typical noble metal micro/nanostructured arrays and their detection applications, including three kinds of periodic metal arrays: Au hierarchically micro/nanostructured particle arrays, Ag nanoparticle–built hollow microsphere (nanoshell) arrays, and standing Ag nanoplate–built hollow microsphere arrays. Such periodic arrays exhibit much better SERS performances than those of the nanoparticle films, showing strong SERS activity, high EF, good structural stability, and excellent measurement reproducibility. The SERS performances could be tunable by structural control of the arrays. The good SERS performances originate from structural parameter-dependent coupling effects and good structural uniformity of the arrays. Such metal micro/nanostructured periodic arrays have exhibited good practicability in trace detection of some molecules such as cyanides.

REFERENCES

1. Xu, X.; Li, H.; Hasan, D.; Ruoff, R. S.; Wang, A. X.; and Fan, D. L. 2013. Near-field enhanced plasmonic-magnetic bifunctional nanotubes for single cell bioanalysis. *Adv. Funct. Mater.* 23: 4332–8.
2. Blackie, E. J.; Le Ru, E. C.; and Etchegoin, P. G. 2009. Single-molecule surface-enhanced Raman spectroscopy of nonresonant molecules. *J. Am. Chem. Soc.* 131: 14466–72.
3. Blackie, E. J.; Le Ru, E. C.; Meyer, M.; and Etchegoin, P. G. 2007. Surface enhanced Raman scattering enhancement factors: A comprehensive study. *J. Phys. Chem. C* 111: 13794–803.
4. Nie, S. and Emory, S. R. 1997. Probing single molecules and single nanoparticles by surface-enhanced Raman scattering. *Science* 275: 1102–6.
5. Le Ru, E. C.; Meyer, M.; and Etchegoin, P. G. 2006. Proof of single-molecule sensitivity in surface enhanced Raman scattering (SERS) by means of a two-analyte technique. *J. Phys. Chem. B* 110: 1944–8.
6. Lu, L.; Randjelovic, I.; Capek, R.; Gaponik, N.; Yang, J.; Zhang, H.; and Eychmuller, A. 2003. Controlled fabrication of gold-coated 3D ordered colloidal crystal films and their application in surface-enhanced Raman spectroscopy. *Chem. Mater.* 17: 5731–6.
7. Duan, G.; Cai, W.; Luo, Y.; Li, Y.; and Lei, Y. 2006. Hierarchical surface rough ordered Au particle arrays and their surface enhanced Raman scattering. *Appl. Phys. Lett.* 89: 181918.
8. Sun, F.; Cai, W.; Li, Y.; Duan, G.; Nichols, W. T.; Liang, C.; Koshizaki, N.; Fang, Q.; and Boyd, I. W. 2005. Laser morphological manipulation of gold nanoparticles periodically arranged on solid supports. *Appl. Phys. B* 81: 765–8.
9. Gaponenko, S. V. 2002. Effects of photon density of states on Raman scattering in mesoscopic structures. *Phys. Rev. B* 65: 140303.
10. Garcia-Vidal, F. J. and Pendry, J. B. 1996. Collective theory for surface enhanced Raman scattering. *Phys. Rev. Lett.* 7: 1163–6.
11. Shalaev, V. M. 2000. *Nonlinear Optics of Random Media*, Springer, New York.

12. Krug, J. T.; Wang, G. D.; Emory, S. R.; and Nie, S. M. 1999. Efficient Raman enhancement and intermittent light emission observed in single gold nanocrystals. *J. Am. Chem. Soc.* 121: 9208–14.
13. Yang, S. K.; Cai, W. P.; Kong, L. C.; and Lei, Y. 2010. Surface nanometer-scale patterning in realizing large-scale ordered arrays of metallic nanoshells with well-defined structures and controllable properties. *Adv. Funct. Mater.* 20: 2527–33.
14. Bohren, C. F. and Huffman, D. R. 1998. *Absorption and Scattering of Light by Small Particles*, Wiley, New York.
15. Prodan, E.; Radloff, C.; Halas, N. J.; and Nordlander, P. 2003. A hybridization model for the plasmon response of complex nanostructures. *Science* 302: 419–22.
16. Lassiter, J. B.; Aizpurua, J.; Hernandez, L. I.; Brandl, D. W.; Romero, I.; Lal, S.; Hafner, J. H.; Nordlander, P.; and Halas, N. J. 2008. Close encounters between two nanoshells. *Nano Lett.* 8: 1212–8.
17. Prodan, E. and Nordlander, P. 2003. Structural tunability of the plasmon resonances in metallic nanoshells. *Nano Lett.* 3: 543–7.
18. Prodan, E.; Nordlander, P. N.; and Halas, J. 2003. Electronic structure and optical properties of gold nanoshells. *Nano Lett.* 3: 1411–5.
19. Sun, Y. and Xia, Y. 2002. Increased sensitivity of surface plasmon resonance of gold nanoshells compared to that of gold solid colloids in response to environmental changes. *Anal. Chem.* 74: 5297–305.
20. Lal, S.; Grady, N.; Kundu, K. J.; Levin, C. S.; Lassiter, J. B.; and Halas, N. J. 2008. Tailoring plasmonic substrates for surface enhanced spectroscopies. *Chem. Soc. Rev.* 37: 898–911.
21. Kneipp, K.; Wang, Y.; Knipp, H.; Perelman, T.; Itzkan, I.; Dasari, R. R.; and Feld, M. S. 1997. Single molecule detection using surface-enhanced Raman scattering (SERS). *Phys. Rev. Lett.* 78: 1667–70.
22. Michaels, A. M.; Nirmal, M.; and Brus, L. E. 1999. Surface enhanced Raman spectroscopy of individual rhodamine 6G molecules on large Ag nanocrystals. *J. Am. Chem. Soc.* 121: 9932–9.
23. Lee, S. J.; Morrill, A. R.; and Moskovits, M. 2006. Hot spots in silver nanowire bundles for surface-enhanced Raman spectroscopy. *J. Am. Chem. Soc.* 128: 2200–1.
24. Emory, S. R.; Haskins, W. E.; and Nie, S. M. 1998. Direct observation of size-dependent optical enhancement in single metal nanoparticles. *J. Am. Chem. Soc.* 120: 8009–10.
25. Liu, G. Q.; Cai, W. P.; Kong, L. C.; Duan, G. T.; Li, Y.; Wang, J. J.; Zuo, G. M.; and Chen, Z. X. 2012. Standing Ag nanoplate-built hollow microsphere arrays controllable structural parameters and strong SERS performances. *J. Mater. Chem.* 22: 3177–84.
26. Zou, X. Q. and Dong, S. J. 2006. Surface-enhanced Raman scattering studies on aggregated silver nanoplates in aqueous solution. *J. Phys. Chem. B.* 110: 21545–50.
27. Lu, L. H.; Kobayashi, A.; Tawa, K.; and Ozaki, Y. S. 2006. Silver nanoplates with special shapes: Controlled synthesis and their surface plasmon resonance and surface-enhanced Raman scattering properties. *Chem. Mater.* 18: 4894–901.
28. Jin, R. C.; Cao, Y. W.; Hao, E.; Métraux, G. S.; Schatz, G. C.; and Mirkin, C. A. 2003. Controlling anisotropic nanoparticle growth through plasmon excitation. *Nature* 425: 487–90.
29. Sau, T. K. and Murphy, C. J. 2004. Room temperature, high-yield synthesis of multiple shapes of gold nanoparticles in aqueous solution. *J. Am. Chem. Soc.* 126: 8648–9.
30. Kim, K. and Yoon, J. K. 2005. Raman scattering of 4-aminobenzenethiol sandwiched between Ag/Au nanoparticle and macroscopically smooth Au substrate. *J. Phys. Chem. B* 109: 20731–6.
31. Khan, M. A.; Hogan, T. P.; and Shanker, B. J. 2009. Gold-coated zinc oxide nanowire-based substrate for surface-enhanced Raman spectroscopy. *J. Raman Spectrosc.* 40: 1539–45.

32. He, H.; Cai, W. P.; Lin, Y. X.; and Chen, B. S. 2010. Au nanochain-built 3D netlike porous films based on laser ablation in water and electrophoretic deposition. *Chem. Commun.* 46: 7223–5.
33. Suzuki, M.; Niidome, Y.; Kuwahara, Y.; Terasaki, N.; Inoue, K.; and Yamada, S. 2004. Surface-enhanced nonresonance Raman scattering from size- and morphology-controlled gold nanoparticle films. *J. Phys. Chem. B* 108: 11660–5.
34. Deckert-Gaudig, T. and Deckert, V. 2009. Ultraflat transparent gold nanoplates: Ideal substrates for tip-enhanced Raman scattering experiments. *Small* 5: 432–6.
35. Xu, C. and Wang, X. 2009. Fabrication of flexible metal-nanoparticle films using graphene oxide sheets as substrates. *Small* 5: 2212–7.
36. Zuev, V. S.; Frantsesson, A. V.; Gao, J.; and Eden, J. G. 2005. Enhancement of Raman scattering for an atom or molecule near a metal nanocylinder: Quantum theory of spontaneous emission and coupling to surface plasmon modes. *J. Chem. Phys.* 122: 214726.
37. Liu, G. Q.; Cai, W. P.; and Liang, C. H. 2008. Trapeziform Ag nanosheet arrays induced by electrochemical deposition on Au-coated substrate. *Cryst. Growth Des.* 8: 2748–52.
38. Wang, H. H.; Liu, C. Y.; Wu, S. B.; Liu, N. W.; Peng, C. Y.; Chan, T. H.; Hsu, C. F.; Wang, J. K.; and Wang, Y. L. 2006. Highly Raman-enhancing substrates based on silver nanoparticle arrays with tunable sub-10 nm gaps. *Adv. Mater.* 18: 491–5.
39. Tessier, P. M.; Velev, O. D.; Kalambur, A. T.; Rabolt, J. F.; Lenhoff, A. M.; and Kaler, E. M. 2000. Assembly of gold nanostructured films templated by colloidal crystals and use in surface-enhanced Raman spectroscopy. *J. Am. Chem. Soc.* 122: 9554–5.
40. Natan, M. 2006. Concluding remarks surface enhanced Raman scattering. *Faraday Discuss.* 132: 321–8.
41. Xu, Z. C.; Chen, X. Q.; Kim, H. N.; and Yoon, J. Y. 2010. Sensors for the optical detection of cyanide ion. *Chem. Soc. Rev.* 39: 127–37.
42. Cho, D. G. and Sessler, J. L. 2009. Modern reaction-based indicator systems. *Chem. Soc. Rev.* 38: 1647–62.
43. Gimeno, N.; Li, X. E.; Durrant, J. R.; and Vilar, R. 2008. Cyanide sensing with organic dyes: Studies in solution and on nanostructured Al_2O_3 surfaces. *Chem. Eur. J.* 14: 3006–12.
44. Palomares, E.; Martínez-Díaz, M. V.; Torres, T.; and Coronado, E. 2006. A highly sensitive hybrid colorimetric and fluorometric molecular probe for cyanide sensing based on a subphthalocyanine dye. *Adv. Funct. Mater.* 16: 1166–70.
45. Anzenbacher, P.; Tyson, Jr, D. S.; Jursiková, K.; and Castellano, F. N. 2002. Luminescence lifetime-based sensor for cyanide and related anions. *J. Am. Chem. Soc.* 124: 6232–3.
46. WHO. 1996. *Guidelines for Drinking-Water Quality*, 2nd edition, World Health Organization, Geneva, Switzerland.
47. Miyaji, H. and Sessler, J. L. 2001. Off-the-shelf colorimetric anion sensors. *Angew. Chem. Int. Ed. Engl.* 40: 154–7.
48. Zhang, Y.; Zhang, D.; and Liu, C. 2006. Novel chemical sensor for cyanides: Boron-doped carbon nanotubes. *J. Phys. Chem. B* 110: 4671–4.
49. Liu, H.; Shao, X.; Jia, M.; Jiang, X.; Li, Z.; and Chen, G. 2005. Selective recognition of sodium cyanide and potassium cyanide by diazo-crown ether-capped Zn-porphyrin receptors in polar solvents. *Tetrahedron Lett.* 61: 8095–100.
50. Tomasulo, M.; Sortino, T. S.; White, A. J. P.; and Raymo, F. M. 2006. Chromogenic oxazines for cyanide detection. *J. Org. Chem.* 71: 744–53.
51. Wu, X. F.; Xu, B. W.; Tong, H.; and Wang, L. X. 2011. Highly selective and sensitive detection of cyanide by a reaction-based conjugated polymer chemosensor. *Macromolecules* 44: 4241–8.
52. Qian, G.; Li, X. Z.; and Wang, Z. Y. 2009. Visible and near-infrared chemosensor for colorimetric and ratiometric detection of cyanide. *J. Mater. Chem.* 19: 522–30.

53. Jin, W. J.; Fernández-Argüelles, M. T.; Costa-Fernández, J. M.; Pereiro, R.; and Sanz-Medel, A. 2005. Photoactivated luminescent CdSe quantum dots as sensitive cyanide probes in aqueous solutions. *Chem. Commun.* 7: 883–5.
54. Touceda-Varela, A.; Stevenson, E. I.; Galve-Gasión, J. A.; Dryden, D. T. F.; and Mareque-Rivas, J. C. 2008. Selective turn-on fluorescence detection of cyanide in water using hydrophobic CdSe quantum dots. *Chem. Commun.* 17: 1998–2000.
55. Badugu, R.; Lakowicz, J. R.; and Geddes, C. D. 2005. Enhanced fluorescence cyanide detection at physiologically lethal levels: Reduced ICT-based signal transduction. *J. Am. Chem. Soc.* 127: 3635–41.
56. Chung, Y. M.; Raman, B.; Kim, D.-S.; and Ahn, K. H. 2006. Fluorescence modulation in anion sensing by introducing intramolecular H-bonding interactions in host–guest adducts. *Chem. Commun.* 2: 186–8.
57. Bozzini, B.; Mele, C.; and Romanello, V. 2006. Time-dependent in situ SERS study of CN^- adsorbed on gold. *J. Electroanal. Chem.* 592: 25–30.
58. Kuncicky, D. M.; Prevo, B. G.; and Velev, O. D. 2006. Controlled assembly of SERS substrates templated by colloidal crystal films. *J. Mater. Chem.* 16: 1207–11.
59. Senapati, D.; Dasary, S. S. R.; Singh, A. K.; Senapati, T.; Yu, H. T.; and Ray, P. C. 2011. A label-free gold-nanoparticle-based SERS assay for direct cyanide detection at the parts-per-trillion level. *Chem. Eur. J.* 17: 8445–51.
60. Premasiri, W. R.; Clarke, R. H.; Londhe, S.; and Womble, M. E. 2001. Determination of cyanide in waste water by low-resolution surface enhanced Raman spectroscopy on sol-gel substrates. *J. Raman Spectrosc.* 32: 919–22.
61. Tessier, P. M.; Christesen, S. D.; Ong, K. K.; Clemente, E. M.; Lenhoff, A. M.; Kaler, E. W.; and Velev, O. D. 2002. On-line spectroscopic characterization of sodium cyanide with nanostructured gold surface-enhanced Raman spectroscopy substrates. *Appl. Spectrosc.* 56: 1524–30.
62. Shelton, R. D.; Haas, J. W.; and Wacher, E. A. 1994. Surface-enhanced Raman detection of aqueous cyanide. *Appl. Spectrosc.* 48: 915–1032.
63. Pettinger, B.; Picardi, G.; Schuster, R.; and Ertl, G. 2003. Surface-enhanced and STM tip-enhanced Raman spectroscopy of CN^- ions at gold surfaces. *J. Electroanal. Chem.* 554–555: 293–9.
64. Tan, S.; Erol, M.; Sukhishvili, S.; and Du, H. 2008. Substrates with discretely immobilized silver nanoparticles for ultrasensitive detection of anions in water using surface-enhanced Raman scattering. *Langmuir* 24: 4765–71.
65. Liu, G. Q.; Cai, W. P.; Kong, L. C.; Duan, G. T.; Li, Y.; Wang, J. J.; and Cheng, Z. X. 2013. Trace detection of cyanide based on SERS effect of Ag nanoplate-built hollow microsphere arrays. *J. Hazard. Mater.* 248–249: 435–41.
66. Freundlich, H. and Heller, W. 1939. The adsorption of cis-and trans-azobenzene. *J. Am. Chem. Soc.* 61: 2228–30.
67. Lou, X. D.; Zhang, L. Y.; Qin, J. G.; and Li, Z. 2008. An alternative approach to develop a highly sensitive and selective chemosensor for the colorimetric sensing of cyanide in water. *Chem. Commun.* 44: 5848–50.
68. Liu, Y. L.; Lv, X.; Zhao, Y.; Liu, J.; Sun, Y. Q.; Wang, P.; and Guo, W. 2012. A Cu(II)-based chemosensing ensemble bearing rhodamine B fluorophore for fluorescence turn-on detection of cyanide. *J. Mater. Chem.* 22: 1747–50.

Index

I

K

N

O

P

R

S

T